T0321377

SMART AUTONOMOUS AIRCRAFT

Flight Control and Planning for UAV

SMART AUTONOMOUS AIRCRAFT

Flight Control and Planning for UAV

Yasmina Bestaoui Sebbane

Université d'Evry, France

CRC Press is an imprint of the
Taylor & Francis Group, an Informa business
A CHAPMAN & HALL BOOK

CRC Press
Taylor & Francis Group
6000 Broken Sound Parkway NW, Suite 300
Boca Raton, FL 33487-2742

Printed on acid-free paper
Version Date: 20151012

International Standard Book Number-13: 978-1-4822-9915-1 (Hardback)

Visit the Taylor & Francis Web site at
http://www.taylorandfrancis.com

and the CRC Press Web site at
http://www.crcpress.com

Dedication

To my family

Contents

Preface

Smart autonomous aircraft have become the new focus of academic research and education, due to their important application potential. They offer new opportunities to perform autonomous missions in field services and tasks such as search and rescue, observation and mapping, emergency and fire fighting, hurricane management.

A smart autonomous aircraft does not require any human intervention during its mission. Without a pilot onboard, an aircraft either must be controlled from the ground by a radio-control ground pilot or it must have its own intelligence to fly autonomously. Autonomy is defined as the ability of a system to sense, communicate, plan, make decisions and act without human intervention. To that end, the goal of autonomy is to teach machines to be smart and act more like humans. Intelligence is necessary for:

1. Mission planning to chart its own course using navigation with guidance and tracking control, while achieving optimal fuel consumption.
2. Path planning and waypoint generation accounting for changing weather and air traffic en route.
3. Mode switching and control reconfiguration decisions for implementation through the use of a flight control system.
4. Staying within the flight envelope of the aircraft.
5. Taking the corrective action to avoid an obstacle or to evade a threat, in the presence of external abnormal conditions.
6. Taking the corrective action by reconfiguring the set of controls to safely continue to fly or land the aircraft, in the presence of internal abnormal conditions.
7. Interpreting the data from a variety of sources to execute these functions.

The objective of this book is to give an interdisciplinary point of view on autonomous aircraft. It aims to develop models and review different methodologies of control and planning used to create smart autonomous aircraft. Some case studies are examined as well.

The topics considered in this book have been derived from the author's research and teaching duties in smart aerospace and autonomous systems over several years. The other part is based on the top literature in the field. This book is primarily geared at advanced graduate students, PhD students, and researchers. It assumes at least an undergraduate-level background in engineering.

Preface

Author

Professor Yasmina Bestaoui Sebbane earned a PhD in control and computer engineering from École Nationale Supérieure de Mecanique, Nantes, France, in 1989 (currently École Centrale de Nantes) and the Habilitation to Direct Research in Robotics, from University of Évry, Évry, France, in 2000.

She has been with the Electrical Engineering Department of the University of Evry since 1999. From 1989 to 1998, she was with the Mechanical Engineering Department of the University of Nantes. From September 1997 to July 1998, she was a visiting associate professor in the Computer Science Department at the Naval Postgraduate School, Monterey, California, USA.

Her research interests include control, planning and decision making of unmanned systems particularly unmanned aerial vehicles and robots. She has authored two other books: *Lighter than Air Robots*, Springer, ISCA 58, 2012 and *Planning and Decision Making for Aerial Robots*, Springer, ISCA 71, 2014.

She also has published eight book chapters, forty journal papers and eighty fully refereed conference papers. She is the coordinator of the Master SAAS: Smart Aerospace and Autonomous Systems, in cooperation with Poznan University of Technology, Poland.

Dr. Yasmina Bestaoui Sebbane is an AIAA senior member and IEEE senior member and was an associate editor of the *IEEE Transactions on Control Systems Technology* from January 2008 to December 2012. She is cited in *Who's Who in the World* and *Who's Who in Engineering*.

1 Introduction

1.1 CURRENT UAV PRESENTATION

An unmanned aircraft is defined by the US Department of Defense [22] as an aircraft that does not carry a human operator and is capable of flight under remote control or autonomous programming. Unmanned aircraft are the preferred alternatives for missions that are characterized as dull, dirty or dangerous. Unmanned aerial vehicles (UAV) provide a great degree of flexibility in mobility and response time [5, 6, 7]. This contributes towards lower mission costs compared to manned aircraft and enables acquisition of information in a time frame not previously possible [36]. The AIAA defines an UAV as an aircraft which is designed or modified not to carry a human pilot and is operated through electronic input initiated by the flight controller or by an onboard autonomous flight management control system that does not require flight controller intervention. During the last decade, significant efforts have been devoted to increasing the flight endurance and payloads of UAV resulting in various UAV configurations with different sizes, endurance levels and capabilities. UAV platforms typically fall into one the following four categories: fixed wing UAV, rotary wing UAV, airships or lighter than air UAV and finally flapping wing UAV. This book is devoted to fixed wing UAV.

The civilian applications can be divided into four categories:

1. Scientific and research related: environmental monitoring [21], climate monitoring, pollution monitoring [26], pollutant estimation [56],
2. Security related: surveillance [16, 37], communications [29], pipeline inspection,
3. Contractor supplied flight services: structure inspection [39, 44, 46], agriculture and farm management [47, 53, 63], bridge monitoring [10],
4. Safety related: weather and hurricane monitoring [55].

Unmanned aircraft typically have complementary sensors that can provide aircraft location along with imagery information and that support mission planning and route following [8, 15, 17]. Both multi-spectral and hyper-spectral cameras are now available for small UAV. The aircraft are used to collect relevant sensor information and transmit the information to the ground control station for further processing. For example, advanced sensors are contributing to precision agriculture which uses the technology to determine crop nutrient levels, water stress, impacts from pests and other factors that affect yield. Advances in platform design, production, standardization of image georeferencing and mosaiking and information extraction work flow are required to provide reliable end products [64].

The characteristics of UAV are:

1. Operation in large, outdoor environments.
2. Motion in 3 dimensions meaning that the planning space must be 4 dimensions.
3. Uncertain and dynamic operating environment.
4. Presence of moving obstacles and environmental forces that affect motion such as winds and changing weather conditions.
5. Differential constraints on movement.

On the basis of gross weight, operational altitude above ground level (AGL) and mission endurance, UAV can be described broadly as:

1. Group 1 : Micro/Mini tactical (< 10 kg)
2. Group 2 : Small tactical (10 to 20 kg)
3. Group 3 : Tactical (< 500 kg)
4. Group 4 : Persistent (> 500 kg and Flight Level $< FL180$)
5. Group 5 : Penetrating (> 500 kg and Flight Level $> FL180$)

Civilian applications deal mainly with group 1 to group 3. **Unmanned aerial systems** (UAS) are systems of systems. These systems make use of a large variety of technologies. These technologies are not always traditionally aviation-related. Some of them are related to robotics, embedded systems, control theory, computer science and technology. Many of these technologies have crossover potential.

Unmanned aircraft rely predominantly on guidance, navigation and control sensors, microprocessors, communication systems and ground station command, control and communications (C3). Microprocessors allow them to fly entire missions autonomously with little human intervention [42]. The principal issues for communication technologies are flexibility, adaptability, security and controllability of the bandwidth, frequency and information/data flows. The key aspects of the off-board command, control and communications are: man-machine interfaces, multi-aircraft command, control and communications, target identification, downsizing ground equipment.

Navigation is concerned with determining where the aircraft is relative to where it should be, guidance with getting the aircraft to the destination and control with staying on track. Situation awareness is used for mission planning and flight mode selection which constitutes the high level control elements. For inhabited aircraft, the pilot provides the intelligence for interpreting the data from a variety of sources to execute these functions. Much of this data is used in pre-flight or pre-mission planning and is updated onboard as the mission proceeds. As the mission segments are executed and abnormal events are encountered, flight mode switching takes place; this constitutes the mid-level control element. On an inhabited aircraft, the pilot flies the aircraft and makes necessary mode switching and control reconfiguration decisions for implementation through the use of the flight control system. This constitutes the low-level control element and is used to execute the smooth transition between modes of flight, for example transition from takeoff to level flight to landing and stay within the **flight envelope** of the aircraft. External abnormal conditions cause the pilot to take corrective actions, such as avoiding a collision or an obstacle or evading a threat. Internal abnormal conditions can also occur, such as a failure or a malfunction of a component onboard the aircraft. Once again, the pilot provides the intelligence to take the corrective action by reconfiguring the set of controls to safely continue to fly or land the aircraft.

Without a pilot onboard the aircraft, the onboard computing architecture must provide the environment for re-usability and reconfigurability. The high level supervisory controller receives mission commands from the command and control post and decomposes them into sub-missions which will then be assigned to connected function modules. Upon reception of start and destination points from the supervisory controller, the route planner generates the *best* route in the form of waypoints for the aircraft to follow. A database of the terrain in the form of a digitized map should be available to the route planner.

Currently, automated functions in unmanned aircraft include critical flight operations, navigation, takeoff and landing and recognition of lost communications requiring implementation of return-to-base procedure [54, 60]. Unmanned aircraft that have the option to operate autonomously today are typically fully preprogrammed to perform defined actions repeatedly and independently of external influence [14]. These systems can be described as self-steering or self-regulating and can follow an externally given path while compensating for small deviations caused by external disturbances. Current autonomous systems require highly structured and predictable environments. A significant amount of manpower is spent directing current unmanned aircraft during mission performance, data collection and analysis and planning and re-planning [9].

An unmanned aircraft system (UAS) is a system whose components include the necessary equipment, network and personnel to control the unmanned aircraft [5]. Unmanned technology innovations are rapidly increasing. Some UAS platforms fly throughout areas of operations for several hours at multiple altitudes. These missions require accurate and timely weather forecasts to improve planning and data collection and to avoid potential weather related accidents [62]. Accurate weather reporting also supports complementary ground and flight planning synchronization.The UAS replaces the onboard pilot's functionality with several distinctive integrated capabilities. From a decision-making perspective, these include a remote-human operator, coupled with an information technology based system of components that enable autonomous functionality. The latter element is significant since it must compensate for the many limitations that arise as a result of operating with a remote pilot [25]. As UAS endurance is increasing, weather predictions must be accurate so that potential weather related incidents can be avoided and coordinated flight and ground operations can be improved [34]. UAS have the potential to be more cost-effective than current manned solutions. UAS can be more effective than manned aircraft at dull, dirty and dangerous tasks. They are also capable of missions that would not be possible with manned aircraft.

1.2 AUTONOMOUS AIRCRAFT

In the context of humans and societies, autonomy is the capacity of a rational individual to make an informed, uncoerced decision and/or to give oneself his own rules according to which one acts in a constantly changing environment [33]. Technical systems that claim to be autonomous will be able to perform the necessary analogies of mind functions that enable humans to be autonomous. Moreover, like in human and animal societies, they will have to abide by rule set or law systems that govern the interaction between individual members and between groups.

Research and development in automation are advancing from a state of automatic systems requiring human control toward a state of autonomous systems able to make decisions and react without human interaction [2]. Advances in technology have taken sensors, cameras and other equipment from the analog to the digital state, making them smaller, lighter and more energy efficient and useful.

Autonomy is a collection of functions that can be performed without direct intervention by the human. Autonomy can be defined as an unmanned system ability of sensing, perceiving, analyzing, communicating, planning, decision-making and acting to achieve its goals [40, 41]. Autonomy is a capability enabled by a set of technologies such as sensing, intelligence, reliability and endurance [50].

Different levels of autonomy can be considered [22, 58]:

1. Human makes all decisions
2. Computer computes a complete set of alternatives
3. Computer chooses a set of alternatives
4. Computer suggests one alternative
5. Computer executes suggestion with approval
6. Human can veto computer's decision within time frame
7. Computer executes then reports to human
8. Computer only reports if asked
9. Computer reports only if it wants to
10. Computer ignores the human.

In today's technology, the last five levels of autonomy have not yet been attained.

An integrated suite of information technologies is required to achieve autonomous capabilities. The four elements of the decision cycle: ***observe, orient, decide, act (OODA)***, are the same actions that any human would perform in achieving an objective in daily life. For an autonomous system, machine-based elements must perform the same functions to achieve a desired result. In that case, the ***observe*** function is carried out by one or more sensors. The ***orient*** function involves the aggregation and fusing of asynchronous data, the contextual assessment of the situation and environment and the development of relevant hypotheses for the decision-making step. The ***decide*** step involves the ability to establish relevant decision criteria, correctly weigh a variety of factors in creating the best decision-making algorithm, accurate determining of timing of action and anticipate the consequences of actions in anticipation of the next decision cycle. The ***act*** function is executed by one or more effectors that interface with the rest of the system [32].

Autonomy is characterized into levels by factors including mission complexity, environmental difficulty and level of human–machine interaction to accomplish the mission. It is their performance in terms of mission success and efficiency, through self-awareness and analysis of their situation, self-learning and self-decision-making, while minimizing human involvement [58]. Autonomy is the ability at run time to operate without the need for a sustained human supervisor [26].

An autonomous aircraft is self-directed by choosing the behavior it has to follow to reach a human-directed goal. Autonomous aircraft may optimize their behavior in a goal-directed manner in unexpected situations. In a given situation, the *best* solution must be found, ensuring accuracy and correctness of a decision-making process through a continual process [30].

The primary goal of UAS regulations is their assurance of safe operations. This goal is quantified as an **equivalent level of safety** (ELOS) with that of manned aircraft. UAS depend on the onboard flight control system and/or the communication link to operate, introducing additional failure modes that may increase the total number of accidents for the same reliability requirement. UAS do not carry passengers and, as a result, the probability of injury and fatalities after an accident is greatly reduced compared with that of general aviation. Primary accidents can be ground impact, mid-air collision, unintended movement. Secondary accidents are falling debris resulting in fatality or injury, damage to property, damage/loss of system, impact on environment, impact on society.

Since failure frequency requirements prescribed for manned aircraft of the same size cannot be used directly, other means to derive such requirements for UAS need to be employed. A different approach frequently used in safety engineering is to define safety constraints for a specific accident based on the desired likelihood of the worst possible outcome, which can in turn be used to determine maximum failure frequency.

The operating environment of the unmanned aircraft is a critical factor in determining the appropriate level of autonomy and the capability to maneuver as needed to accomplish the mission. To be able to fly in any airspace, UAS are required to be certified as airworthy. Airworthiness certification is a core acquisition and engineering process conducted by system safety. It takes into account material, service life and mission requirements within the intended airspace. The level of certification depends on the mission requirements of the system. The aircraft structure, propulsion system, control redundancies, software and control links must all be certified to a certain standard defined by the service's technical airworthiness authority (TAA) [18].

An unmanned aircraft system (UAS) comprises individual system elements consisting of an unmanned aircraft, the control station and any other system elements necessary to enable flight, i.e., command and control link, and launch and recovery elements. There may be multiple control stations, command and control links and launch and recovery elements within a UAS [23].

With no persons onboard the aircraft, the airworthiness objective is primarily targeted at the protection of people and property on the ground. A civil UAS must not increase the risk to people or property on the ground compared with manned aircraft of an equivalent category.

Consideration should be given to the need for continued safe flight and landing. This will bring into question the appropriateness of a flight termination system that brings the unmanned aircraft down immediately when a failure occurs, regardless of location. Emergency sites shall be unpopulated areas. Factors such as gliding capability and emergency electrical power capacity (e.g., in case of loss of power) should be considered in determining the location of emergency sites.

The level of UAS autonomy is likely to have the following impacts of certification issues: human machine interface compliance with air traffic control (ATC) instructions, command and control link integrity, handling of UAS failures and compliance with safety objectives, specific autonomy techniques (e.g., nondeterministic algorithms) but which have to prove safe behavior, collision avoidance, type of airspace, avoidance of noise sensitive areas and objects. A UAS is airworthy if the aircraft and all of its associated elements are in condition for safe operation.

The consideration of UAS flights in the **National Air Space (NAS)** where UAS must operate within the FAA/EASA **Federal Aviation Regulations (FAR)/Joint Aviation Regulation (JAR)** has put a stringent requirement on UAS operations. Users of the UAS must consider the safety of the general public [20, 24]. Sense and avoid (SAA) capability is a technical approach proposed to bridge the gap between the Federal Aviation Regulations/Joint Aviation Regulation requirements for a pilot in the cockpit to see and avoid and the unmanned nature of UAS [38]. The sense and avoid module is a decentralized independent safety assurance system in the unmanned aircraft for immediate to short-term collisions with aircraft and terrain. This module immediately intervenes when the mid-term separation assurance process fails. It uses two types of information: surrounding traffic information and terrain database. Terrain database stores a spatial model of the Earth objects with their locations and heights in a certain resolution. The traffic information is obtained from ADS-B transponders of surrounding aircraft. This module enables both automatic dependent surveillance-broadcast **ADS-B in** and **ADS-B out** applications which allow data transmission between aircraft themselves and ground segments [49].

Airborne sense and avoid systems are focusing on an onboard capability to perform both self-separation and collision avoidance to ensure an appropriate level of safety, to enable autonomous action by the aircraft where the system can identify and react to conflicts [4, 11, 19]. Complex SAA systems may allow for formation flights. Some rules of the air are detect and avoid, air traffic control clearance, 500 feet rule, 1000 feet rule, check weather before flight, check fuel level, check environment, navigation, planning.

By definition, autonomous systems are given goals to achieve and these goals must come from human operators. Furthermore the use of autonomous systems can increase safety and operational effectiveness. It is assumed that the use of autonomous systems on unmanned aircraft is always accompanied by a pilot override function where a remote pilot can take control of the aircraft at any point [59]. Rational agents are well suited for use in autonomous unmanned aircraft because they can provide the overall direction and control for a task or mission and their behavior can be explained by analyzing the beliefs, goals, intentions and plans that define their behavior. Information flows into the flight control system and rational agents from the environment. Both components communicate with the rational agent making abstract decisions about the progress of the mission. These abstract decisions are then passed to the flight control system.

The special feature of an autonomous aircraft is its ability to be goal-directed in unpredictable situations [19, 27]. This ability is a significant improvement in capability compared to the capabilities of automatic systems. The automatic system is not able to initially define the path according to some given goal or to choose the goal that is dictating its path. An autonomous system is able to make a decision based on a set of rules and limitations [54]. Ideally, unmanned aircraft systems should be able to adapt to any environment. Their physical operating environment may vary greatly as they operate in all weather, from low to high altitudes and in airspace that is congested [51].

Small UAS requires careful codesign over both physical and cyber-elements to maximize the system efficiency. Mission objectives and success of the system as a whole are becoming increasingly dependent on appropriate allocation of computational resources balanced against demands of the physical actuation systems. A co-optimization scheme is described that considers trade-offs between costs associated with the physical actuation effort required for control and the computational effort required to acquire and process incoming information [12].

Autonomous aircraft use their hardware and software platforms to complete a given mission in a dynamic and unstructured environment. The architecture of an autonomous aircraft can be seen as a collection of software processes within a software architecture and run on a hardware architecture [41]. Certain key areas of interest for improving technology are autonomy and cognitive behavior, communication systems, interoperability and modularity [22].

Autonomous systems can change their behavior in response to unanticipated events, whereas automatic systems would produce the same outputs regardless of any changes experienced by the system or its surroundings [43]. Another definition of autonomy is given in [5]: a UAS is autonomous if it has the intelligence and experience to react quickly in a reasoned manner to an unplanned situation. For a UAS to operate with autonomy, a measure of artificial intelligence must be present within it plus a readily available source of experience.

Three major levels of autonomy have been identified in unmanned aircraft:

1. Reactive side:
   a. Flight control system, actuator function, engine or propulsion control
   b. Aircraft flight mechanics and air data acquisition
2. Reflective side:
   a. Flight path command and performance envelope protection such as the waypoint following system and the guidance and navigation function
   b. Health manager and fault tolerant control in order to detect and react to possible system failures and malfunctions
3. Decision-making side:
   a. Fault detection and identification
   b. Situation awareness manager
   c. Mission goal manager

The theoretical framework adopted borrows from various disciplines such as aeronautic, automatic control, robotics, computer science and engineering, artificial intelligence, operational research. It integrates algorithms for mission planning, trajectory generation, health monitoring, path following, adaptive control theory, fault tolerant control for fast and robust adaptation. Together, these techniques yield a control architecture that allows meeting strict performance requirements in the presence of complex vehicle dynamics and partial vehicle failures. One of the key requirements is that all maneuvers must be collision free. The temporal and spatial assignments are therefore sometimes separated [61].

Remark 1.1. *Structural mechanics, aerodynamics and propulsion were important to the airplane. Computing power, planning, communications, sensors and other information technology-based capabilities are as important to the unmanned equivalent. The UAS replaces the onboard pilot's functionality with several distinctive integrated capabilities.*

1.3 SMART AUTONOMOUS AIRCRAFT

Smart technology is all about adding intelligence to applications. The goal is to create applications that are smart enough to understand both the concept in which they operate and the data they receive from sensors and automatically use this information to take some type of action.

Smart systems trace their origin to a field of research that envisioned devices and materials that could mimic human muscular and nervous system. The essential idea is to produce a nonbiological system that emulates a biological function. Smart systems consist of systems with sensors and actuators that are either embedded in or attached to the system to form an integral part of it. They respond to stimuli and environmental changes and activate their functions according to these changes. A **smart structure** is a system that incorporates particular functions of sensing and actuation: data acquisition, data transmission, command and control unit, data instructions, action devices [1]. Research on systems with adaptation and learning is being developed. However, higher levels of autonomy that include cognition and reasoning will be required. As stated in [3], everything will, in some sense, be smart; that is every product, every service and every bit of infrastructure will be attuned to the needs of humans it serves and will adapt its behavior to those needs.

A major goal in unmanned aeronautics or aerial robotics is to make autonomous aircraft smarter. Smartness involves the presentation of innovative ideas, approaches, technologies, findings and outcomes of research and development projects.

The development of smart and sustainable vehicles has emerged as one of the most fundamental societal challenges of the next decade. Vehicles should be safer and more environmentally friendly. It is thus important to develop innovative autonomous vehicles and pervasive sensing to monitor the status of the aircraft and the surroundings [45, 48]. A smart structure is a system with actuators, sensors, controllers with built-in intelligence. It involves distributed actuators and microprocessors that analyze the responses from the sensors. It uses integrated control theory to command the actuators to alter system response. For example, the use of smart material actuators has been considered as an effective solution for control surface actuation.

Remark 1.2. *The expression **smart** can be extended to the field of **structural health monitoring** (SHM) where sensor networks, actuators and computational capabilities are used to enable the autonomous aircraft to perform a self-diagnosis with the goal that it can release early warnings about a critical health state, locate and classify damage or even forecast the remaining life term [35].*

Smart materials can be used in vibration control, active shape control, and energy harvesting. These materials have the ability to transduce one form of energy to another which makes them useful as actuators and sensors [13, 31, 52, 57].

The ability to achieve one's goals is a defining characteristic of intelligent behavior [28]. Three topics can be considered:

1. Encoding drives: how the needs of the system are represented.
2. Goal generation: how particular instances of goals are generated from the drives with reference to the current state.
3. Goal selection: how the system determines which goal instances to act on.

Thus, a smart autonomous aircraft can be characterized

1. by being smaller, lighter, faster and more maneuverable and maintainable,
2. by sensing the presence of wind and characterizing its effect on aircraft performance, stability and control,
3. by independent path planning functionality,
4. by improved guidance technologies,
5. by quickly assessing the information from the multi-sensor information,
6. by operating the protection system, providing aircraft envelope protection and adapting the flight controls when some degradation in performance and control can be anticipated,
7. by activating and managing the protection system and providing the autopilot with feedback on the system status,
8. by modifying the aircraft **flight envelope** by use of the flight control system to avoid conditions where flight could potentially be uncontrollable,
9. by adapting the control system to maintain safe flight with the reduced flight envelope,
10. by automatically generating a **flight plan** that optimizes multiple objectives for a predefined mission goal,
11. by optimizing altitude transitions through weather,
12. by delivering actionable intelligence instead of raw information by an increased system-sensor automation,
13. by improving mission performance with situational awareness and weather sensing,
14. by using advanced airborne sense-and-avoid technologies,
15. by detecting, tracking and identifying the time critical targets,
16. by constructing a mission plan with uncertain information satisfying different and possibly conflicting decision objectives,
17. by applying verification methods to these algorithms.

A perspective emerges in this book through a discussion of mission capabilities unique to unmanned aerial systems and an explanation of some processes used to develop those capabilities to achieve improved efficiency, effectiveness and survivability.

1.4 OUTLINE OF THE BOOK

This book is organized as follows:

1. Chapter 1: **Introduction**: This chapter explains the book's purpose and scope. It examines the current unmanned aerial vehicles while also presenting autonomous systems and smart aircraft.
2. Chapter 2: **Modeling**: The main topics presented in this chapter are reference frames and coordinate systems followed by kinematic and dynamic models for airplanes. In this book, flight is restricted to the atmosphere. Flight models involve high order nonlinear dynamics with state and input constraints. For translational dynamics, it is customary to either completely neglect the rotational dynamics by assuming instantaneous changes of attitude or to treat the aircraft attitude dynamics as an actuator for the navigational control system. The presentation of six degrees of freedom nonlinear equations of motion for a fixed wing aircraft over a flat, non rotating Earth modeled by twelve state equations, follows. More innovative modeling approaches such as Takagi–Sugeno modeling, fuzzy modeling and linear hybrid automation are then considered and mission tools presented. Finally the model of the atmosphere is introduced.
3. Chapter 3: **Flight Control**: In this chapter, airplane control problems are considered and solved, in order to achieve motion autonomy. A common control system strategy for an aircraft is a two loops structure where the attitude dynamics are controlled by an inner loop, and the position dynamics are controlled by an outer loop. In the low level control system, algorithms depend on the type of aircraft, its dynamical model, the type of control design used and finally the inputs and sensors choice. Autonomous aircraft can encounter a wide range of flight conditions in an atmospheric flight and, in some cases, disturbances can be as strong as the small aircraft's own control forces. Topics related to classical linear and nonlinear control methods are first presented. Then fuzzy control is introduced followed by filtering. Case studies are presented to illustrate the theoretical methods presented in this chapter.
4. Chapter 4: **Flight Planning**: Flight planning is defined as finding a sequence of actions that transforms some initial state into some desired goal state. This chapter begins with path and trajectory planning: trim trajectories followed by maneuvers without wind. The optimal approach can be used to realize the minimum time trajectory or minimum energy to increase the aircraft's endurance. Trajectory generation refers to determining a path in free configuration space

between an initial configuration of the aircraft and a final configuration consistent with its kinematic and dynamic constraints. Zermelo's problem is then considered; it allows the study of aircraft's trajectories in the wind. In the middle of the chapter, guidance and collision/obstacle avoidance are considered. Planning trajectories is a fundamental aspect of autonomous aircraft guidance. It can be considered as a draft of the future guidance law. The guidance system can be said to fly the aircraft on an invisible highway in the sky by using the attitude control system to twist and turn the aircraft. Guidance is the logic that issues the autopilot commands to accomplish certain flight objectives. Algorithms are designed and implemented such that the motion constraints are respected while following the given command signal. Flight planning is also the process of automatically generating alternate paths for an autonomous aircraft, based on a set of predefined criteria, when obstacles are detected in the way of the original path. Aircraft operate in a three-dimensional environment where there are static and dynamic obstacles and they must avoid turbulence and storms. As obstacles may be detected as the aircraft moves through the environment or their locations may change over time, the trajectory needs to be updated and satisfy the boundary conditions and motion constraints. Then, mission planning is introduced by route optimization and fuzzy planning. Case studies are presented to illustrate the methods presented in this chapter.

5. Chapter 5: **Flight Safety**: In the first part of the chapter, situation awareness is introduced. Integrated navigation systems are important components. Situational awareness is used for low level flight control and for flight and mission planning which constitute the high level control elements. Data are coming from different kinds of sensors, each one being sensitive to a different property of the environment, whose data can be integrated to make the perception of an autonomous aircraft more robust and to obtain new information otherwise unavailable. Due to the uncertainty and imprecise nature of data acquisition and processing, individual informational data sources must be appropriately integrated and validated. A part of these data is used in pre-flight or pre-mission planning and is updated onboard as the mission proceeds. The navigation problem is to find a desired trajectory and its open loop steering control. An integrated navigation system is the combination of an onboard navigation solution providing position, velocity and attitude as derived from accelerometer and gyro-inertial sensors. This combination is accomplished with the use of diverse Kalman filter algorithms as well as Monte Carlo approaches. SLAM and geolocation are also presented. In autonomous aircraft, the onboard control system must be able to interpret the meaning of the system health information and decide on the appropriate course

of action. This requires additional processing onboard the unmanned platform to process and interpret the health information and requires that the health monitoring and vehicle control systems be capable of integration. The benefit of integrating systems health monitoring with the command and control system in unmanned systems is that it enables management of asset health by matching mission requirements to platform capability. This can reduce the chance of mission failures and loss of the platform due to a faulty component. Integrated system health monitoring is investigated, presenting some diagnostic tools and approaches. As there are uncertainties in the information, usually several scenarios are considered and a trade-off solution is offered. The smart autonomous aircraft must be able to overcome environmental uncertainties such as modeling errors, external disturbances and an incomplete situational awareness. In the third part of the chapter, fault tolerant flight control is considered for LTI and LPV formulations followed by model reference adaptive control and maneuver envelope determination. The last part of the chapter concerns a fault tolerant planner detailing trim state, reliability analysis, safety analysis of obstacle avoidance and mission decision.

6. Chapter 6: **General Conclusions**: Some general conclusions are given while also surveying the potential future environment of smart autonomous aircraft.

REFERENCES

1. Akhras, G. (2008): *Smart materials and smart systems for the future*, Canadian Military Journal, pp. 25–32.
2. Angulo, C.; Goods, L. (2007): *Artificial Intelligence Research and Development*, IOS Press.
3. Astrom, K. J.; Kumar, P. R. (2014): *Control: a perspective*, Automatica, vol. **50**, pp. 3–43.
4. Atkins, E. M.; Abdelzaher, T. F.; Shin, K.G.; Durfee, E.H. (2001): *Planning and resource allocation for hard real time fault tolerant plan execution*, Autonomous Agents and Multi-Agents Systems, Vol. **4**, pp. 57–78.
5. Austin, R. (2010): *Unmanned Aircraft Systems: UAVs Design, Development and Deployement*, AIAA Press.
6. Beard, R.; Kingston, D., Quigley, M.; Snyder, D.; Christiansen, R.; Johnson, W.; McLain, T.; Goodrich, M. (2005): *Autonomous vehicle technologies for small fixed-wing UAVs*, AIAA Journal of Aerospace Computing, Information and Communication. Vol. **2**, pp. 92–108.
7. Beard, R.; McLain, T. (2012): *Small Unmanned Aircraft: Theory and Practice*, Princeton University Press.
8. Belta, C.; Bicchi, A.; Egersted, M.; Frazzoli, E.; Klavins, E.; Pappas, G. (2007): *Symbolic planning and control of robot motion*, IEEE Robotics and Automation Magazine, vol. **14**, pp. 61–70.

9. Bestaoui, Y.; Dicheva, S. (2010): *3D flight plan for an autonomous aircraft*, 48th AIAA Aerospace Sciences Meeting Including the New Horizons Forum and Aerospace Exposition, AIAA paper 2010–415.
10. Bestaoui, Y. (2011): *Bridge monitoring by a lighter than air robot*, 49th AIAA Aerospace Sciences Meeting, Orlando, Fl, paper AIAA 2011–81.
11. Bollino, K. P.; Lewis, L. R.; Sekhavat, P.; Ross, I. M. (2007): *Pseudo-spectral optimal control: a clear road for autonomous intelligent path planning* , AIAA Infotech@aerospace conference, Rohnert Park, Ca, USA, AIAA 2007–2831.
12. Bradley, J. M.; Atkins, E. M. (2014): *Cyber-physical optimization for unmanned aircraft systems*, AIAA Journal of Aerospace Information Systems, vol. **11**, pp. 48–59.
13. Braggs, M. B.; Basar T.; Perkins W. R.; Selig M. S.; Voulgaris P. C.; Melody J. W. (2002): *Smart icing system for aircraft icing safety*, 40th AIAA Aerospace Sciences Meeting and Exhibit, paper AIAA 2002–813.
14. Budiyono, A.; Riyanto, B; Joelianto, E. (2009): *Intelligent Unmanned Systems: Theory and Applications*, Springer.
15. Bullo, F.; Cortes, J.; Martinez, S. (2009): *Distributed Control of Robotic Networks*; Princeton series in Applied Mathematics.
16. Cevic, P.; Kocaman, I.; Akgul, A. S.; Akca, B. (2013): *The small and silent force multiplier: a swarm UAV electronic attack*, Journal of Intelligent and Robotic Systems, vol. **70**, pp. 595–608.
17. Choset, H.; Lynch, K.; Hutchinson, S.; Kantor, G.; Burgard, W.; Kavraki, L.; Thrun, S. (2005): *Principles of Robot Motion, Theory, Algorithms and Implementation*, The MIT Press.
18. Clothier, R. A.; Palmer, J. L.; Walker, R. A., Fulton, N. L. (2011): *Definition of an airworthiness certification framework for civil unmanned aircraft system*, Safety Science, vol. **49**, pp. 871–885.
19. Dadkhah, N.; Mettler, B. (2012): *Survey of motion planning literature in the presence of uncertainty: considerations for UAV guidance*, Journal of Intelligent and Robotics Systems, vol. **65**, pp. 233–246.
20. Dalamagkidis, K.; Valavanis, K.; Piegl, L. (2010): *On Integrating Unmanned Aircraft Systems into the National Airspace System*, Springer.
21. Dimitrienko, A. G.; Blinov, A. V.; Novikov, V. N. (2011): *Distributed smart system for monitoring the state of complex technical objects*, Measurement Techniques, vol. **54**, pp. 235–239.
22. Department of Defense (USA) (2013): *Unmanned systems integrated roadmap: FY 2013–2038*, ref. number 14–S–0553 (129 pages).
23. EASA (2013): *Airworthiness certification of unmanned aircraft system (UAS)*, European aviation safety agency, report E. Y013–01.
24. Florio, F. (2006): *Air Worthiness*, Elsevier.
25. Francis, M. S. (2012): *Unmanned air systems: challenge and opportunity*, AIAA Journal of Aircraft, vol. **49**, pp. 1652–1665.
26. Fregene, K. (2012): *Unmanned aerial vehicles and control: Lockheed martin advanced technology laboratories*, IEEE Control System Magazine, vol. **32**, pp. 32–34.
27. Goerzen, C.; Kong, Z.; Mettler, B., (2010): *A survey of motion planning algorithms from the perspective of autonomous UAV guidance*, Journal of Intelligent and Robotics Systems, vol. **20**, pp. 65–100.

28. Hawes, N. (2011): *A survey of motivation frameworks for systems*, Artificial Intelligence, vol. **175**, pp. 1020–1036.
29. Jawhar, I.; Mohamed, N.; Al-Jaroodi, J.; Zhang, S. (2014): *A framework for using UAV for data collection in linear wireles sensor networks*, Journal of Intelligent and Robotics Systems, vol. **6**, pp. 437–453.
30. Jarvis, P. A.; Harris, R.; Frost, C. R. (2007): *Evaluating UAS autonomy operations software in simulation*, AIAA Unmanned Conference, paper AIAA 2007–2798.
31. Jeun, B. H.; Whittaker, A. (2002): *Multi-sensor information fusion technology applied to the development of smart aircraft*, Lockheed Martin Aeronautical Systems Marietta Ga report, pp. 1–10.
32. Jonsson, A. K. (2007): *Spacecraft autonomy: intelligence software to increase crew, spacecraft and robotics autonomy*, AIAA Infotech@ Aerospace Conference and Exhibit, paper AIAA 2007–2791.
33. Knoll, A. (2010): *The dawn of the age of autonomy*, report TUM-I144, Institut fur informatik, Technische Universitat Munchen.
34. Krozel, J.; Penny, S.; Prete, J.; Mitchell, J. S. (2007): *Automated route generation for avoiding deterministic weather in transition airspace.* AIAA Journal of Guidance, Control and Dynamics, vol. **30**, pp. 144–153.
35. Kurtoglu, T.; Johnson, S. B.; Barszcz, E.; Johnson, J. R.; Robinson, P. I. (2008): *Integrating system health management into the early design of aerospace system using functional fault analysis*, Inter. conference on Prognostics and Health Management.
36. Lam, T.M. (ed) (2009): *Aerial Vehicles*, In-Tech, Vienna, Austria.
37. Lan, G.; Liu, H. T. (2014): *Real time path planning algorithm for autonomous border patrol: Design, simulation and experimentation*, Journal of Intelligent and Robotics Systems, vol. **75**, pp. 517–539, DOI 10.1007/s10846-013-9841-7.
38. Lapierre, L.; Zapata, R. (2012): *A guaranteed obstacle avoidance guidance system*, Autonomous Robots, vol. **32**, pp. 177–187.
39. Li, Z.; Bruggermann, T.S.; Ford, J.S.; Mejias, L.; Liu Y (2012): *Toward automated power line corridor monitoring using advanced aircraft control and multisource feature fusion*, Journal of Field Robotics, vol. **29**, pp. 4–24.
40. Ludington, B.; Johnson, E.; Vachtsevanos, A. (2006): *Augmenting UAV autonomy GTMAX*, IEEE Robotics and Automation Magazine, vol. **21**, pp. 67–71.
41. Musial, M. (2008): *System Architecture of Small Autonomous UAV*, VDM.
42. Nonami, K.; Kendoul, F.; Suzuki, S.; Wang, W.; Nakazawa, D. (2010): *Autonomous Flying Robots: Unmanned Aerial Vehicles and Micro-aerial Vehicles*, Springer.
43. Panella, I. (2008): *Artificial intelligence methodologies applicable to support the decision-making capability onboard unmanned aerial vehicles*, IEEE Bio-inspired, Learning and Intelligent Systems for Security Workshop, pp. 111–118.
44. Quaritsch, M.; Kruggl, K.; Wischonmig-Stud, L.; Battacharya, S.; Shah, M.; Riner, B. (2010): *Networked UAV as aerial sensor network for disaster management applications*, Electrotechnik and Informations technik, vol. **27**, pp. 56–63.
45. Sanchez-Lopez, T.; Ranasinghe, D. C.; Patkai, B.; McFarlane, D. (2011): *Taxonomy, technology and application of smart objects*, Information Systems

Frontiers, vol. **13**, pp. 281–300.

46. Song, B. D.; Kim, J.; Park, H.; Morrison, J. R.; Sjin D. (2014): *Persistent UAV service: an improved scheduling formulation and prototypes of system components*, Journal of Intelligent and Robotics Systems, vol. **74**, pp. 221–232.
47. Stefanakis, D.; Hatzopoulos, J. N.; Margaris, N.; Danalatos, N. (2013): *Creation of a remote sensing unmanned aerial system for precision agriculture and related mapping application*, Conference on American Society for Photogrammetry and Remote Sensing, Baltimore, Md, USA.
48. Suzuki, K. A.; Filho, P. K.; Darrison, J. (2012): *Automata battery replacement system for UAV: analysis and design*, Journal of Intelligent and Robotics Systems, vol. **65**, pp. 563–586.
49. Tarhan, A. F.; Koyuncu, E.; Hasanzade, M.; Ozdein, U.; Inalhan, G. (2014): *Formal intent based flight management system design for unmanned aerial vehicles*, Int. Conference on Unmanned Aircraft Systems, pp. 984–992, DOI 978-1-4799-2376-2.
50. Tavana, M. (2004): *Intelligence flight system (IFSS): a real time intelligent decision support system for future manned spaceflight operations at Mission Control Center*, Advances in Engineering Software, vol. **35**, pp. 301–313.
51. Teinreiro Machado, J. A.; Patkai, B.; Rudas, I. (2009): *Intelligent Engineering Systems and Computational Cybernetics*, Springer.
52. Tewari, A. (2011): *Advanced Control of Aircraft, Spacecrafts and Rockets*, Wiley Aerospace Series.
53. Tokekar, P.; Vander Hook, H.; Muller, D.; Isler, V. (2013): *Sensor planning for a symbiotic UAV and UGV system for precision agriculture*, Technical report, Dept. of Computer Science and Engineering, Univ. of Minnesota, TR13–010.
54. Valavanis, K.; Oh, P.; Piegl, L. A. (2008): *Unmanned Aircraft Systems*, Springer.
55. Van Blyenburgh, P. (2012): *European UAS industry and market issues*, ICAO UAS seminar, Lima, Peru, pp. 1–29.
56. Visconti, G. (2008): *Airborne measurements and climatic change science: aircraft, balloons and UAV*, Mem SAIt, vol. **79**, pp. 849–852.
57. Wang, Y.; Hussein, I. I.; Erwin, R. S. (2011): *Risk-based sensor management for integrated detection and estimation*, AIAA Journal of Guidance, Control and Dynamics, vol. **34**, pp. 1767–1778.
58. Wang, Y.; Liu, J. (2012): *Evaluation methods for the autonomy of unmanned systems*, Chinese Science Bulletin, vol. **57**, pp. 3409-3418.
59. Webster, M.; Cameron, N.; Fisher, M.; Jump, M. (2014): *Generating certification evidence for autonomous unmanned aircraft using model checking and simulation*, AIAA Journal of Aerospace Information System, vol. **11**, pp. 258–278.
60. Wu, F.; Zilberstein, S.; Chen, X. (2011): *On line planning for multi-agent systems with bounded communication*, Artificial Intelligence, vol. **175**, pp. 487–511.
61. Xargay, E.; Dobrokhodov, V.; Kaminer, I.; Pascoal, A.; Hovakimyan, N.; Cao, C. (2012): *Time critical cooperative control of multiple autonomous vehicles*, IEEE Control Systems Magazine, vol. **32**, pp. 49–73.
62. Yanushevsky, R.(2011): *Guidance of Unmanned Aerial Vehicles*, CRC Press.

63. Zarco-Tejada, P. J.; Guillen-Climent, M. L.; Hernandez-Clemente, R.; Catalina, A. (2013): *Estimating leaf carotenoid content in vineyards using high resolution hyperspectral imagery acquired from an unmanned aerial vehicle*, Journal of Agricultural and Forest Meteorology, vol. **171**, pp. 281-294.
64. Zhang, C.; Kavacs, J. M. (2012): *The application of small unmanned aerial system for precision agriculture: a review*, Precision Agriculture, vol. **3**, pp. 693–712.

2 Modeling

ABSTRACT

The main topics presented in this chapter are reference frames and coordinate systems followed by kinematic and dynamic models for airplanes. In this book, flight is restricted to the atmosphere. Flight models involve high order nonlinear dynamics with state and input constraints. For the considered translational dynamics, it is customary to either completely neglect the rotational dynamics by assuming instantaneous changes of attitude or to treat the aircraft attitude dynamics as an actuator for the navigational control system. The presentation of six degrees of freedom nonlinear equations of motion for a fixed wing aircraft over a flat, non rotating Earth modeled by twelve state equations, follows. More innovative modeling approaches such as Takagi–Sugeno modeling, fuzzy modeling and linear hybrid automation are then considered and mission tools presented. Finally the model of the atmosphere is introduced.

2.1 INTRODUCTION

Mathematical models that capture the underlying physical phenomena are at the heart of many control and planning algorithms. **Modeling** is a vast domain of knowledge and only some basic topics concerning smart autonomous aircraft are presented in this chapter.

A natural approach for reducing the complexity of large scale systems places a hierarchical structure on the system architecture. For example, in the common two-layer planning and control hierarchies, the planning level has a coarser system model than the lower control level. Furthermore, in general, an aircraft has two separate classes of control systems: the inner loop and the outer loop. One of the main challenges is the extraction of a hierarchy of models at various levels of abstraction while preserving properties of interest. The notion of abstraction refers to grouping the system states into equivalence classes. A hierarchy can be thought of as a finite sequence of abstractions. Consistent abstractions are property preserving abstractions, which can be discrete or continuous [66]. A fixed-wing aircraft has four actuators: forward thrust, ailerons, elevator and rudder. The aircraft's thrust provides acceleration in the forward direction and the control surfaces exert various moments on the aircraft: rudder for yaw torque, ailerons for roll torque and elevator for pitch torque. The general aircraft's configuration has six dimensions: three for the position and three for the orientation. With six degrees of freedom and four control inputs, the aircraft is an under-actuated system. Traditionally, aircraft dynamic models have been analytically determined from principles such as Newton–Euler laws for rigid-body dynamics. The parameters of these dynamic systems are usually determined through costly and time-consuming wind tunnel testing. These methods, while useful, have limitations when applied to small autonomous aircraft due to several differences which include [62, 69]:

1. Low Reynolds number and airspeed,
2. Increased dynamic rates due to decreased mass and moments of inertia,
3. Dominance of propulsion dynamic forces and moments versus aerodynamic body forces and moments,
4. Asymmetric or atypical designs,
5. Aerobatic maneuvers not possible by manned aircraft.

More information about classical aircraft modeling can be found in some textbooks such as [4, 18, 27, 38, 74, 104]. More innovative approaches for modeling small autonomous aircraft are also presented in this chapter.

2.2 REFERENCE FRAMES AND COORDINATE SYSTEMS

Frames are models of physical objects consisting of mutually fixed points while coordinate systems are mathematical abstracts. Frames are models of physical references whereas **coordinate systems** establish the association with Euclidean space [104]. A reference frame is a set of rigidly related points that can be used to establish distances and directions. An inertial frame is a reference frame in which Newton–Euler laws apply [4, 58]. There are coordinated systems in which the phenomena of interest are most naturally expressed.

The precise definition of a number of **Cartesian coordinate reference frames** is fundamental to the process of aircraft performance, navigation, flight and mission planning [72]. For aircraft performance, position and velocity are considered with respect to the atmosphere. For navigation, flight and mission planning, position and velocity are considered with respect to the Earth. Depending on the objective, there may be some particular coordinate system in which the position and velocity make sense. There are many different reference frames useful for navigation, guidance and control, such as heliocentric, inertial, Earth, geographical frame, body frame, wind frame [17, 33, 36, 41, 94]. Each frame is an orthogonal right-handed coordinate frame axis set. **Earth** can considered to be flat, round or oblate, rotating or not [27, 38, 86, 97].

Remark 2.1. *As this book is restricted to atmospheric flight, only some reference frames are detailed in this chapter.*

2.2.1 INERTIAL FRAME

An **inertial reference frame (i-frame)** is a coordinate frame in which Newton's laws of motion are valid. It is neither rotating nor accelerating. The location of the origin may be any point that is completely unaccelerated (inertial) such as the Great Galactic Center.

2.2.2 EARTH-CENTERED REFERENCE FRAME

For navigation over the Earth, it is necessary to define axis sets which allow the inertial measurements to be related to the cardinal directions of the Earth. Cardinal directions have a physical significance when navigating in the vicinity of the Earth.

2.2.2.1 Geocentric-inertial frame

The **geocentric-inertial frame (gi-frame)** has its origin at the center [72] of the Earth and axes which are non-rotating with respect to the fixed stars, defined by the axes $O_{x_i}, O_{y_i}, O_{z_i}$ with O_{z_i} coincident with the Earth's polar axis (which is assumed to be invariant in direction). It is an inertial reference frame which is stationary with respect to the fixed stars, the origin of which is located at the center of the Earth. It is also called **Earth-centered inertial frame** (ECIF) in [72].

2.2.2.2 Earth frame

The **Earth frame (e-frame)** has its origin at the center of the Earth and axes which are fixed with respect to the Earth, assumed to be a uniform sphere, defined by the axes $O_{x_e}, O_{y_e}, O_{z_e}$. The axis O_{x_e} lies along the Greenwich meridian with the Earth's equatorial plane. The axis O_{z_e} is along the Earth's polar axis. The Earth frame rotates with respect to the inertial frame, at a rate ω_{ec} about the axis O_z. This frame is also called **Earth-centered Earth fixed frame** (ECEF) in [72].

2.2.3 GEOGRAPHIC FRAME

2.2.3.1 Navigation frame

The **navigation frame (n-frame)** is a local geographic frame which has its origin, P, fixed to any arbitrary point that may be free to move relative to the Earth, assumed to be a uniform sphere. These axes are aligned with the directions of North, East and the local vertical Down **(NED)** or East, North and Up **(ENU)**, both defined as right-handed reference frames. A grid blanketing the Earth's surface and determining any point on Earth consists of lines of longitude and latitude. Longitude is divided into different meridians from 0 to 180 degrees with the positive direction starting at the Greenwich meridian in an Easterly direction. Latitude is measured from the Equator, positive to the North from 0 to 90 degrees and negative South.

When going from the Earth-frame to a geographic frame, the convention is to perform the longitude rotation first, then the latitude. The coordinate transformation matrix of geographic frame with respect to Earth is:

$$\mathbf{R}_{GE} = \begin{pmatrix} -\sin\lambda\cos\ell & -\sin\lambda\sin\ell & \cos\lambda \\ -\sin\ell & \cos\ell & 0 \\ -\cos\lambda\cos\ell & -\cos\lambda\sin\ell & -\sin\lambda \end{pmatrix} \tag{2.1}$$

where λ represent the latitude and ℓ the longitude.

The turn rate of the navigation frame with respect to the Earth fixed frame ω_{en}, is governed by the motion of the point P, with respect to the Earth, often referred to as the **transport rate**.

2.2.3.2 Tangent plane frame

The **tangent plane frame (t-frame)** is a geographic system with its origin on the Earth's surface: a frame translating with the vehicle center of mass with North, East and Down **(NED)** fixed directions, which moves with it. These axes are aligned with the directions of North, East and the local vertical Down **(NED)** or East, North and Up **(ENU)**, both defined as right-handed reference frames.

The **flight path coordinate system** relates the velocity vector of the aircraft with respect to the Earth to the geographic system via two angles: the flight path angle γ and the heading angle χ. The **heading angle** χ is measured from North to the projection of V (the aircraft velocity relative to the wind) in the local tangent plane and the flight path angle γ takes vertically up to V.

The transformation matrix is given by:

$$\mathbf{R}_{VG} = \begin{pmatrix} \cos\gamma & 0 & -\sin\gamma \\ 0 & 1 & 0 \\ \sin\gamma & 0 & \cos\gamma \end{pmatrix} \begin{pmatrix} \cos\chi & \sin\chi & 0 \\ -\sin\chi & \cos\chi & 0 \\ 0 & 0 & 1 \end{pmatrix} = \tag{2.2}$$

$$= \begin{pmatrix} \cos\gamma\cos\chi & \cos\gamma\sin\chi & -\sin\gamma \\ -\sin\chi & \cos\chi & 0 \\ \sin\gamma\cos\chi & \sin\gamma\sin\chi & \cos\gamma \end{pmatrix}$$

The geographic local horizon **line of sight** (LOS) is defined in terms of **azimuth** and **elevation** relative to the geographic frame: NED frame.

2.2.4 BODY FRAME

In the **body-fixed frame (b-frame)**, the origin and axes of the coordinate system are fixed with respect to the nominal geometry of the aircraft. The axes can be aligned with aircraft reference direction or aircraft stability axes system. As the aircraft has, in general, a plane of symmetry then x_B and z_B lie in that plane. The x_B direction is aligned with the principal axis of symmetry of the aircraft. The origin of the b-frame is defined at an invariant point such as the center of volume, the center of mass, of the aircraft.

Remark 2.2. *In aircraft dynamics, the* ***instantaneous acceleration center of rotation*** *(IACR) of an aircraft is the point on the aircraft that has zero instantaneous acceleration. The instantaneous acceleration center of rotation of a rigid body is related to, but distinct from, the center of rotation [99].*

Direction cosines matrices, quaternions, Euler angles can describe finite rotations between frames. Rigid body attitude is often represented using three or four parameters. A 3×3 direction cosine matrix of Euler parameters can be used to describe the orientation of the body achieved by three successive rotations with respect to some fixed frame reference. Assembled with the three position coordinates $\eta_1 = (x, y, z)^T$, they allow the description of the situation of an aircraft [37, 38, 97, 104]. Unit quaternions and the axis angle representation use four parameters to represent the attitude. Three parameters representation of attitude include the Euler angles or Rodrigues parameters. These three parameter sets can be viewed as embedded subsets of $\mathbb{R}^3$, thus allowing methods of analysis that are suited to the Euclidean space $\mathbb{R}^3$. The usual minimal representation of orientation is given by a set of three Euler angles: roll ϕ, pitch θ and yaw ψ.

The body-fixed frame is an orthogonal reference frame whose set of axes is aligned with the roll, pitch and yaw axes of the aircraft. The position of all points belonging to the rigid body with respect to the inertial-fixed frame can be completely defined by knowing the orientation of a body-fixed frame to the aircraft body and the position of its origin with respect to the reference frame.

Adopting this formulation, the rotation matrix $\mathbf{R}$ can be written as a function of $\eta_2 = (\phi, \theta, \psi)^T$ given by:

$$\mathbf{R}_{GB} = \mathbf{R}(\eta_2) = \mathbf{R}_z(\psi)\mathbf{R}_y(\theta)\mathbf{R}_x(\phi) \tag{2.3}$$

with

$$\mathbf{R}_x(\phi) = \begin{pmatrix} 1 & 0 & 0 \\ 0 & \cos\phi & -\sin\phi \\ 0 & \sin\phi & \cos\phi \end{pmatrix}$$

$$\mathbf{R}_y(\theta) = \begin{pmatrix} \cos\theta & 0 & \sin\theta \\ 0 & 1 & 0 \\ -\sin\theta & 0 & \cos\theta \end{pmatrix}$$

$$\mathbf{R}_z(\psi) = \begin{pmatrix} \cos\psi & -\sin\psi & 0 \\ \sin\psi & \cos\psi & 0 \\ 0 & 0 & 1 \end{pmatrix}$$

This transformation, also called the direction cosine matrix (DCM), can be expressed as:

$$\mathbf{R}_{GB} = \mathbf{R}_{BG}^T = \mathbf{R}(\eta_2) =$$

$$\begin{pmatrix} \cos\psi\cos\theta & \mathbf{R}_{12} & \mathbf{R}_{13} \\ \sin\psi\cos\theta & \mathbf{R}_{22} & \mathbf{R}_{23} \\ -\sin\theta & \cos\theta\sin\phi & \cos\theta\cos\phi \end{pmatrix} \tag{2.4}$$

where the following notation is used: $\mathbf{R}_{ij}$ represents the element on the ith line and the jth column of $\mathbf{R}_{GB}$:

$$\begin{aligned} \mathbf{R}_{12} &= -\sin\psi\cos\phi + \cos\psi\sin\theta\sin\phi \\ \mathbf{R}_{13} &= \sin\psi\sin\phi + \cos\psi\sin\theta\cos\phi \\ \mathbf{R}_{22} &= \cos\psi\cos\phi + \sin\psi\sin\theta\sin\phi \\ \mathbf{R}_{23} &= -\cos\psi\sin\phi + \sin\psi\sin\theta\cos\phi \end{aligned}$$

The Euler angles can be obtained by the following relations:

$$\phi = atan2(\mathbf{R}_{32}, \mathbf{R}_{33}) \tag{2.5}$$

$$\theta = \arcsin(-\mathbf{R}_{31}) \tag{2.6}$$

$$\psi = atan2(\mathbf{R}_{21}, \mathbf{R}_{11}) \tag{2.7}$$

The function $atan2(y, x)$ computes the arctangent of the ratio y/x. It uses the sign of each argument to determine the quadrant of the resulting angle.

2.2.5 WIND FRAME

The **wind frame (w-frame)** is a body carried coordinate system in which the x_w axis is in the direction of the velocity vector of the aircraft relative to the air flow. The z_w axis is chosen to lie in the plane of symmetry of the aircraft and the y_w axis to the right of the plane of symmetry. In the aircraft wind axes system, the aerodynamic forces and moments on an aircraft are produced by its relative motion with respect to the air and depend on the orientation of the aircraft with respect to the airflow. Therefore, two orientation angles with respect to the relative wind are needed to specify the aerodynamic forces and moments: the angle of attack α and the side slip angle β.

The transformation matrix of wind with respect to body coordinates is given by:

$$\mathbf{R}_{WB} = \mathbf{R}_{BW}^T = \begin{pmatrix} \cos\alpha\cos\beta & \sin\beta & \sin\alpha\cos\beta \\ -\cos\alpha\sin\beta & \cos\beta & -\sin\alpha\sin\beta \\ -\sin\alpha & 0 & \cos\alpha \end{pmatrix} \tag{2.8}$$

Remark 2.3. *In flight dynamics problems, the instantaneous interaction of the airframe with the air mass is interesting. The atmosphere can be thought of as a separate Earth fixed system to which appropriate extra components are added to account for winds, gusts and turbulence.*

2.2.6 SPECIAL EUCLIDEAN GROUP

The special Euclidean group is formed by the group of rigid body transformations [37]. The trajectories of an aircraft evolve on the special Euclidean group of order 3, $SE(3) = \mathbb{R}^3 \times SO(3)$, which includes the position and the orientation of the rigid body.

Definition 2.1. ***Special Orthogonal Matrix Group SO(3)**: The rotation matrix* $\mathbf{R}$ *being an orthogonal matrix, the set of such matrices constitutes the special orthogonal matrix group* $SO(3)$ *defined as:*

$$SO(3) = \left\{\mathbf{R} \in \mathbb{R}^{3\times 3}, \mathbf{R}\mathbf{R}^T = \mathbf{R}^T\mathbf{R} = I_{3\times 3}, \det(\mathbf{R}) = 1\right\} \tag{2.9}$$

For a matrix $\mathbf{R} \in \mathbb{R}^{3\times 3}$, the symmetric and anti-symmetric projection operations are:

$$P_s(\mathbf{R}) = \frac{1}{2}\left(\mathbf{R} + \mathbf{R}^T\right) \tag{2.10}$$

$$P_a(\mathbf{R}) = \frac{1}{2}\left(\mathbf{R} - \mathbf{R}^T\right) \tag{2.11}$$

The configuration of the aircraft can be completely defined by associating the orientation matrix and the body-fixed frame origin position vector, $\eta_1 = (x, y, z)^T$, with respect to the inertial frame using homogeneous matrix formulation as:

$$A_M = \begin{pmatrix} \mathbf{R}(\eta_2) & \eta_1 \\ 0_{3\times 3} & 1 \end{pmatrix} \tag{2.12}$$

Definition 2.2. ***Special Euclidean Group SE(3)**: The special Euclidean group of rigid-body transformations in three dimensions* $SE(3)$ *is defined by:*

$$SE(3) = \left\{ A_M | A_M = \begin{pmatrix} \mathbf{R}(\eta_2) & \eta_1 \\ 0_{3\times 3} & 1 \end{pmatrix} \quad \mathbf{R} \in SO(3) \quad \eta_1 \in \mathbb{R}^3 \right\} \tag{2.13}$$

Theorem 2.1

The set $SE(3)$ is a Lie group. This group consists of pairs $(\mathbf{R}, b)$, $\mathbf{R}$ being a rotation matrix and b a vector with a binary operation o:

$$(\mathbf{R}_1, b_1)o(\mathbf{R}_2, b_2) = (\mathbf{R}_1\mathbf{R}_2, \mathbf{R}_1 b_2 + b_1) \tag{2.14}$$

and inverse

$$(\mathbf{R}, b)^{-1} = \left(\mathbf{R}^{-1}, -\mathbf{R}^{-1}b\right) \tag{2.15}$$

■

Definition 2.3. ***Lie Algebra:*** *The Lie algebra of the $SO(3)$ group is denoted by $so(3)$ and is given by*

$$so(3) = \{\mathbf{R} \in \mathbb{R}^{3\times3}, \mathbf{R} = -\mathbf{R}^T\} \tag{2.16}$$

$\mathbf{R}$ *being an anti-symmetric matrix.*

Given a curve $C(t) : [-a, a] \rightarrow SE(3)$, an element $S(t)$ of the Lie algebra $\mathtt{se}(3)$ can be associated to the tangent vector $\dot{C}(t)$ at an arbitrary configuration $A(t)$ by:

$$S(t) = A_M^{-1}(t)\dot{A_M}(t) = \begin{pmatrix} Sk(\Omega) & \mathbf{R}^T\dot{\eta}_1 \\ 0 & 0 \end{pmatrix} \tag{2.17}$$

where

$$Sk(\Omega) = \mathbf{R}^T(t)\dot{\mathbf{R}}(t) \tag{2.18}$$

is a 3×3 skew symmetric operator on a vector defined by:

$$Sk(\Omega) = Sk((p, q, r)^T) = \begin{pmatrix} 0 & -r & q \\ r & 0 & -p \\ -q & p & 0 \end{pmatrix} \tag{2.19}$$

such that

$$\forall x, y \in \mathbb{R}^3 : Sk(y)x = y \times x \tag{2.20}$$

The rotation matrix derivative can be deduced as:

$$\dot{\mathbf{R}}(t) = Sk(\Omega)\mathbf{R}(t) \tag{2.21}$$

A particular parametrization of the vector space of anti-symmetric or skew-symmetric matrices $so(3)$ group is the angle-axis parametrization given by:

$$\mathbf{R}(\theta, U) = \mathbf{I}_{3\times3} + \sin\theta Sk(U) + (1 - \cos\theta)Sk(U)^2 \tag{2.22}$$

where $U \in S^2$ denotes the axis of rotation and $\theta \in [0, \pi]$ denotes the rotation angle [14].

A curve on $SE(3)$ physically represents a motion of the rigid body. If $(V(t), \Omega(t))$ is the pair corresponding to $S(t)$, then V is the linear velocity of the origin O_m of the body frame with respect to the inertial frame, while Ω physically corresponds to the angular velocity of the rigid body.

2.3 KINEMATIC MODELS

Kinematics introduces time and models the motion of vehicles without consideration of forces.

2.3.1 TRANSLATIONAL KINEMATICS

There are two critical frames of reference: an inertial frame, which is fixed with respect to the ground (in this case, a North-East-Down frame), and the air relative frame which is the aircraft motion relative to the surrounding air. Air-relative flight path angle γ is the climb of the velocity vector with respect to the wind. Similarly, the air relative heading χ is the heading with respect to lateral wind. The bank angle σ is the rotation of the lift around the velocity vector. Wind is defined in inertial space and represents the motion of the atmosphere relative to the ground fixed inertial frame.

The derivation of kinematic equations involves three velocity concepts: inertial velocity, local velocity and wind-relative velocity. The aircraft equations of motion are expressed in a velocity coordinate frame attached to the aircraft, considering the velocity of the wind $W = \begin{pmatrix} W_N \\ W_E \\ W_D \end{pmatrix}$, components of the wind velocity in the inertial frame. The position of the aircraft is assumed to be described in the local coordinate system **NED** with unit vectors **n, e, d** pointing respectively, North, East and Down.

The translational kinematics of an aircraft taking into account the wind effect can thus be expressed by the following equations:

$$\begin{aligned} \dot{x} &= V\cos\chi\cos\gamma + W_N \\ \dot{y} &= V\sin\chi\cos\gamma + W_E \\ \dot{z} &= -V\sin\gamma + W_D \end{aligned} \tag{2.23}$$

The aircraft relative velocity vector is defined by the airspeed V. The variables x, y, z are the aircraft inertial coordinates. The x, y directions are chosen such that the $x - y$ plane is horizontal, the x-direction is aligned with the principal axis of symmetry of the aircraft and the z-direction is descending vertically.

2.3.2 SIX DEGREES OF FREEDOM FORMULATION

The kinematic relationship between the different velocities are given by:

$$\begin{pmatrix} \dot{\eta}_1 \\ \dot{\eta}_2 \end{pmatrix} = \mathbf{RV} = \begin{pmatrix} \mathbf{R}_{GB} & 0_{3\times 3} \\ 0_{3\times 3} & \mathbf{J}(\eta_2) \end{pmatrix} \begin{pmatrix} V \\ \Omega \end{pmatrix} \tag{2.24}$$

where the matrix $\mathbf{R}_{GB}$ has been previously defined in equation (2.4), while the matrix $\mathbf{J}(\eta_2)$ is defined by:

$$\mathbf{J}(\eta_2) = \begin{pmatrix} 1 & 0 & -\sin\theta \\ 0 & \cos\phi & \sin\phi\cos\theta \\ 0 & -\sin\phi & \cos\phi\cos\theta \end{pmatrix}^{-1}$$

$$= \begin{pmatrix} 1 & \sin\phi\tan\theta & \cos\phi\tan\theta \\ 0 & \cos\phi & -\sin\phi \\ 0 & \frac{\sin\phi}{\cos\theta} & \frac{\cos\phi}{\cos\theta} \end{pmatrix} \tag{2.25}$$

Both the linear velocity $V = (u, v, w)^T$ and angular velocity $\Omega = (p, q, r)^T$ are expressed in the body-fixed frame. This matrix $\mathbf{J}(\eta_2)$ presents a singularity for $\theta = \pm\frac{\pi}{2}$.

Remark 2.4. *Euler angles are kinematically singular since the transformation from their time rate of change to the angular vector is not globally defined.*

The relationship between the body-fixed angular velocity vector $\Omega = (p, q, r)^T$ and the rate of change of the Euler angles $\dot{\eta}_2 = (\dot{\phi}, \dot{\theta}, \dot{\psi})^T$ can be determined by resolving the Euler rates into the body fixed coordinate frame:

$$\begin{aligned} p &= \dot{\phi} - \dot{\psi}\sin\theta \\ q &= \dot{\theta}\cos\phi + \dot{\psi}\sin\phi\cos\theta \\ r &= -\dot{\theta}\sin\phi + \dot{\psi}\cos\phi\cos\theta \end{aligned} \tag{2.26}$$

2.3.3 SPECIAL EUCLIDEAN SPACE

In the special Euclidean space, the kinematic model of an aircraft can thus be formulated as:

$$\dot{\mathbf{R}} = \mathbf{R}Sk(\omega) = -Sk(\omega)\mathbf{R} \tag{2.27}$$

where $\mathbf{R} \in SO(3)$ is the rotation matrix that describes the orientation of the body-frame relative to the inertial frame. This equation is known as **Poisson equation.**

2.3.4 CURVATURE AND TORSION

Without wind effect, the kinematic model is given by the following equations:

$$\dot{x} = V\cos\chi\cos\gamma \tag{2.28}$$

$$\dot{y} = V\sin\chi\cos\gamma \tag{2.29}$$

$$\dot{z} = -V\sin\gamma \tag{2.30}$$

Two **non-holonomic constraints** can thus be deduced [8]:

$$\begin{aligned} \dot{x}\sin\chi - \dot{y}\cos\chi = 0 \\ \{\dot{x}\cos\chi + \dot{y}\sin\chi\}\sin\gamma + \dot{z}\cos\gamma = 0 \end{aligned} \tag{2.31}$$

Using the Frenet-Serret formulation, **curvature** κ can be deduced [7, 79]:

$$\kappa = (\dot{\gamma}^2 + \dot{\chi}^2 cos^2\gamma)^{-1/2} \tag{2.32}$$

as well as **torsion** τ

$$\tau = \frac{\dot{\chi}\ddot{\gamma}\cos\gamma + 2\dot{\chi}\dot{\gamma}^2\sin\gamma - \dot{\gamma}\ddot{\chi}\cos\gamma}{\dot{\gamma}^2 + \dot{\chi}^2\cos^2\gamma} + \frac{-\dot{\gamma}\dot{\chi}^2\cos\chi\cos\gamma\sin^2\gamma\sin\chi + \dot{\chi}^3\sin\gamma\cos^2\gamma}{\dot{\gamma}^2 + \dot{\chi}^2\cos^2\gamma} \tag{2.33}$$

Remark 2.5. *The Frenet frame equations are pathological when the curve is perfectly straight or when the curvature vanishes momentarily.*

2.4 DYNAMIC MODELS

Dynamics is the effect of force on mass and flight dynamics is the study of aircraft motion through the air, in 3D. Newton's and Euler's laws are used to calculate their motions. Newton's second law governs the translation degrees of freedom and Euler's law controls the attitude dynamics. Both must be referenced to an inertial reference frame, which includes the linear and angular momenta and their time derivative [104]. In classical vector mechanics, the acceleration of a rigid body is represented by means of the linear acceleration of a specified point in the body and an angular acceleration vector which applies to the whole body.

With P the linear momentum vector and f the external force vector, the time rate of change of the linear momentum equals the external force:

$$[f]_I = \left[\frac{dP}{dt}\right]_I \tag{2.34}$$

The time derivative is taken with respect to the inertial frame I. If the reference frame is changed to the aircraft's body frame B, Newton's law can be written as

$$\begin{aligned}[f]_B &= \left[\tfrac{dP}{dt}\right]_B + [\Omega]_B \times [P]_B = \left[\tfrac{dP}{dt}\right]_B + [Sk(\Omega)]_B [P]_B = \\ &= \mathbf{R}_{BI}\left[\tfrac{dP}{dt}\right]_B + [Sk(\Omega)]_B [P]_B\end{aligned} \tag{2.35}$$

where $Sk(\Omega)$ is the skew symmetric form of Ω expressed in body coordinates, $\mathbf{R}_{BI}$ is the transformation matrix of the body coordinates with respect to the inertial coordinates.

The dynamics of position is given by:

$$m\dot{V} + \Omega \times mV = F \tag{2.36}$$

where m is the mass of the aircraft and V is the linear velocity of the aircraft.

The dynamics of orientation is given by the **Euler equation**:

$$\mathbf{I}\dot{\Omega} + \Omega \times \mathbf{I}\Omega = M \tag{2.37}$$

where Ω is the angular velocity of the body relative to the inertial frame, $\mathbf{I}$ is the inertia matrix and M is the torque.

When the wind is included, the velocity of the aircraft center of gravity with respect to the air is given by:

$$V_R = V_B - \mathbf{R}_{BG} \begin{pmatrix} W_N \\ W_E \\ W_D \end{pmatrix} \tag{2.38}$$

When this equation is added to the flat Earth equation, the wind components must be supplied as inputs. Then, V_R rather than V_B must be used in the calculation of aerodynamic forces and moments [55, 73, 74, 84].

As an alternative to analytical method and wind tunnel testing, system identification provides a method for developing dynamic system models and identifying their parameters [39]. System identification is the process of determining a mathematical model of a dynamic system by analyzing the measured input signals and output states of the system [44]. It uses the inputs and states to develop a model that describes the relationship between the input signal and the response of the system. Then the autopilot performs maneuvers designed to excite the autonomous aircraft dynamics. The signal given to the control surfaces and actuators is recorded. The actual deflection of the control surfaces can also be recorded. Then various sensors record the current state of the small autonomous aircraft: linear and angular accelerations, linear and angular velocities, positions, aerodynamic angles and angles relative to the Earth's surface [6].

There are six main elements to system identification:

1. Input signals
2. Data collection
3. Selection of the model structure
4. Selection of the system identification method
5. Optimization of the model using system identification method
6. Model structure and test data.

In this section, aerodynamics forces/moments, point mass models and six degrees of freedom models are presented due to their usefulness for control, guidance and navigation.

2.4.1 AERODYNAMICS FORCES AND MOMENTS

The aerodynamic forces and moments generated on an airplane are due to its geometric shape, attitude to the flow, airspeed and to the properties of the ambient air mass through which it is flying, air being a fluid. Its pressure, temperature, density, viscosity and speed of sound at the flight altitude are important properties [64]. The aerodynamics forces are defined in terms of dimensionless coefficients; the flight dynamic pressure and a reference area are follows:

$$\begin{array}{rl} X = C_x \overline{q} S & \textbf{Axial force} \\ Y = C_y \overline{q} S & \textbf{Side force} \\ Z = C_z \overline{q} S & \textbf{Normal force} \end{array} \tag{2.39}$$

where $\overline{q}$ is the dynamic pressure, S wing surface area, C_x, C_y, C_z are the x-y-z forces' coefficients.

The components of the aerodynamic moments are also expressed in terms of dimensionless coefficients, flight dynamic pressure reference area S and a characteristic length ℓ, as follows:

$$\begin{array}{rl} L = C_l \overline{q} S \ell & \textbf{Rolling moment} \\ M = C_m \overline{q} S \ell & \textbf{Pitching moment} \\ N = C_n \overline{q} S \ell & \textbf{Yawing moment} \end{array} \tag{2.40}$$

where C_l, C_m, C_n are the roll, pitch and yaw moment coefficients.

For airplanes, the reference area S is taken as the wing platform area and the characteristic length ℓ is taken as the wing span for the rolling and yawing moment and the mean chord for the pitching moment.

The aerodynamic coefficients $C_x, C_y, C_z, C_l, C_m, C_n$ are primarily a function of Mach number M, Reynolds number $\Re$, angle of attack α and side-slip angle β and are secondary functions of the time rate of change of angle of attack and side-slip angle and the angular velocity of the airplane. These coefficients are also dependent on control surface deflections; otherwise, the airplane would not be controllable. They are also dependent on other factors such as engine power level, configuration effects (e.g., landing gear, external tanks,) and ground proximity effects. Because of the complicated functional dependence of the aerodynamic coefficient, each coefficient is modeled as a sum of components that are, individually, functions of fewer variables.

The aircraft is required to be capable of operating in all weather conditions. One of the most important factors to be considered is the strength of winds. Flying in constant wind has little influence on the aerodynamic characteristics of the aircraft as the forces on the aircraft are related to the relative/local air movements. Winds also affect the aircraft climb performance relative to the ground [45].

The resultant aerodynamic force produced by the motion of the aircraft through atmosphere is resolved into components along the wind axes. The component along the x-axis is called the drag D. It is in the opposition to the velocity and resists the motion of the aircraft. The component along the z-axis (perpendicular to the aircraft velocity) is called the lift L. The lift is normal to an upward direction and its function is to counteract the weight of the aircraft. It is the lift that keeps the airplane in the air. The third component, along the y-axis, is a side force that appears when the velocity of a symmetrical airplane is not in the plane of symmetry, i.e., when there is a side-slip angle [36].

2.4.1.1 Negligible side-slip angle

The lift and drag can be expressed as a function of a non dimensional drag coefficient C_L, C_D in the form:

$$\begin{array}{c} L = \frac{1}{2}\rho V^2 S C_L(V,\alpha) \\ D = \frac{1}{2}\rho V^2 S C_D(V,\alpha) \end{array} \tag{2.41}$$

where S is the aerodynamic reference area and ρ is the density of the air.

The **drag** coefficient is estimated using the common approximation where the effective drag coefficient C_D is the sum of parasitic $C_{D,0}$ and lift induced $C_{D,i}$ drag components. Induced drag is a function of the lift coefficient C_L, the aspect ratio A_R and the efficiency factor e:

$$C_D(V,\alpha) = C_{D,0} + C_{D,i} = C_{D,0} + \frac{C_L^2}{\pi A_R e} = C_{D,0} + KC_L^2 \tag{2.42}$$

The density ρ expressed in Kg/m^3 is given by the following relation

$$\rho = \frac{P}{RT} \tag{2.43}$$

where the **Mach number**

$$M = \frac{V}{a} \tag{2.44}$$

represents the ratio of the velocity with the sonic speed $a = \sqrt{\gamma_{air} RT}, (m/s^2)$, R is gas constant, $\gamma_{air} = 1.4$ is the ratio of specific heat for air, T is the temperature of the air while the **dynamic pressure** is given by

$$\bar{q} = \frac{\rho}{2} V^2 \tag{2.45}$$

Replacing the lift coefficient with the load factor, the drag D can be computed as:

$$D = \frac{1}{2}\rho V S C_{D_0} + 2K \frac{L^2}{\rho V^2 S} \tag{2.46}$$

The aerodynamic forces and moments acting on the aircraft can be expressed in terms of the non dimensional forces and moments through multiplication by a dimensionalizing factor [24]. The forces and moments are therefore given by:

$$\begin{pmatrix} X \\ Y \\ Z \end{pmatrix} = \bar{q}S \begin{pmatrix} \cos\alpha & 0 & -\sin\alpha \\ 0 & 1 & 0 \\ \sin\alpha & 0 & \cos\alpha \end{pmatrix} \begin{pmatrix} -C_D \\ C_Y \\ C_L \end{pmatrix} \tag{2.47}$$

$$\begin{pmatrix} L \\ M \\ N \end{pmatrix} = \bar{q}S \begin{pmatrix} b_{ref} C_l \\ C_{ref} C_m \\ b_{ref} C_n \end{pmatrix} \tag{2.48}$$

where C_L, C_Y, C_D are the lift, side-force and drag coefficients, respectively, b_{ref} the wingspan, C_{ref} the mean aerodynamic chord. The non-dimensional force coefficient C_L, C_Y, C_D and moment coefficients C_l, C_m, C_n are functions of the control inputs as well as the aircraft state.

The aerodynamic efficiency is the lift to drag ratio:

$$E = \frac{L}{D} = \frac{C_L}{C_D} \tag{2.49}$$

Lift and drag normally increase when α is increased.

2.4.1.2 Nonnegligible side-slip angle

The effective **angle of attack** α_e and **side-slip angle** β_e for the aircraft include the components of wind. The aircraft **lift** force L, **drag** force D and **side-force** C are defined as functions of these angles as follows:

$$\begin{aligned} L &= \bar{q}SC_L(\alpha_e, \beta_e) \\ D &= \bar{q}SC_D(\alpha_e, \beta_e) \\ C &= \bar{q}SC_c(\alpha_e, \beta_e) \end{aligned} \tag{2.50}$$

where the aerodynamic force coefficients C_L, C_D, C_c are transformations of the aerodynamic force components between the wind axes and the aircraft body axes given by the following equations:

$$\begin{pmatrix} D' \\ C' \\ L' \end{pmatrix} = \mathbf{R} \begin{pmatrix} D \\ C \\ L \end{pmatrix} \tag{2.51}$$

with $\mathbf{R} = \mathbf{R}_1\mathbf{R}_2$

$$\mathbf{R}_1 = \begin{pmatrix} \cos\alpha\cos\beta & \sin\beta & \sin\alpha\cos\beta \\ -\cos\alpha\sin\beta & \cos\beta & -\sin\alpha\sin\beta \\ -\sin\alpha & 0 & \cos\alpha \end{pmatrix} \tag{2.52}$$

$$\mathbf{R}_2 = \begin{pmatrix} \cos\alpha_e\cos\beta_e & -\cos\alpha_e\sin\beta_e & -\sin\alpha_e \\ \sin\beta_e & \cos\beta_e & 0 \\ \sin\alpha_e\cos\beta_e & -\sin\alpha_e\sin\beta_e & \cos\alpha_e \end{pmatrix} \tag{2.53}$$

with

$$\bar{q} = \frac{1}{2}\rho V^2 \tag{2.54}$$

$$\begin{aligned} C_L &= C_x \sin\alpha - C_z \cos\alpha \\ C_D &= -C_x \cos\alpha\cos\beta - C_y \sin\beta - C_z \sin\alpha\cos\beta \\ C_c &= C_x \cos\alpha\sin\beta - C_y \cos\beta + C_z \sin\alpha\sin\beta \end{aligned} \tag{2.55}$$

The aerodynamics of an airframe and its controls make a fundamental contribution to the stability and control characteristics of the aircraft. It is usual to incorporate aerodynamics in the equations of motion in the form of aerodynamic stability and control derivatives. An important aspect of flight dynamics is concerned with the proper definition of aerodynamic derivatives as functions of common aerodynamic parameters.

2.4.2 TRANSLATIONAL DYNAMICS: COORDINATED FLIGHT

In studies of the guidance system of an aircraft, **point-mass models** have usually been employed. In these models, attitude control systems are incorporated and through these autopilots, the attitude is assumed to be quickly controlled. Therefore, in many cases, the aircraft is approximated by a point mass and only its translational motion is considered.

Remark 2.6. *The complete set of translational equations of motion should be employed in determining the optimal flight trajectories for long-range navigation.*

In a **coordinated flight**, the flight direction is confined to the plane of symmetry leading to a zero side force. To achieve a coordinated flight and hence the maximum aerodynamic efficiency, an aircraft requires precise attitude control. In the absence of a side force, the only way the aircraft can turn horizontally in a coordinated flight is by banking the wings, that is, tilting the plane of symmetry such that it makes an angle σ, called the **bank angle**, with the local vertical plane. The bank angle is the angle between the aircraft lift and weight force vectors. **Weight** is the force due to gravity and is directed toward the center of mass of the Earth. In this section, a flat Earth model is assumed due to the relatively small scale of the smart aircraft and flight paths. The dynamic model used in this section is an **aerodynamic point mass model**. The applied forces are aerodynamic force (decomposed into lift L and drag D), the thrust and the weight force. The aerodynamic force is a function of the motion of the aircraft relative to the surrounding air and the physical properties of the aircraft (shape, size and surface properties). **Lift** is defined as the component of the aerodynamic force acting perpendicularly to the air-relative velocity vector. Body force due to side-slip is not considered. **Drag** is in the opposite side of the motion [104].

$$m\mathbf{R}^E V_B^E = f_{a,p} + mg \tag{2.56}$$

$f_{a,p}$ represent the sum of the aerodynamic and the propulsion forces. The rotational time derivative is taken with respect to the inertial Earth frame E. Using Euler's transformation, it can be changed to the velocity frame V. The angular velocity is given by $\begin{pmatrix} -\dot{\chi}\sin\gamma \\ \dot{\gamma} \\ \dot{\chi}\cos\gamma \end{pmatrix}$. Thus,

$$\mathbf{R}^V V_B^E + \Omega^{V_E} \left[V_B^E\right]^E = \frac{f_{a,p}}{m} + [g]^V \tag{2.57}$$

The air relative to inertial transformation is denoted as R_{AI} and is made up of the standard rotation transformation matrices denoted as $\mathbf{R}_x, \mathbf{R}_y, \mathbf{R}_z$:

$$\mathbf{R}_{AI} = \mathbf{R}_z(\chi)\mathbf{R}_y(\gamma)\mathbf{R}_x(\sigma) \tag{2.58}$$

The air-relative velocity can be described in terms of the airspeed V and the heading and flight path angle:

$$V_a = \mathbf{R}_{AI}\begin{pmatrix} V \\ 0 \\ 0 \end{pmatrix} = \begin{pmatrix} V\cos\gamma\cos\chi \\ V\cos\gamma\sin\chi \\ -V\sin\gamma \end{pmatrix} \tag{2.59}$$

The **gravity force** is expressed by:

$$f_g = \begin{pmatrix} \cos\gamma\cos\chi & \cos\gamma\sin\chi & -\sin\gamma \\ -\sin\chi & \cos\chi & 0 \\ \sin\gamma\cos\chi & \sin\gamma\sin\chi & \cos\gamma \end{pmatrix}\begin{pmatrix} 0 \\ 0 \\ mg \end{pmatrix} \tag{2.60}$$

The aerodynamic forces f_a and thrust f_p are given in body coordinates:

$$f_{ap} = f_a + f_p = \begin{pmatrix} T\cos\alpha - D \\ (T\sin\alpha + L)\sin\sigma \\ (T\sin\alpha + L)\cos\sigma \end{pmatrix} \tag{2.61}$$

The thrust is labeled as T, the parameters D and L are, respectively, the drag and lift forces, m is the aircraft mass and g is the acceleration due to gravity. α represents the angle of attack.

The **thrust** depends on the altitude z and the throttle setting η by a known relationship $T = T(z, V, \eta)$. Also, it is assumed that the drag is a function of the velocity, the altitude and the lift: $D = D(z, V, L)$. It is assumed that the aircraft will operate close to wing level, steady state flight. Therefore, any uncertainties in drag forces will dominate and be the most influential to the aircraft dynamics [103].

If wind is not considered, the different forces can be written as follows:

$$F_x = m\dot{V} = T\cos\alpha - D - mg sin\gamma \tag{2.62}$$

$$F_y = mV\cos\gamma\dot{\chi} = (L + T\sin\alpha)\sin\sigma \tag{2.63}$$

$$F_z = mV\dot{\gamma} = (L + T\sin\alpha)\cos\sigma - mg\cos\gamma \tag{2.64}$$

Therefore, introducing the wind, the following dynamic model containing only those state variables that concern the control outer-loop design is considered:

$$\begin{aligned}\dot{V} &= -g\sin\gamma + \frac{1}{m}(T\cos\alpha - D) \\ &\quad - \left(\dot{W}_N\cos\gamma\cos\chi + \dot{W}_E\cos\gamma\sin\chi - \dot{W}_D\sin\gamma\right)\end{aligned} \tag{2.65}$$

$$\dot{\chi} = \frac{L + T\sin\alpha}{mV\cos\gamma}(\sin\sigma) + \left(\frac{\dot{W}_N \sin\chi - \dot{W}_E \cos\chi}{V\cos\gamma}\right) \tag{2.66}$$

$$\dot{\gamma} = \frac{T\sin\alpha + L}{mV}\cos\sigma - \frac{g}{V}\cos\gamma - \frac{1}{V}\left(\dot{W}_N \sin\gamma\cos\chi + \dot{W}_E \sin\gamma\sin\chi + \dot{W}_D \cos\gamma\right) \tag{2.67}$$

This point-mass model employs angle of attack α, bank angle σ and thrust T as three control variables. Side-slip angle β is controlled to 0, which is the assumption of a coordinated turn.

Another standard simplified formulation, with the assumption that the angle of attack is small and the wind negligible, is:

$$\dot{V} = \frac{T - D}{m} - g\sin\gamma \tag{2.68}$$

$$\dot{\chi} = g\frac{n_z \sin\phi}{V\cos\gamma} \tag{2.69}$$

$$\dot{\gamma} = \frac{g}{V}(n_z\cos\phi - \cos\gamma) \tag{2.70}$$

where n_z is the load factor.

Definition 2.4. *Load Factor*: *Aircraft load factor n_z is defined as lift L divided by the weight W.*

Differentiating equation (2.23) with respect to time and substituting dynamics of V, χ, γ from equations ((2.41) to (2.46)), with the assumption of a small angle of attack and negligible wind, the dynamics of the position of the aircraft are given by:

$$\ddot{x} = \frac{T}{m}\cos\gamma\cos\chi - \frac{D}{m}\cos\gamma\cos\chi - \frac{L}{m}(\sin\sigma\sin\chi + \cos\sigma\sin\gamma\cos\chi) \tag{2.71}$$

$$\ddot{y} = \frac{T - D}{m}\cos\gamma\sin\chi + \frac{L}{m}(\sin\sigma\cos\chi + \cos\sigma\sin\gamma\sin\chi) \tag{2.72}$$

$$\ddot{z} = g - \frac{T - D}{m}\sin\gamma - \frac{L}{m}\cos\gamma\cos\sigma \tag{2.73}$$

Remark 2.7. *In a coordinated flight, a change in the wind strength or direction manifests itself as changes in the angles of attack and side-slip causing a change in the aerodynamic forces. The* ***ground track*** *is affected even by a steady wind.*

The kino-dynamic equations for the aircraft can be expressed, without wind, as:

$$\dot{x} = V\cos\chi\cos\gamma \tag{2.74}$$

$$\dot{y} = V\sin\chi\cos\gamma \tag{2.75}$$

$$\dot{z} = -V \sin\gamma \tag{2.76}$$

$$\dot{V} = -g\sin\gamma + \frac{1}{m}\left(T\cos\alpha - D\right) \tag{2.77}$$

$$\dot{\chi} = \frac{L + T\sin\alpha}{mV\cos\gamma}\left(\sin\sigma\right) \tag{2.78}$$

$$\dot{\gamma} = \frac{T\sin\alpha + L}{mV}\cos\sigma - \frac{g}{V}\cos\gamma \tag{2.79}$$

These equations are equivalent to:

$$x' = dx/ds = \cos\chi\cos\gamma \tag{2.80}$$

$$y' = dy/ds = \sin\chi\cos\gamma \tag{2.81}$$

$$z' = dz/ds = -\sin\gamma \tag{2.82}$$

$$V' = dV/ds = -\frac{1}{V}g\sin\gamma + \frac{1}{mV}\left(T\cos\alpha - D\right) \tag{2.83}$$

$$\chi' = d\chi/ds = \frac{L + T\sin\alpha}{mV^2\cos\gamma}\left(\sin\sigma\right) \tag{2.84}$$

$$\gamma' = d\gamma/ds = \frac{T\sin\alpha + L}{mV^2}\cos\sigma - \frac{g}{V^2}\cos\gamma \tag{2.85}$$

with $dt = \frac{ds}{V}$, $ds = \sqrt{dx^2 + dy^2 + dz^2}$,

$$\psi = \arctan\left(\frac{dy}{dx}\right) = \arctan\left(\frac{y'}{x'}\right) \tag{2.86}$$

$$\gamma = \arctan\left(\frac{dz}{\sqrt{dx^2 + dy^2}}\right) = \arctan\left(\frac{z'}{\sqrt{x'^2 + y'^2}}\right) \tag{2.87}$$

It is also possible to express as:

$$\psi' = \frac{1}{1 + \left(\frac{y'}{x'}\right)^2}\frac{y''x' - y'x''}{x'^2} = \frac{y''x' - y'x''}{x'^2 + y'^2} \tag{2.88}$$

$$\gamma' = \frac{z''x'^2 + z''y'^2 - z'x''x' - z'y''y'}{\sqrt{x'^2 + y'^2}} \tag{2.89}$$

$$t^*(s) = \int_{s_0}^{s} dt = \int_{s_0}^{s} \frac{ds}{V^*(s)} \qquad s_0 \le s \le s_f \tag{2.90}$$

$$\phi^*(s) = -\arctan\left(\frac{\cos\gamma(s)\psi'(s)}{\gamma'(s) + g\frac{\cos\gamma(s)}{V^{*2}(s)}}\right) \tag{2.91}$$

Airplane maneuverability is limited by several constraints such as stall speed and minimum controllable velocity, engine limitations, structural limitations [73, 74].

Physical limitations mean that the maximum specific lift is limited by maximum lift coefficient $C_{L,max}$ and load factor constraints (n_{min}, n_{max}):

$$|C_L| \leq C_{L,max} \qquad n_{min} \leq n_z \leq n_{max} \tag{2.92}$$

The velocity is saturated to prevent stall as follows:

$$0 < V_{stall} \leq V \leq V_{max} \tag{2.93}$$

The **flight path angle** rate is saturated to prevent stall and ensure commanded paths stay within physical limits.

$$|\dot{\gamma}| \leq \dot{\gamma}_{max} \tag{2.94}$$

2.4.2.1 Aircraft at constant altitude

To maintain a level flight or constant altitude, the lift produced by the aircraft must equal the weight of the aircraft. For a given altitude and airspeed, this implies a required value of the lift coefficient:

$$C_L = 2\frac{mg}{\rho V^2 S} \tag{2.95}$$

The drag for level flight is:

$$D = C_{D_0}\frac{\rho V^2 S}{2} + 2\frac{(mg)^2}{\pi A \rho V^2 S} \tag{2.96}$$

where A is the aspect ratio.

The model of an aircraft at constant altitude can be expressed as:

$$\begin{aligned} \dot{x} &= V\cos\chi \\ \dot{y} &= V\sin\chi \\ \dot{V} &= U_1 - aV^2 - \tfrac{b}{V^2}\sin\chi \\ \dot{\chi} &= U_2 \end{aligned} \tag{2.97}$$

Differentiating once, the following relations can be obtained:

$$\ddot{x} = U_1\cos\left(atan2(\dot{x},\dot{y})\right) - \left(a\left(\dot{x}^2+\dot{y}^2\right) + \frac{b}{\dot{x}^2+\dot{y}^2}\right)\cos\left(atan2(\dot{x},\dot{y})\right) - \dot{y}U_2 \tag{2.98}$$

$$\ddot{y} = \left(U_1 - a\left(\dot{x}^2+\dot{y}^2\right) - \frac{b}{\dot{x}^2+\dot{y}^2}\right)\sin\left(atan2(\dot{x},\dot{y})\right) + \dot{x}U_2 \tag{2.99}$$

2.4.2.2 Aircraft in a vertical flight

The model often adopted to describe the aircraft motion is that of a point mass with three degrees of freedom commonly used for **trajectory prediction** [91]. The equations describe the motion of the aircraft center of mass. The equations of motion for symmetric flight in a vertical plane with thrust parallel to the aircraft aerodynamic velocity are the following:

$$\begin{aligned} \dot{x} &= V\cos\gamma \\ \dot{z} &= -V\sin\gamma \\ m\dot{V} &= T - D(V,z,L) - mg\sin\gamma \\ mV\dot{\gamma} &= L - mg\cos\gamma \end{aligned} \tag{2.100}$$

where V is the aerodynamic velocity modulus, γ is the path angle, m is the aircraft mass, x, z are the horizontal distance and the altitude, respectively, g is the gravity acceleration, T, L, D are, respectively, the thrust, the lift and the aerodynamic drag; the controls are T and L.

2.4.3 TRANSLATIONAL DYNAMICS: NONCOORDINATED FLIGHT

The origins of all of the following coordinates are taken at the aircraft center of mass.

$$\begin{aligned} \dot{x} &= V\cos\chi\cos\gamma + W_N \\ \dot{y} &= V\sin\chi\cos\gamma + W_E \\ \dot{z} &= -V\sin\gamma + W_D \end{aligned} \tag{2.101}$$

The assumption of negligible side-slip angle is relaxed in this approach. In this point mass model, with the control inputs taken as T, α, β, σ, the equations of motion of the aircraft are given as follows [43]:

$$\begin{aligned} \dot{V} &= -g\sin\gamma + \tfrac{1}{m}\left(T\cos\alpha\cos\beta - D'\right) + \\ &- \left(\dot{W}_N\cos\gamma\cos\chi + \dot{W}_E\cos\gamma\sin\chi - \dot{W}_D\sin\gamma\right) \end{aligned} \tag{2.102}$$

$$\begin{aligned} \dot{\chi} = \tfrac{1}{mV\cos\gamma}&\left(T(\sin\alpha\sin\sigma - \cos\sigma\cos\alpha\sin\beta) + L'\sin\sigma - C'\cos\sigma\right) + \\ &+\tfrac{1}{V\cos\gamma}\left(\dot{W}_N\sin\chi - \dot{W}_E\cos\chi\right) \end{aligned} \tag{2.103}$$

$$\begin{aligned} \dot{\gamma} = \tfrac{1}{mV}&\left(T(\sin\alpha\cos\sigma + \sin\alpha\cos\alpha\sin\beta) + (L'\cos\sigma + C'\sin\sigma)\right) - \tfrac{1}{V}g\cos\gamma + \\ &-\tfrac{1}{V}\left(\dot{W}_N\sin\gamma\cos\chi + \dot{W}_E\sin\gamma\sin\chi + \dot{W}_D\cos\gamma\right) \end{aligned} \tag{2.104}$$

The effective angle of attack, the effective side-slip angle, the aircraft relative velocity and its three components and the dynamic pressure are given as follows:

$$\alpha_e = \arctan\left(\frac{V\sin\alpha\cos\beta - W_D}{V\cos\alpha\cos\beta - W_N}\right) \tag{2.105}$$

$$\beta_e = \arcsin\left(\frac{V\sin\beta - W_E}{V_a}\right) \tag{2.106}$$

$$\begin{aligned} V_{ax} &= V\cos\alpha\cos\beta - W_N \\ V_{ay} &= V\sin\beta - W_E \\ V_{az} &= V\sin\alpha\cos\beta - W_D \\ V &= \sqrt{V_{ax}^2 + V_{ay}^2 + V_{az}^2} \end{aligned} \tag{2.107}$$

$$\bar{q} = \frac{1}{2}\rho V^2 \tag{2.108}$$

$$\begin{aligned} C_L &= C_x\sin\alpha - C_z\cos\alpha \\ C_D &= -C_x\cos\alpha\cos\beta - C_y\sin\beta - C_z\sin\alpha\cos\beta \\ C_c &= C_x\cos\alpha\sin\beta - C_y\cos\beta + C_z\sin\alpha\sin\beta \end{aligned} \tag{2.109}$$

The parameters C_x, C_y, C_z are the aerodynamic coefficients in the x, y, z directions.

Remark 2.8. *This model employs angle of attack, side-slip angle, bank angle and thrust as four control variables and can introduce wind and active side-slip angle control precisely.*

2.4.4 SIX DEGREES OF FREEDOM DYNAMICS

The nonlinear equations of motion for a fixed wing aircraft over a flat, non-rotating Earth can also be modeled by twelve state equations. Dynamic modeling of a fixed wing aircraft typically involves establishment of the flow/stability frame (whose coordinates are the angle of attack α and angle of side slip β) [82]. In the stability frame, aerodynamic coefficients can be calculated and aerodynamic forces (lift, drag and side forces) and aerodynamic moments (in pitch, roll and yaw) can be determined with respect to the body frame.

The equations of motion for the rigid-body model have appeared in many textbooks [4, 82, 104].

2.4.4.1 General formulation

Translational motions can be expressed as

$$\begin{pmatrix} \dot{x} \\ \dot{y} \\ \dot{z} \end{pmatrix} = \mathbf{R}_{BG}\begin{pmatrix} u \\ v \\ w \end{pmatrix} \tag{2.110}$$

$$\dot{u} = rv - qw - g\sin\theta + \frac{\bar{q}S}{m}C_x + \frac{X_T}{m} \tag{2.111}$$

$$\dot{v} = pw - ru + g\cos\theta\sin\phi + \frac{\bar{q}S}{m}C_y + \frac{Y_T}{m} \tag{2.112}$$

$$\dot{w} = qu - pv + g\cos\theta\cos\phi + \frac{\bar{q}S}{m}C_z + \frac{Z_T}{m} \tag{2.113}$$

where u, v, w are the three aircraft velocity components, C_x, C_y, C_z are the aerodynamic force coefficients in the aircraft body axes x_b, y_b, z_b, S is the wing area and $\bar{q}$ is the dynamic pressure, the matrix $\mathbf{R}_{BG}$ being given by relation (2.4).

The rotational dynamic model can be formulated as:

$$\mathbf{I}\dot{\Omega} = \mathbf{I}\Omega \times \Omega + M \tag{2.114}$$

where M are the applied moments such as the moment created by the gravity, the aerodynamic moments, the propulsion and control moments, the aerologic disturbances.

In another form, the dynamic equations can be written as:

$$\dot{V}_B = -Sk(\Omega)V_B + \mathbf{R}_{BE}g_0 + \frac{F_B}{m} \tag{2.115}$$

$$\dot{\Omega} = -\mathbf{I}^{-1}Sk(\Omega)I\Omega + \mathbf{I}^{-1}T_B \tag{2.116}$$

F_B and T_B are, respectively, the applied force and torque on the aircraft center of gravity.

The rotational equations are detailed as:

$$\dot{p} = \frac{I_{xz}(I_x - I_y + I_z)}{I_xI_z - I_{xz}^2}pq - \frac{I_z(I_z - I_y) + I_{xz}^2}{I_xI_z - I_{xz}^2}qr + \frac{\bar{q}Sb}{I_x}C_l \tag{2.117}$$

$$\dot{q} = \frac{I_z - I_x}{I_y}pr + \frac{I_{xz}}{I_y}\left(r^2 - p^2\right) + \frac{\bar{q}S\bar{c}}{I_y}C_m \tag{2.118}$$

$$\dot{r} = \frac{(I_x - I_y)I_x + I_{xz}^2}{I_xI_z - I_{xz}^2}pq - I_{xz}\frac{I_x - I_y + I_z}{I_xI_z - I_{xz}^2}qr + \frac{\bar{q}Sb}{I_z}C_n \tag{2.119}$$

$$\dot{\phi} = p + (q\sin\phi + r\cos\phi)\tan\theta \tag{2.120}$$

$$\dot{\theta} = q\cos\phi - r\sin\phi \tag{2.121}$$

$$\dot{\psi} = \frac{1}{\cos\theta}(q\sin\phi + r\cos\phi) \tag{2.122}$$

where I_x, I_y, I_z, I_{xz} are the components of the aircraft inertial matrix, b is the wing span aerodynamic chord and C_l, C_m, C_n are rolling, pitching and yawing aerodynamic moment coefficient, respectively.

Remark 2.9. *Parametric uncertainty can be included by adjusting the aircraft parameters as well as scaling the non dimensional coefficients.*

2.4.4.2 Poincare formulation

In the Euler angle, the model can also be generated in the form of Poincaré equations:

$$\begin{gathered} \dot{Q} = \mathbf{V}(Q)P \\ \mathbf{M}(Q)\dot{P} + \mathbf{C}(Q,P)P + \mathbf{F}(P,Q,U) = 0 \end{gathered} \tag{2.123}$$

where $Q = (\phi, \theta, \psi, x, y, z)^T$ is the generalized coordinate vector, $P = (p, q, r, u, v, w)^T$ is the quasi-velocity vector. The key parameters in the formulation are the kinematic matrix $\mathbf{V}$, the inertia matrix $\mathbf{M}(Q)$ and the gyroscopic matrix $\mathbf{C}$ [47]. The force function $\mathbf{F}(P,Q,U)$ includes all of the aerodynamic, engine and gravitational forces and moments. Ultimately, the engine and aerodynamic forces depend on the control inputs U. The kinematics and dynamics are combined to obtain the state equations:

$$\dot{X} = f(X, U, \mu) \tag{2.124}$$

where the state equation $X = (\phi, \theta, \psi, x, y, z, p, q, r, u, v, w)^T$, the control $U = (T, \delta_e, \delta_a, \delta_r)^T$ and the parameter $\mu \in \mathbb{R}^k$ is an explicitly identified vector of distinguished aircraft parameters.

Multivariate orthogonal function modeling can be applied to wind tunnel databases to identify a generic global aerodynamic model structure that could be used for many aircraft. A successful generic model structure must meet several requirements for practicality. It should be global in the sense that it is valid over a large portion of the flight envelope for each aircraft. The method should also be formulated in a manner that enables a fundamental understanding of the functional dependencies [32].

2.4.4.3 Longitudinal model

Longitudinal models are used for a variety of purposes, from performance analysis to automatic control system design [82]. The longitudinal variables are $X_{long} = (x, z, u, w, \theta, q)^T$.

The longitudinal dynamics of a rigid aircraft can be derived from the six degrees of freedom model by restricting motion to the longitudinal variables. When written in flight path coordinates, the longitudinal model takes the following form:

$$\dot{x} = V \cos\gamma \tag{2.125}$$

$$\dot{z} = -V \sin\gamma \tag{2.126}$$

$$\dot{\theta} = q \tag{2.127}$$

$$\dot{V} = \frac{1}{m}\left(T\cos\alpha - D - mg_D\sin\gamma\right) \tag{2.128}$$

$$\dot{\gamma} = \frac{1}{mV}\left(T\sin\alpha + L - mg_D\cos\gamma\right) \tag{2.129}$$

$$\alpha = \theta - \gamma \tag{2.130}$$

$$\dot{q} = \frac{M}{I_y} \tag{2.131}$$

$$M = \frac{1}{2}\rho V^2 S_c C_M + \frac{1}{2}\rho V^2 S_c C_z \tag{2.132}$$

A common alternative representation for the longitudinal model uses flight path angle as a state variable in place of pitch attitude:

$$m\dot{V}_T = F_T\cos(\alpha + \alpha_T) - D - mg_D\sin\gamma \tag{2.133}$$

$$m\dot{\gamma}V_T = F_T\sin(\alpha + \alpha_T) + L - mg_D\cos\gamma \tag{2.134}$$

$$\dot{\alpha} = q - \dot{\gamma} \tag{2.135}$$

$$\dot{q} = \frac{m}{I_y} \tag{2.136}$$

The variables V_T, α, β describe the magnitude and direction of the relative wind. The thrust vector lies in the $x_b - z_b$ plane but is inclined at an angle α_T to the fuselage reference line, so that positive α_T corresponds to a component of thrust in the negative z_b direction. In load factor coordinates, the resultant aerodynamic force posseses lift and drag as its two components:

$$f_a = \bar{q}S\begin{pmatrix}-C_D & 0 & -C_L\end{pmatrix}^T \tag{2.137}$$

2.4.4.4 Lateral model

The lateral variables are $X_{lat} = (y, v, \phi, \psi, r)^T$. The lateral equations are given as:

$$\dot{y} = u_N\sin\psi + v\cos\phi\cos\psi \tag{2.138}$$

$$\dot{\psi} = r\cos\phi \tag{2.139}$$

$$\dot{\phi} = p \tag{2.140}$$

$$\dot{v} = \frac{Y_b}{m} + g \sin\phi - r u_N \tag{2.141}$$

$$\dot{p} = \frac{I_{zz} L + I_{xz} N}{I_{xx} I_{zz} - I_{xz}^2} \tag{2.142}$$

$$\dot{r} = \frac{I_{xz} L + I_{xx} N}{I_{xx} I_{zz} - I_{xz}^2} \tag{2.143}$$

2.4.5 UNCERTAINTY

Aircraft have to deal with typical types of uncertainties [101]:

1. Uncertainty in aircraft dynamics and limited precision in command tracking,
2. Uncertainty in the knowledge of the environment (fixed and mobile obstacle locations, collision avoidance),
3. Disturbances in the operational environment (wind, atmospheric turbulence),
4. Uncertainty in configuration information.

These uncertainties are often introduced by approximations made while deriving physical models from first principles, unforeseen increases in system complexity during operation, time variations, nonlinearities, disturbances, measurement noises, health degradation, environmental uncertainties. Another uncertainty is flexible aircraft effects, such as structural modes and resulting vibrations.

Uncertainty can arise in various ways. **Structural uncertainty** refers to an imprecise knowledge of the system dynamics, whereas **input uncertainty** is associated with exogenous signals: disturbances that steer the aircraft away from its nominal behavior. Many methods have been developed for the construction of uncertainty sets. Some of these methods are purely deterministic and are based on the derivation of the upper bounds on the set of parameters that are consistent with data, whereas other methods allow the user to express preferences in terms of a coherent risk measure [29].

1. The **average approach** can be used when a more structured and probabilistic point of view in the description of the uncertainty is adopted. This framework is often adopted when uncertainty is associated with disturbance signals, although the average approach can also be used for structural uncertainty.
2. The **worst case approach** is a suitable design methodology when the uncertainty level is moderate enough that the solution to the problem secures an adequate performance for all uncertainty outcomes. For other problems, the uncertainty is much larger; the approach becomes too conservative.

3. The **modulating robustness approach** consists of using probability to quantify the chance that a certain performance specification is attained. This approach can be used in the context of flight control. One way to evaluate the probability that a certain performance is attained is through randomization and explicit bounds on the number of samples. Methods can be used to solve feasibility problems with probability specifications, to quantify the chance that a certain performance is attained.

Uncertainty assessment consists of three steps:

1. uncertainty characterization by percentage uncertainty with confidence level,
2. uncertainty analysis using error propagation techniques,
3. sensitivity analysis based on iterative binary search algorithms.

Parametric uncertainty on a dynamical system is represented as a vector q in which each element represents a parameter whose nominal value is considered as uncertain. There are mainly three types of uncertainty:

Box:

$$q_{i_{min}} \leq q_i \leq q_{i_{max}} \tag{2.144}$$

Sphere:

$$||q_i||_2 \leq r \tag{2.145}$$

Diamond:

$$||q_i||_\infty \leq r \tag{2.146}$$

Generally, uncertainty is defined as a connected set.

Different norms of a signal can be defined.

Norm H_1

$$\|u\|_1 = \int_0^{+\infty} |u(t)|dt \tag{2.147}$$

Norm H_2

$$\|u\|_2 = \sqrt{\int_0^{+\infty} u(t)^2 dt} \tag{2.148}$$

Norm H_∞

$$\|u\|_\infty = sup_{t \geq 0}|u(t)|dt \tag{2.149}$$

Uncertainty can be dealt with in a number of ways:

1. **Generative probabilistic graphical models** describe how the data are generated via classical distribution functions. Depending on the problem, metric learning techniques can be considered for generative models.

2. **Discriminative models** do not attempt to model data generation, but focus on learning how to discriminate between data belonging to different categories or classes, in recognizing actions based on a limited training set of examples.
3. **Imprecise-probabilistic models** assume the data are probabilistic but insufficient to estimate a precise probability distribution.

There are three major approaches to the treatment of uncertainty: **Bayesian probabilities, pattern matching** and treating **uncertainty** as a problem solving task [51, 71]. Uncertainty is a key property of existence in the physical world: the environment can be stochastic and unpredictable, physical sensors provide limited noisy and inaccurate information, physical effectors produce limited, noisy and inaccurate action and often models are simplified.

If the position of the aircraft is considered as a probability density function, the $3D$ Gaussian probability density function is:

$$\Pr(X) = \frac{1}{2\pi\sqrt{\det \mathbf{P}}} \exp\left(-\frac{1}{2}(X - \mu_X)^T \mathbf{P}^{-1}(X - \mu_X)\right) \tag{2.150}$$

where $\mu_X \in \mathbb{R}^3$ is the mean of X and $\mathbf{P} \in \mathbb{R}^{3\times3}$ is the covariance matrix. Such an ellipsoid represents the positional uncertainty. An ellipse corresponds to the uncertainty about position. A scalar measure of total uncertainty is the volume of the ellipsoid.

The focus can be put on the uncertainty modeling for an aerodynamic model because the aircraft model is a complex wing body configuration. The general aeroelastic equations of motion can be written as [98]:

$$\mathbf{M}q + \mathbf{M}_c\delta + \mathbf{C}\dot{q} + \mathbf{K}q = F_q + F_\delta \tag{2.151}$$

The modal based **double lattice method** (DLM) is a widely used unsteady aerodynamic computational method. Under the double lattice method framework, the general aerodynamic force F_q in the frequency domain is represented by:

$$F_q = \frac{1}{2}\rho V^2 \Phi_p^T S C_p(Ma, ik) q \tag{2.152}$$

$C_p(Ma, ik) \in \mathbb{R}^{n_a \times n}$ represents the dynamic pressure coefficients on aerodynamic elements due to harmonic modal motion, at a given reduced frequency k. The expression for F_δ is similar to this equation.

With the consideration of the calculation uncertainty or error for $C_p(Ma, ik)$, it can be rewritten as:

$$C_p = C_{p_0} + W_{cpl} \times \Delta_{cp} \times W_{cpr} \tag{2.153}$$

W_{cpr} and W_{cpl} scale the pressure perturbation so that Δ_{cp} satisfies $||\Delta_{cp}|| \leq 1$. Under the framework, Δ_{cp} is assumed to be a diagonal complex valued matrix accounting for the uncertainty in dynamic pressure on selected aerodynamic elements. The key to establishing an adequate aerodynamic uncertainty model is to determine the left and right weighting matrices in equation (2.153).

Robust optimization involves optimizing the mean of some desired observables and minimizing the variance in those observables. Thus, robust optimization is inherently multi-objective even if there is only a single main observable to optimize. Its variance must be minimized, resulting in at least two objectives for any robust optimization problem. Multi-objective optimization does not usually produce a single solution or design point, but rather a set of non-dominated solutions. For an optimization problem with two objectives, two design points (1 and 2) are non-dominated with respect to each other if one has a superior value for one objective by an inferior value for the other. For example, when minimizing lift to drag ratio (L/D) and minimizing its variance, two solutions are non dominated if $(L/D)_1 > (L/D)_2$ and $\sigma_1 > \sigma_2$. Non-dominated solutions are often represented as a Pareto set that shows trade-offs between two or more objectives. Because robust optimization involves optimizing both the mean and variance of one or more observables, the underlying system must compute such information from similar input information. For each objective function evaluation, the underlying system accepts Gaussian random variables with a defined mean and variance, representing geometric uncertainty in the edge thickness of an airfoil and computed mean and variance information for the airfoil's (L/D). Obtaining the variance information required involves modeling a stochastic system. There are several methods for modeling stochastic systems of which Monte Carlo simulations (MCS) are the most accurate and straightforward but it usually requires on the order of thousands of deterministic runs to obtain accurate output statistics for a single design point. Optimization routines usually require several hundreds or thousands of design point evaluations. If Monte Carlo simulations are used to obtain the mean and variance the computational cost becomes prohibitive. Another solution is to find an efficient means of obtaining variance information for a system and non intrusive polynomial chaos is a promising candidate. A single objective function evaluation still requires several deterministic runs, but this is on the order of tenths rather than thousands making robust optimization computationally feasible. Problems of sequential resource allocation in stochastic environments are considered in [31].

2.5 MULTI-MODEL APPROACH

2.5.1 GLOBAL REPRESENTATION FROM LOCAL MODELS

The aircraft motion is supposed to be described by the model:

$$\begin{aligned} \dot{X}(t) &= f(X(t), U(t)) \\ Y(t) &= h(X(t), U(t)) \end{aligned} \tag{2.154}$$

The evolution and output functions f, h assumed unknown [87], the objective is to find a multi-model representation describing the behavior of the aircraft in the flight envelope $\mathbb{D}$. This domain can be subdivided into N local domains noted $\mathbb{D}_i$ such that $\mathbb{D} = \cup_i \mathbb{D}_i$. On each of these domains, the aircraft motion can be modeled as:

$$\begin{aligned} \dot{X}(t) &= f_i(X(t), U(t)) \\ Y(t) &= h_i(X(t), U(t)) \qquad i = 1 \dots N \end{aligned} \tag{2.155}$$

such that

$$\forall \zeta \in \mathbb{D}_i \qquad \begin{aligned} \|f - f_i\| &< M_f \\ \|h - h_i\| &< M_h \end{aligned} \tag{2.156}$$

The scalars M_f, M_h represent the maximum bounds on the errors on f, h.

Let $\mu_i(\zeta) > 0$, a function representing the validity of the local model i on the domain

$$\mu_i(\zeta) : \mathbb{D} \to [0, 1] \tag{2.157}$$

such that $\mu_i(\zeta) \approx 1$ for $\zeta \in \mathbb{D}_i$ and converging rapidly to zero outside $\mathbb{D}_i$. Different functions can be used:

1. Gaussian:

$$\mu_i(\zeta) = \prod_{j=1}^{dim(\zeta)} \exp\left(-\frac{1}{2}\left(\frac{\zeta_j - m_i^j}{\sigma_i^j}\right)^2\right) \tag{2.158}$$

2. Triangular:

$$\mu_i(\zeta) = \prod_{j=1}^{dim(\zeta)} \max\left(\min\left(\frac{\zeta_j - a_i^j}{b_i^j - a_i^j}, \frac{c_i^j - \zeta_j}{c_i^j - b_i^j}\right), 0\right) \tag{2.159}$$

3. Trapezoidal:

$$\mu_i(\zeta) = \prod_{j=1}^{dim(\zeta)} \max\left(\min\left(\frac{\zeta_j - a_i^j}{b_i^j - a_i^j}, 1, \frac{c_i^j - \zeta_j}{c_i^j - b_i^j}\right), 0\right) \tag{2.160}$$

with $\zeta = (\zeta_1, \zeta_2, \dots, \zeta_{dim(\zeta)})^T$.

To obtain the multi-model in the general form:

$$\begin{aligned}\dot{X}(t) &= \hat{f}(X(t), U(t)) \\ Y(t) &= \hat{h}(X(t), U(t))\end{aligned} \tag{2.161}$$

such that

$$\forall \zeta \in \mathbb{D}_i \qquad i = 1 \dots N \qquad \begin{aligned}\hat{f} &= f_i \\ \hat{h} &= h_i\end{aligned} \tag{2.162}$$

The following expressions must thus be respected:

$$\forall \zeta \in \mathbb{D} \qquad \hat{f} = \sum_{i=1}^{N} f_i \varpi_i(\zeta) \tag{2.163}$$

$$\forall \zeta \in \mathbb{D} \qquad \hat{h} = \sum_{i=1}^{N} h_i \varpi_i(\zeta) \tag{2.164}$$

$$\forall \zeta \in \mathbb{D} \qquad i = 1 \dots N \qquad \varpi_i(\zeta) = \frac{\mu_i(\zeta)}{\sum_{j=1}^{N} \mu_j(\zeta)} \tag{2.165}$$

where $\varpi_i(\zeta)$ is the interpolation function.

2.5.2 LINEAR APPROXIMATE MODEL

The nonlinear state–space equations (2.111) to (2.122) can be formulated as follows:

$$f\left(\dot{X}, X, U\right) = 0 \tag{2.166}$$

where X represents the state-space variable while U is the control variable. Their definitions depend on the model used.

The linear model state-space formulation is the following:

$$\mathbf{E}\dot{X} = \mathbf{A}'X + \mathbf{B}'U \tag{2.167}$$

where

$$\mathbf{E} = \frac{\partial f}{\partial \dot{X}}|_{X=X_{ref}, U=U_{ref}} \tag{2.168}$$

$$\mathbf{A}' = \frac{\partial f}{\partial X}|_{X=X_{ref}, U=U_{ref}} \tag{2.169}$$

$$\mathbf{B}' = \frac{\partial f}{\partial U}|_{X=X_{ref}, U=U_{ref}} \tag{2.170}$$

This general formulation can be used to express linear longitudinal and lateral models.

The longitudinal and lateral dynamics for a fixed wing aircraft are given by the linear equations:

$$\dot{u} = X_u u + X_w w + X_q q + X_\theta \theta + X_{\delta_e} \delta_e + X_t T \tag{2.171}$$

$$\dot{v} = Y_v v + Y_p p - (u_0 - Y_r) r + Y_\phi \phi + Y_\psi \psi + Y_{\delta_a} \delta_a + Y_{\delta_r} \delta_r \tag{2.172}$$

$$\dot{w} = Z_u u + Z_w w + Z_q q + Z_\theta \theta + Z_{\delta_e} \delta_e + Z_t T \tag{2.173}$$

$$\dot{p} = L_v v + L_p p + L_r r + L_{\delta_a} \delta_a + L_{\delta_r} \delta_r \tag{2.174}$$

$$\dot{q} = M_u u + M_w w + M_q q + M_\theta \theta + M_{\delta_e} \delta_e + M_t T \tag{2.175}$$

$$\dot{r} = N_v v + N_p p + N_r r + N_{\delta_a} \delta_a + N_{\delta_r} \delta_r \tag{2.176}$$

where $\delta_{inputs} = (\delta_e, \delta_a, \delta_r, T)$ are control inputs corresponding to the elevator, aileron, rudder deflection angles and thrust, respectively. The stability and control derivatives used in this dynamic linear model are derived from a nonlinear UAV model using linearization. Therefore, these derivatives depend on the physical parameters and aerodynamic coefficient parameters which are $X_u, X_w, Z_u, Z_w, X_t, Z_{\delta_e}, Y_v, Y_{\delta_a}, Y_{\delta_r}$ being inversely proportional to the aircraft mass m, X_t, M_{δ_e} proportional to aerodynamic coefficients of $C_{\delta_t}, C_{m\delta_e}$, respectively.

Stability investigations are an important part of any vehicle design. They require the linearization of the equations of motion in order to take advantage of linear stability criteria. The eigenvalues of the linear equations serve to indicate frequency and damping.

2.5.2.1 Linear longitudinal model

Linearized longitudinal models are classically used, where the state variable is given by $X = \begin{pmatrix} \alpha \\ q \\ v_T \\ \theta \end{pmatrix}$ and the control variable is given by $U = \begin{pmatrix} \delta_e \\ \delta_t \end{pmatrix}$ while the matrices are given by:

$$\mathbf{E} = \begin{pmatrix} V_{T_e} - Z_{\dot{\alpha}} & 0 & 0 & 0 \\ -M_{\dot{\alpha}} & 1 & 0 & 0 \\ 0 & 0 & 1 & 0 \\ 0 & 0 & 0 & 1 \end{pmatrix} \tag{2.177}$$

$$\mathbf{A}' = \begin{pmatrix} Z_\alpha & V_{T_e} + Z_q & Z_V - X_{T_V} \sin(\alpha_e + \alpha_T) & -g_D \sin \gamma_e \\ M_\alpha + M_{T_\alpha} & M_q & M_V + M_{T_V} & 0 \\ X_\alpha & 0 & X_V + X_{T_V} \cos(\alpha_e + \alpha_T) & -g_D \cos \gamma_e \\ 0 & 1 & 0 & 0 \end{pmatrix} \tag{2.178}$$

$$\mathbf{B}' = \begin{pmatrix} Z_{\delta_e} & -X_{\delta_t}\sin(\alpha_e + \alpha_T) \\ M_{\delta_e} & M_{\delta_T} \\ X_{\delta_e} & X_{\delta_T}\cos(\alpha_e + \alpha_T) \\ 0 & 0 \end{pmatrix} \tag{2.179}$$

where the aerodynamic derivatives constants are $X, Y, Z, L, M, N, \alpha, \beta, \delta_e$, $\delta_t, \delta_a, \delta_r$. More information about their derivation can be found in [82].

2.5.2.2 Linear lateral model

In the linear lateral model state-space formulation, the state variable is given by $X = \begin{pmatrix} \beta \\ \phi \\ p_s \\ q_s \end{pmatrix}$ and the control variable is given by $U = \begin{pmatrix} \delta_a \\ \delta_r \end{pmatrix}$ while the matrices are given by:

$$\mathbf{E} = \begin{pmatrix} V_{T_e} & 0 & 0 & 0 \\ 0 & 1 & 0 & 0 \\ 0 & 0 & 1 & 0 \\ 0 & 0 & 0 & 1 \end{pmatrix} \tag{2.180}$$

$$\mathbf{A}' = \begin{pmatrix} Y_\beta & g_D\cos\theta_e & Y_p & Y_r - V_{T_e} \\ 0 & 0 & \cos\gamma_e/\cos\theta_e & \sin\gamma_e/\cos\theta_e \\ L_\beta & 0 & L_p & L_r \\ N_\beta & 0 & N_p & N_r \end{pmatrix} \tag{2.181}$$

$$\mathbf{B}' = \begin{pmatrix} Y_{\delta_a} & Y_{\delta_r} \\ 0 & 0 \\ L_{\delta_a} & L_{\delta_r} \\ N_{\delta_a} & N_{\delta_r} \end{pmatrix} \tag{2.182}$$

2.5.2.3 Linear translational model

The dynamic translational model of an aircraft can be expressed as:

$$\begin{aligned} \dot{x} &= V\cos\gamma\cos\chi \\ \dot{y} &= V\cos\gamma\sin\chi \\ \dot{z} &= -V\sin\gamma \end{aligned} \tag{2.183}$$

$$\dot{V} = \frac{T - D - mg\sin\gamma}{m} \tag{2.184}$$

$$\dot{\chi} = \frac{L\sin\sigma}{mV\cos\gamma} \tag{2.185}$$

$$\dot{\gamma} = \frac{L\cos\sigma - mg\cos\gamma}{mV} \tag{2.186}$$

taking into account the fact that:

$$
\begin{aligned}
L &= n_z mg \\
C_L &= \frac{2n_z mg}{\rho V^2 S} \\
C_D &= C_{D_0} + KC_L^2 \\
D &= \frac{1}{2}\rho V^2 S C_{D_0} + K\frac{4n_z^2 m^2 g^2}{\rho^2 V^4 S^2}
\end{aligned} \tag{2.187}
$$

The dynamic model can also be rewritten as:

$$
\begin{aligned}
\dot{x} &= V\cos\gamma\cos\chi \\
\dot{y} &= V\cos\gamma\sin\chi \\
\dot{z} &= -V\sin\gamma
\end{aligned} \tag{2.188}
$$

$$
\dot{V} = \frac{T}{m} - aV^2 - b\frac{n_z^2}{V^2} - g\sin\gamma \tag{2.189}
$$

$$
\dot{\chi} = \frac{n_z g\sin\sigma}{V\cos\gamma} \tag{2.190}
$$

$$
\dot{\gamma} = g\frac{n_z\cos\sigma - \cos\gamma}{V} \tag{2.191}
$$

with the parameters $a = \frac{\rho S C_{D_0}}{2m}$, $b = \frac{4mg^2 K}{\rho^2 S^2}$. The following choice is made for the state variable $X = (x, y, z, V, \chi, \gamma)^T$ and the control input $u = (\frac{T}{m}, n_z, \sigma)^T$

The equilibrium trajectories can thus be obtained for:

$$
\begin{aligned}
u_{1r} &= aV_r^2 + b\frac{\cos^2\gamma_r}{V_r^2} + g\sin\gamma_r \\
u_{2r} &= n_{zr} = \cos\gamma_r \\
u_{3r} &= \sigma_r = 0
\end{aligned} \tag{2.192}
$$

The linear translational model is given by:

$$
\dot{X} = \mathbf{A}X + \mathbf{B}U \tag{2.193}
$$

The Jacobian matrices evaluated at the reference trajectories and inputs give for the state matrix:

$$
\mathbf{A} = \begin{pmatrix}
0 & 0 & 0 & \cos\gamma\cos\chi & -V\cos\gamma\sin\chi & -V\sin\gamma\cos\chi \\
0 & 0 & 0 & \cos\gamma\sin\chi & V\cos\gamma\cos\chi & -V\sin\gamma\sin\chi \\
0 & 0 & 0 & -\sin\gamma & 0 & -V\cos\gamma \\
0 & 0 & 0 & -2aV + b\frac{n_z^2}{V^3} & 0 & -g\cos\gamma \\
0 & 0 & 0 & 0 & 0 & 0 \\
0 & 0 & 0 & 0 & 0 & g\frac{\sin\gamma}{V}
\end{pmatrix} \tag{2.194}
$$

and for the input matrix

$$\mathbf{B} = \begin{pmatrix} 0 & 0 & 0 \\ 0 & 0 & 0 \\ 0 & 0 & 0 \\ 1 & -2b\frac{n_z}{V^2} & 0 \\ 0 & 0 & g/V \\ 0 & g/V & 0 \end{pmatrix} \tag{2.195}$$

The properties of these linear models will be studied in the following chapter.

2.5.3 LINEAR PARAMETER VARYING MODEL: TAKAGI–SUGENO FORMULATION

In many cases, it is not enough to simply linearize the system around the equilibrium point because the variable may vary in a wide range far away from the equilibrium point. Such a local model describes the system's input/output (I/O) relationship in a small region around the given point. In contrast, a global model describes the system input/output relationship for the entire input space. In particular, a nonlinear global model can often be approximated by a set of linear local models via rule based function approximation. The Takagi–Sugeno (TS) approach generalizes the classical partitioning by allowing a sub-region to partially overlap with neighboring sub-regions [83].

Takagi–Sugeno models allow the applications of powerful learning techniques for their identification from data [3]. This model is formed using a set of rules to represent a nonlinear system as a set of local affine models which are connected by membership functions. This modeling method presents an alternative technique to represent complex nonlinear systems and reduces the number of rules in modeling higher order nonlinear systems.

Remark 2.10. *Takagi–Sugeno models are proved to be universal function approximators as they are able to approximate any smooth nonlinear functions to any degree of accuracy in any convex compact region. This result provides a theoretical foundation for applying Takagi–Sugeno models to represent complex nonlinear systems.*

Several results have been obtained concerning the identification of Takagi–Sugeno models. They are based upon two principal approaches:

1. The first one is to linearize the original nonlinear system in various operating points when the model of the system is known.
2. The second is based on the input-output data collected from the original nonlinear system when its model is unknown.

The Takagi–Sugeno model is a specific mathematical form of representing nonlinear dynamics. There are at least two methods for constructing such a model from general nonlinear state space dynamics:

1. The first approach, **sector nonlinearity**, is a method that results in a Takagi–Sugeno model that is a mathematically exact representation of the nonlinear system equations.
2. The second approach, **local approximation**, to fuzzy partition spaces produces a Takagi–Sugeno model that can capture the dynamics of any nonlinearity that the designer who has insight into the model desires, and therefore can capture the most prominent nonlinear effects of the system while ignoring minor effects. This results in a simpler, less computationally intensive model and, for this reason, is generally preferred in practice.

Definition 2.5. ***Sector Nonlinearity:*** *A continuous mapping $\phi : \mathbb{R} \rightarrow \mathbb{R}$ is said to be a sector nonlinearity in $[\alpha, \beta]$ if $\phi(0) = 0$ and $\alpha U^2 \leq U\phi(U) < \beta U^2, \forall U \neq 0$.*

To generate the Takagi–Sugeno model, the system dynamics are rewritten as:

$$\begin{aligned} \dot{X} &= \mathbf{A}(X,U)X(t) + \mathbf{B}(X,U)U(t) \\ Y(t) &= \mathbf{C}X(t) \end{aligned} \tag{2.196}$$

where the $\mathbf{A}(X,U), \mathbf{B}(X,U)$ matrices are nonlinear and can depend on any way on the states and the inputs of the system.

The idea behind the Takagi–Sugeno model is to label the model's nonlinearities of interest as **premise variables**. At any operating condition within a predefined range, the premise variables can be evaluated. To capture and evaluate all the premise variables, the Takagi–Sugeno model defines a set of if-then rules corresponding to each premise variable's minimum and maximum over the predefined range of interest and every combination between premise variables. Each rule has a corresponding linear system associated to it representing its combination of minimum and maximum bounds of the premise variables. These rules for the linear system are then weighted through membership functions and finally summed. The result is equivalent to the nonlinear system and although the Takagi–Sugeno model is composed of a set of weighted summed linear systems, the result is nonlinear because the weightings are functions of the states and inputs.

The Takagi–Sugeno formulation is a combination of linear models that can represent satisfactorily a nonlinear model with a compromise of complexity and error. It can be described through the following polytopic form:

$$\dot{X}(t) = \sum_{i=1}^{r} \varpi_i(\zeta(t)) \left(\mathbf{A}_i X(t) + \mathbf{B}_i U(t)\right) \tag{2.197}$$

$$Y(t) = \sum_{i=1}^{r} \varpi_i(\zeta)\mathbf{C}_i X(t) \tag{2.198}$$

$X(t), Y(t)$ are the state and output vectors, $\zeta = (\zeta_1(t), \zeta_2(t), \dots, \zeta_p(t))^T$ the premise vector which may depend on the state vector and $\mathbf{A}_i, \mathbf{B}_i, \mathbf{C}_i$ are constant matrices. The rules are denoted with r and their number is equal to $2^{|\zeta|}$, where $|\zeta|$ denotes the number of nonlinearities considered. In addition, ϖ_i are normalized weighted functions which are defined by:

$$\varpi_i(\zeta(t)) = \frac{\mu_i(\zeta(t))}{\sum_{i=1}^{r} \mu_i(\zeta(t))} \qquad \mu_i(\zeta(t)) = \prod_{j=1}^{p} M_{ij}(\zeta_j(t)) \tag{2.199}$$

$M_{ij}(\zeta(t)) \in [0,1]$ denotes the ranking of the membership functions of the premise variables $\zeta_j(t) \in M_{ij}$ and $\varpi_i(\zeta(t))$ satisfies the convex sum property for all t. The nonlinear system is modeled as a Takagi–Sugeno model in a compact region of the state space variable using the sector nonlinearity approach.

There are four important steps involved in order to develop the equivalent Takagi–Sugeno models to the aircraft represented as a nonlinear system. Those are:

1. Definition of the premise variable of the nonlinear system and calculation of their compact region.
2. Determination of the membership functions.
3. Determination of the number of rules involved and each rule with respect to possible associations between the premise variables.
4. Calculation of $\mathbf{A}_i, \mathbf{B}_i, \mathbf{C}_i$ with respect to the compact region of the premise variables

This approach guarantees an exact approximation; however, it is often difficult to find global sectors for the general nonlinear system. Thus local sectors are utilized since variables in the nonlinear system can be bounded. The latter means that every bounded nonlinear term (premise variable) is decomposed in a convex combination of its bounds. In essence, the compact region for the premise variables should be known a priori in order for the previous to hold [76].

Remark 2.11. *In some cases, a division by different state variables of the system is necessary to obtain the LPV form. Consequently, some additional conditions are necessary for the divisions: these state variables must be different from zero. To solve this problem, a translation may be realized: if $[-\alpha_1, \alpha_2]$ is the variation domain of one of these variables X_1, where $\alpha_1, \alpha_2 > 0$, the following translation can be realized:*
$\bar{X}_d = X_d + \alpha_1 + \epsilon$ *thus* $\bar{X}_d \in [\epsilon, \alpha_1 + \alpha_2 + \epsilon], \epsilon > 0$
Replacing X_d by $\bar{X}_d$, the appropriate LPV form can be obtained.

2.5.3.1 Longitudinal model

Since the number of rules needed to create a Takagi–Sugeno model is at least 2^p, where p is the number of premise variables, there is a trade-off between having a mathematically approximate Takagi–Sugeno model with fewer premise variables, which is quicker to calculate and more feasible to find Lyapunov functions for stability and control purposes. For these reasons, only the most prominent nonlinearities that influence the angular rates of the aircraft are modeled in the Takagi–Sugeno model. The other nonlinear terms are linearized [68].

In the first and second rows of the **A** matrix, there has been a second order approximating of the gravity terms [12]. This was done because the zeroth order of gravity and the trimmed lift force cancel taking a second order approximation of these terms while canceling the zeroth order terms eliminates need to have constants in the equations and it keeps sufficient accuracy in the flight envelope. If the pitch rate is defined in the premise variables:

$$\zeta_1 = \frac{SC_{mac}}{I_y}\bar{q}C_{M_\alpha} \qquad \zeta_2 = \frac{SC_{mac}}{2VI_y}\bar{q}C_{M_q} \qquad \zeta_3 = \frac{SC_{mac}}{I_y}\bar{q}C_{M_{\delta_e}} \tag{2.200}$$

with $U = \delta_e, X = \begin{pmatrix} V \\ \alpha \\ q \\ \theta \\ h \end{pmatrix}$, then the matrix $\mathbf{A}(X,U)$ is given by:

$$\mathbf{A}(X,U) = \begin{pmatrix} 0 & \mathbf{A}_{12} & 0 & \mathbf{A}_{14} & 0 \\ 0 & \mathbf{A}_{22} & 1 & \mathbf{A}_{24} & 0 \\ 0 & \mathbf{A}_{32} & \mathbf{A}_{33} & 0 & 0 \\ 0 & 0 & 1 & 0 & 0 \\ 0 & \mathbf{A}_{52} & 0 & \mathbf{A}_{54} & 0 \end{pmatrix} \tag{2.201}$$

with the following matrix elements:

$$\begin{aligned} \mathbf{A}_{12} &= g\cos\theta\frac{\sin\alpha}{\alpha} - \frac{F_T\alpha}{m} - \frac{S}{m}\bar{q}C_{D_\alpha} \\ \mathbf{A}_{14} &= -g\cos\alpha\frac{\sin\theta}{\theta} \\ \mathbf{A}_{22} &= -g\frac{\alpha}{2} + g\theta - \frac{F_T}{mV}\frac{\sin\alpha}{\alpha} - \frac{S}{mV}\bar{q}C_{L_\alpha} \\ \mathbf{A}_{24} &= -g\frac{\theta}{2} \end{aligned}$$

$$\mathbf{A}_{32} = \frac{SC_{mac}}{I_y}\bar{q}C_{M_\alpha}$$
$$\mathbf{A}_{33} = \frac{SC_{mac}C_{mac}}{2VI_y}\bar{q}C_{M_q}$$
$$\mathbf{A}_{52} = -V\cos\theta\frac{\sin\alpha}{\alpha}$$
$$\mathbf{A}_{54} = V\cos\alpha\frac{\sin\theta}{\theta}$$

The matrix $\mathbf{B}(X, U)$ is given by the following matrix:

$$\mathbf{B}(X,U) = \begin{pmatrix} -\frac{S}{m}\bar{q}C_{D_e} \\ -\frac{S}{mV}\bar{q}C_{L_e} \\ \frac{SC_{mac}}{I_y}\bar{q}C_{M_e} \\ 0 \\ 0 \end{pmatrix} \tag{2.202}$$

In choosing which nonlinear terms to incorporate, all terms that feed into the pitch rate are taken. The reason to do this is that longitudinal stabilization is controlled by commanding the pitch rate of the aircraft. It is thus most important to control the pitch rate most precisely over the largest nonlinear region as needed.

In [57], another point of view is taken. Because the full set of aerodynamic coefficients is deemed too complex for modeling and control design, a low-fidelity model of the aircraft is obtained through a two-steps simplification study. First, an analytical study of the importance of each stability derivative with respect to the nominal value of the total aerodynamic coefficient is performed. Second, open loop time simulations are performed to refine the low fidelity model and to ascertain the validity of the final reduced set.

2.5.3.2 Lateral model

The lateral Takagi–Sugeno model is constructed by choosing the premise variables that influence the determination of the roll rate as well as the most prominent nonlinearity in the yaw rate: the rudder input term. To incorporate all the nonlinear terms that do so, the following premise variables can be defined:

$$\begin{aligned} \zeta_1 &= I_z C_{L_\beta} + I_{xz} C_{N_\beta} \\ \zeta_2 &= I_z C_{L_{\delta_a}} + I_{xz} C_{N_{\delta_a}} \\ \zeta_3 &= I_z C_{L_{\delta_r}} + I_{xz} C_{N_{\delta_r}} \\ \zeta_4 &= I_{xz} C_{L_{\delta_r}} + I_x C_{N_{\delta_r}} \end{aligned} \tag{2.203}$$

with $U = \begin{pmatrix} \delta_a \\ \delta_r \end{pmatrix}, X = \begin{pmatrix} \beta \\ p \\ r \\ \phi \\ \psi \end{pmatrix}$,

with the following parameters $C_1 = \frac{S\bar{q}_T}{mV_T}$, $C_2 = Sb\bar{q}_T \frac{1}{I_x I_z - I_{xz}^2}$.

The dynamics are for a trimmed velocity $V_T = V(trim)$ and angle of attack $\alpha_T = \alpha(trim)$. The state matrix $\mathbf{A}(X,U)$ is given by:

$$\mathbf{A}(X,U) = \begin{pmatrix} \mathbf{A}_{11} & \mathbf{A}_{12} & \mathbf{A}_{13} & \mathbf{A}_{14} & 0 \\ \mathbf{A}_{21} & \mathbf{A}_{22} & \mathbf{A}_{23} & 0 & 0 \\ \mathbf{A}_{31} & \mathbf{A}_{32} & \mathbf{A}_{33} & 0 & 0 \\ 0 & 1 & 1 & 0 & 0 \\ 0 & 0 & \cos\phi & 0 & 0 \end{pmatrix} \tag{2.204}$$

with its matrix elements:

$$\begin{aligned}
\mathbf{A}_{11} &= C_1 C_{Y_\beta} \\
\mathbf{A}_{12} &= \sin\alpha_T \\
\mathbf{A}_{13} &= -\cos\alpha_T \\
\mathbf{A}_{14} &= \frac{g}{V_T}\cos\beta\frac{\sin\beta}{\beta} \\
\mathbf{A}_{21} &= C_2\left(I_z C_{L_\beta} + I_{xz} C_{N_\beta}\right) \\
\mathbf{A}_{22} &= \frac{C_2 b}{2V_T}\left(I_z C_{L_p} + I_{xz} C_{N_p}\right) \\
\mathbf{A}_{23} &= \frac{C_2 b}{2V_T}\left(I_z C_{L_r} + I_{xz} C_{N_r}\right) \\
\mathbf{A}_{31} &= C_2\left(I_{xz} C_{L_\beta} + I_x C_{N_\beta}\right) \\
\mathbf{A}_{32} &= \frac{C_2 b}{2V_T}\left(I_{xz} C_{L_p} + I_x C_{N_p}\right) \\
\mathbf{A}_{33} &= \frac{C_2 b}{2V_T}\left(I_{xz} C_{L_r} + I_x C_{N_r}\right)
\end{aligned}$$

The control matrix $\mathbf{B}(X,U)$ is given by:

$$\mathbf{B}(X,U) = \begin{pmatrix} C_1 C_{Y_{\delta_a}} & C_1 C_{Y_{\delta_r}} \\ C_2\left(I_z C_{L_{\delta_a}} + I_{xz} C_{N_{\delta_a}}\right) & C_2\left(I_z C_{L_{\delta_r}} + I_{xz} C_{N_{\delta_r}}\right) \\ C_2\left(I_{xz} C_{L_{\delta_a}} + I_x C_{N_{\delta_a}}\right) & C_2\left(I_{xz} C_{L_{\delta_r}} + I_x C_{N_{\delta_r}}\right) \\ 0 & 0 \\ 0 & 0 \end{pmatrix} \tag{2.205}$$

2.5.3.3 Multi-model approach for tracking error model for an aircraft at constant altitude

The aircraft model can be represented with the Takagi–Sugeno model. The inference performed via the Takagi–Sugeno model is an interpolation of all the relevant linear models. The degree of relevance becomes the weight in the interpolation process.

Remark 2.12. *Even if the rules in a Takagi–Sugeno model involve only linear combinations of the model inputs, the entire model is nonlinear.*

The kinematic equations of motion of an aircraft at constant altitude are given by

$$\begin{aligned}\dot{x} &= V\cos\chi \\ \dot{y} &= V\sin\chi \\ \dot{\chi} &= \omega\end{aligned} \tag{2.206}$$

The reference trajectory satisfies also:

$$\begin{aligned}\dot{x}_r &= V_r\cos\chi_r \\ \dot{y}_r &= V_r\sin\chi_r \\ \dot{\chi}_r &= \omega_r\end{aligned} \tag{2.207}$$

In 2D, the error posture model can be written as:

$$\begin{pmatrix} e_x \\ e_y \\ e_\chi \end{pmatrix} = \begin{pmatrix} \cos\chi & \sin\chi & 0 \\ -\sin\chi & \cos\chi & 0 \\ 0 & 0 & 1 \end{pmatrix}\begin{pmatrix} x_r - x \\ y_r - y \\ \chi_r - \chi \end{pmatrix} \tag{2.208}$$

Differentiating,

$$\begin{pmatrix} \dot{e}_x \\ \dot{e}_y \\ \dot{e}_\chi \end{pmatrix} = \begin{pmatrix} \omega e_y - V + V_r\cos e_\chi \\ -\omega e_x + V_r\sin e_\chi \\ \omega_r - \omega \end{pmatrix} \tag{2.209}$$

The considered input is $U = \begin{pmatrix} V \\ \omega \end{pmatrix}$. If an anticipative action is chosen such as $U = U_B + U_F$ with $U_F = \begin{pmatrix} V_r\cos e_\chi \\ \omega_r \end{pmatrix}$ then system (2.209) can be written as:

$$\begin{pmatrix} \dot{e}_x \\ \dot{e}_y \\ \dot{e}_\chi \end{pmatrix} = \begin{pmatrix} 0 & \omega_r & 0 \\ -\omega_r & 0 & V_r\frac{\sin e_\chi}{e_\chi} \\ 0 & 0 & 0 \end{pmatrix}\begin{pmatrix} e_x \\ e_y \\ e_\chi \end{pmatrix} + \begin{pmatrix} -1 & e_y \\ 0 & -e_x \\ 0 & -1 \end{pmatrix} U_B \tag{2.210}$$

This nonlinear model can be used for the development of control laws with observers. It is possible with this methodology to prove the stability of the global structure of the control [34].

The errors are assumed to be bounded by:

$$|e_x| \leq e_{max} \qquad |e_y| \leq e_{max} \qquad |e_\chi| \leq \frac{\pi}{2} rad \tag{2.211}$$

while the limits of the inputs are given by:

$$0 < V_{stall} \leq V_r \leq V_{max} \qquad |\omega_r| \leq \omega_{max} \tag{2.212}$$

Considering the model of the tracking error, four nonlinearities appear:

$$n_1 = \omega_r \qquad n_2 = V_r \frac{\sin e_\chi}{e_\chi} \qquad n_3 = e_y \qquad n_4 = e_x \tag{2.213}$$

The Takagi–Sugeno model is obtained by a nonlinear sector approach. As there are four nonlinearities, then $r = 2^4 = 16$ Takagi–Sugeno sub-models can be obtained:

$$\dot{e}(t) = \sum_{i=1}^{16} \varpi_i(\zeta(t)) \left(\mathbf{A}_i e(t) + \mathbf{B}_i U_B(t)\right) \tag{2.214}$$

with the state and control matrices:

$$\mathbf{A}_i = \begin{pmatrix} 0 & -\epsilon_i^1 \omega_{r,max} & 0 \\ \epsilon_i^1 \omega_{r,max} & 0 & \mu_i \\ 0 & 0 & 0 \end{pmatrix} \tag{2.215}$$

$$\mathbf{B}_i = \begin{pmatrix} -1 & \epsilon_i^2 e_{max} \\ 0 & \epsilon_i^3 e_{max} \\ 0 & -1 \end{pmatrix} \tag{2.216}$$

where:

$$\epsilon_i^1 = \left\{ \begin{array}{cc} +1 & 1 \leq i \leq 8 \\ -1 & \text{Otherwise} \end{array} \right\} \tag{2.217}$$

$$\epsilon_i^2 = \left\{ \begin{array}{cr} +1 & i \in \{3,4,7,8,11,12,15,16\} \\ -1 & \text{Otherwise} \end{array} \right\} \tag{2.218}$$

$$\epsilon_i^3 = (-1)^{i+1} \tag{2.219}$$

$$\mu_i = \left\{ \begin{array}{cc} \frac{2}{\pi} V_{r,min} & 1 \leq i \leq 4 \text{ and } 9 \leq i \leq 12 \\ V_{r,max} & \text{Otherwise} \end{array} \right\} \tag{2.220}$$

The membership functions $\varpi_i, i = 1, \dots, 16$ for the Takagi–Sugeno models are:

$$\begin{array}{ll} \varpi_1 = \omega_{01}\omega_{02}\omega_{03}\omega_{04} & \varpi_2 = \omega_{01}\omega_{02}\omega_{03}\omega_{14} \\ \varpi_3 = \omega_{01}\omega_{02}\omega_{13}\omega_{04} & \varpi_4 = \omega_{01}\omega_{02}\omega_{13}\omega_{14} \\ \varpi_5 = \omega_{01}\omega_{12}\omega_{03}\omega_{04} & \varpi_6 = \omega_{01}\omega_{12}\omega_{03}\omega_{14} \\ \varpi_7 = \omega_{01}\omega_{12}\omega_{13}\omega_{04} & \varpi_8 = \omega_{01}\omega_{12}\omega_{13}\omega_{14} \\ \varpi_9 = \omega_{11}\omega_{02}\omega_{03}\omega_{04} & \varpi_{10} = \omega_{11}\omega_{02}\omega_{03}\omega_{14} \\ \varpi_{11} = \omega_{11}\omega_{02}\omega_{13}\omega_{04} & \varpi_{12} = \omega_{11}\omega_{02}\omega_{13}\omega_{14} \\ \varpi_{13} = \omega_{11}\omega_{12}\omega_{03}\omega_{04} & \varpi_{14} = \omega_{11}\omega_{12}\omega_{03}\omega_{14} \\ \varpi_{15} = \omega_{11}\omega_{12}\omega_{13}\omega_{04} & \varpi_{16} = \omega_{11}\omega_{12}\omega_{13}\omega_{14} \end{array} \tag{2.221}$$

with:

$$\omega_{01} = \frac{\omega_{r,max} - \omega_r}{\omega_{r,max} - \omega_{r,min}} \qquad \omega_{11} = 1 - \omega_{01} \tag{2.222}$$

$$\omega_{02} = \frac{V_{r,max} - V_r \frac{\sin e_\chi}{e_\chi}}{V_{r,max} - V_{r,min} \frac{\sin \pi/2}{\pi/2}} \qquad \omega_{12} = 1 - \omega_{02} \tag{2.223}$$

$$\omega_{03} = \frac{e_{max} - e_x}{2e_{max}} \qquad \omega_{13} = 1 - \omega_{03} \tag{2.224}$$

$$\omega_{04} = \frac{e_{max} - e_y}{2e_{max}} \qquad \omega_{14} = 1 - \omega_{04} \tag{2.225}$$

under the hypothesis of controllability for all (A_i, B_i).

2.5.3.4 Multi-model approach for tracking error model for a 3D aircraft

The kinematic equations of motion of an aircraft in 3D are given by

$$\begin{aligned} \dot{x} &= V \cos\gamma \cos\chi \\ \dot{y} &= V \cos\gamma \sin\chi \\ \dot{z} &= -V \sin\gamma \\ \dot{\chi} &= \omega_1 \\ \dot{\gamma} &= \omega_2 \end{aligned} \tag{2.226}$$

The reference trajectory satisfies also:

$$\begin{aligned} \dot{x}_r &= V_r \cos\gamma_r \cos\chi_r \\ \dot{y}_r &= V_r \cos\gamma_r \sin\chi_r \\ \dot{z}_r &= -V_r \sin\gamma_r \\ \dot{\chi}_r &= \omega_{1r} \\ \dot{\gamma}_r &= \omega_{2r} \end{aligned} \tag{2.227}$$

In 3D, the error posture model can be written as:

$$\begin{pmatrix} e_x \\ e_y \\ e_z \\ e_\chi \\ e_\gamma \end{pmatrix} = \begin{pmatrix} E_1 & E_2 & -\sin\gamma & 0 & 0 \\ -\sin\chi & \cos\chi & 0 & 0 & 0 \\ E_3 & E_4 & \cos\gamma & 0 & 0 \\ 0 & 0 & 0 & 1 & 0 \\ 0 & 0 & 0 & 0 & 1 \end{pmatrix} \begin{pmatrix} x_r - x \\ y_r - y \\ z_r - z \\ \chi_r - \chi \\ \gamma_r - \gamma \end{pmatrix} \tag{2.228}$$

where

$$E_1 = \cos\gamma \cos\chi$$

$$E_2 = \cos\gamma \sin\chi$$

$$E_3 = \sin\gamma \cos\chi$$

$$E_4 = \sin\gamma \sin\chi$$

Differentiating,

$$\begin{pmatrix} \dot{e}_x \\ \dot{e}_y \\ \dot{e}_z \\ \dot{e}_\chi \\ \dot{e}_\gamma \end{pmatrix} = \begin{pmatrix} e_1 \\ e_2 \\ e_3 \\ \omega_{1r} - \omega_1 \\ \omega_{2r} - \omega_2 \end{pmatrix} \tag{2.229}$$

where

$$e_1 = \omega_1 \cos\gamma e_y - \omega_2 e_z - V + V_r \cos\gamma \cos\gamma_r \cos e_\chi + V_r \sin\gamma \sin\gamma_r$$

$$e_2 = -\omega_1 \cos\gamma e_x - \omega_1 \sin\gamma e_z + V_r \cos\gamma_r \sin e_\chi$$

$$e_3 = \omega_1 \sin\gamma e_y + \omega_2 e_x + V_r \sin\gamma \cos\gamma_r \cos e_\chi - V_r \cos\gamma \sin\gamma_r$$

The considered input is $U = \begin{pmatrix} V \\ \omega_1 \\ \omega_2 \end{pmatrix}$. If an anticipative action is chosen such as $U = U_B + U_F$ with

$$U_F = \begin{pmatrix} V_r \cos\gamma \cos\gamma_r \cos e_\chi + V_r \sin\gamma \sin\gamma_r \\ \omega_{1r} \\ \omega_{2r} \end{pmatrix} \tag{2.230}$$

then system (2.229) can be written as:

$$\begin{pmatrix} \dot{e}_x \\ \dot{e}_y \\ \dot{e}_z \\ \dot{e}_\chi \\ \dot{e}_\gamma \end{pmatrix} = \mathbf{A}_{lin} \begin{pmatrix} e_x \\ e_y \\ e_z \\ e_\chi \\ e_\gamma \end{pmatrix} + \mathbf{B}_{lin} U_B \tag{2.231}$$

where

$$\mathbf{A}_{lin} = \begin{pmatrix} 0 & \omega_{1r}\cos\gamma & -\omega_{2r} & 0 & 0 \\ -\omega_{1r}\cos\gamma & 0 & -\omega_{1r}\sin\gamma & \mathbf{A}_{24} & 0 \\ \omega_{2r} & \omega_{1r}\sin\gamma & 0 & \mathbf{A}_{34} & \mathbf{A}_{35} \\ 0 & 0 & 0 & 0 & 0 \\ 0 & 0 & 0 & 0 & 0 \end{pmatrix} \tag{2.232}$$

with its elements

$$\mathbf{A}_{24} = V_r \cos\gamma_r \frac{\sin e_\psi}{e_\chi}$$

$$\mathbf{A}_{34} = V_r \cos\gamma_r \sin\gamma \frac{\cos e_\chi}{e_\chi}$$

$$\mathbf{A}_{35} = -V_r \sin\gamma_r \cos\gamma \frac{1}{e_\chi}$$

and

$$\mathbf{B}_{lin} = \begin{pmatrix} -1 & \cos\gamma e_y & -e_z \\ 0 & -e_x\cos\gamma + e_z\sin\gamma & 0 \\ 0 & \sin\gamma e_y & e_x \\ 0 & -1 & 0 \\ 0 & 0 & -1 \end{pmatrix} \tag{2.233}$$

This nonlinear model is used for the development of control laws with observers. The errors are assumed to be bounded by:

$$|e_x| \leq e_{max} \qquad |e_y| \leq e_{max} \qquad |e_z| \leq e_{max} \qquad |e_\chi| \leq \frac{\pi}{2} rad \qquad |e_\gamma| \leq \frac{\pi}{2} rad \tag{2.234}$$

while the limits of the inputs are given by:

$$V_{stall} \leq V_r \leq V_{max} \qquad |\omega_{1r}| \leq \omega_{max} \qquad |\omega_{2r}| \leq \omega_{max} \tag{2.235}$$

Considering the model of the tracking error, eleven nonlinearities appear:

$$\begin{array}{c} n_1 = \omega_{1r}\cos\gamma \qquad n_2 = \omega_{1r}\sin\gamma \qquad n_3 = \omega_{2r} \\ n_4 = V_r\cos\gamma_r \frac{\sin e_\psi}{e_\chi} \qquad n_5 = V_r\cos\gamma_r\sin\gamma\frac{\cos e_\chi}{e_\chi} \\ n_6 = V_r\sin\gamma_r\cos\gamma\frac{1}{e_\chi} \qquad n_7 = \cos\gamma e_y \qquad n_8 = \sin\gamma e_y \\ n_9 = e_x \qquad n_{10} = e_h \qquad n_{11} = e_x\cos\gamma + e_z\sin\gamma \end{array} \tag{2.236}$$

The Takagi–Sugeno model is obtained by a nonlinear sector approach. As there are eleven nonlinearities, then $r = 2^{11} = 2048$ Takagi–Sugeno sub-models can be obtained:

$$\dot{e}(t) = \sum_{i=1}^{2048} \varpi_i(Z(t))\left(\mathbf{A}_i e(t) + \mathbf{B}_i U_B(t)\right) \tag{2.237}$$

with the state and control matrices:

$$\mathbf{A}_i = \begin{pmatrix} 0 & \omega_{max} & -\omega_{max} & 0 & 0 \\ -\omega_{max} & 0 & -\omega_{max} & V_r\frac{\sin e_\psi}{e_\chi} & 0 \\ \omega_{max} & \omega_{max} & 0 & V_r\frac{\cos e_\chi}{e_\chi} & -V_r\frac{1}{e_\chi} \\ 0 & 0 & 0 & 0 & 0 \\ 0 & 0 & 0 & 0 & 0 \end{pmatrix} \tag{2.238}$$

$$\mathbf{B}_i = \begin{pmatrix} -1 & \cos\gamma e_y & -e_h \\ 0 & -e_x\cos\gamma - e_z\sin\gamma & 0 \\ 0 & \sin\gamma e_y & e_x \\ 0 & -1 & 0 \\ 0 & 0 & -1 \end{pmatrix} \tag{2.239}$$

The number of sub-models being too important, the implementation is not an easy task. In this case, only the more influential nonlinearities should be kept while the others should be linearized by first order approximation. The following approximation is used: $\cos e_\chi \approx 1$.

The following equation model is thus obtained.

$$\begin{pmatrix} \dot{e}_x \\ \dot{e}_y \\ \dot{e}_z \\ \dot{e}_\chi \\ \dot{e}_\gamma \end{pmatrix} = \begin{pmatrix} \omega_1 \cos\gamma e_y - \omega_2 e_z - V + V_r \cos e_\gamma \\ -\omega_1 \cos\gamma e_x - \omega_1 \sin\gamma e_z + V_r \cos\gamma_r \sin e_\chi \\ \omega_1 \sin\gamma e_y + \omega_2 e_x - V_r \sin e_\gamma \\ \omega_{1r} - \omega_1 \\ \omega_{2r} - \omega_2 \end{pmatrix} \tag{2.240}$$

The considered input is $U = \begin{pmatrix} V \\ \omega_1 \\ \omega_2 \end{pmatrix}$. Then an anticipative action is chosen such as $U = U_B + U_F$ with

$$U_F = \begin{pmatrix} V_r \cos e_\gamma \\ \omega_{1r} \\ \omega_{2r} \end{pmatrix} \tag{2.241}$$

This error model will be retained in the sequel.

Then the following system is obtained:

$$\dot{e} = \mathbf{A}_{line} e + \mathbf{B}_{line} U + \begin{pmatrix} \cos e_\gamma \\ \sin e_\gamma \cos\gamma_r \\ -\sin e_\gamma \\ 0 \\ 0 \end{pmatrix} V_r \tag{2.242}$$

where

$$\mathbf{A}_{line} = \begin{pmatrix} 0 & \omega_{1r}\cos\gamma & -\omega_{2r} & 0 & 0 \\ -\omega_{1r}\cos\gamma & 0 & -\omega_{1r}\sin\gamma & V_r\cos\gamma_r & 0 \\ \omega_{2r} & \omega_{1r}\sin\gamma & 0 & 0 & -V_r \\ 0 & 0 & 0 & 0 & 0 \\ 0 & 0 & 0 & 0 & 0 \end{pmatrix} \tag{2.243}$$

$$\mathbf{B}_{line} = \begin{pmatrix} -1 & 0 & 0 \\ 0 & 0 & 0 \\ 0 & 0 & 0 \\ 0 & -1 & 0 \\ 0 & 0 & -1 \end{pmatrix} \tag{2.244}$$

Three nonlinearities appear:

$$n_1 = \omega_{1r} \cos\gamma \qquad n_2 = \omega_{1r} \sin\gamma \qquad n_3 = \omega_{2r} \tag{2.245}$$

Then there will exist $p = 2^3 = 8$ sub-models. Taking into account the limitations on n_1, n_2, n_3 and the mutual exclusion between $\sin\gamma$ and $\cos\gamma$ to attain their maximum, the following four state matrices can be used.

$$\mathbf{A}_l = \begin{pmatrix} 0 & \epsilon_i\omega_{1r} & -\omega_{2r} & 0 & 0 \\ -\epsilon_i\omega_{1r} & 0 & 0 & V_r\cos\gamma_r & 0 \\ \omega_{2r} & 0 & 0 & 0 & -V_r \\ 0 & 0 & 0 & 0 & 0 \\ 0 & 0 & 0 & 0 & 0 \end{pmatrix} \tag{2.246}$$

$\epsilon_i = \pm 1$, or

$$\mathbf{A}_l = \begin{pmatrix} 0 & 0 & -\omega_{2r} & 0 & 0 \\ 0 & 0 & -\epsilon_j\omega_{1r} & V_r\cos\gamma_r & 0 \\ \omega_{2r} & \epsilon_j\omega_{1r} & 0 & 0 & -V_r \\ 0 & 0 & 0 & 0 & 0 \\ 0 & 0 & 0 & 0 & 0 \end{pmatrix} \tag{2.247}$$

$\epsilon_j = \pm 1$, and

$$\mathbf{B}_l = \begin{pmatrix} -1 & 0 & 0 \\ 0 & 0 & 0 \\ 0 & 0 & 0 \\ 0 & -1 & 0 \\ 0 & 0 & -1 \end{pmatrix} \tag{2.248}$$

Model validation refers to assessing the predictive accuracy of a model with experimental data. Existing techniques can generally be placed into two categories [23]:

1. A **model perturbation** is identified that accounts for the discrepancy between simulation and experiment. If the perturbation belongs to an allowable set, the result is deemed not to invalidate the model. This approach mostly relies on linear matrix inequality optimization and provides rigorous conclusions about model quality.
2. The second method relies on **statistical analysis** of the estimated output error between simulation and experiment.

A method based on the gap metric is presented in [23] as a technique to validate aircraft models using flight data. The gap metric is a generalization of the statistical validation metric: the **Theil inequality coefficient**. It allows a comparison for an identified aircraft LTI model to flight data and the derivation of a set of robustness requirements for closed-loop control.

The attraction of LPV systems is the possibility of determining a global model over an extended region of the operating parameter space on data in a limited region of the operating parameter space. For an aircraft simulation, this allows the prediction and extrapolation of the flight characteristics into regions of the operating parameter space where no actual data were previously available [10].

2.5.4 FUZZY MODELING

Fuzzy logic is a form of soft (approximate) computing that allows for better tolerance to imprecision through use of fuzzy **membership functions** (MF). It is possible to model complex systems with fuzzy logic; each point in the variable space or universe has a degree of membership to a fuzzy set on the interval $[0, 1]$. A fuzzy membership function can also be used to model uncertainty as a possible distribution. There are three stages in a fuzzy system, which evaluates disjoint IF-THEN rules of the form:

$$\text{IF antecedent THEN consequent} \tag{2.249}$$

2.5.4.1 Type 1 Mamdani approach

Fuzzy logic is a modification of boolean or crisp logic which allows approximate and common sense reasoning in the absence of *true* or *false* certainty. In crisp logic, set membership is *all* or *nothing*. In contrast fuzzy logic allows partial membership of sets, known as fuzzy sets, and forms the basis of fuzzy systems. Fuzzy systems can deal with partial truth and incomplete data and are capable of producing accurate models of how systems behave in the real world, particularly when appropriate conventional system models are not available [63]. The system operates when inputs are applied to the rules consisting of the current values of appropriate membership functions. Once activated, each rule will fire and produce an output which will also be a partial truth value. In the final stage, the outputs from all the rules are combined, in some way, and converted into a single crisp output value.

In summary, a fuzzy system consists of the following:

1. A **set of inputs**,
2. A **fuzzification** system, for transforming the raw inputs into grades of memberships of fuzzy sets,
3. A set of **fuzzy rules**,
4. An **inference** system to activate the rules and produce their outputs,
5. A **defuzzification** system to produce one or more crisp outputs.

A variation of fuzzy sets is rough sets. The basic idea is to take concepts and decision values and create rules for upper and lower boundary approximations of the set. With these rules, a new object can easily be classified into one of the regions. Rough sets are especially helpful in dealing with vagueness and uncertainty in decision situations and for estimating missing data.

Mamdani-type inference expects the output membership functions to be fuzzy sets. After the aggregation process, there is a fuzzy set for each output variable that needs defuzzification. It is possible to use a single spike as the output membership function rather than a distributed fuzzy set. This set of output is sometimes known as a singleton output membership function and it can be thought of as a pre-defuzzified fuzzy set. It enhances the efficiency of the defuzzification process because it greatly simplifies the computation required by the more general Mamdani method, which finds the centroid of a 2D function [76].

Definition 2.6. ***Fuzzy Set:*** *Given a set* $\mathbb{X}$*, a conventional Type 1 fuzzy set* $\mathbb{A}$ *defined on* $\mathbb{X}$ *is given by a 2D membership function, also called Type 1 membership function. The primary membership function, denoted by* $\mu_{\mathbb{A}}(X)$*, is a crisp number in* $[0,1]$*, for a generic element* $X \in \mathbb{X}$*. Usually, the fuzzy set* $\mathbb{A}$ *is expressed as a 2-tuple given by:*

$$\mathbb{A} = \{(X, \mu_{\mathbb{A}}(X)) \,|\forall X \in \mathbb{X}\} \tag{2.250}$$

An alternative representation of the fuzzy set $\mathbb{A}$ is also found in the literature [35] as

$$\mathbb{A} = \int_{X \in \mathbb{X}} \mu_{\mathbb{A}}(X) dX \tag{2.251}$$

where $\int$ denotes the union of all admissible X.

Definition 2.7. ***Fuzzy Binary Relation:*** *A relation* $R \in \mathbb{U} \times \mathbb{W}$ *is referred to as a fuzzy binary relation from* $\mathbb{U}$ *to* $\mathbb{W}$*,* $R(X,Y)$ *is the degree of relation between* X *and* Y *where* $(X,Y) \in \mathbb{U} \times \mathbb{W}$*. If for each* $X \in \mathbb{U}$*, there exists* $Y \in \mathbb{W}$ *such that* $R(X,Y) = 1$*, then* R *is referred to as a fuzzy relation from* $\mathbb{U}$ *to* $\mathbb{W}$*. If* $\mathbb{U} = \mathbb{W}$*, then* R *is referred to as a fuzzy relation on* $\mathbb{U}$*.* R *is referred to as a reflexive fuzzy relation if* $R(X,X) = 1, \forall X \in \mathbb{U}$*, symmetric if* $R(X,Y) = R(Y,X), \forall X,Y \in \mathbb{U}$*, transitive fuzzy relation if* $R(X,Y) \leq \vee_{Y \in \mathbb{U}} (R(X,Y) \wedge R(Y,Z)), \forall X, Z \in \mathbb{U}$*.*

For example, the designed autonomous navigation controller can consist of six fuzzy logic modules for the control of altitude, yaw angle, roll angle, pitch angle and motion along x, y, z axes using the error and the rate of change of these errors [25]. There are five membership functions for each input set as:

1. NB: Negative Big
2. N: Negative
3. Z: Zero
4. P: Positive
5. PB: Positive Big

There are seven membership functions for each output set as:

1. NB: Negative Big
2. NM: Negative Mean
3. N: Negative
4. Z: Zero
5. P: Positive
6. PM: Positive Mean
7. PB: Positive Big

The membership function used for the input and output are Gaussian:

$$\mu_{A_i} = \exp\left(\frac{(c_i - X)^2}{2\sigma_i^2}\right) \tag{2.252}$$

Mean-area defuzzification method can be selected for the controllers [50].

2.5.4.2 Fuzzy estimation of Takagi–Sugeno models

The fuzzy Takagi–Sugeno model is an effective tool for uncertain nonlinear system approximation and has been applied successfully to control theory and techniques [46]. The Takagi–Sugeno system consists of *if-then* rules with fuzzy rule antecedent and consequent, where fuzzy rule consequent is specified as an affine function. More precisely, for m dimensional input variable $X = (X_1, X_2, \dots, X_n)^T \in \mathbb{R}^n$, the ith fuzzy rule is presented as

$$\begin{array}{c} R^i : \text{ IF } X_1 \text{ is } A_{i1}, X_2 \text{ is } A_{i2}, \dots, X_n \text{ is } A_{in} \text{ THEN} \\ Y = \varpi_{i0} + \varpi_{i1}X_1 + \varpi_{i2}X_2 + \dots + \varpi_{in}X_n; (i = 1, 2, \dots, r) \end{array} \tag{2.253}$$

Bell-shaped membership function can be employed to represent the fuzzy linguistic proposition $A_{ij}(X_j)$ such that:

$$\mu_{\mathbb{A}_{ij}}(X_j) = \exp\left[-\left(\frac{X_j - c_{ij}}{\sigma_{ij}}\right)^2\right] \tag{2.254}$$

where c_{ij} and σ_{ij} denote the mean and the variance of the corresponding bell-shaped membership function, respectively. Other shapes can also be used.

With weighted-average defuzzifier, the output function of the Takagi–Sugeno system is derived as:

$$\hat{Y} = \sum_{i=1}^{r} \Phi_i(X) l_i(X) \tag{2.255}$$

where $\Phi_i(X)$ denotes the firing strength of the ith fuzzy rule with respect to the input variable X by:

$$\Phi_i(X) = \frac{\mu_i(X)}{\sum_{k=1}^{r} \mu_k(X)} \tag{2.256}$$

where

$$\mu_i(X) = \prod_{j=1}^{n} \mu_{A_{ij}}(X_j) \quad (2.257)$$

and

$$l_i(X) = \varpi_{i0} + \varpi_{i1}X_1 + \varpi_{i2}X_2 + \cdots + \varpi_{in}X_n = \left[1, X^T\right]^T \varpi_i = X_e\varpi_i \quad (2.258)$$

is the corresponding fuzzy rule consequent where $\varpi_i = (\varpi_{i0}, \varpi_{i1}, \ldots, \varpi_{in})^T$ indicates the column vector of parameters for the ith rule consequent part. Let the input-output dataset be:

$$\mathbb{D} = \left\{ \left(X_k^T, Y_k^T\right)^T : X_k = [X_{k1}, X_{k2}, \ldots, X_{kn}]^T, k = 1, \ldots, N \right\} \quad (2.259)$$

where X_k and Y_k represent the kth n-dimensional input variable and output variable, respectively. $\mathbb{M}_{\mathbb{D}}^r$ is the Takagi–Sugeno system with r fuzzy rules, which is learned from data set $\mathbb{D}$.

Let $\hat{Y} = \left[\hat{Y}_1, \hat{Y}_2, \ldots, \hat{Y}_N\right]$ be the output of the Takagi–Sugeno system $\mathbb{M}_{\mathbb{D}}^r$,

$$\hat{Y} = \sum_{i=1}^{r} \Phi_i \varpi_i = \Phi\varpi \quad (2.260)$$

where $\Phi \in \mathbb{R}^{N \times r(n+1)}$ consists of r blocks

$$\Phi_i = diag(\Phi_i(X_1), \Phi_i(X_2), \ldots, \Phi_i(X_N))X_e \quad r = 1, \ldots, r \quad (2.261)$$

Φ is called the dictionary of the Takagi–Sugeno System $\mathbb{M}_{\mathbb{D}}^r$ and its component Φ_i is the sub-dictionary of the ith fuzzy rule. The Takagi–Sugeno fuzzy system dictionary has a natural block structure. Each block is associated with one fuzzy rule sub-dictionary. In this sense, fuzzy model output $\hat{Y}$ can be expressed as a linear combination of the fuzzy rule sub-dictionaries. A fuzzy clustering method can be used to identify Takagi–Sugeno models, including identification of the number of rules and parameters of membership functions, and identification of parameters of local linear models by using a least squares method. The goal is to minimize the error between the Takagi–Sugeno models and the corresponding original nonlinear system [24]. Another method of interval fuzzy model identification can also be used. This method combines a fuzzy identification methodology with linear programming. This results in lower and upper fuzzy models or a fuzzy model with lower and upper parameters. An approach to fuzzy modeling using the **relevance vector learning mechanisms** (RVM) based on a kernel-based Bayesian estimation has also been proposed. The main concern is to find the best structure of the Takagi–Sugeno fuzzy model for modeling nonlinear dynamic systems with measurement error. The number of rules and the parameter values of membership functions can be found as optimizing the marginal likelihood of the relevance vector learning mechanisms in the proposed **fuzzy inference system** (FIS) [15, 46, 50].

Establishing the Takagi–Sugeno system on the basis of data may lead to a nonlinear programming problem because it is quite technical to determine the optimum fuzzy rules including **antecedent membership functions** (AMF) and the corresponding consequent parameters. Some numerical optimization approaches such as neuro-fuzzy method via gradient-descent optimization techniques, genetic algorithms by balancing the trade-off between model complexity and accuracy and Levenberg-Marquardt algorithms have been investigated and widely used for fuzzy system modeling [42].

Two important steps are necessary for **Takagi–Sugeno fuzzy system** (TSFS) modeling based on data [85]:

1. Determining rule membership functions which divide the input space into a number of regions.
2. Estimating the rule consequent parameter vector, which is used to describe the system's behavior in each region.

Besides the situations in which experts can provide fuzzy rules with some prior knowledge, statistics and **clustering techniques** are naturally and extensively exploited to partition the input space and determine the membership functions of fuzzy rule antecedents. Many well known clustering techniques are widely used such as k-means algorithm, fuzzy c-means algorithm and its extensions, subtractive clustering algorithms, vector quantization (QV) approach. After **fuzzy rule antecedents** have been determined, the estimation of consequent parameters can be viewed as a linear regression problem in the product space of the given Input-Output data [54].

Given an input-output data set, the Takagi–Sugeno system can be identified with a minimal number of rules which simultaneously process sparse consequent parameters. To this end, block-structured information existing in the Takagi–Sugeno model are taken into account and block structured sparse representation to the framework of the Takagi–Sugeno system identification are extended. Block-structured sparse representation as a successor of traditional sparse representation is introduced in the **least absolute shrinkage and selection operator** (LASSO). It provides a regression model where many blocks of the regression coefficients with small contribution would shrink exactly to zero while keeping high prediction accuracy. The main important rules are selected while the redundant ones are eliminated. As a result, the Takagi–Sugeno system is established with a minimal number of rules that also possess a minimal number of non zero consequent parameters [90].

2.5.4.3 Type 2 fuzzy systems

Type 1 fuzzy sets handle the uncertainties associated with the fuzzy system inputs and outputs by using precise and crisp membership functions which the user believes would capture the uncertainties [40]. Once the type 1 membership functions have been chosen, the fact that the actual degree of membership itself is uncertain is no longer modeled in type 1 fuzzy sets. It is assumed that a given input results in a precise single value of membership. However, the linguistic and numerical uncertainties associated with dynamic unstructured environments cause problems in determining the exact and precise antecedents and consequent membership functions during the fuzzy logic system design. As a consequence, the designed type 1 fuzzy sets can be sub-optimal under specific environment and operation conditions. This can cause degradation in the fuzzy system performance which can result in poor control.

Definition 2.8. ***Type 2 fuzzy set:*** *A type 2 fuzzy set $\tilde{A}$ is characterized by a 3D membership function, also called T2 membership function, which itself is fuzzy. The T2 membership function is usually denoted by $\mu_{\tilde{A}}(X, U)$ where $X \in \mathbb{X}$ and $U \in \mathbb{J}_x \subseteq [0, 1]$ where $0 \leq f_x(U) \leq 1$. The amplitude of a secondary membership function is called secondary grade of membership, $\mathbb{J}_x$ being the primary membership of X.*

Definition 2.9. ***Footprint of uncertainty:*** *Uncertainty in the primary membership of a Type 2 fuzzy set $\tilde{A}$ is represented by a bounded region, called footprint of uncertainty (FOU) which is defined as the union of all primary membership:*

$$FOU(\tilde{A}) = \bigcup_{X \in \mathbb{U}} \mathbb{J}_x \tag{2.262}$$

If all the secondary grades of a Type 2 fuzzy set $\tilde{A}$ are equal to 1, i.e.,

$$\mu_{\tilde{A}}(X, U) = 1, \forall X \in \mathbb{X}, \forall U \in \mathbb{J}_x \subseteq [0, 1] \tag{2.263}$$

then $\tilde{A}$ is called interval type 2 fuzzy logic systems (IT2FLS). The footprint of uncertainty is bounded by two curves called the lower and upper membership functions denoted by $\underline{\mu}_{\tilde{A}}, \overline{\mu}_{\tilde{A}}$ at all X, respectively, take up the minimum and maximum of the membership functions of the embedded T1 fuzzy sets of the footprint of Uncertainty.

A **type 2 fuzzy set** is characterized by a fuzzy membership function, i.e., the membership value, for each element of this set is a fuzzy set in $[0, 1]$, unlike a type 1 fuzzy set where the membership grade is a crisp number in $[0, 1]$. The membership functions of type 2 fuzzy sets are 3D and include a **footprint of uncertainty**. It is the third dimension of type 2 fuzzy sets and the footprint of uncertainty that provide additional degrees of freedom that make it possible to model and handle uncertainties. Hence, type 2 fuzzy logic systems have the potential to overcome the limitations of type 1 fuzzy logic system [40, 59].

Type reduction (TR) followed by defuzzification is commonly used in **interval type 2 fuzzy logic systems** (IT2FLS)[59]. Some direct approaches to defuzzification bypass type reduction, the simplest of which is the **Nie–Tan** (NT) direct defuzzification method. The original interval type 2 fuzzy logic system requires both type reduction and defuzzification where type reduction which projects an interval type 2 fuzzy logic system into an interval of numbers is accomplished using iterative **Kernel–Mendel**(KM) or **Enhanced Kernel–Mendel** (EKM) algorithm. In such an interval type 2 fuzzy logic system, after computing the firing interval for each rule, fired rules can be aggregated in two different ways:

1. **Aggregation A1**: Compute interval type 2 fired-rule output sets; then take the union of those sets leading to a single interval type 2 fuzzy logic system, reduce that interval type 2 fuzzy logic system by means of centroid type reduction and finally defuzzify the resulting type reduced set by computing the average value of the two end points of the type reduced set.
2. **Aggregation A2**: Go directly to type reduction by mean of center of sets, height or modified height type reduction and then defuzzify the resulting type-reduced set by computing the average value of the two end points of the type reduced set.

Mini-max uncertainty bounds (UB) are developed that provide closed form formulas for the lower and upper bounds of both end points of the type-reduced set. Defuzzification is simple after the bounds have been computed; it is just the average of the two end points of the approximate type reduced set: uncertainty bounds and defuzzification.

A closed form type reduction and defuzzification method has also been proposed; it makes use of equivalent type 1 membership grades. The basic idea is to first find an equivalent type 1 membership grade, a firing strength to replace the firing interval, after which those firing strengths are used in center of sets defuzzification. Finding the equivalent type 1 membership grade being complicated, the closed form type reduction and defuzzification method method can only be used as a replacement for **aggregation A2**.

In the case of a **centroid computation** for an interval type 2 fuzzy logic system, fired rule output sets in an interval type 2 fuzzy logic system are combined by means of the union operation to give one aggregated interval type 2 fuzzy logic system: $\mathbb{A}$. The defuzzified value of $\mathbb{A}$ is computed in two steps:

1. Compute the centroid of $\mathbb{A}$, which is an interval set, $C(\mathbb{A})$.
2. Defuzzify $C(\mathbb{A})$.

Interval type 2 fuzzy logic system is a bivariate function on the Cartesian product: $\mu : \mathbb{X} \times [0,1] \to [0,1]$ where $\mathbb{X}$ is the universe for the primary variable X of $\mathbb{A}$. The point-valued representation of $\mathbb{A}$ is:

$$\mathbb{A} = \{(X,U), \mu_A(X,U) = 1, \forall X \in \mathbb{X}, \forall U \in [0,1]\} \tag{2.264}$$

The 2D support of $\mu_{\mathbb{A}}$ is called the footprint of uncertainty (FOU) of $\mathbb{A}$, i.e.,

$$FOU(\mathbb{A}) = \{(X, U) \in \mathbb{X}, \times [0,1], \mu_{\mathbb{A}} > 0\} \tag{2.265}$$

$FOU(\mathbb{A})$ is bounded by lower and upper bounding membership functions (MF) which are denoted by $\underline{\mu}_{\mathbb{A}}$ and $\overline{\mu}_{\mathbb{A}}$, respectively, where:

$$\overline{\mu}_{\mathbb{A}} = sup\{U \in [0,1], \mu_{\mathbb{A}}(X, U) > 0\} \tag{2.266}$$

and

$$\underline{\mu}_{\mathbb{A}} = inf\{U \in [0,1], \mu_{\mathbb{A}}(X, U) > 0\} \tag{2.267}$$

The primary membership function of $\mathbb{A}$ at X is the interval $\left[\underline{\mu}_{\mathbb{A}}, \overline{\mu}_{\mathbb{A}}\right]$ i.e.,

$$J_X = \left[\underline{\mu}_{\mathbb{A}}, \overline{\mu}_{\mathbb{A}}\right] \tag{2.268}$$

An embedded type 1 fuzzy system i.e., $\mathbb{A}_e^j$, is a fuzzy set whose range is a subset of $[0,1]$ determined by $\mu_{\mathbb{A}}$:

$$A_e^j = \{(X, U(X)), X \in \mathbb{X}, U \in \mathbb{J}_X\} \tag{2.269}$$

where the primary variable X is sampled at N values $X_1, \ldots, X_N$ and, at each of these values, its primary membership is sampled at M_i values $\mu_{i1}, \ldots, \mu_{iM_i}$; then there will be $n_A = \prod_{i=1}^{N} M_i$ embedded type 1 fuzzy systems that are contained within $FOU(\mathbb{A})$.

The wavy slide representation for $\mathbb{A}$ is:

$$FOU(\mathbb{A}) = \bigcup_{j=1}^{n_A} \mathbb{A}_e^j \tag{2.270}$$

The centroid $C(\mathbb{A})$ of the type 1 fuzzy system $\mathbb{A} \in \mathbb{X} = \{\mathbb{X}_1, \mathbb{X}_2, \ldots, \mathbb{X}_N\}$ is defined as:

$$C(\mathbb{A}) = \frac{\sum_{i=1}^{N} X_i \mu_A(x_i)}{\sum_{i=1}^{N} \mu_A(X_i)} \tag{2.271}$$

The centroid $C(\mathbb{A})$ of an interval type 2 fuzzy system $\mathbb{A}$ is the union of the centroids of all its embedded type 1 fuzzy systems A_e^j.

$$C(\mathbb{A}) = \bigcup_{j=1}^{n_A} C(A_e^j) = [C_l(\mathbb{A}), C_r(\mathbb{A})] \tag{2.272}$$

where

$$C_l(\mathbb{A}) = min_{\forall A_e^j} C_{\mathbb{A}}(A_e^j) \tag{2.273}$$

and

$$C_r(\mathbb{A}) = max_{\forall A_e^j} C_{\mathbb{A}}(A_e^j) \tag{2.274}$$

In the case of a continuous Kernel–Mendel algorithm, for a continuous domain of support, C_l and C_r can be addressed as:

$$C_l = min_{\zeta \in [a,b]} \frac{\int_a^{\zeta_l} x\overline{\mu}_{\mathbb{A}}(X)dX + \int_{\zeta_l}^b x\underline{\mu}_{\mathbb{A}}(X)dX}{\int_a^{\zeta_l} \overline{\mu}_{\mathbb{A}}(X)dX + \int_{\zeta_l}^b \underline{\mu}_{\mathbb{A}}(X)dX} \tag{2.275}$$

and

$$C_r = max_{\zeta \in [a,b]} \frac{\int_a^{\zeta_l} x\overline{\mu}_{\mathbb{A}}(x)dx + \int_{\zeta_l}^b x\underline{\mu}_{\mathbb{A}}(x)dx}{\int_a^{\zeta_l} \overline{\mu}_{\mathbb{A}}(X)dX + \int_{\zeta_l}^b \underline{\mu}_{\mathbb{A}}(X)dX} \tag{2.276}$$

ζ_l^*, ζ_r^* the solutions to (2.275) and (2.276) are called switch points and are computed by continuous Kernel–Mendel (CKM) or Continuous Enhanced Kernel–Mendel (CEKM).

Transforming the centroid computation into root finding problems: the optimization problems that are associated with the continuous Kernel–Mendel (and continuous enhanced Kernel–Mendel) can be transformed into equivalent root finding problems. It is those root findings that define the Nie–Tan and Kernel–Mendel and defuzzification architectures.

Using the Kernel–Mendel method and defuzzification, the defuzzified value of $\mathbb{A}, m()$ is the center of $C(\mathbb{A})$, i.e.,

$$m = \frac{1}{2}(c_l + C_r) \tag{2.277}$$

In the Nie–Tan method, the average is first computed, i.e., C_i, of the lower membership function and the upper membership function of $\mathbb{A}$ at each X_i:

$$C_i = \frac{1}{2}\left(\underline{\mu}_{\mathbb{A}}(X_i) + \overline{\mu}_{\mathbb{A}}(X_i)\right), i = 1, \ldots, N \tag{2.278}$$

Each C_i is a spike that is located at $X = X_i$. Then, one computes the center of gravity (COG) of the N spikes:

$$C_{NT} = \frac{\sum_{i=1}^N X_i C_i}{\sum_{i=1}^N C_i} \tag{2.279}$$

Equation (2.279) shows that the Nie–Tan formulation of the crisp output of an interval type 2 fuzzy logic system depends only on the lower and upper bound of its $FOU(\mathbb{A})$ [35, 40, 59].

2.5.5 LINEAR HYBRID AUTOMATON

A typical aircraft dynamics consists of several flight modes, for example constant velocity (CV), coordinated turn (CT) and so on. The **stochastic linear hybrid system (SLHS)** model is well suited for describing such dynamics, with each discrete state matched to a flight mode, and each pair (A_Q, B_Q)

models the discrete-time continuous state evolution corresponding to different flight modes. When there is no information available about the aircraft intent, the mode transition is decoupled from the continuous state dynamics. Thus, the Markov jump transition model can be used to describe the aircraft flight mode transition [49].

2.5.5.1 Stochastic linear hybrid system

The stochastic linear hybrid system (SLHS) model contains a set of discrete-time continuous state models, each matched to a discrete state. At each time k, the discrete-time state dynamics is described by the following difference equation:

$$X(k) = \mathbf{A}_{Q(k)}X(k-1) + \mathbf{B}_{Q(k)}\tilde{W}(k) \tag{2.280}$$

where $X(k)$ is the state vector, $Q(k) \in \mathbb{Q} = 1, 2, .., n_d$ is the discrete state at time k, $\mathbb{Q}$ is a finite set of all the discrete states, $\mathbf{A}_Q$ and $\mathbf{B}_Q$ are the system matrices with appropriate dimensions, corresponding to each discrete state $Q \in \mathbb{Q}$, and $\tilde{W}(k) \in \mathbb{R}^p$ is the white Gaussian process noise with zero mean and covariance $\tilde{Q}(k)$. The initial state is assumed to be independent of $\tilde{W}(k)$ for any k.

There are two types of discrete transition models in the stochastic linear hybrid system:

1. The first type is the **Markov jump transition model**. The discrete state transition history is a realization of a homogeneous Markov chain. The finite state space of the Markov chain is the discrete state space $\mathbb{Q}$. Suppose at each time k, the probability vector is given by $\pi(k) = (\pi_1, \ldots, \pi_{n_d}(k))^T$ where each entry of the vector $\pi_i(k)$ denotes the probability that the system's true discrete state is k. Then at the next time step, the probability vector is updated as

$$\pi(k+1) = \Gamma\pi(k) \tag{2.281}$$

where a constant matrix Γ is the Markov transition matrix with

$$\sum_j \Gamma_{ij} = 1 \tag{2.282}$$

Γ_{ij} denotes the scalar component in the ith row and the j^{th}column in the Markov transition matrix Γ. The discrete transition is independent of the continuous dynamics.
2. The second type is the **state-dependent transition model**. The discrete state transition is governed by

$$Q(k+1) = \tilde{\mathbf{A}}\left(Q(k), X(k), \theta\right) \tag{2.283}$$

where $\theta \in \Theta = \mathbb{R}^l$ and $\tilde{\mathbf{A}}$ is the discrete state transition function defined as

$$\tilde{\mathbf{A}}(i, X, \theta) = j \quad \textit{if } [X^T \;\; \theta^T]^T \in G(i,j) \tag{2.284}$$

where $G(i,j)$ is the guard condition.

A specific kind of the guard condition is the stochastic linear guard condition

$$G_{ij} = \left\{ \begin{bmatrix} X \\ \theta \end{bmatrix} | X \in \mathbb{X}, \theta \in \Theta, \mathbf{L}_{ij} \begin{bmatrix} X \\ \theta \end{bmatrix} + b_{ij} \leq 0 \right\} \tag{2.285}$$

where $\theta \approx N(\theta, \bar{\theta}, \Sigma_\theta)$ is a l-dimensional Gaussian random vector with mean $\bar{\theta}$ and covariance Σ_θ representing uncertainties in the guard condition; $\mathbf{L}_{ij}$ is a $\xi \times (n+l)$ matrix, b_{ij} is a constant ξ dimensional vector and ξ is the dimension of the vector inequality. Here, a vector inequality $Y \leq 0$ means that each scalar element of Y is non positive.

2.5.5.2 State transition

In this paragraph, the state dependent transition, **guard condition**, is mainly used to model the flight mode transition with intent information (or flight plan). To reach its destination, the unmanned aircraft needs to perform a series of given maneuvers following the given routes from one way point to another. An aircraft changes its flight mode when it reaches a way point, which can be interpreted as aircraft intent information. With the given aircraft intent information, the flight mode changes can be modeled as the discrete state transitions of a hybrid system. The aircraft's flight mode depends on which segment it currently stays in. Such partitions can be mathematically described by the guard conditions. The flight mode transition is dependent on the current state of the aircraft and can be modeled by the state dependent discrete state transition function [44].

In real applications, the flight mode transition does not happen exactly at the way point but around it, due to various uncertainties. The **stochastic linear guard condition** is designed to model such uncertainty in the flight mode transition. The design parameter θ is a Gaussian random variable to account for the uncertainties. It can be chosen according to the position of the given way point and Σ_θ is chosen according to the aircraft's navigation capability: for aircraft with high navigational accuracy, Σ_θ is small, whereas Σ_θ is large for aircraft with poor navigational accuracy.

The aircraft's dynamics can be modeled by a linear hybrid automaton in which each control location corresponds to one of the phases of the flight plan. The activities of each control location contain functions which describe the evolution of the aircraft's position on a tangent plane coordinate system. Assuming that the speed during each phase of the flight is constant, the transitions connecting the different control locations are guarded by tests belonging to one of two categories:

1. **Tests on the aircraft's position**. This type of test is used when the end of the nth flight phase is defined by passing waypoint $WP_{n+1} = (x_{n+1}, y_{n+1}, z_{n+1})$, specified by coordinates in the tangent plane coordinate system.
2. **Tests on a clock**. This type of test is used when the duration of the nth phase of the flight plan is specified, instead of its final waypoint. The clock used has to be initialized when entering control location n.

The model accepts constant values for the wind's speed and direction as well as ranges of minimum/maximum values.

2.5.5.3 Specific fuel consumption

In general, the specific fuel consumption of an aircraft is a function of several variables and flight conditions which makes the resulting system a nonlinear hybrid one. In order to be analyzed, this system must be approximately modeled by a **linear hybrid automaton**. This is accomplished by the rate translation technique which replaces the nonlinear fuel consumption variable by a piecewise linear variable that approximates the nonlinear one. A model for the specific fuel consumption s_{fc} reflects its dependency on the aircraft's mass and flight condition:

$$s_{fc} = \dot{f} = g_u(m)$$

where $m = m_e + f$ is the total aircraft's mass with m_e being the empty aircraft's mass, f the current fuel mass onboard and $U \in \{bestrange, endurance, maxspeed, \dots\}$ a parameter determining the flight condition. Under such conditions, g_u is a non-positive function which decreases when m grows, approximated by:

$$g_u(m) = g_u(m_e) + h_u f$$

where $g_u(m_e)$ is the nominal specific fuel consumption for flight condition u and $h_u f$ its increment due to the current amount of fuel f in the aircraft, h_u being a negative constant.

Considering a flight plan as the composition of several phases, each of them corresponding to a constant flight condition, the fuel consumption of an unmanned aircraft can be modeled by a nonlinear hybrid automaton such that location P_i models a phase with fuel consumption dynamics given by $s_{fc} = k_{1_i} + k_{2_i} f$ with k_{1_i}, k_{2_i} being non-positive constants corresponding to $g_u(m_e)$ and h_u for location i and flight condition u. Also, at each location P_i, f is constrained to be in the interval $[0, F]$ where F is the initial amount of fuel onboard the aircraft.

2.6 MISSION TOOLS

The main purpose of a smart autonomous aircraft is to provide a platform for the payload to operate from. The smart autonomous aircraft enables sensors to be used in the desired location. Hence the design of the smart autonomous aircraft and its operational requirements are highly influenced by the payload parameters. Accurately capturing payload performance and its influence on smart autonomous aircraft design at an early stage allows a more effective overall system solution.

2.6.1 SENSORS

Autonomous aircraft can incorporate different kinds of sensors, each one sensitive to a different property of the environment, whose data can be integrated to make the perception of the aircraft more robust and to obtain new information. Sensors now represent one of the largest cost items in a smart autonomous aircraft and are necessary for navigation and mission. Main requirements for sensors are:

1. Integration and environmental requirements
    a. Integration on aircraft (size, weight)
    b. Environmental requirements for aircraft
2. Operational requirements for sensors, depending on the mission
    a. Atmospheric propagation of electromagnetic radiations
    b. Disruptive conditions for in route, approach/takeoff or on the runway (rain, cloud, fog)
    c. Image processing requirements for sensors
    d. Data bases' requirements for sensors
    e. Terminal area localization and trajectory management requirements

Sensors' state of the art can be summarized as follows [11]:

1. Optronic sensors
2. Passive sensors
    a. day light camera: CCD and CMOS
    b. Low light level camera for night vision
    c. Infrared camera
3. Active system
    a. Lidar laser radar (obstacle detection)
    b. Active laser imaging (disruptive conditions, fog, rain)
4. Radar sensors
    a. Millimeter wave radar (all weather conditions for medium range 5 km and good resolution)
    b. SAR (long range 100km, poor resolution)
5. Enhanced vision system (EVS)

Inertial sensors measure rotation rate and acceleration both of which are vector valued variables [33, 60, 61, 77].

1. Gyroscopes are sensors for measuring rotation. Rate gyroscopes measure rotation rate and displacement gyroscopes (also called whole-angle gyroscopes) measure rotation angles
2. Accelerometers are sensors for measuring accelerations. However, accelerometers cannot measure gravitational acceleration.

Airspeed data are provided by pitot tubes; however the measurements can be affected by the aircraft hull.

An inertial navigation system estimates aircraft velocity, orientation and position with respect to the inertial reference frame. It has no external inputs such as radio signals. Inertial navigation uses gyroscopes and accelerometers to maintain an estimate of the position, velocity, attitude and attitude rates of the aircraft on which the **inertial navigation system** (INS) is carried. An INS consists of the following:

1. An **inertial measurement unit** (IMU) or inertial reference unit (IRU) containing in general three accelerometers and three gyroscopes. These sensors are rigidly mounted to a common base to maintain the same relative orientation.
2. Navigation computers calculate the gravitational acceleration (not measured by accelerometers) and doubly integrate the net acceleration to maintain an estimate of the position of the aircraft.

Data coming from different kind of sensors must be integrated. Data fusion for navigation has been characterized by the use of probabilistic techniques such as Kalman filters, Gauss approximation methods, vector maps, grid maps. These quantitative methods can manage incomplete and inaccurate information, but at a high computational cost and they usually obtain a description of the world which is more accurate for the task.

During flight, the aircraft angular rates p, q, r can be measured by rate gyroscopes. The three orthogonally mounted gyroscopes measure the components of the angular velocity Ω and use the following discrete matricial equation:

$$\mathbf{R}(k+1) = \delta_t Sk(\Omega)\mathbf{R}(k) + \mathbf{R}(k) \tag{2.286}$$

to continuously update the estimated orientation matrix $\mathbf{R}_B^E$ of the aircraft body-fixed frame with respect to the inertial reference frame. δ_t is the increment in time. The measured acceleration a^B of the aircraft body frame is rotated into the inertial frame:

$$a^I = \mathbf{R}_B^I a^B \tag{2.287}$$

and then can be integrated twice to update the estimate of the aircraft's position in the inertial frame.

The angle of attack α and the side-slip angle β can be measured by a spherical flow direction sensor. This sensor uses the differential pressure measurements from several pressure ports on a sphere mounted to the tip of the aircraft to determine the direction of the flow.

The **Global Positioning System** (GPS) currently comprises around thirty active satellites orbiting the Earth in six planes. A GPS receiver works by measuring the time of travel of radio signals from four or more satellites whose orbital position is encoded in the GPS signal. With four known points in space and four measured time delays, it is possible to compute the position of the aircraft and the time. If the GPS signals are received after reflecting off some surface, the distance traveled is longer and this will introduce an error in the position estimate [19]. Variations in the propagation speed of radiowaves through the atmosphere is the main cause of errors in the position estimate. However, these errors vary slowly with time and are approximately constant over large areas. This allows the error to be measured at a reference station and transmitted to compatible nearby receivers that can offset the error: this is the **differential global positioning system** (DGPS)

Electronic sensors of many types and sizes have emerged to provide the necessary functionalities. Situation awareness/understanding is an attribute required for successful aircraft operations. More information about avionics can be found in some textbooks such as [33].

There is a need to develop a simulation tool that is simple enough to be used during the early design phase where timing is critical. The tool should be computationally cheap to allow quick evaluation of design spaces and competing designs. Ideally an agent based modeling and simulation approach supports creating a complex operational environment through simple agent definition.

There are a few commercial **UAV simulators** available and the numbers continue to grow as the use of UAV becomes more popular. Most of these simulators are developed to replicate the state-of-the-art training and operation procedures for current type UAV. The tele-operated system can be made up of five major parts: the motion platform, the aerial platform, the onboard sensors including wireless operation, the PC to remote control (RC) circuit and the ground station.

Wireless sensor network (WSN) is characterized by the dense deployment of sensor nodes that are connected wirelessly to observe physical phenomena. The main advantages of the wireless sensor network include its low cost, rapid deployment, self-organization and fault tolerance. The wireless sensor network has become essential to unmanned systems for environment and situation awareness, based on which intelligent control and decision can be achieved, in a fully autonomous manner. For example, a ground static sensor network can often be deployed for target detection, classification and tracking. A group of smart autonomous aircraft can form a wireless mobile sensor network to undertake cooperative search, exploration and surveillance.

Intensive research and development work has been done for the wireless sensor network, covering from sensor platform development, wireless communication and networking, signal and information processing, as well as to network performance evaluation and design. However, in unmanned systems, there still are a number of challenges in the underlying wireless sensor network, induced by limited network resources (such as node energy and network bandwidth), sensor heterogeneity, large scale deployment and distributed mechanism. For example, state estimation, signal processing and sensor fusion are constrained by the limited sensor resources and network resources, and the associated algorithms should be jointly designed with the sensor and network resource management; Consensus is desirable for distributed estimation with only local information exchanges, but trade-off should be made between the consensus performance and the communication and computation overhead needed; Big data processing approaches are to be sought to deal with the huge amount of sensor data which usually have low quality due to data loss, large data forwarding delay and compression operation forced by the hard bandwidth constraint.

2.6.2 CAMERA MODEL

The main purpose of the current UAV is to gather intelligence by the use of cameras to capture imagery of a location of interest. Digital imaging cameras use a collection of individual detectors to form an image. The individual detectors are arranged in a rectangular array known as the focal plane array. The **field of view** (FOV) at the focal plane array is the angular view of the focal plane. In general, the field of view characteristics are determined by the sensors' designers and forwarded as inputs to UAV manufacturers [78].

One of the fundamental parameters that govern image quality is **ground sample distance** (GSD). It is a function of focal lane array, optics and collection geometry:

$$GSD_H = 2\tan\left(\frac{FOV_H}{2Pix_H}\right)R \tag{2.288}$$

where GSD_H refers to the horizontal ground sample distance, FOV_H is the horizontal field of view, Pix_H is the number of horizontal sensor pixels and R is the slant range. Similarly, the vertical ground sample distance can be given as:

$$GSD_V = \frac{2\tan\left(0.5FOV_V Pix_H\right)}{\cos\left(\theta_{look}\right)}R \tag{2.289}$$

where θ_{look} is the look angle. But the look angle and the slant range are dependent on the UAV operational parameters such as altitude and velocity. The slant range between the UAV and the object of interest is found by:

$$R = \sqrt{h^2 + GR^2} \tag{2.290}$$

where h is the UAV altitude and GR is the ground range from the UAV to the target.

The ground sample distance acts as a metric to identify the performance of the camera module, but it does not act as a metric to identify the quality of the image. Image quality prediction is arbitrary and based on empirical approaches. It defines three levels of object discrimination, namely detection, recognition and identification.

Definition 2.10. ***Detection*** *is the probability that an imagery feature is recognized to be part of a general group. Recognition is the discrimination of the target class.* ***Identification*** *is the discrimination of the target type.*

Determining the probability of detection, recognition and identification is based on the sensor resolution. Targets are replaced by black and white stripes each of which constitutes a cycle. The total number of cycles for a target of given dimensions is:

$$d_c = \sqrt{W_{target} + H_{target}} \tag{2.291}$$

where d_c is the target characteristic dimension and W_{target}, H_{target} are the target width and height (as viewed by the camera), respectively; the number of cycles across the target is given by:

$$N = \frac{d_c}{2GSD_{avg}} \tag{2.292}$$

More information about cameras can be found in [9, 28].

2.6.3 SOFTWARE

Autonomous aircraft face a rapid growth in the complexity of needs and requirements for aircraft responsible for multiple tasks, able to coordinate and developed in a way that guarantees safety. Safety can be verified and certified for a use in a human environment. A parallel evolution similar in the area of real time embedded systems has spread justifying the emerging field of *cyber-physical* systems which mirror a similar increase in complexity. Other areas are also emerging, such as autonomic computing, sharing same scientific objectives in the design and implementation of their software architectures.

Software architectures in these areas seek globally to integrate a computerized control system in a context engagement with the real world, with other treatment information increasingly heavy (mapping, planning, data analysis, learning, etc.). The experience of autonomous aircraft shows that achieving of correct and reliable software architectures easily reusable and scalable remains a challenge because it is difficult to get a good decoupling between sensors, control algorithms and actuators. Taking into account the time and various constraints led to robotic systems tightly coupled. The characteristics of the sensors and actuators as well as the specification features of autonomous aircraft eventually dictate specific solutions in terms of algorithms, architectures, middleware configurations and system. Solutions are costly to develop and then difficult to change.

The themes covered, always within the context of these software architecture, are not limited to:

1. Definition of software architectures,
2. Programming languages and programming abstractions,
3. Specification, modeling and validation, especially according to their impact and constraints on architecture's software,
4. Aspects of software engineering directly related architecture's software: design, development, testing, validation, management life cycle,
5. Composition, coordination, inter-connection of these systems.

Remark 2.13. *The National air space is a rather complex and heterogeneous system in which different types of airborne and ground based entities need to interact and collaborate efficiently often under tight schedules. Each of those entities follows its own sets of behavioral rules and is characterized by its own dynamics and operates in a highly dimensional and constantly changing environment. Hence system modeling in a classical sense may become infeasible for such large and diversified architecture. The use of agent-based modeling and simulation is well suited to describe these types of systems as this approach takes advantage of the concept of decentralization by focusing on each agent's microscopic behavior rather than attempting to macroscopically model the entire framework's dynamics which instead is left to emerge from the agent to agent interactions [13].*

Typically unmanned aircraft that are designed to fly in urban settings have to cope with computationally costly algorithms. The fusion of all sensors for state estimation, flight control and mission management is already complex, although these are only the basic modules. Furthermore, obstacle sensing has to be included and environmental mapping integrated, which creates a model of the environment and its impact on the mission. onboard path planning is necessary if the area of interest could cause a deviation from the predefined flight software which include, among others, the flight controller, mission management, sensor fusion, path planning and the required middleware. The software has to be distributable to different computational units optionally [21].

Remark 2.14. *Beyond the responsibility of responding to unexpected system faults, the software enabled control (SEC) program is also charged with making smart autonomous aircraft more agile without exceeding critical flight parameters. Improved performance of smart autonomous aircraft is expected to be achieved when such vehicles are endowed with levels of autonomy that will allow them to operate safely and robustly under external and internal disturbances, to be able to accommodate fault conditions without significant degradation of their performance, to adapt to unexpected events and to coordinate/cooperate among themselves to accomplish mission objectives.*

2.6.3.1 Petri nets

A software modeling approach using Petri nets is presented in [56] to design, develop and verify/validate small software. The key elements differentiating hardware and software development include identifying defects, visualizing architecture and testing. Visualizing software architecture is normally realized by creating functional blocks, timing, class and numerous other diagrams. Software testing is approached by attempting to identify and test all failure cases post-design, subsequently rectifying found defects. Hardware testing involves the creation of virtual model simulations that are used to identify design flaws and subsequently improve designs before fabrication and physical tests. Petri nets are often used to verify system properties, simulate concurrent systems and analyze temporal attributes of systems.

Petri nets are a graphical and mathematical modeling tool. They can be used as a communication tool similar to other modeling notations like state-transition diagrams and entity-relationship diagrams. As a mathematical tool, it can set up mathematical models that govern the behavior of systems. Those systems can be asynchronous, concurrent, parallel, stochastic or nondeterministic. Using a **Petri net** formalism allows to visualize the structure of the rules based system, making the model more legible and easier to understand. It allows also to express the behavior of the system in mathematical form.

The term **fuzzy Petri net** (FPN) has been used to describe Petri nets that use their formalism to implement fuzzy reasoning algorithms. In fuzzy Petri nets, places can represent propositions, transitions can represent rules and tokens can represent truth values.

Fuzzy reasoning Petri nets (FRPN) can also be proposed where the properties of the Petri net are further defined [53]:

1. If a place represents a truth degree, it can have at most one token.
2. Fuzzy reasoning Petri nets are conflict-free nets as rules may share propositions.
3. Tokens are not removed from the input places after it fires.
4. Complementary arcs do not inhibit the firing of a transition if its place has a token.

2.6.3.2 Middleware

Middleware designed to describe sensing, control and computational communications of components within unmanned systems enables the creation of clean interfaces between the low-level mechanics of such systems and the higher-level logic designed to control them. In [96], a modeling environment is presented with a domain-specific ontology for autonomous systems, capable of generating software necessary for inter-computer communications according to existing autonomous systems middleware standards. Meta-models are used to specify the domain-specific modeling language to model the messages used, the interfaces between components and some of the functionality of the components that transmit and receive messages. The generated code supports the high data rates expected in autonomous systems that use lossy message compression. Tests for the messaging infrastructure are also generated for the messages. Also, using this research, this code generation process can be extended to any component based platform with a similar ontology.

Software verification for highly automated unmanned aircraft is constrained by certification standards and regulatory rules. As a top level view, the processes and tools that were established for the software development, verification and validation must be presented [89]. Automated tests must drive the development of the mission planning, mission management and sensor fusion system. In the UAV community, there is a high interest in UAV airworthiness certification and especially software development and verification as a problem. Compared to conventional manned aircraft, the software may be more complex and larger parts may be safety critical. Autonomy software is complex and especially adaptive and it is an extreme case for conventional verification and validation approaches but with formal techniques (run time monitoring, static analysis, model checking and theorem proving), there exist some new methodologies for the validation and verification of autonomous software, although the required level of expertise to master these techniques increase significantly.

In [30], the focus is on value-driven design and how advances in software engineering, rationale capture, operational simulation and rapid prototyping can be leveraged to create an integrated design suite that allows the rapid design and manufacture of low-cost civilian UAV. Value-driven design is a movement that is using economic theory to transform system engineering to better use optimization to improve the design of large systems. The different modules that can be used are: concept design, CAD, design viewer, manufacturing, operational simulation, value model, aerodynamics, structures, cost, design rationale, reliability, test flight telemetry, mission simulation.

2.6.4 HUMAN SUPERVISORY MODELING

In most current UAV, the role of the human operator has shifted from an aircraft controller to a mission manager and automation supervisor. A variety of ground and onboard automation systems relieve the operator of high frequent tasks such as flight control and support him in tasks such as flight guidance and sensor management functions of aided/automated target recognition. These systems support the operator in clearly defined tasks during mission execution. Higher cognitive tasks such as planning, decision making and problem solving are still solely left to the human. The operator controls the aircraft using the available automation and orders discrete and rather abstract commands, such as waypoint navigation and supervises their execution. The human only intervenes in the case of unforeseen events [16].

One application in human supervisory control considers the case where the human operators act as mission managers overseeing high level aspects of the mission (such as resource planning, scheduling and generating new strategies). Mathematical models for human operators interacting with autonomous aircraft have been developed using a queuing framework; where external tasks/events arrive according to an underlying stochastic process and the human supervisor, modeled as a server, has to service this stream of tasks/events. These queuing based mathematical models can be used to infer limits of performance on the human operators as well as used for a sensitivity analysis to parameters (such as variation of the task arrival rate) via the use of discrete event simulation.

Event modeling is a method of intelligent analyzing of streams of information (data, percepts) about things that happen (events), and deriving conclusions from them. The goal is to identify meaningful events and respond to them appropriately and quickly. The complexity of the event is defined as both the complexity of the modeled physical phenomenon (fire, weather, chemical reaction, biological process) as well as the heterogeneity of the data (digital images, percepts, sensory data, natural language, semi-structured and structured data). In addition, the emphasis should be placed on the intelligent aspect of these models. This means that systems should semi-autonomously perceive their environment and take action.

An important feature is the ability for operators to conclude that there is insufficient information to make a good decision and require an additional visit to look at the target. These so-called re-looks are important in minimizing collateral damage and reducing errors, and have been studied in the context of optimal stopping problems and inspection problems. Single autonomous aircraft re-look problems in the presence of fuel constraints have been also presented. In the first approaches, the operator is primarily modeled with a stationary confusion matrix, while the true error rates may depend on the actual search time. Furthermore, multi-video stream visual search tasks with the possibility of re-look can be much more complex than single UAV counterparts, since the multi-autonomous aircraft aspect requires a fine balance between planning for the other vehicles and understanding how to re-allocate the vehicles to gain additional information, all the while under intensive time pressure [65, 103].

2.6.4.1 Operator modeling

Queuing models for human operators were originally proposed in the context of air traffic control, where the human operator is treated as a serial controller, capable of handling one complex task at a time. Operator models were extended in the context of human supervisory control of multiple autonomous aircraft to account for operator workload and situational awareness. A simplified queuing model can be described as follows: the tasks/events are generated by a Poisson process with rate $\hat{\lambda}$ and the human operator, with possible help from the **decision support system** (DSS), services the tasks at a rate $\hat{\lambda}_e$. In complex tasks, operators may dedicate themselves only to a single task, allowing the incoming tasks to accumulate in the queue. A key concern in supervisory control is the role of vigilance-like effects whereby operator performance degrade with time, and this is a critical consideration for human supervisors of autonomous aircraft missions. Data collected in previous human-in-the-loop experiments have shown that detection probability in the search task can degrade with increased search time dedicated to the task. The estimate of detection probability $\hat{P}_d$ can be modeled using a logistic regression of the form

$$\hat{P}_d = \frac{1}{1 + \exp\left(\hat{\beta}^T \hat{t}\right)} \tag{2.293}$$

where $\hat{t} = [1, t_s]$, t_s being the total search time, and $\hat{\beta} = [\beta_0, \beta_1]$ is the vector of parameters of the logistic regression. Using this insight, there are certain thresholds beyond which the operator performance degrades and it may be beneficial to temporarily abandon the current search task and possibly relook at it later.

2.6.4.2 Retrial queuing model

In order to account for re-look in the queuing model, this paragraph formulates this problem as a retrial. A retrial queue model of the human operator treats the human as a server and if the server is available, the task can be serviced immediately. If the server is not available, the task is inserted in a so-called orbit pool to be serviced at a later time. In this setting, the orbit pool could represent the list of re-queued targets but also targets waiting to be processed.

This model assumes that new tasks arrive in the system as a Poisson arrival with rate $\hat{\lambda}$. The model further assumes that new tasks are serviced by the operator at a rate $\hat{\lambda}_e$. For queuing models, $\hat{\lambda}_e$ has a strong dependence on numerous factors, but the two principal drivers are operator search time and vehicle routing policy. Thus, if an operator is not efficient in the search task and does not route the vehicles along efficient paths, the service rate can be potentially much lower than the arrival rate $\hat{\lambda}_e << \hat{\lambda}$, leading to queue instability, where the number of outstanding targets will grow unbounded over time.

The re-queuing policy is one of the most important features of the re-look task, and yet remains to be completely identified from experimental data. However, insight can be gained from re-queuing models with Bernoulli feedback where the operators perform re-looks (i.e., re-queues the targets) with some probability. Serious lags have been exposed in determining a possible cause of air accidents because implications on flight safety lie deep in flight data that cannot be gained ahead of time. A human strategy method is presented in [100] to generate a very large sample of flight data for safety analysis.

2.7 ATMOSPHERE

The Earth's atmosphere is the gaseous envelope surrounding the planet. It figures centrally in transfers of energy between the Sun, the Earth and deep space [75]. It also figures in transfer of energy from one region of the globe to another. The Earth's atmosphere is related closely to ocean and surface processes.

Since the Earth rotates around its own axis and moves around the Sun, it is convenient to use a non inertial reference frame fixed on Earth when atmospheric disturbances are analyzed [55]. This reference frame rotates once a day. The movement of air parcel is measured relatively to the surface of the Earth. The equations that adequately govern the motion of an air parcel in this framework takes on the form:

$$\left(\frac{dV}{dt}\right)_{rotating} = \frac{F}{m} - \Omega_{Earth} \times (\Omega_{Earth} \times r) - 2\Omega_{Earth} \times V \qquad (2.294)$$

where Ω_{Earth} is the vector representing the Earth's rotation about its own axis, V is the velocity of the air parcel in this rotating frame, r is the position vector of the air parcel from the center of the Earth, m the mass of the air parcel and F the net force acting on the air parcel.

The equation giving the velocity is:

$$V_{inertial} = V_{relative} + \Omega_{Earth} \times r \qquad (2.295)$$

The motions of the atmosphere are complex and can vary according to a number of factors. The atmosphere is always in motion due to combined effects of planetary rotation and thermodynamic variations in local atmospheric temperature and pressure, leading to strong horizontal air currents. If unaccounted for, winds can cause large errors in the position of airplanes, which can lead to a shortfall in range and fuel starvation. Therefore, careful periodic observations of the wind velocities above selected weather stations along the route, yielding periodic winds aloft data at a series of altitudes, are essential for accurate navigation [1, 2, 22, 70]. Adverse weather affects the operations of aircraft in a wide variety of ways and often very differently in the different phases of flight. Thus a wide range of specific meteorological factors can be identified as being potentially hazardous [26, 28, 80]. Winds are driven by forces [92]. These forces can be:

1. pressure gradient forces, which are due to a change in the air pressure over a distance,
2. centrifugal and Coriolis forces due to the Earth's rotation about its axis of rotation,
3. frictional forces due to the Earth's surface: roughness or structures. Frictional forces are strongest within the region near and adjacent to the Earth's surface, which is called boundary layer region

The surface winds typically, going from the Equator to the Pole, have an East-West-East pattern. The surface winds are stronger in the Southern hemisphere than in the Northern hemisphere because in the former the surface drag is weaker because of the relative lack of continental land masses and topography [92].

Because the atmosphere is a fluid system, it is capable of supporting a wide spectrum of motions. They range from turbulent eddies of a few meters to circulations with dimensions of the Earth itself. Two frameworks can be used to describe the atmospheric behavior [75]:

1. The **Eulerian description** represents atmospheric behavior in terms of field properties such as the instantaneous distributions of temperature, motions and constituents.
2. The **Lagrangian description** represents atmospheric behavior in terms of properties of individual air parcels, e.g., in terms of their instantaneous positions, temperature and constituent concentrations.

Remark 2.15. *The surface winds are not explained by thermal wind balance. Indeed, unlike the upper level winds, they must be maintained against the dissipating effects of friction. The large-scale mid-latitude circulation of the atmosphere is a turbulent flow. This turbulence is neither fully developed nor isotropic.*

Definition 2.11. ***Turbulence*** *is a state of fluid flow in which the instantaneous velocities exhibit irregular and apparently random fluctuations so that in practice only statistical properties can be recognized and subjected to analysis.*

In the troposphere, altitude < 11000 m, the temperature $\tilde{T}(K)$ and the pressure $\tilde{P}(Pa)$ can be, respectively, approximated by:

$$\tilde{T} = 288.15 - 0.0065\ z \qquad \tilde{P} = 101325\left(\frac{\tilde{T}}{288.15}\right)^{5.2559} \tag{2.296}$$

where the altitude z is expressed in meters.

Shape of the mean velocity profile of the wind can be defined on the degree of surface roughness. Deviations of the air flow velocity from the mean air flow velocity, like gusts, waves and fluctuations, have various discrete or continuous modes over a wide range of amplitude, frequencies, wave numbers, length, scales and time scales [70].

The performance characteristics of an aircraft depend on the properties of the atmosphere through which it flies. As the atmosphere is continuously changing with time, it is impossible to determine airplane performance parameters precisely without first defining the state of the atmosphere [64]. The Earth's atmosphere is a gaseous envelope surrounding the planet. The relative percentages of the constituents remain essentially the same up to an altitude of 90 Km owing primarily to atmospheric mixing caused by winds and turbulence.

2.7.1 GUSTS AND WIND SHEAR

The atmosphere is in a continuous state of motion [64]. The winds and wind gusts created by the movement of atmospheric air masses can degrade the performance and flying qualities of an aircraft. In addition, the atmospheric gusts impose structural loads that must be accounted for in the structural design of an aircraft. The movement of atmospheric air masses is driven by solar heating, the Earth's rotation and various chemical, thermodynamic and electromagnetic processes.

Wind gusts model presented in [95] takes into account the link between high gusts and lower frequency fluctuations in wind speed which are represented by observations of hourly mean speed which are themselves generated by large-scale meteorological processes.

Wind shear is a type of space and time-dependent airflow vector field. Wind shear is an atmospheric phenomenon that occurs within thin layers separating two regions where the predominant air flows are different, either in speed, in direction or in both speed and direction. The air layer between these regions usually presents a consistent gradient in the flow field [5]. The air masses in the troposphere are in constant motion and the region is characterized by unsteady or gusting winds and turbulence. Wind shear is the variation of the wind vector in both magnitude and direction. Wind shears are created by the movement of air masses relative to one another or to the Earth's surface. Thunderstorms, frontal systems and the Earth's boundary layer all produce wind shear to aircraft flying at low altitudes.

If there is a local variation for the wind vector, there exists a wind shear. Wind shears have always been considered as a potential hazard for aviation safety [67]. Wind shear is defined as a local variation of the wind vector. It can also be defined either as a spatial or temporal variation of wind speed or direction. The variations in wind speed and direction are measured in the vertical and horizontal directions [64]. A vertical wind shear is one in which the wind speed and direction varies with changing altitude; horizontal wind shear refers to variations along some horizontal distance. Wind shears are created by the movement of air masses relative to one another or to the Earth's surface. Thunderstorms, frontal systems and the Earth's boundary layer all produce wind shear profiles, which, at times, can be hazardous to aircraft flying at low altitudes. The strong gusts/fronts associated with thunderstorms are created by **downdrafts** within the storm system. As the downdrafts approach the ground, they turn and move outward along the Earth's surface. The wind shear produced by the gust front can be quite severe. The wind shear created by a frontal system occurs at the transition zone between two different air masses. The wind shear is created by the interaction of the winds in the air masses. If the transition zone is gradual, the wind shear will be small. However, if the transition zone is small, the conflicting wind speeds and directions of the air masses can produce a very strong wind shear. The surface boundary layer also produces wind shear. The shape of the profile is determined primarily by local terrain and atmospheric condition. Additional problems arise when there is an abrupt change in surface roughness in addition to internal boundary layers and when the direction of the wind varies with altitude.

To analyze the influence of wind shear, the magnitude of the shear can be expressed in terms of the change of wind speed with respect to altitude $\frac{du_W}{dz}$ where a positive wind shear is one which increases with increasing altitude. The most severe type of wind shear is a downburst.

Definition 2.12. ***Downburst*** *is a descending column of cold air that spreads out in radial direction upon hitting the ground [67]. Downbursts are subdivided into microburst and macroburst according to their horizontal scale of damaging winds.* ***Macroburst*** *is a large downburst with its outburst winds extending in excess of* 4000 *m in horizontal dimension. An intense macroburst often causes widespread tornado-like damage. Damaging winds lasting* 5 *to* 30 *minutes could be as high as* 60 *m/s.* ***Microburst*** *is a small downburst with its outburst winds extending radially* 4000 *m or less but contains speed changes larger than* 10 *m/s within itself. Outburst winds of a microburst may have up to* 70 *m/s intensity.*

Remark 2.16. *Because of severe wind gradients, a microburst is the most hazardous downburst that brings about catastrophic conditions to aircraft flying through it, especially for small ones.*

The adopted microburst model is the one elaborated by Vicray [67, 93]. This analytical model is an axisymmetric steady-state model that uses shaping functions to satisfy the flow mass continuity as well as the boundary layer effects. The burst velocity components at a point P in this coordinate system are given by:

$$W_x = \frac{\tilde{\lambda} x}{2}\left[e^{c_1(z/z_m)} - e^{c_2(z/z_m)}\right] e^{\frac{2-(x^2+y^2)^{\alpha}/r_p^{2\alpha}}{2\alpha}} \tag{2.297}$$

$$W_y = \frac{\tilde{\lambda} y}{2}\left[e^{c_1(z/z_m)} - e^{c_2(z/z_m)}\right] e^{\frac{2-(x^2+y^2)^{\alpha}/r_p^{2\alpha}}{2\alpha}} \tag{2.298}$$

$$\begin{aligned} W_z = &-\tilde{\lambda}\left\{\tfrac{z_m}{c_1}\left[e^{c_1(z/z_m)} - 1\right] - \tfrac{z_m}{c_2}\left[e^{c_2(z/z_m)} - 1\right]\right\} \\ &\times\left[1 - \tfrac{(x^2+y^2)^{\alpha}}{r_p^{2\alpha}}\right] e^{\frac{2-(x^2+y^2)^{\alpha}/2r_p^{2\alpha}}{2\alpha}} \end{aligned} \tag{2.299}$$

In this model, the horizontal and vertical wind velocities are both functions of horizontal and vertical coordinates. Intensity of vertical velocity W_z increases with altitude and decreases away from the burst center. The horizontal wind intensity increases with altitude up to an altitude that corresponds to the most severe horizontal velocity z_m and then decreases with further increase in altitude. Usually z_m is located between 200 m and 1000 m above the ground [67]. Furthermore, the horizontal velocity grows radially from the shaft reaching its maximum at a radius 1.2 times the downburst shaft radius. Thus the model is fully defined by specifying only four parameters: the radial distance from burst center r_p, the altitude above the ground z_m, the magnitude of the maximum outflow velocity u_m and the value of the shaping variable α. Its value is approximately 2. The scale factor $\tilde{\lambda}$ can then be determined from

$$\tilde{\lambda} = \frac{2u_m}{r_p(e^{c_1} - e^{c_2})e^{1/2\alpha}} \tag{2.300}$$

where $c_1 = -0.15$ and $c_2 = -3.2175$ [67].

There are three definitions of altitude: **absolute, geometric and geo-potential** related to each other in the following manner:

$$h_a = h_G + R_0 \tag{2.301}$$

h_a is absolute altitude, h_G geometric altitude and R_0 is radius of the Earth.

$$h_G = \frac{R_0}{R_0 - h} h \tag{2.302}$$

h is the geo-potential altitude.

Most of the known models of wind and temperature consider the fluctuations around the mean field (measured or produced by numerical meteorological prediction) as a Gaussian random field. The covariance structure of the field has a direct impact on the adverse effect of the uncertainties upon the collision risk level and it represents the main characteristic of various models.

2.7.2 TURBULENCE

A perturbation model with stochastic white noise properties can be used. The velocity variations in a turbulent flow can be decomposed into a mean and a fluctuating part. The size or scale of the fluctuations vary from small wavelengths of the order of 10^{-2} m to wavelengths of the order of 10^{+3} m. Atmospheric turbulence being a random phenomenon, it can only be described in a statistical way.

To predict the effect of atmospheric disturbances on aircraft response requires a mathematical model of atmospheric turbulence which is a random process and the magnitude of the gust fields can only be described by statistical parameters. The properties of atmospheric turbulence include homogeneity and stationarity.

Definition 2.13. ***Homogeneity:*** *The property of homogeneity means that the statistical properties of turbulence are the same throughout the region of* *interest; stationarity implies that the statistical properties are independent of time.*

Atmospheric turbulence is typically modeled as a filtered Gaussian with white noise. The three components of turbulence (i.e., longitudinal, lateral and vertical) are modeled independently. There are two spectral forms of random continuous turbulence used to model atmospheric turbulence for aircraft response: **Dryden** and **Von Karman**.

The velocity field within the atmosphere varies in both space and time in a random manner. This random velocity field is called atmospheric turbulence. The mathematical model can be proposed as follows:

$$u = W_h \frac{ln(z/z_0)}{ln(W_h/z_0}), 1 \leq z \leq 300 \text{ m} \tag{2.303}$$

where u is the mean wind speed, W_h is the measured speed at an altitude h, z is the aircraft altitude and z_0 is a constant depending on the flight phase.

The **Dryden** turbulence mode is one commonly used atmospheric turbulence model. It specifies the spectra of the three components of turbulence as follows.

$$\Phi_{u_g}(\omega_s) = \sigma_u^2 \frac{2L_u}{\pi} \frac{1}{1 + (L_u \omega_s)^2} \tag{2.304}$$

$$\Phi_{v_g}(\omega_s) = \sigma_v^2 \frac{L_v}{\pi} \frac{1 + 3(L_v \omega_s)^2}{(1 + (L_v \omega_s)^2)^2} \tag{2.305}$$

$$\Phi_{w_g}(\omega_s) = \sigma_w^2 \frac{L_w}{\pi} \frac{1 + 3(L_w \omega_s)^2}{(1 + (L_w \omega_s)^2)^2} \tag{2.306}$$

The spectra are given in terms of spatial frequency, which is converted to temporal frequency ω_s by multiplying by the speed of the aircraft.

The power spectral density for the turbulence velocities modeled by **Von Karman** is given by:

$$\Phi_{u_g} = \sigma_u^2 \frac{2L_u}{\pi} \frac{1}{(1 + (1.339 L_u \omega_s)^2)^{5/6}} \tag{2.307}$$

$$\Phi_{v_g} = \sigma_v^2 \frac{2L_v}{\pi} \frac{1 + \frac{8}{3}(1.339 L_v \omega_s)^2}{(1 + (1.339 L_v \omega_s)^2)^{11/6}} \tag{2.308}$$

$$\Phi_{w_g} = \sigma_w^2 \frac{2L_w}{\pi} \frac{1 + \frac{8}{3}(1.339 L_w \omega_s)^2}{(1 + (1.339 L_w \omega_s)^2)^{11/6}} \tag{2.309}$$

where σ_g is the root mean square intensity of the gust component, ω_s is the spatial frequency defined by $\frac{2\pi}{\lambda}$ where λ is the wavelength of a sinusoidal component and L is the scale of the turbulence. The subscripts u, v, w refer to the gust components. The scales and intensities of atmospheric turbulence depend on altitude and the type of turbulence, i.e., clear air (high or low altitude) and thunderstorm turbulence.

For an airplane passing through a gust field, it is assumed that the encountered turbulence is independent of time (i.e., the turbulence is stationary). The relationship between the spatial and temporal frequency is given by:

$$\omega_s = \frac{\omega}{V_0} \tag{2.310}$$

where ω is in rad/s and V_0 is the velocity of the aircraft relative to the air mass it is passing through.

At medium to high altitudes (above 610 m) the turbulence is assumed to be isotropic. The characteristic lengths and the intensities in each direction are equal to each other. A typical characteristic is 530 m. Intensities can be charted as a function of altitude. Moderate turbulence has a root-mean-square intensity of about 3 m/s at 610 m, decreasing roughly linearly to near zero at 18000 m.

Whereas lateral turbulence has little effect on the speed on an aircraft, longitudinal turbulence has a direct effect on airspeed. Longitudinal turbulence with a spectrum matching that given in equation (2.304) can be obtained by passing white noise through a filter of the form:

$$\sigma_u = \sqrt{\frac{2L_u}{U_u}} \frac{U_u}{L_u s + U_u} \tag{2.311}$$

Vertical turbulence has an indirect effect on airspeed.

Aircraft can experience the effects of severe mechanical turbulence. When the vortices are almost stationary in location relative to the surface terrain and topography below, such as the lee of a high cliff, they are referred to as standing vortices, or, more usual when there is a downwind addition to any **standing vortex** in the form of a zone of transitory traveling vortices they often rotate in different rotations, forming a part of a **Karman vortex train**.

When the atmosphere is stably stratified, the occurrence of strong winds generates predominantly only mechanical turbulence, but as soon as the vertical structure of the atmosphere becomes thermally and dynamically unstable, as the daytime solar heating of the air layer is in contact with the ground, thermal turbulence generated by warm and buoyant updrafts and cold compensatory downdrafts is added. Whereas mechanical turbulence is often characterized by a steady succession of vortices which fall within a limited size range, the strength and character of thermal turbulence are much less predictable, especially when the turbulence is just beginning and thermals of all sizes rise up from a wide variety of terrain features. There is normally a lot of possible variation in terms of the linear spacing of thermals and their size, shape and vertical speeds [102].

Remark 2.17. *Existing UAS platforms currently do not lend themselves to autonomous operation within complex highly variable aerodynamic environments. As such, there is a need for an accurate high fidelity airflow model to help reduce the risk of failure of urban UAS missions. This is specially important in urban missions. Urban aerodynamics are exceptionally complicated because of complex interactions between geometry, physical conditions and varying meteorology and possible urban canyons [20].*

Estimation of atmospheric turbulence field encountered by an aircraft continues to be an interesting problem [81]. In meteorology, turbulence has represented a challenge due to its fine spatial and temporal scale. In numerical meteorology, wind with horizontal dimensions ranging from 5 to several hundred kilometers can be resolved adequately. However, to accurately diagnose aircraft normal local accelerations, resolved scales as small as 50 m are needed. In fact, spatial resolution should be well below the aircraft length [52].

From another point of view, an aircraft itself is in fact a big sensor in the atmospheric environment. Penetrating a turbulent air zone, the aircraft responds in a definite way depending on the imposed wind field and aircraft aerodynamic characteristics. Therefore, from the response, the input including the turbulence can be identified. As the atmosphere turbulence is random in nature, it is composed of a wide spectrum of frequencies. Therefore, the response data can only provide the estimation of the significant components of the low frequency part of atmospheric turbulence, the wind. A normal force coefficient is primarily dependent linearly on the angle of attack with much smaller contributions from other parameters such as the elevator and pitch rate. Combinations of high wind speeds, thermal and mechanical turbulence, vertical thermal instability, heavy precipitation and lightning risks can all be found in severe storms, making them priorities for avoidance at all costs by aircraft [48].

2.8 CONCLUSION

This chapter has presented classical modeling approaches for autonomous aircraft with the presentation of reference frames, kinematics and dynamics of airplanes. Traditionally, aircraft dynamic models have been analytically determined from principles such as Newton–Euler laws for rigid-body dynamics. These approaches are followed by the presentation of less conventional techniques such as Takagi–Sugeno modeling, fuzzy logic modeling and linear hybrid automaton. Then mission tools are presented such as sensors, cameras, simulation tools and software human supervisory modeling. The presentation of the atmosphere modeling closes the chapter of modeling. Particular attention is given to mathematical models useful for analysis, simulation and evaluation of control, planning and safety algorithms for autonomous aircraft.

REFERENCES

1. Abdulwahah, E.; Hongquan, C. (2008): *Aircraft response to atmospheric turbulence at various types of the input excitation*, Space Research Journal, vol. **1**, pp. 17–28.
2. AIAA (2010): *Guide to reference and standard atmosphere models*, AIAA report G003C–2010.
3. Al-Hadithi, B.; Jimenez, A.; Matia, F. (2012): *A new approach to fuzzy estimation of Takagi–Sugeno model and its applications to optimal control of nonlinear systems*, Applied Soft Computing, vol. **12**, pp. 280–290.
4. Anderson, J. (1991): *Aircraft Performance and Design*, McGraw–Hill.
5. Bencatel, R.; Kabamba, P.; Girard, A. (2014): *Perpetual dynamic soaring in linear wind shear*, AIAA Journal of Guidance, Control and Dynamics, vol. **37**, pp. 1712–1716, DOI 10.2514/1.G000425.
6. Berchtold, M.; Riedel, T.; Decker, C; Laerhoeven, K. (2008): *Gath-Geva specification and genetic generalization of Takagi-Sugeno-Kang fuzzy models*, IEEE Int. Conference on Systems, Man and Cybernetics, pp. 595–600.
7. Bestaoui, Y. (2011): *3D curves with a prescribed curvature and torsion for an aerial robot*, Int. Journal of Computer Applications, vol. **41**, pp. 269–274.
8. Bestaoui, Y. (2012): *Lighter Than Air Robots*, Springer.
9. Braunl, T. (2008): *Embedded Robotics*, Springer.
10. Buchholz, M.; Larrimore, W. E. (2013): *Subspace identification of an aircraft linear parameter varying flutter model*, American Control Conference, Washington, DC, pp. 2263–2267.
11. Burkholder, J. O.; Tao, G. (2011): *Adaptive detection of sensor uncertainties and failures*, AIAA Journal of Guidance, Control and Dynamics, vol. **34**, pp. 1065–1612.
12. Butler, E. J.; Wang, H. O.; Burken, J. J. (2011): *Takagi–Sugeno fuzzy model based flight control and failure stabilization*, AIAA Journal of Guidance, Control and Dynamics, vol. **34**, pp. 1543–1555.
13. Calanni Fraccone, G.C.; Valenzuela-Vega, R.; Siddique, S.; Volovoi, V.; Kirlik, A. (2014): *Nested modeling of hazards in the national air space system*, AIAA Journal of Aircraft, vol. **50**, pp. 370–377, DOI: 10.2514/1.C031690.
14. Casan, P.; Sanfelice, R.; Cunha, R.; Silvestre, C. (2012): *A landmark based controller for global asymptotic stabilization on SE(3)*, IEEE Conference on Decision and Control, Hawai, pp. 496–501.
15. Chuang, C. C.; Su, S. F.; Chen, S. S. (2001): *Robust TSK fuzzy modeling for function approximation with outliers*, IEEE Transactions on Fuzzy System, vol. **9**, pp. 810–821.
16. Claub, S.; Schulte, A. (2014): *Implications for operator interactions in an agent supervisory control relationship*, Int. Conference on Unmanned Aircraft Systems, pp. 703–714, DOI 978-1-4799-2376-2.
17. Conway, B. A. (2010): *Spacecraft Trajectory Optimization*, Cambridge Press.
18. Cook, M. V (2007): *Flight Dynamics Principles*, Elsevier.
19. Corke, P. (2011): *Robotics, Vision and Control*, Springer.
20. Cybyk, B. Z.; McGrath, B. E.; Frey, T. M.; Drewry, D. G.; Keane, J. F. (2014): *Unsteady airflows and their impact on small unmanned air systems in urban environments*, AIAA Journal of Aerospace Information Systems, vol. **11**, pp. 178–194.

21. Dauer, J. C.; Goorman, L.; Torens, C. (2014): *Steps towards scalable and modularized flight software for unmanned aircraft system*, Int. Journal of Advanced Robotic System, vol. **11**, pp. 81–88, DOI: 10.5772/58363.
22. Davidson, P. A; Kaneda, Y.; Moffatt, K.; Sreenivasan, K.R. (2011): *Voyage Through Turbulence*, Cambridge University Press.
23. Dorobantu, A.; Balas, G. J.; Georgiou, T. T. (2014): *Validating aircraft models in the gap metric*, AIAA Journal of Aircraft, vol. **51**, pp. 1665–1672, DOI 10.2514/1.C032580.
24. Dydek, Z. T.; Annaswamy, A.; Lavretsky, R. (2010): *Adaptive control and the NASA X-15-3 flight revisited*, IEEE Control System Magazine, vol. **30**, pp. 32–48.
25. Erginer, B.; Altug, E. (2012): *Design and implementation of a hybrid fuzzy logic controller for a quad-rotor VTOL vehicle*, Int. Journal of Control, Automation and Systems, vol.**11**, pp. 61–70.
26. Etele, J. (2006): *Overview of wind gust modelling with application to autonomous low level UAV control*, Contract report, DRDC Ottawa, CR 2006–211.
27. Etkin, B. (2000): *Dynamics of Atmospheric Flight*, Dover Publishing.
28. Gao, Z.; Gu., H.; Liu, H. (2009): *Real time simulation of large aircraft flying through microburst wind field*, Chinese Journal of Aeronautics, vol. **22**, pp. 459–466
29. Garatti, S.; Campi, M. C. (2013): *Modulating robustness in control design*, IEEE Control System Magazine, vol. **33**, pp. 37–51.
30. Garissen, D.; Quaranta, E.; Ferraro, M.; Schumann, B.; van Schaik, J.; Bolinches, M.; Gisbert, I.; Keane, A. (2014): *Value-based decision environment vision and application*, AIAA Journal of Aircraft, vol. **51**, pp. 1360–1372.
31. Girard, A. R.; Larba, S. D.; Pachter, M.; Chandler, P. R. (2007): *Stochastic dynamic programming for uncertainty handling in UAV operations*, American Control Conference, pp. 1079–1084.
32. Grauer, J. A.; Morelli, E. A. (2014): *Generic global aerodynamic model for aircraft*, AIAA Journal of Aircraft, vol. **52**, pp. 13–20, DOI: 10.2514/C032888.
33. Gruwal, M.; Weill, L. R.; Andrews, A. (2001): *Global Positionning Systems, Inertial Navigation and Integration*, Wiley.
34. Guechi, E.; Lauber, J.; Dambine, M.; Klancar, G.; Blazic, S. (2010): *PDC control design for nonholonomic wheeled mobile robots with delayed outputs*, Journal of Intelligent and Robotic Systems, vol. **60**, pp. 395–414.
35. Halder, A.; Konar, A.; Mandal, R.; Chakraborty, A.; Bhownik P.; Pal N. T.; Nagan, A. K. (2013): *General and interval type 2 fuzzy face space approach to emotion recognition*, IEEE Transactions on Systems Man and Cybernetics, vol. **43**, pp. 587–597.
36. Hale, F. J. (1984): *Introduction to Aircraft Performance, Selection and Design*, Wiley.
37. Heard, W. B. (2006): *Rigid Body Mechanics*, Wiley.
38. Henderson, D. M. (2006): *Applied Cartesian Tensors for Aerospace Simulations*, AIAA Press.
39. Hoffer, N.; Coopmans, C.; Jensen, A.; Chen, Y. (2014): *A survey and categorization of small low-cost UAV systems identification*, Journal of Intelligent and Robotic Systems, vol. **74**, pp. 129–145.

40. Hogras, H.; Wagner, C. (2012): *Towards the wide spread use of type 2 fuzzy logic systems in real world applications*, IEEE Computational Intelligence Magazine, vol. **7**, pp. 14–24.
41. Holzapfel, F.; Theil, S.(eds) (2011): *Advances in Aerospace Guidance, Navigation and Control*, Springer.
42. Hossain, M. A.; Shill, P. C.; Sarker, B., Murose, K. (2011): *Optimal fuzzy model construction with statistical information using genetic algorithm*, Int. Journal of Computer Science and Information Technology, vol. **3**, pp. 241–257.
43. Imado, F.; Heike, Y.; Kinoshita, T. (2011): *Research on a new aircraft point mass model*, AIAA Journal of Aircraft, vol. **48**, pp. 1121–1130.
44. Jategaonkar, R. V. (2006): *Flight Vehicle System Identification*, AIAA Press.
45. Jenkinson, L. R.; Marchman J. F. (2003): *Aircraft Design Projects*, Butterworth-Heinemann.
46. Kharrati, H.; Khanmohammadi, S.; Pedrycz, W.; Alizadeh, G. (2012): *Improved polynomial fuzzy modeling and controller with stability analysis for nonlinear dynamical system*, Mathematical problems in engineering, DOI. 10.1155/2012/273631.
47. Kwatny, H. G.; Dongno, J. E.; Chang, B. C.; Bajpar, G.; Yasar, M.; Belcastro, C. (2013): *Nonlinear analysis of aircraft loss of control*, AIAA Journal of Guidance, Control and Dynamics, vol. **36**, pp. 149–162.
48. Lawrance, N.R.; Sukkarieh, S.(2011): *Autonomous exploration of a wind field with a gliding aircraft*, AIAA Journal of Guidance, Control and Dynamics, vol. **34**, pp. 719–733.
49. Liu, W., Hwang I. (2011): *Probabilistic trajectory prediction and conflict detection for air traffic control*, AIAA Journal of Guidance, Control and Dynamics, vol. **34**, pp. 1779–1789.
50. Liu, M.Y., Xu F. (2013): *Intelligent digital redesign of fuzzy model based UAV controllers*, 32nd Chinese Control Conference, Xian, China, pp. 2251–2255.
51. Lopez, I., Sarigul-Klijn N. (2010): *A review of uncertainty in flight vehicle structural damage monitoring, diagnosis and control: challenges and opportunities*, Progress in Aerospace Sciences, vol. **46**, pp. 247–273.
52. Lorenz, R. D. (2001): *Flight power scaling of airplanes, airships and helicopters: application to planetary exploration*, AIAA Journal of Aircraft, vol. **38**, pp. 208–214.
53. Lundell, M.; Tang, J.; Nygard, K. (2005): *Fuzzy Petri net for UAV decision-making*, IEEE International Symposium on Collaborative Technologies and Systems, pp. 347–352.
54. Luo, M.; Sun, F.; Liu, H. (2013): *Hierarchical structured sparse representation for Takagi–Sugeno fuzzy system identification*, IEEE Transactions on Fuzzy Systems, vol. **21**, pp. 1032–1043.
55. Mak, M. (2011): *Atmospheric Dynamics*, Cambridge University Press.
56. Malott, L.; Palangpou, P.; Pernicka, H.; Chellapan, S. (2014): *Small spacecraft software modeling: a Petri net based approach*, AIAA Journal of Aerospace Information System, DOI 10.2514/1.I010168.
57. Marcos, A.; Galas, G. J. (2004): *Development of linear parameter varying models for aircraft*, AIAA Journal of Guidance, Control and Dynamics, vol. **27**, pp. 218–228.

58. Marsden, J.; Ratiu, T. S. (1999): *Introduction to Mechanics and Symmetry*, Springer-Verlag.
59. Mendel, J.; Liu, X. (2013): *Simplified interval type 2 fuzzy logic system*, IEEE Transactions on Fuzzy systems, pp. 1056–1069.
60. Moir, A.; Seabridge, A. (2006): *Civil Avionics Systems*, Wiley.
61. Moir, A.; Seabridge, A. (2007): *Military Avionics Systems*, AIAA Press.
62. Mueller, T. J.; Kellogg, J. C.; Ifju, P. G.; Shkarayev, T. (2006): *Introduction to the Design of Fixed Wing Micro Air Vehicles*, AIAA Press.
63. Mumford, C. L.; Jain, L. C. (eds) (2009): *Computational Intelligence, Collaboration, Fusion and Emergence*, Springer.
64. Nelson, R. (1989): *Flight Stability and Automatic Control*, McGraw Hill.
65. Ollero, A.; Maza, I. (2007): *Multiple Heterogeneous Unmanned Aerial Vehicle*, Springer.
66. Pappas, G. J.; Simic, S. (2002): *Consistent abstractions of affine control systems*, IEEE transactions on Automatic Control, vol. **47**, pp. 745–756.
67. Pourtakdoust, S. H.; Kiani, M.; Hassanpour, A. (2011): *Optimal trajectory planning for flight through microburst wind shears*, Aerospace Science and Technology, vol. **15**, pp. 567–576.
68. Poussot-Vassal C., Roos C. (2011): *Flexible aircraft reduced order LPV model generation from a set of large scale LTI models*, American Control Conference, pp. 745–750, DOI 978-1-4577-0081-1.
69. Raymer, D. P.(2006): *Aircraft Design: a Conceptual Approach*, AIAA Press.
70. Riahi, D. N. (2005): *Mathematical Modeling of Wind Forces*, Taylor and Francis.
71. Riley, M.; Grandhi, R. (2011): *Quantification of modeling in aeroelastic analyzes*, AIAA Journal of Aircraft, vol. **48**, pp. 866–876.
72. Rogers, R. M. (2007): *Applied Mathematics in Integrated Navigation Systems*, AIAA Press.
73. Roskam, J.; Lan, C. T. (1997): *Airplane Aerodynamics and Performance*, DAR Corporation.
74. Roskam, J. (2001): *Airplane Flight Dynamics and Automatic Control*, DAR Corporation.
75. Salby, M. L. (2011): *Physics of the Atmosphere and Climate*, Cambridge University Press.
76. Salman, S. A.; Puttige, V. R.; Anavatti, S. (2006): *Real time validation and comparison of fuzzy identification and state-space identification for a UAV platform*, IEEE Int. Conference on Control Applications, Munich, Germany, pp. 2138–2143.
77. Schetzen, M. (2006): *Airborne Doppler Radar*, AIAA Press.
78. Schuman, B.; Ferraro, M.; Surendra, A.; Scanlan, J. P.; Fangohr, H. (2014): *Better design decisions through operational modeling during the early design phases*, AIAA Journal of Aerospace Information System, vol **11**, pp. 195–210.
79. Selig, J. M. (1996): *Geometric Methods in Robotics*, Springer.
80. Seube, N.; Moitie, G.; Leitman, B. (2000): *Aircraft takeoff in wind shear: a viability approach*, Set Valued Analysis, vol. **8**, pp. 163–180.
81. Sheu, D.; Lan, C.T. (2011): *Estimation of turbulent vertical velocity from nonlinear simulations of aircraft response*, AIAA Journal of Aircraft, vol. **48**, pp. 645–651.

82. Stevens, B. L., Lewis,F. L. (2007): *Aircraft Control and Simulation*, Wiley.
83. Takagi, T.; Sugeno, M. (1985): *Fuzzy identification of systems and its applications to modeling and control*, IEEE Transactions on Systems, Man and Cybernetics, vol. **1**, pp. 116–132.
84. Talay, T. A. (1975): *Introduction to the Aerodynamics of Flight*, NASA Report.
85. Tanaka, K.; Ohtake, H.; Tanaka, M.; Wang, H.O. (2012): *A Takagi–Sugeno fuzzy model approach to vision based control of a micro helicopter*, IEEE Conference on Decision and Control, Maui, HI, pp. 6217–6222.
86. Titterton, D. H.; Weston, J. L. (2004): *Strapdown Inertial Navigation Technology*, AIAA Press.
87. Toscano, R. (2007): *Robust synthesis of a PID controller by uncertain multimodel approach*, Information Sciences, vol. **177**, pp. 1441–1451.
88. Toscano, R.; Lyonnet, P. (2009): *Robust PID controller tuning based on the heuristic Kalman algorithm*, Automatica, vol. **45**, pp. 2099–2106.
89. Torens, C., Adolf, F. M.; Goorman, L. (2014): *Certification and software verification considerations of autonomous unmanned aircraft*, AIAA Journal of Information Systems, vol. **11**, pp. 649–664, DOI 10.2514/1.I010163.
90. Ursem, R. K. (2003): *Models for evolutionary algorithms and their applications in system identification and control optimization*, PhD thesis, Univ. of Aarhus, Denmark.
91. Valenzuela, A.; Rivas, D. (2014): *Optimization of aircraft cruise procedures using discrete trajectory patterns*, AIAA Journal of Aircraft, DOI 10.2514/1.C032041.
92. Vallis, G. K. (2012): *Atmospheric and Oceanic Fluid Dynamics: Fundamentals and Large Scale Circulation*, Cambridge University Press.
93. Vicroy, D. (1992): *Assessment of microburst models for downdraft estimation*, AIAA Journal of Aircraft, vol. **29**, pp. 1043–1048.
94. Vinh, N. X. (1993): *Flight Mechanics of High Performance Aircraft*, Cambridge Aerospace series.
95. Walshaw, D.; Anderson, C. (2000): *A model for extreme wind gusts*, Journal of the Royal Statistical Society, vol. **49**, pp. 499–508.
96. Whitsitt, S.; Sprinkle, J. (2013): *Modeling autonomous systems*, AIAA Journal of Aerospace Information System, vol. **10**, pp. 396–412.
97. Wie, B. (1998): *Space Vehicle Dynamics and Control*, AIAA Press.
98. Wu, Z.; Dai, Y.; Yang, C.; Chen, L. (2013): *Aeroelastic wind tunnel test for aerodynamic uncertainty model validation*, AIAA Journal of Aircraft, vol. **50**, pp. 47–55.
99. Yan, J.; Hoagg, J. B.; Hindman, R. E.; Bernstein D. S. (2011): *Longitudinal aircraft dynamics and the instantaneous acceleration center of rotation*, IEEE Control Systems Magazine, vol. **31**, pp. 68–92.
100. Yin, T.; Huang, D.; Fu, S. (2014): *Human strategy modeling via critical coupling of multiple characteristic patterns*, AIAA Journal of Aircraft, vol. **52**, pp. 617–627, DOI 10.2514/1.C032790.
101. Yokoyama, N. (2012): *Path generation algorithm for turbulence avoidance using real-time optimization*, AIAA Journal of Guidance, Control and Dynamics, vol. **36**, pp. 250–262.
102. Zhao, Y.J. (2009): *Extracting energy from downdraft to enhance endurance of Uninhabited Aerial Vehicles*. AIAA Journal of Guidance, Control and Dynamics, vol. **32**, pp. 1124–1133.

103. Zhou, Y.; Pagilla, P. R.; Ratliff, R. T. (2009): *Distributed formation flight control using constraint forces*, AIAA Journal of Guidance, Control and Dynamics, vol. **32**, pp. 112–120.
104. Zipfel, P. H. (2007): *Modeling and Simulation of Aerospace Vehicle Dynamics*, AIAA Education series, Reston, VA.

3 Flight Control

ABSTRACT

In this chapter, airplane control problems are considered and solved, in order to achieve motion autonomy. A common control system strategy for an aircraft is a two loops structure where the attitude dynamics are controlled by an inner loop, and the position dynamics are controlled by an outer loop. In the low level control system, algorithms depend on the type of aircraft, its dynamical model, the type of control design used and finally the inputs and sensors choice. The difficulty is that disturbances can be as strong as the unmanned aircraft's own control forces. In this chapter, topics related to classical linear and nonlinear control methods are first presented. Then fuzzy control is used.

3.1 INTRODUCTION

For successful and efficient operation, flight requires control of position, attitude and velocity. Control law design can only be performed satisfactorily if a set of design requirement or performance criteria is available. **Classical control theory** provides a standard approach:

1. Model the dynamics of the aircraft.
2. Represent the task in terms of a state error.
3. Design a control algorithm to drive the state error to zero.
4. Measure all the states if possible.
5. Estimate the system state online.
6. Input the state/output estimate into the control algorithm to close the loop.

UAV are dynamical systems that can be classified as **under-actuated mechanical systems**. An under-actuated mechanical system has fewer control inputs than degrees of freedom. **Under-actuated** systems control is an important challenge.

Two approaches are usually employed for ground track control of UAV. In the first approach, the guidance and control design problems are separated into an outer-loop guidance and an inner-loop control design problem [89]:

1. **An outer loop controller** is designed to track the given position commands and reach the desired positions given the engine thrust, the lift and the bank angle as the control inputs. Another possibility is the engine thrust, the angle of attack and the bank angle. It results in the aircraft's navigation from one position to another relative to a fixed frame. The desired position and velocity can be stored online at

discrete times serving as nominal values. The difference between the actual position and velocity and the nominal ones produces acceleration commands in order to correct the errors.

2. **An inner loop controller** is designed to follow the desired attitude generated from the outer loop, using the control surface of the aileron, elevator and rudder. It results in control of aircraft's orientation. Continuous control laws using Euler angles cannot be globally defined; thus, these representations are limited to local attitude maneuvers. Since attitude control problems are nonlinear control problems, they can be categorized as [16]:
   a. **Local attitude control** issues address changes in the rigid body attitude and angular velocity that lie in the open neighborhood of the desired attitude and angular velocity.
   b. **Global attitude control** issues arise when arbitrary changes in the rigid body attitude and angular velocity are allowed. No a priori restrictions are placed on the possible rotational motion.

In the second approach, the guidance and control problems are addressed together in an integrated and unified framework. The UAV flight controller is designed to stabilize the attitude of an aircraft by holding a desired orientation and position. It also provides the means for an aircraft to navigate by tracking a desired trajectory/path. Different control techniques have been used to design flight controllers ranging from linear to nonlinear control algorithms:

1. The problem of **path following** is making an aircraft converge to and follow a desired spatial path, while tracking a desired speed profile that may be path dependent. The temporal and spatial assignments can be therefore separated.
2. The aim of the **trajectory following control** system is guiding the aircraft to stay on a given reference trajectory. Due to unconsidered disturbances and simplifications made in the model used for guidance, the aircraft will not follow the trajectory without error feedback. Therefore a control system for trajectory error reduction must be designed.

Remark 3.1. *In an* ***atmospheric flight****, smart autonomous aircraft can encounter a wide range of flight conditions. In some cases, disturbances can be as strong as the aircraft's own control forces.*

The main challenges of flight control and planning include:

1. Generation of a feasible trajectory taking into account aircraft's dynamics and capabilities.
2. Tight integration between guidance and control to ensure safety, maximizing performance and flexibility.
3. Satisfaction of performance criteria and terminal state constraints.

An **autonomous flight control system** must simultaneously account for changing weather and possible collision avoidance, while achieving optimal fuel consumption [2]. Depending on the mission, a set of design requirements may be different. For example, in a **dynamic maneuvering situation**, the autopilot is mainly concerned with the control forces that must be exerted and the resulting six degrees of freedom translational and angular accelerations. In a task requiring **precision tracking**, the evaluation of the control system is more influenced by landmarks and the response of the aircraft to turbulence [85].

Flight models involve high order nonlinear dynamics with state and input constraints. In the last decades, control for these systems was developed using the linearization approach. A lot of work has been devoted to the control and state estimation of these linear systems. The trend now is to use nonlinear control approaches as well as computational intelligence methods [5, 9, 37, 40, 48, 72].

3.2 LINEAR CONTROL METHODS

Linear control methods are generally divided into:

1. Linear time invariant (LTI) (linearization around an operating point)
2. Linear time variant (LTV) (linearization around a trajectory)
3. Linear parameter variant (LPV) (multi-model approach)

3.2.1 PROPERTIES OF LINEAR SYSTEMS

Main properties of linear systems are stabilizability, controllability and observability:

3.2.1.1 Controllability and observability

The following linear system is considered:

$$\begin{aligned} \dot{X} &= \mathbf{A}X + \mathbf{B}U \\ Y &= \mathbf{C}X \end{aligned} \tag{3.1}$$

The variables X, U, Y are, respectively, the state variable, the input or control variable and the output or measure variable. The matrices $\mathbf{A}, \mathbf{B}, \mathbf{C}$ are, respectively, the state matrix, the control matrix and the output matrix.

The following theorems give the Kalman conditions for controllability and observability.

Theorem 3.1

Controllability The linear system (3.1) is controllable if and only if the controllability matrix $\mathbf{Co} = \left[\mathbf{B}, \mathbf{AB}, \mathbf{A}^2\mathbf{B} \dots \mathbf{A}^{n-1}\mathbf{B}\right]$ has rank n. The pair $(\mathbf{A}, \mathbf{B})$ is said to be controllable. ■

Definition 3.1. ***Reachability*** *addresses whether it is possible to reach all points in the state space in a transient fashion. Computation of the states can be carried out by a dynamical system called an observer.*

Theorem 3.2

Observability The linear system (3.1) is Observable if and only if the Observability matrix $\mathbf{Ob} = \begin{bmatrix} \mathbf{C} \\ \mathbf{CA} \\ \mathbf{CA}^2 \\ \vdots \\ \mathbf{CA}^{n-1} \end{bmatrix}$ has rank n. The pair $(\mathbf{A}, \mathbf{C})$ is said to be observable. ■

3.2.1.2 Stability of a linear system

The system (3.1) is said to be stable if the eigenvalues of the matrix $\mathbf{A}$ have negative real parts. The matrix is said to be Hurwitz.

Remark 3.2. *Many different approaches to stability exist such as local stability, asymptotic stability, bounded input bounded output stability, absolute stability [72].*

An alternative method exists. The Lyapunov method is a useful technique for determining stability and is at the basis of many linear and nonlinear control methods such as backstepping and sliding mode. Some definitions and theorems are introduced in the sequel before the application of the Lyapunov method to an autonomous linear system.

Definition 3.2. ***Positive Definite Function and Candidate Function:*** *A function $\tilde{V} : \mathbb{R}^n \longrightarrow \mathbb{R}$ is called positive definite if $\tilde{V}(0) = 0, \tilde{V}(X) > 0, \forall X \in \mathbb{R}^n - \{0\}$ and $\tilde{V}$ is called a Lyapunov function candidate if it is positive definite and radially unbounded. If $\tilde{V}$ is differentiable, then the vector $\tilde{V}_X(X) = \nabla\tilde{V}(X)$ denotes the derivative of $\tilde{V}$ with respect to X.*

Theorem 3.3

Lyapunov Stability Theorem: Let $\tilde{V}$ be a non negative function on $\mathbb{R}^n$ and let $\dot{\tilde{V}}$ represent the time derivative of $\tilde{V}$ along trajectories of the system dynamics:

$$\dot{X} = f(X) \qquad X \in \mathbb{R}^n \tag{3.2}$$

$$\dot{\tilde{V}} = \frac{\partial \tilde{V}}{\partial X}\frac{dX}{dt} = \frac{\partial \tilde{V}}{\partial X} f(X) \tag{3.3}$$

Let $B_R = B_R(0)$ be a ball on radius R around the origin. If there exists $R > 0$ such that $\tilde{V}$ is positive definite and $\dot{\tilde{V}}$ is negative semi-definite for all $X \in B_R$, then $X = 0$ is locally stable in the sense of Lyapunov.

If $\tilde{V}$ is positive definite and $\dot{\tilde{V}}$ is negative definite in B_R, then $X = 0$ is locally asymptotically stable. ■

Remark 3.3. *A* ***Lyapunov function*** $\tilde{V} : \mathbb{R} \to \mathbb{R}$ *is an energy-like function that can be used to determine the stability of a system. If a non negative function that always decreases along trajectories of the system can be found, it can be concluded that the minimum of the function is a locally stable equilibrium point.*

The following stability analysis is useful in the derivation of the linear matrix inequalities in the linear parameter varying systems.

The following autonomous linear system is considered:

$$\dot{X} = \mathbf{A}X \tag{3.4}$$

where $\mathbf{A}$ is an invertible matrix. This system has a unique equilibrium point $X = 0$. To study its stability, the following quadratic form is introduced:

$$\tilde{V}(X) = X^T \mathbf{P} X \tag{3.5}$$

where $\mathbf{P}$ is a constant symmetric definite positive function. The derivative of this function can be written as:

$$\dot{\tilde{V}}(X) = X^T \left(\mathbf{A}^T\mathbf{P} + \mathbf{P}\mathbf{A}\right) X \tag{3.6}$$

Hence, the origin is an equilibrium point globally asymptotically stable if there exists a symmetric definite matrix $\mathbf{P}$ such that:

$$\mathbf{A}^T\mathbf{P} + \mathbf{P}\mathbf{A} < 0 \tag{3.7}$$

or equivalently:

$$\mathbf{A}^T\mathbf{P} + \mathbf{P}\mathbf{A} + \mathbf{Q} = 0 \tag{3.8}$$

where $\mathbf{Q}$ is a symmetric definite positive matrix.

Quadratic stability implies that the equilibrium point $X = 0$ is uniformly asymptotically stable.

Remark 3.4. *The principal difficulty is that there is no general rule allowing to find a Lyapunov function for any system.*

The origin is uniformly stable and hence exponentially stable for the following linear time-varying system:

$$\dot{X} = \mathbf{A}(t)X \tag{3.9}$$

assumed to be uniformly completely observable from the output:

$$Y(t) = \mathbf{C}(t)X(t) \tag{3.10}$$

if there exists a $\delta > 0$ such that the observability Grammian:

$$\mathbf{W}_o(t, t+\delta) = \int_t^{t+\delta} \Phi^T(\tau, t)\mathbf{C}^T(\tau)\mathbf{C}(\tau)\Phi(\tau, t)d\tau \geq k\mathbf{I}_{n\times n}, \forall t \tag{3.11}$$

is uniformly positive definite and if there exists a uniformly positive definite bounded matrix $\mathbf{P} \in \mathbb{R}^{n\times n}$ satisfying:

$$k_1\mathbf{I}_{n\times n} \leq \mathbf{P}(t) \leq k_2\mathbf{I}_{n\times n} \tag{3.12}$$

$$\dot{\mathbf{P}}(t) = \mathbf{A}^T(t)\mathbf{P}(t) + \mathbf{P}\mathbf{A}(t) + \mathbf{C}^T(t)\mathbf{C}(t) \tag{3.13}$$

with $(\mathbf{A}(t), \mathbf{C}(t))$ a uniformly completely observable pair [72].

If a pair $(\mathbf{A}(t), \mathbf{C}(t))$ is a uniformly completely observable that its observability Grammian is uniformly positive definite then for any $\mathbf{K}(t) \in \mathbb{R}^{n\times n}$, $((\mathbf{A}(t) + \mathbf{K}\mathbf{C}(t)), \mathbf{C}(t))$ is also uniformly completely observable.

3.2.2 LINEAR APPROACHES FOR LTI MODELS

Traditional methods for flight control design typically use nested single-input-single-output (SISO) control loops and strongly structured control architectures. These methods are based on detailed aircraft system analysis and exploit paths with weak coupling to obtain good results for conventional flight control design. Autopilots have been designed using these methods. However, multivariable methods such as optimal control and robust control design methods are state of the art for more complex flight control tasks under coupled and/or uncertain system dynamics. Three large groups of control design methodologies are optimal and adaptive control design methods as well as robust control design methods [74].

Definition 3.3. ***Robustness*** *is a property that guarantees that essential functions of the designed system are maintained under adverse conditions in which the model no longer accurately reflects reality.*

Different linear control techniques have been used for the linear models of an aircraft such as longitudinal and lateral models presented in Chapter 2. More information on aircraft linear control can be found in some textbooks such as [9, 12, 18, 64, 89].

3.2.2.1 PID control

The simplest control law is the **proportional, integral, derivative** law (PID). It necessitates three direct feedback gains:

1. **P**: On the measurement deviation (error to be minimized),
2. **I**: On its integral (to counter any disturbances),
3. **D**: On its derivative (to provide damping).

The control law has the following form:

$$U(t) = \mathbf{K}_P e(t) + \mathbf{K}_I \int_0^t e(\tau)d\tau + \mathbf{K}_V \frac{de(t)}{dt} \tag{3.14}$$

where $e(t)$ represents the error $e(t) = R(t) - Y(t), R(t)$ being the reference signal and $Y(t)$ is the measured signal; the parameters $\mathbf{K}_P, \mathbf{K}_I, \mathbf{K}_V$ are the gain diagonal matrices.

The advantages of PID law are that tuning is easily linked to physics and objectives, while the drawbacks are lack of optimization and filtering must be in series. However, a potential improvement such as **transverse acceleration feedback** can be used.

3.2.2.2 Classical methods

Many state-space methods have been used for the design of the autopilots such as the direct approach, the pole placement, the eigenvalue assignment by output feedback, the linear quadratic regulator and the adaptive approach.

3.2.2.2.1 Direct approach

One of the earliest control methods is the direct approach where a static control law can be proposed:

$$U(t) = \mathbf{K}_r R(t) + \mathbf{K}_x X(t) \tag{3.15}$$

where $\mathbf{K}_x$ is the feedback gain matrix and $\mathbf{K}_r$ is the feed-forward gain matrix. These matrices must be determined so that the closed-loop control system has the following form:

$$\begin{aligned} \dot{X} &= \mathbf{A}_d X + \mathbf{B}_d U \\ Y &= \mathbf{C}X \end{aligned} \tag{3.16}$$

The matrix $\mathbf{B}$ being generally not invertible, the following gain matrices can be proposed:

$$\mathbf{K}_x = \mathbf{B}^T \left(\mathbf{B}\mathbf{B}^T\right)^{-1} (\mathbf{A}_d - \mathbf{A}) \tag{3.17}$$

$$\mathbf{K}_r = \mathbf{B}^T \left(\mathbf{B}\mathbf{B}^T\right)^{-1} \mathbf{B}_d \tag{3.18}$$

This method is seldom used because of the difficulty of the choice of the matrices $\mathbf{A}_d, \mathbf{B}_d$.

3.2.2.2.2 *Pole placement*

Theorem 3.4

Pole placement: If the pair $(\mathbf{A}, \mathbf{B})$ of system (3.1) is completely controllable, then for any set of eigenvalues $\lambda_1, \lambda_2, \dots, \lambda_n \in \mathbb{C}$ symmetric with respect to the real axis, there exists a matrix $\mathbf{K} \in \mathbb{R}^{n \times n}$ such that the eigenvalues of $\mathbf{A} + \mathbf{BK}$ are precisely $\lambda_1, \lambda_2, \dots, \lambda_n$. ■

In this method, the following control law is proposed

$$U(t) = -\mathbf{K}_x X(t) \tag{3.19}$$

If all the eigenvalues $\lambda_1, \dots, \lambda_n$ are given then using the characteristic polynomial:

$$\lambda(s) = \det\left(s\mathbf{I} - A + \mathbf{BK}_x\right) \tag{3.20}$$

Identification or other methods such as Bass–Gura method allow to solve n equations giving the elements of the matrix $\mathbf{K}_x$.

If only $m < n$ eigenvalues are imposed, the matrix $\mathbf{K}_x$ can be decomposed into the product of two vectors:

$$\mathbf{K}_x = bd^T \qquad b \in \mathbf{R}^m \qquad d \in \mathbf{R}^n \tag{3.21}$$

The closed-loop system is then given by:

$$\dot{X} = \mathbf{A}X - \mathbf{B}(bd^T)X = \mathbf{A}X - \varphi\varpi \tag{3.22}$$

with $\varphi = \mathbf{B}b$ and $\varpi = d^T X$ with the requirement that the pair $(\mathbf{A}, \varphi)$ is controllable. The control law can be proposed:

$$U = -b\left(k_1\varpi_1^T + k_2\varpi_2^T + \dots + k_m\varpi_m^T\right)X \tag{3.23}$$

with

$$k_j = \frac{\prod_{i=1}^{m}\left(\lambda_i^d - \lambda_j\right)}{\varphi\varpi_j \prod_{i=1, i\neq j}^{m}\left(\lambda_i - \lambda_j\right)} \tag{3.24}$$

It imposes the pole placement λ_i^d while the other eigenvalues remain unchanged. The ϖ_j are the m eigenvectors of $\mathbf{A}^T$ associated with the eigenvalues $\lambda_1, \dots, \lambda_m$.

3.2.2.2.3 *Eigenvalue assignment by output feedback*

If only some of the state space variables are measurable, the following controller can be proposed:

$$\dot{\hat{X}} = \mathbf{A}\hat{X} + \mathbf{B}U + \mathbf{L}\left(Y - \mathbf{C}\hat{X}\right) = \left(\mathbf{A} - \mathbf{BK} - \mathbf{LC}\right)\hat{X} + \mathbf{L}Y \tag{3.25}$$

with

$$U = -\mathbf{K}\hat{X} \tag{3.26}$$

The closed loop system can be assigned arbitrary roots if the system is reachable and observable.

3.2.2.2.4 *Linear quadratic regulator*

A well-known classical control method is the **linear quadratic regulator** (LQR). The linear quadratic regulator is one of the most commonly used method to solve optimal control algorithms. The optimal control problem can be expressed as:

$$\min \left(J = \frac{1}{2} \int_0^T \left(X^T \mathbf{Q} X + U^T \mathbf{R} U \right) dt \right) \tag{3.27}$$

subject to

$$\dot{X} = \mathbf{A}X + \mathbf{B}U \tag{3.28}$$

where the weight matrix $\mathbf{Q}$ is a symmetrical semi-definite positive matrix and $\mathbf{R}$ is a symmetrical definite positive weight matrix. The following control law can be proposed:

$$U(t) = -\mathbf{K}X \tag{3.29}$$

with

$$\mathbf{K} = \mathbf{R}^{-1}\mathbf{B}^T\mathbf{P} \tag{3.30}$$

where $\mathbf{P}$ is the solution of the **Riccati** equation:

$$\dot{\mathbf{P}} = \mathbf{PA} + \mathbf{A}^T\mathbf{P} + \mathbf{Q} - \mathbf{PBR}^{-1}\mathbf{B}^T\mathbf{P} \tag{3.31}$$

When $T \to \infty$, the following equation must be solved:

$$\mathbf{PA} + \mathbf{A}^T\mathbf{P} + \mathbf{Q} - \mathbf{PBR}^{-1}\mathbf{B}^T\mathbf{P} = \mathbf{0} \tag{3.32}$$

Algebraic equations are then solved.

3.2.2.3 Adaptive approach

Adaptive control is a leading methodology intended to guarantee stable high-performance controllers in the presence of uncertainty [28, 36, 100]. A distinction is made between indirect adaptive control and direct adaptive control.

1. **Indirect adaptive control** involves two stages: first an estimate of the plant model is generated online. Once the model is available, it is used to generate the controller parameters.

a. **Model reference adaptive control** (MRAC) relies on a reference model and works on minimizing the tracking error between aircraft output and reference output. With model reference indirect adaptive control, it is feasible to achieve three important goals: trim value, adjustment for the inputs and outputs and closed-loop tracking of autopilot commands.
b. **Self-tuning control** (STC) focuses on adapting the PID control gains of the controller by making use of the estimated parameter values. This method is known to be more flexible.

2. **Direct adaptive control approach** estimates the controller parameters directly in the controller, instead of estimating a plant model. This can be done via two main approaches: output error and input error.

Remark 3.5. *The indirect adaptive control is preferable due to its flexibility and its property of being model based.*

Model reference adaptive control (MRAC) is an approach that provides feedback controller structures and adaptive laws for control of systems with parameter uncertainties to ensure closed-loop signals boundedness and asymptotic tracking of independent reference signals despite uncertainties in the system parameters. To develop an adaptive state feedback controller, it is necessary to first solve the related non adaptive control problem assuming the plant parameters are known, so that an ideal fixed state feedback controller can be obtained. This ideal or nominal controller will be used as a part of a priori knowledge in the design of the adaptive control scheme. The existence of such a nominal controller is equivalent to a set of matching equations. A parameter adaptation scheme must thus be used [28].

The uncertain system given by:

$$\dot{X}(t) = \mathbf{A}X(t) + \mathbf{B}\left(U(t) + \Delta(X(t))\right) \tag{3.33}$$

where $X(t) \in \mathbb{R}^n$ the state vector is available for feedback, $U(t) \in \mathbb{R}^m$ the control input is restricted to the class of admissible controls consisting of measurable functions, $\mathbf{A} \in \mathbb{R}^{n \times n}$, $\mathbf{B} \in \mathbb{R}^{n \times m}$ are known matrices and Δ is a matched uncertainty [100]. The pair $(\mathbf{A}, \mathbf{B})$ is assumed to be controllable.

Definition 3.4. *The uncertainty is said to be matched with the control input because the disturbance enters through the same input channel as the control.*

The reference model is given by:

$$\dot{X}_m(t) = \mathbf{A}_m X_m(t) + \mathbf{B}_m R(t) \tag{3.34}$$

where $X_m(t) \in \mathbb{R}^n$ is the reference state vector, $R(t) \in \mathbb{R}^r$ is a bounded continuous reference input, $\mathbf{A}_m \in \mathbb{R}^{n \times n}$ is Hurwitz (the real parts of its eigenvalues are negative) and $\mathbf{B}_m \in \mathbb{R}^{n \times r}$ with $r \leq m$. Since $R(t)$ is bounded, it follows that X_m is uniformly bounded for all $X_m(0)$. The matched uncertainty in (3.33) can be linearly parametrized as:

$$\Delta(X) = \mathbf{W}^T \varpi(X) + \epsilon(X) \qquad |\epsilon(X)| \leq \epsilon^*, X \in \mathbb{D}_x \tag{3.35}$$

where $\varpi : \mathbb{R}^n \rightarrow \mathbb{R}^s$ is a known vector of basis functions of the form $\varpi(X) = (\varpi_1(X), \varpi_2(X), \ldots, \varpi_s(X))^T \in \mathbb{R}^s$, $\mathbf{W} \in \mathbb{R}^{s \times m}$ is the unknown constant weight matrix, $\epsilon : \mathbf{R}^n \rightarrow \mathbb{R}^m$ is the residual error and $\mathbb{D}_x \subseteq \mathbb{R}^n$ is a sufficiently large compact set.

The following feedback control law is considered:

$$U(t) = U_m(t) - U_{ad}(t) \tag{3.36}$$

where $U_m(t)$ is a nominal feedback control given by:

$$U_m(t) = \mathbf{K}_1 X(t) + \mathbf{K}_2 R(t) \tag{3.37}$$

where $\mathbf{K}_1 \in \mathbb{R}^{m \times n}, \mathbf{K}_2 \in \mathbb{R}^{m \times r}$ are nominal control gains such that $\mathbf{A} + \mathbf{B}\mathbf{K}_1$ is Hurwitz and $U_{ad}(t)$ is the adaptive feedback control component given by:

$$U_{ad}(t) = \hat{\mathbf{W}}^T \varpi(X(t)) \tag{3.38}$$

where $\hat{\mathbf{W}} \in \mathbb{R}^{s \times m}$ is an estimate of $\mathbf{W}$ satisfying the weight update law:

$$\dot{\hat{\mathbf{W}}} = \gamma \left(\varpi(X(t)) e^T(t) \mathbf{P}\mathbf{B} + \dot{\hat{\mathbf{W}}}_m(t) \right), \gamma > 0 \tag{3.39}$$

where $e(t) = X(t) - X_m(t)$ is the state tracking error, $\mathbf{P} \in \mathbb{R}^{n \times n}$ is the positive definite solution of the Lyapunov equation:

$$\mathbf{A}_m^T \mathbf{P} + \mathbf{P}\mathbf{A}_m + \mathbf{Q} = 0 \tag{3.40}$$

for any $\mathbf{Q} = \mathbf{Q}^T$ and $\dot{\hat{\mathbf{W}}}_m \in \mathbb{R}^{s \times m}$ is a modification term such as

$$\dot{\hat{\mathbf{W}}}_m = -\sigma \hat{\mathbf{W}} \qquad \text{for } \sigma \text{ modification} \tag{3.41}$$

or

$$\dot{\hat{\mathbf{W}}}_m = -\sigma |e(t)| \hat{\mathbf{W}} \qquad \text{for e modification} \tag{3.42}$$

σ is a positive fixed gain.

If $\mathbf{A}_m, \mathbf{B}_m$ are chosen so that:

$$\mathbf{A}_m = \mathbf{A} + \mathbf{B}\mathbf{K}_1 \qquad \mathbf{B}_m = \mathbf{B}\mathbf{K}_2 \tag{3.43}$$

then theorems exist that provide sufficient conditions under which the closed-loop system errors $e(t)$ are uniformly ultimately bounded for the σ and e modifications cases [36, 100].

Remark 3.6. *The development of* **envelope protection methods** *for UAV is in the context of the adaptive flight control system [98]. Envelope protection is the task of monitoring and ensuring aircraft operations within its limits. Recent advances in the flight control system enable autonomous maneuvering that can challenge a UAV flight envelope. The options available for envelope protection are either to limit the aircraft inputs to the automatic flight control system (AFCS) or to limit the actuator commands from the automatic flight control system. Typically, conservative hard limits are set in a UAV flight control channels as maximum and minimum allowable command inputs.*

In reality, true command limits are complex functions of highly nonlinear aircraft dynamics and therefore vary with flight condition and/or aircraft configuration. The goal of automatic flight envelope protection is not only to enable the aircraft to safely operate within its envelope, but also to do so without restricting the aircraft to a smaller region of its operational envelope, hence making use of the aircraft's full flight envelope. Therefore, an effective automatic envelope protection system will reduce the compromise between safety and performance, thus improving the overall confidence of safe operations of UAV, especially during aggressive maneuvering close to their operational limits.

3.2.2.4 Robust control method for LTI models

In modeling for robust control design, an exactly known nominal plant is accompanied by a description of plan uncertainty that is a characterization of how the true plant might differ from the nominal one. This uncertainty is then taken into account during the design process [23].

Linear quadratic Gaussian (LQG) or H_2, weighted H_2 and H_∞ are widely used for robust linear control. The linear, quadratic, gaussian (LQG) controller combines a linear quadratic estimator with a linear quadratic regulator. It was originally intended for systems disturbed by white Gaussian noise, susceptible to parameter variation uncertainty [99]. Optimality is guaranteed under mean-free, uncorrelated, white noise excitation of known variance in terms of the state, control-input or output signal variances. It is time-related and is a physical approach to control. The updating **Kalman** gains are tuned decoupled while the command gains **Riccati** equations are solved, minimizing quadratic criteria. Filtering objectives are not taken directly into account in the synthesis; filtering must be in series.

The design model, which consists only of the controlled modes, are given by:

$$\begin{aligned} \dot{X} &= \mathbf{A}X + \mathbf{B}U + \nu \\ Y &= \mathbf{C}X + \vartheta \end{aligned} \tag{3.44}$$

where ν, ϑ denote zero-mean white noise. The controller design problem can be formulated as an LQG design problem with the following objective function:

$$J = \lim_{t\to\infty} \left(E\left[X^T \mathbf{Q} X + U^T \mathbf{R} U \right] \right) \tag{3.45}$$

where $E[\,]$ denotes the expected value, the weighting matrices are such that $\mathbf{Q} = \mathbf{Q}^T \geq 0$ and $\mathbf{R} = \mathbf{R}^T > 0$. The design parameters are the LQ regulator weighting matrices as well as the Kalman–Bucy filter weighting matrices $\mathbf{V} = \mathbf{V}^T \geq 0$ and $\mathbf{W} = \mathbf{W}^T > 0$. A sufficient condition for stability in the presence of unstructured additive uncertainty is:

$$\bar{\sigma}\left[\Delta\mathbf{P}\mathbf{H}\left(\mathbf{I}+\mathbf{P}_0\mathbf{H}\right)^{-1}\right] < 1, \forall \text{ real } \omega \tag{3.46}$$

where $\mathbf{H}(s)$ denotes the controller transfer function matrix and $\bar{\sigma}$ is the largest singular value of the matrix [39]. The uncontrolled mode dynamics are represented by an additive uncertainty $\Delta\mathbf{P}$; $\mathbf{P}_0$ represents the design model. An upper bound on the magnitude $\bar{\sigma}\left[\Delta\mathbf{P}\right]$ can be obtained to form an uncertainty envelope [39].

The design process is summarized as follows:

1. **Step 1**: Select design parameters $\mathbf{Q}, \mathbf{R}, \mathbf{V}, \mathbf{W}$ to obtain a nominal LQG controller for the design model that gives satisfactory closed-loop eigenvalues and frequency response.
2. **Step 2**: Apply the robustness test in the presence of additive uncertainty. If the test fails, adjust the weighting matrices and go back to step 1.
3. **Step 3**: Repeat until satisfactory performance and robustness are obtained.

For example, for lateral control in [74], the Dutch roll mode should be strongly damped for all parameter variations. A coordinated turn control is required and loads in turbulence should be improved during maneuvers and in turbulence.

The dynamics of the aircraft are given by:

$$\begin{aligned} \dot{X} &= \mathbf{A}X + \mathbf{B}U + \mathbf{B}_g W_g \\ Y &= \mathbf{C}X + \mathbf{D}U \end{aligned} \tag{3.47}$$

W_g is the gust velocity and $\mathbf{B}_g$ the related input matrix. The plant transfer matrix $\mathbf{G}(s)$ mapping $\left(U^T W_g^T\right)^T$ to Y is given by:

$$\mathbf{G}(s) = \left(\begin{array}{c|c} \mathbf{A} & [\mathbf{B}, \mathbf{B}_g] \\ \mathbf{C} & [\mathbf{D}, \mathbf{0}] \end{array}\right) \tag{3.48}$$

The signals to be minimized are collected in $Z = \left[Y^T U^T\right]^T$. The normalized generalized plant

$$\mathbf{P}(s) = \begin{pmatrix} \mathbf{P}_{11}(s) & \mathbf{P}_{12}(s) \\ \mathbf{P}_{21}(s) & \mathbf{P}_{22}(s) \end{pmatrix} \tag{3.49}$$

has a minimal state-space realization

$$\begin{aligned} \dot{X} &= \mathbf{A}_p X + \mathbf{B}_{p1} U + \mathbf{B}_{p2} W_g \\ Z &= \mathbf{C}_{p1} X + \mathbf{D}_{p11} U + \mathbf{D}_{p12} W_g \\ Y &= \mathbf{C}_{p2} X + \mathbf{D}_{p21} U + \mathbf{D}_{p22} W_g \end{aligned} \tag{3.50}$$

which combines the aircraft model, the Dryden filter and the weighting functions.

The design aircraft model not only includes a nominal model of the dynamics of the aircraft but also reflects the scaling of the variables and may include augmentation dynamics such as integrators that the designer has appended to the aircraft model to meet special command following and disturbance rejection performance specifications [6]. The LQG design methodology seeks to define the MIMO compensator so that the stability-robustness and performance specifications are met to the extent possible.

In [3], this LQG technique is used to establish a nominal performance baseline for the vertical acceleration control of an aircraft. The aircraft is assumed to be subjected to severe wind gusts causing undesirable vertical motion. The purpose of the controllers is to reduce the transient peak loads on the aircraft caused by the gusts. The gust load alleviation system uses motion sensor feedback to derive the aircraft control surfaces in order to attenuate aerodynamics loads induced by wind gusts.

The H_∞ approach to control systems is very appropriate for the optimization of stability and disturbance rejection robustness properties, while the linear quadratic Gaussian type of cost function is often a more practical criterion for minimizing tracking errors or control signal variations due to reference input changes.

The synthesis of the optimal H_∞ controller is to solve the following optimization problem:

$$\inf_{\mathbf{K}\text{stabilizing}} \|F_l(\mathbf{P},\mathbf{K})\| = \inf_{\mathbf{K}\text{stabilizing}} \sup_{\omega} \bar{\sigma}\left(F_l(\mathbf{P},\mathbf{K})(j\omega)\right) \tag{3.51}$$

where the lower **linear fractional transformation** (LFT) $F_l(\mathbf{P},\mathbf{K})$ is the transfer function matrix from the exogenous input into the panelist output Z and $\bar{\sigma}$ denotes the largest singular value. Let $\mathbf{K}_0$ be the corresponding optimal controller. Then it possesses the following all-pass property:

$$\bar{\sigma}\left(F_l(\mathbf{P},\mathbf{K})(j\omega)\right) = constant, \forall\omega \tag{3.52}$$

The principle of linear fractional transformation representation is to build an uncertain system in the form of a feedback between an augmented invariant system in which parameters are perfectly known and an uncertainty block grouping the various uncertainties. This allows to separate what is perfectly known from what is not.

H_∞ law is a frequency and global method: tuning a control law (all in a row) at a given time by seeking to minimize the ∞ norm (i.e., the max.) of a transfer between disturbing inputs and weighted outputs of an enhanced model. It offers a good handling of trade-offs between the various objectives such as filtering in the synthesis. However, it has a delicate management of time aspects [79]. Therefore to quantitatively demonstrate design trade-offs, the simultaneous treatment of both H_2 and H_∞ performance criteria becomes indispensable [96].

3.2.3 GAIN SCHEDULING

In the design of aircraft control systems, it is important to realize that the rigid body equations are only an approximation of the nonlinear aircraft dynamics. An aircraft has also **flexible modes** that are important at high frequencies, being potentially destabilizing. As the aircraft changes its equilibrium flight conditions, the linearized rigid body model describing its perturbed behavior changes. This parameter variation is a low-frequency effect that can also act to destabilize the system. To compensate for this variation, suitable controller gains must be determined for linearized models at several design equilibrium points to guarantee stability for actual flight conditions near that equilibrium point. Thus, it is important to design controllers that have **stability robustness**, which is the ability to provide stability in spite of modeling errors due to high frequency unmodeled dynamics and plant parameter variations [85].

Gain scheduling is a practical method of coping with known plant nonlinearities.

Definition 3.5. *Gain scheduling is the process of varying a set of controller coefficients according to the current value of a scheduling signal.*

Given a set of flight condition variables such as Mach number, angle of attack, dynamic pressure, center of gravity location ..., control loop gain settings yielding the best **handling qualities** at each individual flight condition can be directly selected. Given the control loop gain settings for best handling qualities, one of the earliest papers [56] presented a procedure for determining which simplified gains programs would yield the best performance for their level of complexity.

Conventional gain scheduling is carried out by taking linearization of the nonlinear plant at a few selected points in the aircraft's flight envelope.

The nonlinear model is considered:

$$\dot{X} = f(X, U), X \in \mathbb{R}^n, U \in \mathbb{R}^m \tag{3.53}$$

The standard approach is to take linearization at several design points throughout the flight envelope [67]:

$$\delta\dot{X} = \frac{\partial f}{\partial X}|_i \delta X + \frac{\partial f}{\partial U}|_i \delta U = \mathbf{A}_i \delta X + \mathbf{B}_i \delta U \tag{3.54}$$

where i represents the evaluation at the ith design points and $\delta X, \delta U$ are perturbations in the system from the design point. A linear control strategy can be implemented using the linearized system to achieve the desired closed-loop characteristics. Gain scheduling is performed between the design points by interpolating the gains to effect a smoothly varying set of gains throughout the flight envelope [97]. Subsequently, either the model linearization or resultant controller gains are scheduled between the design points by using simple interpolation methods. In practice, a scheduling variable usually is based on the physics of the situation and on particular characteristics of the model. Simple curve fitting approaches are used. In fact, linear interpolation seems to be the standard approach [70].

Conventional **eigenstructure assignment** can be used to place the eigenvalues of the linearized system around a predetermined finite set of trim or equilibrium conditions. Eigenstructure assignment can be used to select a set of feedback gains $\mathbf{K}$ that guarantees the desired closed-loop dynamic properties. This method places closed-loop poles of the linearized system:

$$(\mathbf{A}_i - \mathbf{B}_i\mathbf{K})\,\nu_i^d = \lambda_i^d \nu_i^d \tag{3.55}$$

The eigenstructure is assigned through the set of n self-conjugate distinct complex eigenvalues of the linearized closed-loop system, λ_i^d and the n distinct desired eigenvectors of the linearized closed-loop system ν_i^d.

Remark 3.7. ***Gain scheduled controllers*** *are effective for tackling the changes of aircraft dynamics [73]. The classical method selects the design points, designs LTI controllers at the design points and connects all the LTI controllers to cover the admissible envelope. The provided airspeed data include some uncertainties because they are calculated from the measured dynamic and static pressures by using pitot tubes and the measurements are affected by the hull of the aircraft, inaccuracies due to the limited resolution of onboard sensors. Thus the gain scheduled controller should have robustness against uncertainties. The onboard actuators have large uncertainties and the flight controller should have robustness against them. One objective is to design a flight controller that realizes a model-matching property under wind gusts as well as the uncertainties related to the onboard sensors and actuators in the admissible speed range.*

3.2.4 RECEDING HORIZON APPROACH FOR LTV MODELS

Receding horizon approach is a well-known control method presented in this paragraph for linear time varying aircraft models. The linear time-varying differential equation of the controlled system can be expressed as:

$$\dot{X} = \mathbf{A}(t)X + \mathbf{B}(t)U \qquad X(0) = X_0 \tag{3.56}$$

The system (3.56) is assumed to be **uniformly completely controllable**. The receding horizon control problem at any fixed time $t \geq 0$ is defined to be an optimal control problem in which the performance index:

$$J = \int_t^{t+T} \left(X^T(\tau)\mathbf{Q}(\tau)X(\tau) + U^T(\tau)\mathbf{R}(\tau)U(\tau)\right) d\tau \tag{3.57}$$

is minimized for some chosen $\delta \leq T \leq \infty$ (δ is a positive constant), subject to linear time-varying dynamical system (3.56) and with the constraint:

$$X(t+T) = 0 \tag{3.58}$$

where $\mathbf{Q}, \mathbf{R}$ are the weighting matrices, $\mathbf{Q}$ is a non-negative definite symmetric matrix and $\mathbf{R}$ is a positive definite symmetric matrix.

The main goal of receding horizon control is to solve the optimal control $U_{opt}(t)$ for the preceding interval in the interval $[t, t+T]$ with the current state $X(t)$ as the initial condition, where only the first data of $U_{opt}(t)$ are employed as the current control input to the system and the rest of the control $U_{opt}(t)$ is discarded [65].

For the next instantaneous time t, the preceding solution process is repeated and the control input is recomputed:

$$U^*(t) = -\mathbf{R}^{-1}(t)\mathbf{B}^T(t)\mathbf{P}^{-1}(t, t+T)X(t) \tag{3.59}$$

where $\mathbf{P}(t, t+T)$ satisfies the **matrix Riccati differential equation** at any time $\tau \in (t, t+T) = (t, t_T)$:

$$\begin{aligned} \dot{\mathbf{P}}(\tau, t_T) &= \mathbf{P}(\tau, t_T)\mathbf{A}(\tau) + \mathbf{A}^T(\tau)\mathbf{P}(\tau, t_T) + \\ &+ \mathbf{Q}(\tau) - \mathbf{P}(\tau, t_T)\mathbf{B}(\tau)\mathbf{R}^{-1}(\tau)\mathbf{B}^T(\tau)\mathbf{P}(\tau, t_T) \end{aligned} \tag{3.60}$$

with the boundary condition

$$\mathbf{P}(t_T, t_T) = 0 \tag{3.61}$$

For solving the receding horizon control problems, many methods have been proposed to avoid the online integration of the **Riccati** differential equation, such as Lu's method and Legendre and Jacobi pseudospectral methods [47]. Most of these methods transform the receding horizon control problem into a quadratic programming or linear equation problem. Another technique can be proposed to construct an efficient sparse numerical approach for solving the receding horizon control problem online [65].

3.2.5 LINEAR PARAMETER VARYING MODELS

The Takagi–Sugeno (TS) model theory has proven useful in the description of nonlinear dynamic systems as a means of blending of models obtained by local analysis. Such descriptions are referred to as **model-based systems**. In addition, the Takagi–Sugeno approach can be used for the synthesis of gain scheduled controllers [33].

A **linear parameter varying model** is a linear time-varying system whose matrices depend on a vector of time-varying parameters that are either measured in real time or estimated using some known scheduling functions. This modeling framework is appealing because by embedding the nonlinearities in the varying parameters of the model that depend on some exogenous signals (for example system states), powerful linear analysis and design tools can be applied to nonlinear systems [69].

3.2.5.1 Principle

The LPV controller design starts with the conversion of the aircraft nonlinear model into an LPV model:

1. The nonlinear model is converted into a **linear parameter varying model** with affine dependence on the parameter vector that varies within a bounded region and depends on the system states through a known relation based on measured signals. These characteristics guarantee that the trajectories of the LPV model are trajectories of the nonlinear model. Some examples of aircraft models have been given in Chapter 2.
2. This affine model is converted into a polytopic model that is used to formulate the control problem in terms of **linear matrix inequalities** (LMI) allowing the convex optimization theory to be employed in the solution of these LMI applying numerical algorithms available in the LMI solvers [90].
3. For the controller synthesis, the parameter is assumed to vary freely taking arbitrary values in the bounded region.

The validity of the local model i is indicated by a function $\mu_i(\zeta(t)) : \mathbb{D} \rightarrow [0,1]$. The multi-model made of local linear models can be written as:

$$\dot{X}(t) = \sum_{i=1}^{r} \varpi_i(\zeta(t)) \left(\mathbf{A}_i X(t) + \mathbf{B}_i U(t)\right) \tag{3.62}$$

$$Y(t) = \sum_{i=1}^{r} \varpi_i(\zeta(t)) \left(\mathbf{C}_i X(t) + \mathbf{D}_i U(t)\right) \tag{3.63}$$

with

$$\varpi_i(\zeta(t)) = \frac{\mu_i(\zeta(t))}{\sum_{i=1}^{r} \mu_i(\zeta(t))} \tag{3.64}$$

The number of rules $r = 2^{|\zeta|}$ denotes the number of nonlinearities considered.

Instead of designing a single LPV controller for the entire parameter spaces in a conventional LPV synthesis, LPV controllers can be synthesized for parameter subspaces which are overlapped with each other. Then, these LPV controllers are blended into a single LPV controller over the entire parameter space. Thus the performance of the closed-loop system with the blended controller is preserved when parameter trajectories travel over the overlapped parameter subspaces [78].

A nonlinear control law $U(t)$ must be determined to stabilize the multi-model with the general law:

$$U(t) = \sum_{j=1}^{r} \varpi_j(\zeta(t)) \mathbf{K}_j X(t) \tag{3.65}$$

In the closed form, the following relationship is obtained:

$$\dot{X}(t) = \left(\sum_{i=1}^{r} \varpi_i(\zeta(t))\mathbf{A}_i + \left(\sum_{i=1}^{r} \varpi_i(\zeta(t))\mathbf{B}_i \right) \left(\sum_{j=1}^{r} \varpi_j(\zeta(t))\mathbf{K}_j \right) \right) X(t) \tag{3.66}$$

of the form $\dot{X} = f(X)$ with $f(0) = 0$.

The following Lyapunov function candidate is used:

$$\tilde{V}(X) = X^T \mathbf{P} X \tag{3.67}$$

where $\mathbf{P}$ is a symmetrical definite positive matrix. The origin is an equilibrium point globally asymptotically stable if $\dot{\tilde{V}}(X) < 0$. It can be demonstrated that

$$\dot{\tilde{V}}(X) \leq \sum_{i,j} X^T \left(\mathbf{P}\left(\mathbf{A}_i + \mathbf{B}_i\mathbf{K}_j\right) + \left(\mathbf{A}_i + \mathbf{B}_i\mathbf{K}_j\right)^T \mathbf{P} \right) X(t) \tag{3.68}$$

This inequality is verified if $\forall i = 1, \dots, r, \forall j = 1, \dots, r$:

$$\mathbf{P}\left(\mathbf{A}_i + \mathbf{B}_i\mathbf{K}_j\right) + \left(\mathbf{A}_i + \mathbf{B}_i\mathbf{K}_j\right)^T \mathbf{P} < 0 \tag{3.69}$$

Definition 3.6. ***Linear Matrix Inequality:*** *A linear matrix inequality is an expression of the form*

$$\mathbf{F}_0 + \mathbf{F}_1\Theta_1 + \cdots + \mathbf{F}_d\Theta_d \leq 0 \tag{3.70}$$

where $\Theta_i, i = 1, \dots, d$ are real variables, $F_i, i = 1, \dots, d$ are $n \times n$ symmetric matrices and $\mathbf{A} \leq 0$ means that $\mathbf{A}$ is a negative semi-definite matrix; that is, $z^T\mathbf{A}z \leq 0, \forall z \in \mathbb{R}^n$.

Remark 3.8. ***Linear matrix inequalities techniques*** *have been developed for LPV systems whose time-varying parameter vector ϖ varies within a polytope Θ. In this kind of LPV system, the state matrices range in a polytope of matrices defined as the convex hull of a finite number r of matrices. Each polytope vertex corresponds to a particular value of the scheduling variable ϖ.*

The simplest polytopic approximation relies on bounding each parameter by an interval. This approximation is known as the **bounding box approach**.

Definition 3.7. *A* ***quasi-LPV*** *system is defined as a linear time-varying plant whose state space matrices are fixed functions of some vector of varying parameters ϖ that depend on the state variables [58].*

3.2.5.2 Parallel distributed compensator

A crucial task for autonomous flying is tracking a predefined trajectory when wind disturbances are present [44]. For tracking purposes, formulating the problem with respect to the error posture rather than the inertial coordinates model of the aircraft can aid the design of the control methodology. Particularly, the reference and the current posture are used for the generation of the error posture of the aircraft. Among the advantages of using such a model, an important one is that there can be created any kind of reference trajectory to be tracked, even discontinuous ones.

In this section, the error posture model and the control methodology design to track a predefined trajectory are described. Additionally, its equivalence is shown with respect to a Takagi–Sugeno model which is used in the control methodology. The control methodology used involves a **parallel distributed compensation** (PDC) law to be computed via the use of linear matrix inequality (LMI) techniques:

$$\begin{aligned}\dot{x} &= V\cos\gamma\cos\chi + W\cos\gamma_w\cos\chi_w \\ \dot{y} &= V\cos\gamma\sin\chi + W\cos\gamma_w\sin\chi_w \\ \dot{z} &= -V\sin\gamma + W\sin\gamma_w \\ \dot{\chi} &= \omega_1 \\ \dot{\gamma} &= \omega_2\end{aligned} \tag{3.71}$$

In the error posture model, the reference and the current posture used are denoted as $P_{ref} = (x_{ref}, y_{ref}, z_{ref}, \chi_{ref}, \gamma_{ref})^T$ and $P_c = (x_c, y_c, z_c, \chi_c, \gamma_c)^T$, respectively. The reference position is the position of the aircraft to be tracked and the current is the real one calculated at each time step. The error model can be determined by applying a transformation of P_{ref} in the local frame with origin P_c. Hence the tracking error is governed by:

$$P_e = \mathbf{T}_e(P_{ref} - P_c) \tag{3.72}$$

where, in 3D, the matrix $\mathbf{T}_e$ can be written as:

$$\mathbf{T}_e = \begin{pmatrix} \cos\gamma\cos\chi & \cos\gamma\sin\chi & -\sin\gamma & 0 & 0 \\ -\sin\chi & \cos\chi & 0 & 0 & 0 \\ \sin\gamma\cos\chi & \sin\gamma\sin\chi & \cos\gamma & 0 & 0 \\ 0 & 0 & 0 & 1 & 0 \\ 0 & 0 & 0 & 0 & 1 \end{pmatrix} \tag{3.73}$$

The error posture model can be used for tracking purposes.
Differentiating,

$$\begin{pmatrix} \dot{e}_x \\ \dot{e}_y \\ \dot{e}_z \\ \dot{e}_\chi \\ \dot{e}_\gamma \end{pmatrix} = \begin{pmatrix} e_1 \\ e_2 \\ e_3 \\ \omega_{1r} - \omega_1 \\ \omega_{2r} - \omega_2 \end{pmatrix} \tag{3.74}$$

where

$$e_1 = \omega_1 \cos\gamma e_y - \omega_2 e_z - V + V_r \cos\gamma \cos\gamma_r \cos e_\chi + V_r \sin\gamma \sin\gamma_r$$

$$e_2 = -\omega_1 \cos\gamma e_x - \omega_1 \sin\gamma e_z + V_r \cos\gamma_r \sin e_\chi$$

$$e_3 = \omega_1 \sin\gamma e_y + \omega_2 e_x + V_r \sin\gamma \cos\gamma_r \cos e_\chi - V_r \cos\gamma \sin\gamma_r$$

The considered input is $U = \begin{pmatrix} V \\ \omega_1 \\ \omega_2 \end{pmatrix}$. If an anticipative action is chosen such as $U = U_B + U_F$ with

$$U_F = \begin{pmatrix} V_r \cos\gamma \cos\gamma_r \cos e_\chi + V_r \sin\gamma \sin\gamma_r \\ \omega_{1r} \\ \omega_{2r} \end{pmatrix} \tag{3.75}$$

U_B is a feedback control action vector. The control U_B is calculated in the analysis by the use of parallel distributed compensation (PDC) control. The control law used is designed to guarantee stability within a compact region of space and the gains are calculated through the use of LMI conditions.

Then system (3.74) can be written as:

$$\begin{pmatrix} \dot{e}_x \\ \dot{e}_y \\ \dot{e}_z \\ \dot{e}_\chi \\ \dot{e}_\gamma \end{pmatrix} = \mathbf{A}_{lin} \begin{pmatrix} e_x \\ e_y \\ e_z \\ e_\chi \\ e_\gamma \end{pmatrix} + \mathbf{B}_{lin} U_B \tag{3.76}$$

where

$$\mathbf{A}_{lin} = \begin{pmatrix} 0 & \omega_{1r}\cos\gamma & -\omega_{2r} & 0 & 0 \\ -\omega_{1r}\cos\gamma & 0 & -\omega_{1r}\sin\gamma & V_r\cos\gamma_r & 0 \\ \omega_{2r} & \omega_{1r}\sin\gamma & 0 & 0 & -V_r \\ 0 & 0 & 0 & 0 & 0 \\ 0 & 0 & 0 & 0 & 0 \end{pmatrix} \tag{3.77}$$

$$\mathbf{B}_{lin} = \begin{pmatrix} -1 & 0 & 0 \\ 0 & 0 & 0 \\ 0 & 0 & 0 \\ 0 & -1 & 0 \\ 0 & 0 & -1 \end{pmatrix} \tag{3.78}$$

The controller used for the Takagi–Sugeno system is the **parallel distributed compensation** (PDC). In the PDC design, every control law is used, as the aircraft model is designed from the rules of the Takagi–Sugeno model. It involves the same sets in the premise part with the Takagi–Sugeno model fed. Hence for the model described by equation (3.74), the ith control rule layout is equivalent to the following:

Rule i : If $\zeta_1(t)$ is M_{i1} AND ... AND $\zeta_p(t)$ is M_{ip} THEN $U_B = -\mathbf{K}_i X(t)$ for $i = 1, ..., r$ where $r = dim(\zeta)$.

Thus the layout falls under the category of state feedback laws [52]. The overall state feedback law to be used in the design has the form of:

$$U_B = -\frac{\sum_{i=1}^{r} \mu_i(\zeta(t))\mathbf{K}_i\mathbf{X}(t)}{\sum_{i=1}^{r} \mu_i(\zeta(t))} = -\sum_{i=1}^{r} \varpi_i(\zeta(t))\mathbf{K}_i\mathbf{X}(t) \tag{3.79}$$

The matrices $\mathbf{K}_i$ are the local feedback gains that are calculated in order to guarantee global stability of the system. The latter is performed through the use of LMI conditions which guarantee global stability in the consequent compact regions of the premise parts.

3.2.5.3 Integral action

The LPV or quasi-LPV model dynamics of the aircraft is given by:

$$\dot{X}(t) = \mathbf{A}(\varpi)X(t) + \mathbf{B}(\varpi)U(t) \tag{3.80}$$

If the varying plant matrices $\mathbf{A}(\varpi)$, $\mathbf{B}(\varpi)$ are assumed to depend affinely on the parameter ϖ, they can be written as:

$$\mathbf{A}(\varpi) = \mathbf{A}_0 + \sum_{i=1}^{m} \varpi_i \mathbf{A}_i \qquad \mathbf{B}(\varpi) = \mathbf{B}_0 + \sum_{i=1}^{m} \varpi_i \mathbf{B}_i \tag{3.81}$$

The time-varying parameter vector $\varpi(t)$ is assumed to lie in a specified bounded compact set and is assumed to be available for the controller design. In order to introduce a tracking facility in the proposed scheme, the aircraft states are first augmented with integral action states given by:

$$\dot{X}_r(t) = R(t) - \mathbf{C}_c X(t) \tag{3.82}$$

where $R(t)$ is the reference to be tracked and $\mathbf{C}_c$ is the controlled output distribution matrix gain. The augmented LPV system becomes:

$$\begin{pmatrix} \dot{X}_r(t) \\ \dot{X}(t) \end{pmatrix} = \begin{pmatrix} 0 & -\mathbf{C}_c \\ 0 & \mathbf{A}(\varpi) \end{pmatrix} \begin{pmatrix} X_r(t) \\ X(t) \end{pmatrix} + \begin{pmatrix} 0 \\ \mathbf{B}(\varpi) \end{pmatrix} U(t) + \begin{pmatrix} \mathbf{I}_{n\times n} \\ 0 \end{pmatrix} R(t) \tag{3.83}$$

Two objectives must be met: the first relates to achieving a good nominal performance for all the admissible values of ϖ and the second is to satisfy the closed-loop stability condition for which the small gain theorem is required for the controller to guarantee stability in the face of faults or failures [1].

Theorem 3.5

If for the switched polytopic system

$$\begin{aligned} \dot{X} &= \mathbf{A}(\varpi(t))X(t) + \mathbf{B}(\varpi(t))U(t) \\ U(t) &= \mathbf{K}(t)\left(R(t) - X(t)\right) \end{aligned} \tag{3.84}$$

where

$$\mathbf{A}(\varpi(t)) = \sum_{i=1}^{r} \varpi_i(t)\mathbf{A}_i \qquad \mathbf{B}(\varpi_i(t)) = \sum_{i=1}^{r} \varpi_i(t)\mathbf{B}_i \tag{3.85}$$

with

$$\varpi_i(t) > 0 \qquad \sum_{i=1}^{r} \varpi_i(t) = 1$$

and the following assumptions hold:

1. The vertices of any polytopic subsystem, i.e., the subsystem of the corresponding locally overlapped switched system, satisfy the linear matrix equalities:

$$\left(\mathbf{A}_i - \mathbf{B}_i\mathbf{K}_j\right)^T \mathbf{P}_j + \mathbf{P}_j\left(\mathbf{A}_i - \mathbf{B}_i\mathbf{K}_j\right) + \mathbf{Q}_{ij} = 0 \tag{3.86}$$

   where $\mathbf{P}_j, \mathbf{Q}_{ij}$ are appropriately dimensioned positive symmetric matrices.
2. The gain scheduled controller of every polytopic subsystem is synthesized as:

$$\mathbf{K}_j = \mathbf{B}_j^+(\varpi_i(t))\left(\mathbf{B}_i\mathbf{K}_i\right) \tag{3.87}$$

   where $\mathbf{B}_j^+(\varpi_i(t))$ is the Moore–Penrose inverse of the full row rank matrix $\mathbf{B}_j(\varpi_i(t)$.

Then, the system is uniformly input to state bounded under any piecewise bounded command input. ■

The proof of this theorem can be found in [1].

Remark 3.9. *A linear parameter varying approach for designing a constant output feedback controller for a linear time-invariant system with uncertain parameters is presented in [77]. It achieves minimum bound on either the H_2 or H_∞ performance level. Assuming that the uncertain parameters reside in a given polytope a parameter-dependent Lyapunov function is described which enables the derivation of the required constant gain via a solution of a set of **linear matrix inequalities** that correspond to the vertices of the uncertain polytope.*

3.2.5.4 Gain scheduling

For LPV gain scheduled methods, arbitrarily fast variation for parameters is allowed by using a common Lyapunov function which guarantees the closed-loop stability. A switched polytopic system is established to describe the aircraft dynamics within the full flight envelope. Every polytopic subsystem represents the system dynamics in a part of the flight envelope and its vertices are the subsystems of a locally overlapped switched system which describes dynamics on operating points within this part of the flight envelope. For every polytopic subsystem, a gain scheduled subcontroller is achieved by interpolating between the state-feedback controllers on vertices. Gain scheduled controller with respect to the full flight envelope is composed of these gain scheduled subcontrollers [30].

The considered aircraft model is a nonlinear continuous and continuously differentiable system of the form:

$$\dot{X} = f(X, U) \tag{3.88}$$

where $X \in \mathbb{R}^n, U \in \mathbb{R}^m, f : \mathbb{R}^n \times \mathbb{R}^m \rightarrow \mathbb{R}^n$.

The aim is to design a controller capable of following some desired trajectories (X_r, U_r) where X_r is a differentiable slowly varying state trajectory and U_r is the nominal input necessary to follow the unperturbed X_r state. A subset $\mathbb{XU} \subset \mathbb{R}^{n+m}$ of the system's state and input spaces is defined as a bound of the possible state and input values. A set of operating points is also defined as $\{(X_i, U_i) \in \mathbb{XU}, i \in I\}$; I is the set of all positive integers that form a regular grid J in the trajectory space.

Linearization of (3.88) about all the points in J yields:

$$\mathbf{A}_i = \frac{\partial f}{\partial X}|_{(X_i,U_i)} \quad \mathbf{B}_i = \frac{\partial f}{\partial U}|_{(X_i,U_i)} \tag{3.89}$$

resulting in perturbed dynamics about the linearization points given by

$$\begin{aligned} \dot{X} &= \mathbf{A}_i(X - X_i) + \mathbf{B}_i(U - U_i) + f(X_i, U_i) \\ &= \mathbf{A}_i\bar{X} + \mathbf{B}_i\bar{U} + d_i \\ d_i &= f(X_i, U_i) - (\mathbf{A}_i\bar{X} + \mathbf{B}_i\bar{U}) \end{aligned} \tag{3.90}$$

When the linearized systems (3.90) are interpolated through a Takagi–Sugeno model, a nonlinear approximation of (3.88) is obtained through:

$$\dot{X} \approx \hat{f}(X, U) = \sum_{i \in I} \varpi_i(X, U)(\mathbf{A}_i X + \mathbf{B}_i U + d_i) \tag{3.91}$$

A control law for system (3.91) is also designed as a gain scheduling controller based on a Takagi–Sugeno model. Under the hypothesis of controllability for all $(\mathbf{A}_i, \mathbf{B}_i), \forall i \in I$, and being all of the states measured, full-state feedback linear control laws can be synthesized and interpolated through a Takagi–Sugeno system yielding

$$U = U_r + \sum_{j \in I} \varpi_j(X,U)\mathbf{K}_j\left(X - X_r\right)) \tag{3.92}$$

In equations (3.91) to (3.92), the expressions ϖ represent the Takagi–Sugeno linear membership functions relating the input variables to the domain described by *if-then-else* rules consequent. The system membership functions are chosen such as they constitute a convex sum over the input range $\mathbb{XU}$.

Substituting (3.92) in (3.91), the closed-loop perturbed system dynamics become:

$$\dot{X} - \dot{X}_r = \sum_i \varpi_i(X,U)\left[\mathbf{A}_i + \mathbf{B}_i\left(\sum_j \varpi_j(X,U)\mathbf{K}_j\right)\right](X - X_r) + \epsilon \tag{3.93}$$

where

$$\epsilon = \sum_i \varpi_i(X,U)\left[\mathbf{A}_i X_r + \mathbf{B}_i U_r + d_i\right] - \dot{X}_r \tag{3.94}$$

The term $\sum_i \mathbf{A}_i X_r$ is added and subtracted so that the matrix $\sum_i \varpi_i(X,U)\left(\mathbf{A}_i + \mathbf{B}_i\left(\sum_j \varpi_j(X,U)\mathbf{K}_j\right)\right)$ gives the dynamics of the perturbation from the desired trajectory. Also, from the definition of d_i, ϵ represents the error with respect to X_r as a result of the approximation of the function f with the Takagi–Sugeno model.

The asymptotic stability conditions of the Takagi–Sugeno gain scheduling controller around (X_r, U_r) can be derived.

Definition 3.8. ***Scheduling Indexes:*** *Given the grid point set $\mathbb{J}$ and any linearized dynamics $(\mathbf{A}_i, \mathbf{B}_i), i \in \mathbb{J}, \mathbb{J}_i$ is defined as the set of all indexes m of the neighborhood points of (X_i, U_i) whose controllers $\mathbf{K}_m$ have a non negligible influence over $(\mathbf{A}_i, \mathbf{B}_i)$.*

$\mathbb{J}_i$ contains all points such that

$$\varpi_m(X,U) > 0, \forall (X,U) \in \left\{(X,U) : \varpi_i(\bar{X},U) > 0\right\}$$

Given a generic input state pair $(X_l, U_l), l \notin \mathbb{I}$, the stability property for the tracking error $(X_l - X_r) \to 0$ requires the following assumption:

Condition 1: $(X_i, U_i) \in \mathbb{J}$ is assumed to be the nearest linearization grid point to the operating point $(X_l, U_l) \in \mathbb{J}$. The system $(\mathbf{A}_i, \mathbf{B}_i)$ remains closed-loop stable using a convex combination of controllers $\mathbf{K}_m, m \in \mathbb{J}$.

Condition 1 is verified using the following test guaranteeing that the Takagi–Sugeno modeling of the plant $\hat{f}$ is stable when controlled by the

Takagi–Sugeno controller for all positive controller combinations. The closed-loop system dynamics about (X_i, U_i) are considered

$$\dot{X} = \left[\mathbf{A}_i + \mathbf{B}_i\left(\sum_j \varpi_j(X,U)\mathbf{K}_j\right)\right]X = \sum_j \varpi_i(X,U)(\mathbf{A}_i + \mathbf{B}_i\mathbf{K}_j)X \tag{3.95}$$

obtained by a convex combination of controllers $\mathbf{K}_m, m \in \mathbb{J}$. Equation (3.95) has the form of a polytopic differential inclusion, where the vertices are the matrices $(\mathbf{A}_i + \mathbf{B}_i\mathbf{K}_j), j \in \mathbb{J}_i$. Stability of all $(\mathbf{A}_i + \mathbf{B}_i\mathbf{K}_j), j \in \mathbb{J}_i$ is required.

The stability test is repeated for all grid points obtaining:

$$\forall i \in \mathbb{I}, \forall j \in \mathbb{J}_i, \exists \mathbf{P}_i > 0, (\mathbf{A}_i + \mathbf{B}_i\mathbf{K}_j)^T\mathbf{P}_i + \mathbf{P}_i(\mathbf{A}_i + \mathbf{B}_i\mathbf{K}_j) < 0 \tag{3.96}$$

Inequality (3.96) can be solved using **linear matrix inequalities** (LMI) techniques.

The set of Θ, where (3.70) is satisfied, is a convex set that is an LMI specifies a convex constraint on $\Theta = (\Theta_1, \dots, \Theta_d)^T$. **Linear matrix inequalities** are useful tools to describe constraints arising in systems and control applications.

Remark 3.10. *If the LMI test fails, then the grid $\mathbb{J}$ must be made denser. Furthermore, the LMI test suggests where to add additional linearization points, in order to make the closed-loop system stable.*

Condition 2: The approximation error caused by linearization and successive Takagi–Sugeno modeling with respect to the original nonlinear system is small enough as not to compromise robust stability with respect to structured uncertainties. If the desired closed-loop dynamics are given by

$$\mathbf{A}_d = \mathbf{A}_i + \mathbf{B}_i\mathbf{K}_i \quad \forall i \in I \tag{3.97}$$

then from (3.95)

$$\sum_j \varpi_j(X,U)(\mathbf{A}_i + \mathbf{B}_i\mathbf{K}_j)X = \left(\mathbf{A}_d + \sum_j \varpi_j(X,U)\delta\mathbf{A}_{ij}\right)X \tag{3.98}$$

with

$$\delta\mathbf{A}_{ij} = \mathbf{B}(X_i, U_i)\left[\mathbf{K}(X_i, U_j) - \mathbf{K}(X_i, U_i)\right] \tag{3.99}$$

Using the robust stability theorem, under structured uncertainties, the following matrix

$$\mathbf{A} + \sum_i \varpi_i(X,U)\sum_j \varpi_j(X,U)\delta\mathbf{A}_{ij} \tag{3.100}$$

should be tested for stability.

3.2.5.5 Cost control analysis: Actuator saturation

Aircraft are subject to constraints on input and state. Actuator saturation is one of the most common constraints but sometimes constraints involving input and state variables are also useful to limit the rate of variation of inputs or internal variables.

The Takagi–Sugeno model is given by

$$\begin{aligned} \dot{X} &= \mathbf{A}(\varpi)X + \mathbf{B}(\varpi)U \\ Y &= \mathbf{C}(\varpi)X \end{aligned} \tag{3.101}$$

Given positive definite weighting matrices $\mathbf{W}$ and $\mathbf{R}$ and a quadratic objective function of the form:

$$J = \int_0^\infty \left(Y^T(t)\mathbf{W}Y(t) + U^T\mathbf{R}U(t)\right) dt \tag{3.102}$$

The designed controller minimizes an upper bound of it, defined by a quadratic positive definite function, $\tilde{V} = X^T\mathbf{P}X$, so that

$$J < X_0^T\mathbf{P}X_0 \tag{3.103}$$

This bound holds if the following index J_d is negative:

$$J_d = Y^T\mathbf{W}Y + U^T\mathbf{R}U + \dot{X}^T\mathbf{P}X + X^T\mathbf{P}\dot{X} < 0 \tag{3.104}$$

Substituting in (3.104) the expression of X, Y and defining U by a parallel distributed compensator (PDC) control law of the form [4]:

$$U = \sum_{j=1}^{r} \varpi_i \mathbf{K}_j X = \mathbf{K}(\varpi)X \tag{3.105}$$

The double summation condition (3.106) is obtained:

$$\sum_{i=1}^{r}\sum_{j=1}^{r} \varpi_i(t)\varpi_j(t)X(t)^T\mathbf{Q}_{ij}X(t) \leq 0 \tag{3.106}$$

where $\mathbf{Q}_{ij}$ is defined as:

$$\mathbf{Q}_{ij} = \frac{1}{2}\begin{pmatrix} 2\mathbf{D}_{ij} & \mathbf{P}^{-1}\mathbf{C}_i^T & -\mathbf{M}_j & \mathbf{P}^{-1}\mathbf{C}_j^T & -\mathbf{M}_i \\ \mathbf{C}_i\mathbf{P}^{-1} & -\mathbf{W}^{-1} & 0 & 0 & 0 \\ -\mathbf{M}_j & 0 & -\mathbf{R}^{-1} & 0 & 0 \\ \mathbf{C}_j\mathbf{P}^{-1} & 0 & 0 & -\mathbf{W}^{-1} & 0 \\ -\mathbf{M}_i & 0 & 0 & 0 & -\mathbf{R}^{-1} \end{pmatrix} \tag{3.107}$$

$$\mathbf{D}_{ij} = \mathbf{A}_i\mathbf{P}^{-1} + \mathbf{P}^{-1}\mathbf{A}_i^T - \mathbf{B}_i\mathbf{M}_j - \mathbf{M}_j^T\mathbf{B}_i^T \tag{3.108}$$

where $\mathbf{K}_i = \mathbf{M}_i\mathbf{P}$.

The congruence lemma is used: if $\mathbf{Q} < 0$ then $\mathbf{XQX} < 0$ where $\mathbf{X}$ is also a matrix [32]. The above expression (3.106) is a double summation which is not directly an LMI. The optimizer U^{opt} has a piecewise affine structure of the form:

$$U^{opt} = \mathbf{K}_k X + \mathbf{G}_k \quad \text{for } X \in \Omega_k = \{X, \mathbf{H}_k X \leq l_k\} \tag{3.109}$$

defined in l regions $\Omega_k, k = 1, \dots, l$ which conforms a polytopic partition of the operating domain $\mathbb{X}$.

The objective of this subsection is to design a guaranteed cost optimal control in the presence of constraints on the inputs and the states of the form:

$$U \in \mathbb{U} = \{\Phi U \leq \Delta + \Lambda X\} \tag{3.110}$$

defined in the polytopic operating domain, described by:

$$X \in \mathbb{X} = \{\mathbf{K}X \leq \tau\} \tag{3.111}$$

The main idea is to search a parallel distributed compensator (3.105) that does not need any saturation in the state space. If the initial state is inside the ball $\|X_0\|_2 \leq \gamma$, and the control is bounded by $\|U\|_2 \leq \delta$, the aim is to find a controller that guarantees a bound on the cost function (3.103) which is in turn bounded by some scalar λ such that:

$$J < X_0^T \mathbf{P} X_0 < \lambda \tag{3.112}$$

Considering the initial state zone, condition (3.112) holds if:

$$X_0^T \mathbf{P} X_0 < \lambda \frac{X_0^T X_0}{\gamma^2} \tag{3.113}$$

or

$$\mathbf{P} < \frac{\lambda}{\gamma^2} \mathbf{I}_{n \times n} \tag{3.114}$$

or

$$\gamma^2 \lambda^{-1} \mathbf{I}_{n \times n} < \mathbf{P}^{-1} \tag{3.115}$$

From the constraints in the control action and the PDC controller (3.105) described as $U = \mathbf{K}(\varpi)X$:

$$U^T U = X^T \mathbf{K}(\varpi)^T \mathbf{K}(\varpi) X < \delta^2 \tag{3.116}$$

Since $X^T \mathbf{P} X$ is a Lyapunov function, $X^T \mathbf{P} X < X_0^T \mathbf{P} X_0$ and therefore (3.116) holds if

$$\frac{1}{\delta^2} X^T \mathbf{K}(\varpi)^T \mathbf{K}(\varpi) X < \frac{1}{\lambda} X^T \mathbf{P} X \tag{3.117}$$

The bound condition on the control action can be expressed as the following LMI:

$$\begin{pmatrix} \mathbf{P}^{-1} & \mathbf{M}_i^T \\ \mathbf{M}_i & \lambda^{-1}\delta^2\mathbf{I}_{n\times n} \end{pmatrix} > 0 \tag{3.118}$$

where $\mathbf{M}_i = \mathbf{F}_i\mathbf{P}^{-1}$.

Merging all the conditions, the parallel distributed compensator with the best bound on the performance function that holds the constraints on the control action $|u_j| < \delta_j$ for an initial state inside $\|X_0\|_2 < \gamma$ can be obtained.

3.3 NONLINEAR CONTROL

Numerous techniques exist for the synthesis of control laws for nonlinear systems. Many nonlinear control approaches for the autonomous aircraft can be used such as [19, 50, 60, 103]:

1. Dynamic inversion,
2. Model predictive control,
3. Variable structure robust control (sliding mode),
4. Adaptive control,
5. Feedback linearization,
6. Robust control,
7. Backstepping,
8. Tracking by path curvature.

Their application is more general than that of linear control methods.

3.3.1 AFFINE FORMULATION OF AIRCRAFT MODELS

Some of these nonlinear control methods can be applied to an affine formulation of aircraft models. Affine nonlinear control systems are defined as:

$$\dot{X} = f(X) + \sum_{i=1}^{m} g_i(X)U_i = f(X) + \mathbf{G}U \tag{3.119}$$

$$Y_j = h_j(X), j = 1, \dots, m \qquad \text{or} \qquad Y = h(X) \tag{3.120}$$

where X is the state of the aircraft, U is the control input, Y is the measured output and $f, g_i, i = 1, \dots, m$ are smooth vector fields on $\mathbb{M}$. The constraint set contains an open set of the origin in $\mathbb{R}^m$. Thus $U \equiv 0$ is an admissible control resulting in trajectories generated by the vector field f. The vector field f is usually called the drift vector field and the g_i the control vector fields.

Definition 3.9. ***Lie Bracket:*** *The* ***Lie bracket*** *is defined as*

$$[f, g] = \frac{\partial g}{\partial X} f - \frac{\partial f}{\partial X} g \tag{3.121}$$

while the adjoint notation is given as:

$$\begin{aligned} ad_f^0 g &= g \\ ad_f^1 g &= [f, g] \\ &\vdots \\ ad_f^i g &= \left[f, ad_f^{i-1} g\right] \end{aligned} \tag{3.122}$$

Definition 3.10. ***Lie Derivative:*** *If the system (3.119) is assumed to have the same number of inputs and outputs, the kth Lie derivative of $h_j(X)$ along $f(X)$ is*

$$L_f h_j(X) = \frac{dh_j(X)}{dX} f(X) \tag{3.123}$$

$$L_f^v h_j(X) = \frac{dL_f^{v-1} h_j(X)}{dX} f(X) \qquad \forall v = 2, \dots, k \tag{3.124}$$

Similarly with $1 \leq i \leq m, 1 \leq j \leq m, k \geq 1$, define

$$L_{g_i} L_f^k h_j(X) = \frac{dL_f^k h_j(X)}{dX} g_i(X) \tag{3.125}$$

To compute the normal form of this affine system (3.119)–(3.120), the notion of a vector relative degree of the system is important.

Definition 3.11. ***Vector Relative Degree:*** *The system (3.119) is said to have vector relative degree $(r_1, \dots, r_m)$ at a point $X_0 \in \mathbb{R}^n$ if the following two properties hold:*

1. *The existence of* **the Lie derivatives** *for $X \in$ Neighborhood of X_0*

$$L_{g_i} L_f^k h_j(X) = 0 \qquad \forall 1 \leq i \leq m, 1 \leq k < r_i, 1 \leq j \leq m \tag{3.126}$$

2. *The $m \times m$ matrix*

$$\mathbf{A}(X) = \begin{pmatrix} L_{g_1} L_f^{r_1-1} h_1(X) & \dots & L_{g_m} L_f^{r_1-1} h_1(X) \\ \vdots & \dots & \vdots \\ L_{g_1} L_f^{r_m-1} h_m(X) & \dots & L_{g_m} L_f^{r_m-1} h_m(X) \end{pmatrix} \tag{3.127}$$

is non singular at $X = X_0$.

Remark 3.11. *The above state-space model (3.119) can be represented as an equation graph where the nodes represent the state variables X, the input variables U and the output variables Y are the edges representing the interdependencies among the process variables [29]:*

1. *There is an edge from a node X_k to a node X_f if $\frac{\partial f_f(X)}{\partial X_k} \neq 0$.*
2. *There is an edge from a node U_i to a node X_f if $g_{if}(X) \neq 0$.*

3. There is an edge from a node X_k to a node Y_j if $\frac{\partial h_j(X)}{\partial X_k} \neq 0$.

Based on the rules of placing edges, only the structural forms of f, g_i, h_j are required to construct an equation graph. A path is an open walk of nodes and edges of a graph such that no node is repeated. An ***input-to-output path*** *(IOP) is defined as a path which starts from an input variable node and terminates at an output variable node of the equation graph. The length of a path is the number of edges contained in the path. Once the equation graph is constructed all the input-to-output paths can be easily identified using the algorithm. Then alternative single-input-single-output multi-loop control configurations can be generated by choosing input-to-output path such that any input/output variable node is included in only one input-to-output path.*

Given all the alternative control configurations, they must be evaluated in order to select the one to be employed. One possible criterion that can be used for evaluation is structural coupling, i.e., a coupling among the process variables based on their structural inter-dependencies. This can be captured through the concept of relative degree.

In an equation graph, the relative degree r_{ij} is related to l_{ij}, the length of the shortest ***input-to-output path*** *connecting u_i and y_j as follows:*

$$r_{ij} = l_{ij} - 1 \tag{3.128}$$

Since the relative degree quantifies how direct the effect of the input U_i is on the output Y_j, it can be used as a measure of structural coupling. A ***relative degree matrix (RDM)*** *has as element i, j the relative degree between the input U_i and the output Y_j for the process described by equation (3.119). For a square system, it is possible to search over all the possible pairings to identify m input-output pairings such that the* off-diagonal *(with respect to the rearranged outputs so that the input-output pairs fall on the diagonal) elements are maximized. For non square systems with m inputs and q outputs ($m < q$), in this case, any m-columns of the relative degree matrix can be taken to form an $m \times m$ submatrix. Then the procedure mentioned above can be followed, to identify suboptimal I-O pairings for each combination of the columns and select the one with the maximum sum of the off-diagonal elements. In this case, the relative degrees can be classified into three categories:*

1. *Diagonal:* $r_d = \{r_{ij} \in \mathbb{R}, |i = j\}$
2. *Off-Diagonal:* $r_o = \{r_{ij} \in \mathbb{R}, |i \neq j\} \quad 1 \leq i, j \leq min(m, q)$
3. *Remaining:* $r_r = r - (r_d \cup r_0)$

where $r = \{r_{ij} \in \mathbb{R}\} \, |\forall i, j$.

The problem of identifying the optimal input-output pairings is equivalent to looking for r_o, such that the sum of individual elements of r_0 is maximized.

Almost all of the information in the Lie groups is contained in its Lie algebra and questions about systems evolving on Lie groups could be reduced to their

Lie algebras. Notions such as accessibility, controllability and observability can thus be expressed [51].

Definition 3.12. ***Controllability:*** *The system* (3.119) *to* (3.120) *is said to be controllable if for any two points* X_0 *and* X_f *in* $\mathbb{X}$ *there exists an admissible control* $U(t)$ *defined on some time interval* $[0, T]$ *such that the system* (3.119) *to*(3.120) *with initial condition* X_0 *reaches the point* X_f *in time* T.

Remark 3.12. *Some of the early works on nonlinear controllability were based on linearization of nonlinear systems. It was observed that if the linearization of a nonlinear system at an equilibrium point is controllable, the system itself is locally controllable [13].*

A lot of interesting control theoretic information is contained in the Lie brackets of these vector fields. Chow's theorem leads to the controllability of systems without drift. Considering the affine nonlinear model with drift

$$\dot{X} = f(X) + \sum_{i=1}^{m} g_i(X) U_i \qquad U \in \mathbb{U} \subset \mathbb{R}^m \qquad X \in \mathbb{X} \subset \mathbb{R}^n \tag{3.129}$$

Theorem 3.6

In terms of the **accessibility algebras**, system (3.129) is locally accessible if and only if $dim\,(L(X)) = n, \forall X \in \mathbb{X}$ and locally strongly accessible if $dim\,(L_0(X)) = n, \forall X \in \mathbb{X}$. ■

This condition is often referred to as the **Lie algebra rank condition** (LARC). If system (3.129) satisfies this rank condition, then it is locally accessible.

Definition 3.13. ***Small Time Local Controllability:*** *The system* (3.129) *is small time locally controllable from* X_0*, if* X_0 *is an interior point of* $Neighborhood(X_0)$*, for any* $T > 0$.

Studying controllability of general systems with drift is usually a hard problem [87, 88].

Theorem 3.7

Assuming the vector fields in the set $\{f, g_1, \dots, g_m\}$ are assumed to be real and analytic and the input vector fields g_i are linearly independent of each other, if the drift term $f(X) \neq 0$ is bounded and the vectors $(ad_f)^k (g_i)(X), \forall i \in$

$\{1,\ldots,m\}, k \in \{0,1,\ldots,\}$ together with the vectors $[g_i, g_j](X)$ for all pairs $\forall i,j \in \{1,\ldots,m\}$ span $T_X\mathbb{X}$, then system (3.129) is small time locally controllable from X if the input controls are sufficiently large, i.e., with controls λU where $U \in \mathbb{U}, |U_i| \leq 1, i = 1,\ldots,m$ for some large scalar $\lambda > 0$. ■

The proof of this theorem can be found in [25].

Depending on the chosen formulation two aircraft affine systems can be formulated: the first one being without drift and the second one with drift.

3.3.1.1 Affine formulation without drift

The following kinematic equations can be considered, when wind is neglected:

$$\begin{aligned}
\dot{x} &= V\cos\gamma\cos\chi \\
\dot{y} &= V\cos\gamma\sin\chi \\
\dot{z} &= -V\sin\gamma \\
\dot{\chi} &= \omega_1 \\
\dot{\gamma} &= \omega_2
\end{aligned} \tag{3.130}$$

where the state variable is $X = (x, y, z, \chi, \gamma)^T$, the input variable being $U = (V, \omega_1, \omega_2)^T$ and the state equations as:

$$\dot{X} = \mathbf{G}(X)U = \begin{pmatrix} \cos\gamma\cos\chi & 0 & 0 \\ \cos\gamma\sin\chi & 0 & 0 \\ -\sin\gamma & 0 & 0 \\ 0 & 1 & 0 \\ 0 & 0 & 1 \end{pmatrix} U \tag{3.131}$$

The Lie brackets of model (3.130) can be calculated as

$$g_4 = [g_1, g_2] = \begin{pmatrix} \cos\gamma\sin\chi \\ -\cos\gamma\cos\chi \\ 0 \\ 0 \\ 0 \end{pmatrix} \qquad g_5 = [g_1, g_3] = \begin{pmatrix} \cos\chi\sin\gamma \\ \sin\gamma\sin\chi \\ \cos\gamma \\ 0 \\ 0 \end{pmatrix} \tag{3.132}$$

The system is not symmetric as $0 < V_{stall} \leq V \leq V_{max}$.

Remark 3.13. *System (3.131) is a regular system with degree of nonholonomy* $= 2$ *as span* $\{g_1, g_2, g_3, [g_1, g_2], [g_1, g_3], [g_2, g_3]\} = \mathbb{R}^5$.

3.3.1.2 Affine formulation with drift

3.3.1.2.1 *Kinematic formulation*

The following equations are considered:

$$\begin{aligned}
\dot{x} &= V\cos\gamma\cos\chi + W_N \\
\dot{y} &= V\cos\gamma\sin\chi + W_E \\
\dot{z} &= -V\sin\gamma + W_D \\
\dot{V} &= U_1 \\
\dot{\chi} &= U_2 \\
\dot{\gamma} &= U_3
\end{aligned} \tag{3.133}$$

or equivalently

$$\dot{X} = f(X) + \mathbf{G}U + D_w = \begin{pmatrix} V\cos\gamma\cos\chi \\ V\cos\gamma\sin\chi \\ -V\sin\gamma \\ 0 \\ 0 \\ 0 \end{pmatrix} + \begin{pmatrix} 0 & 0 & 0 \\ 0 & 0 & 0 \\ 0 & 0 & 0 \\ 1 & 0 & 0 \\ 0 & 1 & 0 \\ 0 & 0 & 1 \end{pmatrix} U + \begin{pmatrix} W_N \\ W_E \\ W_D \\ 0 \\ 0 \\ 0 \end{pmatrix} \tag{3.134}$$

The wind represents the disturbance D_w.

When the accessibility is analyzed using LARC, the following conditions must be verified:

$$V \neq -\left(W_N\cos\gamma\cos\chi + W_E\cos\gamma\sin\chi - W_D\sin\gamma\right)$$

$\gamma \neq \frac{\pi}{2}$, $|W| < V$ and $0 < V_{stall} \leq V \leq V_{max}$.

3.3.1.2.2 *Kino-dynamic formulation in a coordinated flight*

The following dynamics are used for this derivation:

$$\begin{aligned}
\dot{x} &= V\cos\chi\cos\gamma + W_N \\
\dot{y} &= V\sin\chi\cos\gamma + W_E \\
\dot{z} &= -V\sin\gamma + W_D \\
\dot{V} &= -g\sin\gamma + \frac{T\cos\alpha - D}{m} + \\
&\quad -\left(\dot{W}_N\cos\gamma\cos\chi + \dot{W}_E\cos\gamma\sin\chi - \dot{W}_D\sin\gamma\right) \\
\dot{\chi} &= \frac{(L + T\sin\alpha)\sin\sigma}{mV\cos\gamma} + \left(\frac{\dot{W}_N\sin\chi - \dot{W}_E\cos\chi}{V\cos\gamma}\right) \\
\dot{\gamma} &= \frac{1}{mV}\left((L + T\sin\alpha)\cos\sigma - mg\cos\gamma\right) + \\
&\quad -\frac{1}{V}\left(\dot{W}_N\sin\gamma\cos\chi + \dot{W}_E\sin\gamma\sin\chi + \dot{W}_D\cos\gamma\right)
\end{aligned} \tag{3.135}$$

where $D = \frac{1}{2}\rho V^2 S C_D$. As the controls T, α, σ appear in a non-affine mode, the virtual control $U = (\dot{T}, \dot{\alpha}, \dot{\sigma})^T$ is used in this derivation to obtain an affine system. This is a dynamic extension. The following notations are used: $X_1 = \begin{pmatrix} x \\ y \\ z \end{pmatrix}$, $X_2 = \begin{pmatrix} V \\ \chi \\ \gamma \end{pmatrix}$, $X_3 = \begin{pmatrix} T \\ \alpha \\ \sigma \end{pmatrix}$ and the whole state variable is $X = \begin{pmatrix} X_1 \\ X_2 \\ X_3 \end{pmatrix}$; the affine model without wind can be written under the following form

$$\begin{aligned} \dot{X}_1 &= f_0(X_2) \\ \dot{X}_2 &= f_1(X_2) + f_2(X_2, X_3) \\ \dot{X}_3 &= U \end{aligned} \tag{3.136}$$

where

$$f_0(X_2) = \begin{pmatrix} V\cos\chi\cos\gamma \\ V\sin\chi\cos\gamma \\ -V\sin\gamma \end{pmatrix}, \quad f_1(X_2) = \begin{pmatrix} -g\sin\gamma \\ 0 \\ -\frac{g}{V}\cos\gamma \end{pmatrix}$$

and $f_2(X_2, X_3) = \begin{pmatrix} \frac{T\cos\alpha - D}{m} \\ \frac{(L+T\sin\alpha)\sin\sigma}{mV\cos\gamma} \\ \frac{1}{mV}\left((L+T\sin\alpha)\cos\sigma\right) \end{pmatrix}$.

The complete affine control system can be written as:

$$\dot{X} = f(X) + \mathbf{G}U + D_w \tag{3.137}$$

with the drift

$$f(X) = \begin{pmatrix} V\cos\chi\cos\gamma \\ V\sin\chi\cos\gamma \\ -V\sin\gamma \\ -g\sin\gamma + \frac{T\cos\alpha - D}{m} \\ \frac{1}{mV\cos\gamma}\left((L+T\sin\alpha)\sin\sigma\right) \\ \frac{1}{mV}\left((L+T\sin\alpha)\cos\sigma - mg\cos\gamma\right) \\ 0 \\ 0 \\ 0 \end{pmatrix} \tag{3.138}$$

and the constant matrix

$$\mathbf{G} = \begin{pmatrix} 0 & 0 & 0 \\ 0 & 0 & 0 \\ 0 & 0 & 0 \\ 0 & 0 & 0 \\ 0 & 0 & 0 \\ 0 & 0 & 0 \\ 1 & 0 & 0 \\ 0 & 1 & 0 \\ 0 & 0 & 1 \end{pmatrix} \tag{3.139}$$

The wind effect is considered as a perturbation:

$$D_w = \begin{pmatrix} W_N \\ W_E \\ W_D \\ -\left(\dot{W}_N \cos\gamma \cos\chi + \dot{W}_E \cos\gamma \sin\chi - \dot{W}_D \sin\gamma\right) \\ \left(\dfrac{\dot{W}_N \sin\chi - \dot{W}_E \cos\chi}{V \cos\gamma}\right) \\ -\frac{1}{V}\left(\dot{W}_N \sin\gamma \cos\chi + \dot{W}_E \sin\gamma \sin\chi + \dot{W}_D \cos\gamma\right) \\ 0 \\ 0 \\ 0 \end{pmatrix} \tag{3.140}$$

As the wind disturbance does not enter through the same input channel as the control, this uncertainty is said to be unmatched.

To check the accessibility the following Lie brackets are computed:

$$g_4 = [f, g_1] = \begin{pmatrix} 0 \\ 0 \\ 0 \\ \dfrac{1}{m}(\cos\alpha \sin\sigma) \\ \dfrac{1}{mV \cos\gamma}(\sin\alpha \sin\sigma) \\ \dfrac{1}{mV}(\cos\sigma) \\ 0 \\ 0 \\ 0 \end{pmatrix} \tag{3.141}$$

$$g_5 = [f, g_2] = \begin{pmatrix} 0 \\ 0 \\ 0 \\ -\dfrac{1}{m}\left(T \sin\alpha - \dfrac{\partial D}{\partial \alpha}\right) \\ \dfrac{\sin\sigma}{mV \cos\gamma}\left(T \cos\alpha + \dfrac{\partial L}{\partial \alpha}\right) \\ \dfrac{\cos\sigma}{mV}\left(T \cos\alpha + \dfrac{\partial L}{\partial \alpha}\right) \\ 0 \\ 0 \\ 0 \end{pmatrix} \tag{3.142}$$

and

$$g_6 = [f, g_3] = \begin{pmatrix} 0 \\ 0 \\ 0 \\ \dfrac{\cos\sigma}{mV\cos\gamma}(T\sin\alpha + L) \\ -\dfrac{\sin\sigma}{mV}(T\sin\alpha + L) \\ 0 \\ 0 \\ 0 \end{pmatrix} \tag{3.143}$$

Two other Lie brackets should be calculated in the second order. Then the LARC condition should be checked for accessibility. This would give some additional conditions on the wind.

3.3.1.3 Properties

Some additional properties are necessary for the analysis of these affine models.

Definition 3.14. ***Stabilizability:*** *Let X_0 be an equilibrium point of the control system $\dot{X} = f(X,U)$. The system is said to be asymptotically stabilizable at X_0 if a feedback control $U(X)$ can be found that renders the closed-loop system asymptotically stable.*

Definition 3.15. ***Robust Global Uniform Asymptotic Stability:*** *Let Ω be a compact subset of $\mathbb{R}^n$; suppose that the control input U is chosen as $U = k(X)$. Then the solutions of the closed-loop system for system* (3.129) *are robustly globally uniformly asymptotically stable with respect to Ω (**RGUAS-Ω**) when there exists a regular function β such that for any initial condition $X_0 \in \mathbb{R}^n$, all solutions $X(t)$ of the closed system starting from X_0 exist for all $t \geq 0$ and satisfy $\|X(t)\|_\Omega \leq \beta(\|X_0\|_\Omega, t), \forall t \geq 0$.*

Definition 3.16. ***Robust Stabilizability:*** *The system* (3.129) *is robustly stabilizable (RS) when there exists a control law $U = k(X)$ and a compact set $\Omega \subset \mathbb{R}^n$ such that the solutions of the closed-loop system are robustly globally uniformly asymptotically stable with respect to Ω.*

Definition 3.17. ***Input to State Stability:*** *The system* (3.119) *is input to state stable if there exist comparison functions γ_1 and γ_2 such that $\forall(X_0, U)$, the unique solution X is such that*

$$\|X(t)\| \leq \gamma_1(t, \|X_0\|) + \gamma_2\left(sup_{s\in[0,t]}\|U(s)\|\right) \qquad t \geq 0 \tag{3.144}$$

or

$$\|X(t)\| \leq max\left\{\gamma_1(t, \|X_0\|), \gamma_2\left(sup_{s\in[0,t]}\|u(s)\|\right)\right\} \qquad t \geq 0 \tag{3.145}$$

If system (3.119) is input to state stable, then it has the **global asymptotic stability (GAS)**, **bounded input bounded state (BIBS)** and **convergent input convergent state (CICS)** properties [38]. **Input to state stability** admits a characterization in terms of a Lyapunov-like function.

Theorem 3.8

System (3.119) is **input to state stable** if there exists a smooth function $\tilde{V} : \mathbb{R}^n \times \mathbb{R} \to \mathbb{R}$ and comparison functions $\alpha_1, \alpha_2, \alpha_3, \alpha_4$, such that:

$$\alpha_1 (\|z\|) \leq \tilde{V}(z) \leq \alpha_2 (\|z\|) \tag{3.146}$$

and

$$\nabla \tilde{V}(z)^T g(z, \tilde{V}) \leq -\alpha_3 (\|z\|) + \alpha_4 (\|v\|) \tag{3.147}$$

■

The relation between **closures, interiors and boundaries** is investigated in [27] for the systems generated by square integrable controls.

3.3.1.4 Decoupling by feedback linearization

The **change of coordinates** is often used to transform nonlinear systems in special forms that make easier the interpretation of structural properties. This is the starting point for the solution of many control problems such as feedback linearization, state and output feedback stabilization, output regulation and tracking by system inversion [31, 53].

The principle of this feedback linearization approach is explained for the following generic system:

$$\begin{array}{c} \dot{X} = f(X) + \mathbf{G}(X)U \\ Y = h(X) \end{array} \tag{3.148}$$

The principle is to derive each output Y_i till the input U appears. Assuming that each output Y_i must be derived ρ_i times, the following relations are obtained:

$$Y_i = h_i(X) \tag{3.149}$$

$$\dot{Y}_i = \frac{\partial h_i}{\partial X}\dot{X} = \frac{\partial h_i}{\partial X}\left(f(X) + \mathbf{G}(X)U\right) \tag{3.150}$$

$$\vdots$$

$$\dot{Y}_i^{(\rho_i)} = \frac{\partial h_i^{(\rho_i - 1)}}{\partial X}\dot{X} = \frac{\partial h_i^{(\rho_i)}}{\partial X}\left(f(X) + \mathbf{G}(X)U\right) = h_i^{(\rho_i)} + g_i^{(\rho_i)}U \tag{3.151}$$

The matrix form can then be written:

$$\begin{pmatrix} Y_1^{(\rho_1)} \\ Y_2^{(\rho_2)} \\ \vdots \\ Y_m^{(\rho_3)} \end{pmatrix} = \begin{pmatrix} h_1^{(\rho_1)} \\ h_2^{(\rho_2)} \\ \vdots \\ h_m^{(\rho_3)} \end{pmatrix} + \begin{pmatrix} g_1^{(\rho_1)} \\ g_2^{(\rho_2)} \\ \vdots \\ g_m^{(\rho_3)} \end{pmatrix} U \tag{3.152}$$

If the matrix $\begin{pmatrix} g_1^{(\rho_1)} \\ g_2^{(\rho_2)} \\ \vdots \\ g_m^{(\rho_3)} \end{pmatrix}$ is invertible then the control law can be calculated as:

$$U = \begin{pmatrix} g_1^{(\rho_1)} \\ g_2^{(\rho_2)} \\ \vdots \\ g_m^{(\rho_3)} \end{pmatrix}^{-1} \left(R - \begin{pmatrix} h_1^{(\rho_1)} \\ h_2^{(\rho_2)} \\ \vdots \\ h_m^{(\rho_3)} \end{pmatrix} \right) \tag{3.153}$$

The closed-loop system can thus be written as:

$$Y_i^{(\rho_i)} = R_i \qquad i = 1, \dots, m \tag{3.154}$$

Remark 3.14. *If the sum of the characteristic indexes is less than the degree of the system, then a non observable subsystem has been generated.*

A direct application of decoupling for the aircraft model, for a small angle of attack, is shown next:

$$\dot{V} = \frac{T - D}{m} - g \sin\gamma \tag{3.155}$$

$$\dot{\chi} = g \frac{n_z \sin\phi}{V \cos\gamma} \tag{3.156}$$

$$\dot{\gamma} = \frac{g}{V} (n_z \cos\phi - \cos\gamma) \tag{3.157}$$

Assuming perfect modeling, aircraft dynamics can be feedback linearized with the following control law [34]

$$T = K_v(V_d - V) + mg \sin\gamma + D \tag{3.158}$$

$$n_z \cos\phi = \frac{V}{g} K_\gamma(\gamma_d - \gamma) + \cos\gamma = c_1 \tag{3.159}$$

$$n_z \sin\phi = \frac{V}{g} K_\chi(\chi_d - \chi) \cos\gamma = c_2 \tag{3.160}$$

where V_d, γ_d, χ_d are, respectively, the references of the linear velocity, the flight path angle and the heading angle. The resulting linear system becomes

$$\dot{V} = K_v(V_d - V) \tag{3.161}$$

$$\dot{\chi} = K_\chi(\chi_d - \chi) \tag{3.162}$$

$$\dot{\gamma} = K_\gamma(\gamma_d - \gamma) \tag{3.163}$$

This controller guarantees convergence to zero of the tracking error.

A robust nonlinear control scheme is proposed in [92] for a nonlinear multi-input multi-output (MIMO) system subject to bounded time-varying uncertainty which satisfies a certain integral quadratic constraint condition. The scheme develops a robust feedback linearization approach which uses a standard feedback linearization approach to linearize the nominal nonlinear dynamics of the uncertain nonlinear aircraft and linearizes the nonlinear time-varying uncertainties at an arbitrary point using the mean value theorem. This approach transforms an uncertain nonlinear MIMO system into an equivalent linear uncertain aircraft model with unstructured uncertainty. Finally, a robust mini-max linear quadratic Gaussian control design is proposed for the linearized model. This scheme guarantees the internal stability of the closed-loop system.

3.3.2 INPUT/OUTPUT LINEARIZATION

The aircraft model is expressed by its kinematics:

$$\begin{aligned} \dot{x} &= V\cos\gamma\cos\chi \\ \dot{y} &= V\cos\gamma\sin\chi \\ \dot{z} &= -V\sin\gamma \\ \dot{\chi} &= \omega_1 \\ \dot{\gamma} &= \omega_2 \end{aligned} \tag{3.164}$$

where the input $U = (V, \omega_1, \omega_2)^T$ is used.

The following output is chosen, with the parameter $b \neq 0$

$$\begin{aligned} Y_1 &= x + b\cos\gamma\cos\chi \\ Y_2 &= y + b\cos\gamma\sin\chi \\ Y_3 &= z - b\sin\gamma \end{aligned} \tag{3.165}$$

Differentiating,

$$\begin{pmatrix} \dot{Y}_1 \\ \dot{Y}_2 \\ \dot{Y}_3 \end{pmatrix} = \begin{pmatrix} \cos\gamma\cos\chi & -b\cos\gamma\sin\chi & -b\sin\gamma\cos\chi \\ \cos\gamma\sin\chi & b\cos\gamma\cos\chi & -b\sin\gamma\sin\chi \\ -\sin\gamma & 0 & -b\cos\gamma \end{pmatrix} \begin{pmatrix} V \\ \dot{\chi} \\ \dot{\gamma} \end{pmatrix} \tag{3.166}$$

or

$$\begin{pmatrix} \dot{Y}_1 \\ \dot{Y}_2 \\ \dot{Y}_3 \end{pmatrix} = \mathbf{R}(\chi, \gamma) \begin{pmatrix} U_1 \\ U_2 \\ U_3 \end{pmatrix} \tag{3.167}$$

Calculating the inverse of this matrix $\mathbf{R}(\chi, \gamma)$, as its determinant is equal to $-b^2 \cos\gamma$, gives

$$\begin{aligned} \dot{Y}_1 &= U_1 \\ \dot{Y}_2 &= U_2 \\ \dot{Y}_3 &= U_3 \end{aligned} \tag{3.168}$$

$$V = \dot{x}\cos\gamma\cos\chi + \dot{y}\cos\gamma\sin\chi - \dot{z}\sin\gamma$$

$$\dot{\chi} = \tfrac{1}{b\cos\gamma}\left(-\sin\chi U_1 + \cos\chi U_2\right) \tag{3.169}$$

$$\dot{\gamma} = \tfrac{-\cos\chi\sin\gamma}{b}U_1 + \tfrac{-\sin\chi\sin\gamma}{b}U_2 + \tfrac{-\cos\gamma}{b}U_3$$

An input-output linearization has been obtained as long as $b \neq 0$ and $\gamma \neq \frac{\pi}{2}$.

A simple linear controller of the form:

$$\begin{aligned} U_1 &= \dot{Y}_{1d} + k_1(Y_{1d} - Y_1) \\ U_2 &= \dot{Y}_{2d} + k_2(Y_{2d} - Y_2) \\ U_3 &= \dot{Y}_{3d} + k_3(Y_{3d} - Y_3) \end{aligned} \tag{3.170}$$

guarantees exponential convergence to zero of the Cartesian tracking error, with decoupled dynamics on its three components. The flight path angle and the heading whose evolutions are governed by relations (3.169) are not controlled.

3.3.3 DYNAMIC INVERSION

Advanced flight control systems have been under investigation for decades. However, few advanced techniques have been used in practice. Possibly the most studied method of nonlinear flight control is dynamic inversion. The problem of inversion or exact tracking is to determine the input and initial conditions required to make the output of the aircraft exactly track the given reference. The main idea behind this is to create a detailed nonlinear model of an aircraft's dynamics, invert the model and impose desired handling dynamics to solve for the required actuator commands. Clearly the drawback of this method is that the nonlinear model must be very close to the actual aircraft for the system to be effective.

The application of nonlinear dynamic inversion is demonstrated on the nonlinear single-input-single-output system

$$\begin{aligned} \dot{X} &= f(X) + \mathbf{G}(X)U \\ Y &= h(X) \end{aligned} \tag{3.171}$$

The output to be controlled is differentiated until the input appears explicitly in the expression. Thus:

$$\dot{Y} = \frac{\partial h}{\partial X}\dot{X} = \frac{\partial h}{\partial X}f(X) + \frac{\partial h}{\partial X}\mathbf{G}(X)U \tag{3.172}$$

Assuming that the inverse of $\frac{\partial h}{\partial X}\mathbf{G}(X)$ exists for all $X \in \mathbb{R}^n$, the use of the state feedback

$$U = \left(\frac{\partial h}{\partial X}\mathbf{G}(X)\right)^{-1}\left(U_d - \frac{\partial h}{\partial X}f(X)\right) \tag{3.173}$$

results in an integrator response from the external input U_d to the output Y [63].

As a direct application, the following derivations are presented, using the affine formulation of the kino-dynamic model of the aircraft.

$$\begin{aligned} \dot{X}_1 &= f_0(X_2) \\ \dot{X}_2 &= f_1(X_2) + f_2(X_2, X_3) \\ \dot{X}_3 &= U \end{aligned} \tag{3.174}$$

where $U = (\dot{T}, \dot{\alpha}, \dot{\sigma})^T$ and

$$f_0(X_2) = \begin{pmatrix} V\cos\chi\cos\gamma \\ V\sin\chi\cos\gamma \\ -V\sin\gamma \end{pmatrix} \qquad f_1(X_2) = \begin{pmatrix} -g\sin\gamma \\ 0 \\ -\frac{g}{V}\cos\gamma \end{pmatrix} \tag{3.175}$$

$$f_2(X_2, X_3) = \begin{pmatrix} \frac{T\cos\alpha - D}{m} \\ \frac{(L+T\sin\alpha)\sin\sigma}{mV\cos\gamma} \\ \frac{1}{mV}\left((L+T\sin\alpha)\cos\sigma\right) \end{pmatrix} \tag{3.176}$$

The output chosen is $Y = X_1$. Differentiating this relation, the following equations are obtained

$$\begin{aligned} \dot{Y} &= f_0(X_2) \\ \ddot{Y} &= \mathbf{B}_1\left(f_1(X_2) + f_2(X_2, X_3)\right) \\ Y^{(3)} &= \mathbf{B}_1\left(\dot{f}_1(X_2) + \dot{f}_2(X_2, X_3)\right) + \dot{\mathbf{B}}_1\left(f_1(X_2) + f_2(X_2, X_3)\right) \\ &= f_3(X_2, X_3) + \mathbf{B}_2 U \end{aligned} \tag{3.177}$$

where

$$\mathbf{B}_1 = \begin{pmatrix} \cos\gamma\cos\chi & -V\cos\gamma\sin\chi & -V\sin\gamma\cos\chi \\ \cos\gamma\sin\chi & V\cos\gamma\cos\chi & -V\sin\gamma\sin\chi \\ -\sin\gamma & 0 & -V\cos\gamma \end{pmatrix} \tag{3.178}$$

$\mathbf{B}_2 = \mathbf{B}_1\mathbf{B}_4$ and $f_3 = \left(\dot{\mathbf{B}}_1 + \mathbf{B}_1\left(\mathbf{B}_5 + \mathbf{B}_3\right)\right)(f_1 + f_2)$ with

$$\mathbf{B}_3 = \begin{pmatrix} -\frac{1}{m}\frac{\partial D}{\partial V} & 0 & 0 \\ \frac{\sin\sigma}{mV\cos\gamma}\frac{\partial L}{\partial V} - \frac{(L+T\sin\alpha)\sin\sigma}{mV^2\cos\gamma} & 0 & \frac{(L+T\sin\alpha)\sin\sigma}{mV}\frac{\sin\gamma}{\cos^2\gamma} \\ \frac{\cos\sigma}{mV}\frac{\partial L}{\partial V} - \frac{(L+T\sin\alpha)\cos\sigma}{mV^2} & 0 & 0 \end{pmatrix} \tag{3.179}$$

$$\mathbf{B}_4 = \begin{pmatrix} \frac{\cos\alpha}{m} & \frac{1}{m}\left(-T\sin\alpha - \frac{\partial D}{\partial \alpha}\right) & 0 \\ \frac{\sin\alpha\sin\sigma}{mV\cos\gamma} & \left(\frac{\partial L}{\partial \alpha} + T\cos\alpha\right)\frac{\sin\sigma}{mV\cos\gamma} & \frac{(L+T\sin\alpha)\cos\sigma}{mV\cos\gamma} \\ \frac{\sin\alpha\cos\sigma}{mV} & \frac{\cos\sigma}{mV}\left(\frac{\partial L}{\partial \alpha} + T\sin\alpha\right) & -\frac{(L+T\sin\alpha)\sin\sigma}{mV} \end{pmatrix} \quad (3.180)$$

$$\mathbf{B}_5 = \begin{pmatrix} 0 & 0 & g\cos\gamma \\ 0 & 0 & 0 \\ \frac{g}{V^2}\cos\gamma & 0 & \frac{g}{V}\sin\gamma \end{pmatrix} \quad (3.181)$$

and

$$\dot{\mathbf{B}}_1 = \mathbf{B}_6\dot{V} + \mathbf{B}_7\dot{\chi} + \mathbf{B}_8\dot{\gamma} \quad (3.182)$$

$$\mathbf{B}_6 = \begin{pmatrix} 0 & -\cos\gamma\sin\chi & -\cos\chi\sin\gamma \\ 0 & \cos\gamma\cos\chi & -\sin\gamma\sin\chi \\ 0 & 0 & -\cos\gamma \end{pmatrix} \quad (3.183)$$

$$\mathbf{B}_7 = \begin{pmatrix} -\cos\gamma\sin\chi & -V\cos\gamma\cos\chi & V\sin\chi\sin\gamma \\ \cos\gamma\cos\chi & -V\cos\gamma\sin\chi & -V\sin\gamma\cos\chi \\ 0 & 0 & 0 \end{pmatrix} \quad (3.184)$$

$$\mathbf{B}_8 = \begin{pmatrix} -\sin\gamma\cos\chi & V\sin\gamma\sin\chi & -V\cos\chi\cos\gamma \\ -\sin\gamma\sin\chi & -V\sin\gamma\cos\chi & -V\cos\gamma\sin\chi \\ -\cos\gamma & 0 & V\sin\gamma \end{pmatrix} \quad (3.185)$$

As $\mathbf{B}_2$ is invertible, the following control law can be proposed:

$$U = \mathbf{B}_2^{-1}\left(Y_r^{(3)} - f_3(X_{2r})\right) + \mathbf{K}_v\dot{e} + \mathbf{K}_p e + \mathbf{K}_i\int e dt \quad (3.186)$$

with $e = Y - Y_r$.

Remark 3.15. *The following relation has been used for these calculations:*

$$\dot{f} = \frac{\partial f}{\partial V}\dot{V} + \frac{\partial f}{\partial \chi}\dot{\chi} + \frac{\partial f}{\partial \gamma}\dot{\gamma} + \frac{\partial f}{\partial T}\dot{T} + \frac{\partial f}{\partial \alpha}\dot{\alpha} + \frac{\partial f}{\partial \sigma}\dot{\sigma}$$

Using this approach, the input and initial conditions required to make the output of the aircraft track a given reference Y_r are determined.

3.3.4 CONTROL LYAPUNOV FUNCTION APPROACH

A **control Lyapunov function (CLF)** is a candidate Lyapunov function whose derivative can be made negative point-wise by the choice of control values for the system:

$$\dot{X} = f(X, U) \quad (3.187)$$

If f is continuous and there exists a continuous state feedback such that the point $X = 0$ is a globally asymptotically stable equilibrium of the closed-loop system, then by standard converse Lyapunov theorem, there must exist a control Lyapunov function for the preceding system.

If the function f is affine in the control variable then the existence of a control Lyapunov function is also sufficient for stabilizability via continuous state feedback.

3.3.4.1 Properties

Controller design methods based on control Lyapunov functions have attracted much attention in nonlinear control theory [57, 84]. The following input affine nonlinear system is considered:

$$\dot{X} = f(X) + \mathbf{G}(X)U \tag{3.188}$$

where $X \in \mathbb{R}^n$ is a state vector, $U \in \mathbb{R}^m$ is an input vector, $f : \mathbb{R}^n \to \mathbb{R}^n$, $\mathbf{G} : \mathbb{R}^n \times \mathbb{R}^m \to \mathbb{R}^n$ are continuous mappings and $f(0) = 0$ as presented in relations (3.137) to (3.140).

Definition 3.18. ***Control Lyapunov Function:*** *A continuously differentiable function $\tilde{V} : \mathbb{R}^+ \times \mathbb{R}^n \to \mathbb{R}$ is a control Lyapunov function for system* (3.188) *with input constraints $U \in \mathbb{U} \subset \mathbb{R}^m$ if it is positive definite, decrescent, radially unbounded in X and satisfies:*

$$inf_{U \in \mathbb{U}} \left\{ \frac{\partial \tilde{V}}{\partial t} + \frac{\partial \tilde{V}}{\partial X} \left(f(X) + \mathbf{G}(X)U \right) \right\} \leq -\Upsilon(X) \tag{3.189}$$

$\forall X \neq 0, \forall t \geq 0$ *where $\Upsilon(X)$ is a continuous positive definite function.*

If additionally

$$\dot{\tilde{V}}(X,U) = L_f\tilde{V}(X) + L_{\mathbf{G}}\tilde{V}(X)U \tag{3.190}$$

then a C^1 proper definite function $\tilde{V} : \mathbb{R}^n \to \mathbb{R}^+$ is a control Lyapunov function if and only if the following holds

$$\forall X \in \mathbb{R}^{n*}, L_{\mathbf{G}}\tilde{V} = 0 \Rightarrow L_f\tilde{V} < 0 \tag{3.191}$$

Definition 3.19. ***Small Control Property:*** *A control Lyapunov function $\tilde{V}(X)$ for system* (3.188) *is said to satisfy the small control property (SCP) if for any $\epsilon > 0$, there exists $\delta > 0$ such that $0 \neq \|X\| < \delta \Rightarrow \exists \|U\| < \epsilon$ such that $L_f\tilde{V} + L_{\mathbf{G}}\tilde{V}U < 0$.*

Theorem 3.9

A control Lyapunov function for system (3.188) satisfies the small control property if and only if for any $\epsilon > 0$, there exists $\delta > 0$ such that $\|X\| < \delta$ and $L_{\mathbf{G}}\tilde{V}(X) \neq 0 \Rightarrow L_f\tilde{V}(X) < \epsilon max_{1 \leq j \leq n}|L_{\mathbf{G}}\tilde{V}(X)|$. ■

Sector margins for robustness of the controllers are considered.

Definition 3.20. ***Sector Margin:*** *A state feedback controller* $U : \mathbb{R}^n \Rightarrow \mathbb{R}^m$ *is said to have a sector margin* $[\alpha, \beta]$ *if the origin of the closed-loop system* $\dot{X} = f(X) + \mathbf{G}(X)\phi(U(X))$ *is asymptotically stable where* $\phi(U) = (\phi_1(U), \ldots, \phi_m(U))^T$ *and each* $\phi_j(U_j)$ *is an arbitrary sector nonlinearity in* $[\alpha, \beta]$.

In order to find a control Lyapunov function with bounded input constraints, the partial derivative of $\tilde{V}$ has to be bounded [66]. Physically there exists perturbation terms in system (3.188) due to uncertainties and external disturbances. The issue of uncertainties and disturbances under the input-to-state framework is discussed in the particular case of relations (3.137) to (3.140):

$$\dot{X} = f(X) + \mathbf{G}U + D_w \tag{3.192}$$

which introduces a perturbation term $D_w \in \mathbb{R}^n$ to the nominal system (3.188), as presented in the paragraph *affine formulation with drift*.

Definition 3.21. ***Input to State Stable-Control Lyapunov Function:*** *A continuously differentiable function* $\tilde{V} : \mathbb{R}^+ \times \mathbb{R}^n \to \mathbb{R}$ *is an input to state stable-control Lyapunov function (ISS-CLF) for system* (3.192) *if it is positive-definite, decrescent, radially unbounded in* X *and there exist class* K *functions* Υ *and* ρ *such that*

$$inf_{U \in \mathbb{U}} \left\{ \frac{\partial \tilde{V}}{\partial t} + \frac{\partial \tilde{V}}{\partial X} \left(f(X) + \mathbf{G}U + D_w \right) \right\} \leq -\Upsilon(\|X\|) \tag{3.193}$$

$\forall \|X\| \geq \rho(\|d\|)$

3.3.4.2 Trajectory tracking

The kinematic equations of motion of an aircraft are given by:

$$\begin{aligned} \dot{x} &= V \cos\gamma \cos\chi \\ \dot{y} &= V \cos\gamma \sin\chi \\ \dot{z} &= -V \sin\gamma \\ \dot{\chi} &= \omega_1 \\ \dot{\gamma} &= \omega_2 \end{aligned} \tag{3.194}$$

The reference trajectory satisfies also:

$$\begin{aligned} \dot{x}_r &= V_r \cos\gamma_r \cos\chi_r \\ \dot{y}_r &= V_r \cos\gamma_r \sin\chi_r \\ \dot{z}_r &= -V_r \sin\gamma_r \\ \dot{\chi}_r &= \omega_{1r} \\ \dot{\gamma}_r &= \omega_{2r} \end{aligned} \tag{3.195}$$

Problem 3.1. *For a small initial tracking error, determine the feedback control law* (V_c, χ_c, γ_c) *so that the tracking error:*

$$(e_x, e_y, e_z, e_\chi, e_\gamma) \tag{3.196}$$

converges to a neighborhood about zero.

The following tracking error model has been obtained in chapter 2:

$$\begin{pmatrix} \dot{e}_x \\ \dot{e}_y \\ \dot{e}_z \\ \dot{e}_\chi \\ \dot{e}_\gamma \end{pmatrix} = \begin{pmatrix} \omega_1 \cos\gamma e_y - \omega_2 e_z - V + V_r \cos e_\gamma \\ -\omega_1 \cos\gamma e_x - \omega_1 \sin\gamma e_z + V_r \cos\gamma_r \sin e_\chi \\ \omega_1 \sin\gamma e_y + \omega_2 e_x - V_r \sin e_\gamma \\ \omega_{1r} - \omega_1 \\ \omega_{2r} - \omega_2 \end{pmatrix} \tag{3.197}$$

A feedforward action is used such as:

$$U_F = \begin{pmatrix} V_r \cos\gamma \cos\gamma_r \cos e_\chi + V_r \sin\gamma \sin\gamma_r \\ \omega_{1r} \\ \omega_{2r} \end{pmatrix} \tag{3.198}$$

Let the following candidate Lyapunov function be defined by:

$$\tilde{V} = \frac{1}{2}e_x^2 + \frac{1}{2}e_y^2 + \frac{1}{2}e_z^2 \tag{3.199}$$

and

$$\dot{\tilde{V}} = -Ve_x + V_r \cos e_\gamma e_x + V_r \cos\gamma_r \sin e_\chi e_y - V_r \sin e_\gamma e_z \tag{3.200}$$

by choosing

$$V = k_x e_x + V_r \cos e_\gamma \qquad k_x > 0 \tag{3.201}$$

and

$$e_\chi = \arctan\left(-k_y e_y\right) \qquad k_y > 0 \tag{3.202}$$

$$e_\gamma = \arctan\left(k_z e_z\right) \qquad k_z > 0 \tag{3.203}$$

The derivative of the candidate Lyapunov function becomes negative definite.

Theorem 3.10

Suppose $V_r > 0$, then for any initial condition $e_x(t_0), e_y(t_0), e_z(t_0)$, the tracking control error $(e_x(t), e_y(t), e_z(t), e_\chi(t), e_\gamma(t))$ of the closed-loop (3.199) to (3.203) is uniformly bounded and converges to zero. ■

The closed-loop system has good robustness property against acceptable measurement noise and unmodeled disturbances. Proof of a similar theorem for a constant altitude case can be found in [49].

3.3.4.3 Path tracking

This paragraph details a control method used to guide an autonomous aircraft in the case of tracking a predefined path $p^* = (x^*, y^*, z^*)^T$ even in the presence of bounded disturbances and model uncertainties. To fulfill this requirement, a method based on a Lyapunov controller is employed [14]. The algorithm is based on the postulation of two fictitious particles, one of them belonging to the prescribed path and the other corresponding to a particle of the autonomous aircraft.

Then considering the relative positions of these particles, the method determines:

1. The velocity and orientation that the aircraft particle must follow in order to approach the path particle.
2. The path particle position refreshment in order not to be reached by the aircraft particle.

Consider an ideal particle p without dynamics whose position and speed are, respectively, described in the N-frame by the vectors $p = (x, y, z)^T$ and $\dot{p} = (\dot{x}, \dot{y}, \dot{z})^T$. The speed vector can also be characterized by its magnitude and orientation

$$V_d = \|\dot{p}\|_2 \qquad \chi_d = \arctan\left(\frac{\dot{y}}{\dot{x}}\right) \qquad \gamma_d = \arctan\left(\frac{-\dot{z}}{\dot{x}\cos\chi + \dot{y}\sin\chi}\right) \tag{3.204}$$

The objective of this paragraph is to determine the expressions corresponding to these variables to ensure the convergence of the hypothetical ideal particle towards a desired path. The desired path geometric locus is continuously parametrized by a time-dependent scalar variable ϖ. A path particle p_p is considered in this geometric locus whose instantaneous position in the N-frame is denoted as $p_p(\varpi) = (x_p(\varpi), y_p(\varpi), z_p(\varpi))$. If the purpose is developing a search of a rectilinear target such as a pipeline or a road, then a sinusoidal trajectory can be chosen. In this case, the desired path p_p is defined by choosing:

$$x_p = A\sin(\varpi) \qquad y_p = \varpi \qquad z_p = A\cos(\varpi) \tag{3.205}$$

where A is the width of the corridor where the search is performed. The orientation of this particle is defined by its evolution on the desired path:

$$\chi_p = \arctan\left(\frac{y_p'(\varpi)}{x_p'(\varpi)}\right) \tag{3.206}$$

and

$$\gamma_p = \arctan\left(\frac{-z_p'(\varpi)}{x_p'(\varpi)\sin\chi_p + y_p'(\varpi)\cos\chi_p}\right) \tag{3.207}$$

with

$$\frac{dx_p(\varpi)}{d\varpi} = A\cos(\varpi) \qquad \frac{dy_p(\varpi)}{d\varpi} = 1 \qquad \frac{dz_p(\varpi)}{d\varpi} = -A\sin(\varpi) \tag{3.208}$$

It is useful to define an auxiliary frame attached to p_p and aligned with the path particle orientation p-frame. The angle χ_p determines the transformation matrix from the N-frame to the p-frame, local reference frame:

$$\mathbf{R}_p = \begin{pmatrix} \cos\gamma_p \cos\chi_p & -\sin\chi_p & \sin\gamma_p \cos\chi_p \\ \cos\gamma_p \sin\chi_p & \cos\chi_p & \sin\gamma_p \sin\chi_p \\ -\sin\gamma_p & 0 & \cos\gamma_p \end{pmatrix} \tag{3.209}$$

The new local frame allows to define a positional error vector ϵ composed by the along-track error s, the cross-track error e and the altitude error h: $\epsilon = (s, e, h)^T$. Then in the N-frame, this error vector can be expressed as:

$$\epsilon = \mathbf{R}_p^T (p - p_p(\varpi)) \tag{3.210}$$

Once the expression of the error vector on the $N - frame$ is obtained, the following definite Lyapunov function is defined:

$$\tilde{V} = \frac{1}{2}\epsilon^T \epsilon \tag{3.211}$$

Then, differentiating (3.211) with respect to time along the trajectories of ϵ,

$$\dot{\tilde{V}} = \epsilon^T \dot{\epsilon} = \epsilon^T \left(\dot{\mathbf{R}}_p^T (p - p_p) + \mathbf{R}_p^T (\dot{p} - \dot{p}_p) \right) \tag{3.212}$$

Thus

$$\begin{aligned} \dot{\tilde{V}} = s\left(V_d(\cos(\chi_d)\cos(\chi_p)\cos(\chi_r) + \sin\gamma_p \sin\gamma_d) - V_p\right) + eV_d \cos\gamma_d \sin(\chi_r) + \\ + hV_d \left(\sin\gamma_p \cos\gamma_d \cos\chi_r - \cos\gamma_p \sin\gamma_d\right) \end{aligned} \tag{3.213}$$

where $\chi_r = \chi_d - \chi_p$.

Finally, to ensure (3.213) to be negative, its first term should always be negative. If the following parameter is chosen:

$$\begin{aligned} V_p = V_d(\cos(\chi_d)\cos(\chi_p)\cos(\chi_r) + \sin\gamma_p \sin\gamma_d) + \tilde{\gamma} s + \\ + \tfrac{h}{s} V_d \left(\sin\gamma_p \cos\gamma_d \cos\chi_r - \cos\gamma_p \sin\gamma_d\right) \end{aligned} \tag{3.214}$$

with $\tilde{\gamma}$ an arbitrary positive gain constant, this requirement is verified.

On the other hand, looking for a negative definite expression for the second term, one of the possible options for the selection of χ_r may be:

$$\chi_r = -\arctan\left(\frac{e}{\Delta_e}\right) \tag{3.215}$$

where Δ_e is an upper bounded positive time-varying variable called look-ahead distance, used to shape the convergence of ϵ to zero. Choosing the velocity of the ideal particle as:

$$V_d = \mu\sqrt{e^2 + \Delta_e^2} \tag{3.216}$$

with $\mu > 0$ determines

$$\dot{\tilde{V}}_X = -\tilde{\gamma}s^2 - \mu e^2 < 0 \tag{3.217}$$

The necessary references can be determined that

$$u_d = V_d \cos\gamma_d \cos(\chi_p + \chi_r) \tag{3.218}$$

$$v_d = V_d \cos\gamma_d \sin(\chi_p + \chi_r) \tag{3.219}$$

$$w_d = -V_d \sin\gamma_d \tag{3.220}$$

with

$$\chi_d = \chi_p + \chi_r \tag{3.221}$$

Therefore, the error vector ϵ converges uniformly globally and exponentially to zero when the position of the ideal and path particle is governed by (3.215) to (3.216).

3.3.4.4 Circular path following in wind with input constraints

This section considers the problem of UAV following circular paths to account for winds [7]. This path-following approach takes into account roll and flight path angle constraints. The kinematics are taken as:

$$\begin{aligned} \dot{x} &= V\cos\chi\cos\gamma + W_N \\ \dot{y} &= V\sin\chi\cos\gamma + W_E \\ \dot{z} &= -V\sin\gamma + W_D \end{aligned} \tag{3.222}$$

If a coordinated turn condition is assumed and the roll and pitch dynamics are assumed much faster than the heading and altitude dynamics, respectively, which implies that the roll and flight path angles can be considered as the control variables:

$$\dot{\chi} = \frac{g}{V}\tan\phi_c \qquad \dot{z} = -V\sin\gamma_c + W_D \tag{3.223}$$

where ϕ_c, γ_c are the commanded roll angle and flight path angle with the following constraints:

$$|\phi_c| \leq \phi_{max} < \frac{\pi}{2} \qquad |\gamma_c| \leq \gamma_{max} < \frac{\pi}{2} \tag{3.224}$$

A circular path is described by an inertially referenced center $C = (C_N, C_E, C_D)^T$, a radius $\Xi \in \mathbb{R}$ and a direction $\lambda \in \{-1, +1\}$ as

$$P_{orbit}(C, \Xi, \lambda) = \left\{ r \in \mathbb{R}^3, r = C + \lambda\Xi \left(\cos\varphi, \sin\varphi, 0\right)^T, \varphi \in (0, 2\pi) \right\} \tag{3.225}$$

where $\lambda = 1$ signifies a clockwise circle and $\lambda = -1$ signifies a counterclockwise orbit. The longitudinal controller is assumed to maintain a constant altitude and airspeed. As the altitude is assumed to be constant:

$$\gamma_c = \arcsin\left(\frac{W_D}{V_a}\right) \tag{3.226}$$

with the constraint that $|W_D| \leq V_a$.

The guidance strategy is derived in polar coordinates. Let $d = \sqrt{(p_N - C_N)^2 + (p_E - C_E)^2}$ be the lateral distance from the desired center of the orbit to the UAV and let:

$$\varphi = \arctan\left(\frac{p_E - C_E}{p_N - C_N}\right) \tag{3.227}$$

be the phase angle of the relative position.

Define the wind speed W and wind direction χ_W so that

$$\begin{pmatrix} W_N \\ W_E \end{pmatrix} = W \begin{pmatrix} \cos\chi_W \\ \sin\chi_W \end{pmatrix} \tag{3.228}$$

The kinematics in polar coordinates are therefore given by:

$$\dot{d} = V\cos\gamma\cos(\chi - \varphi) + W\cos(\chi_W - \varphi) \tag{3.229}$$

$$\dot{\varphi} = \frac{V}{d}\cos\gamma\sin(\chi - \varphi) + \frac{W}{d}\sin(\chi_W - \varphi) \tag{3.230}$$

If the commanded roll angle is given by:

$$\phi^c = \left\{ \begin{array}{ll} 0 & \text{if } d < d_{min} \\ -\lambda\phi_{max} & \text{if}(d \geq d_{min}) \text{ and } (\lambda\tilde{\chi} \geq \tilde{\chi}_{max}) \\ \lambda\phi_{max} & \text{if}(d \geq d_{min}) \text{ and } (-\lambda\tilde{\chi} \geq \tilde{\chi}_{max}) \\ \phi_1 & \text{otherwise} \end{array} \right\} \tag{3.231}$$

where

$$\phi_1 = \arctan\left(\lambda\frac{V^2}{gd}\cos\gamma\cos\tilde{\chi} + \sigma_{M_3}\right) \tag{3.232}$$

where:

$$\sigma_{M_3} = \sigma_{M_1}\left(\frac{k_1\dot{\tilde{d}} + \sigma_{M_2}(\Xi)}{\lambda g\cos\tilde{\chi}\cos\gamma + g\frac{W}{V}\sin(\chi - \chi_W)}\right)$$

where $k_1 > 0$, the angles are such that $0 < \phi_{max}, \gamma_{max}, \tilde{\chi}_{max} < \frac{\pi}{2}$, the parameters d_{min} and Ξ satisfy :

$$\frac{V^2 + VW}{g \tan \phi_{max}} < d_{min} < \Xi$$

where σ_{M_i} is the ith saturation function

$$\sigma_{M_i}(U) = \left\{ \begin{array}{cc} M_i & \text{if } U > M_i \\ -M_i & \text{if } U < -M_i \\ U & \text{otherwise} \end{array} \right\} \tag{3.233}$$

The magnitude of the wind is such that :

$$W < V \cos \tilde{\chi}_{max} \cos \gamma_{max} \tag{3.234}$$

and the parameters M_1 and M_2 are given, respectively, by:

$$M_1 = \tan \phi_{max} - \frac{V^2}{d_{min} g} \cos \gamma_{max} \cos \tilde{\chi}_{max} \tag{3.235}$$

and

$$M_2 = \frac{1}{2} M_1 g \left| \cos \tilde{\chi}_{max} \cos \gamma_{max} - \frac{W}{V} \right| \tag{3.236}$$

Then $|\phi^c(t)| \leq \phi_{max}$ and $(d, \dot{d}) \rightarrow (\Xi, 0)$, the error variables being $\tilde{d} = d - \Xi$ and $\tilde{\chi} = \chi - \chi^d$, $\chi^d = \varphi + \lambda \frac{\pi}{2}$. Proof of this discussion can be found in [7].

3.3.4.5 Tracking of moving targets with wind term

The aircraft dynamics expressed relative to a moving target and incorporating a wind term can be expressed as [86]:

$$\begin{aligned} \dot{x} &= u_1 \cos \chi + W_N - V_{xT} \\ \dot{y} &= u_1 \sin \chi + W_E - V_{yT} \\ \dot{\chi} &= u_2 \end{aligned} \tag{3.237}$$

where χ is the heading, u_1 is the commanded airspeed, u_2 the commanded heading rate, with the following constraints:

$$0 < V_{stall} \leq u_1 \leq V_{max} \qquad |u_2| \leq \omega_{max} \tag{3.238}$$

$[W_N, W_E]^T$ are components of the constant wind velocity and $[V_{xT}, V_{yT}]^T$ are components of the constant inertial target velocity [81]. The wind and target velocity both affect the dynamics additively [86] and this allows to combine both effects into one variable:

$$T_x = V_{xT} - W_N \qquad T_y = V_{yT} - W_E \tag{3.239}$$

T_x, T_y are treated as unknown constants and the a priori availability of an upper bound T^* is assumed, satisfying

$$max(T_x, T_y) \leq T^* \tag{3.240}$$

which encompasses worst-case combined effect of wind and target velocities.

The relative distance of the target is $d = \sqrt{x^2 + y^2}$ and r_d is the desired standoff radius of the circular orbit. If the polar coordinates are represented by (d, ϕ) then :

$$\cos\phi = \frac{d^2 - r_d^2}{d^2 + r_d^2} \qquad \sin\phi = \frac{2dr_d}{d^2 + r_d^2} \tag{3.241}$$

The aircraft moves around the standoff circle with a constant angular velocity $\dot{\varphi} = \frac{u_0}{r_d}$ where u_0 is the constant nominal vehicle airspeed.

The following controller can be proposed:

$$\begin{aligned} u_1 \cos\chi &= -u_0 \cos(\varphi - \phi) + \hat{T}_x - v_s \sin\varphi \\ u_1 \sin\chi &= -u_0 \sin(\varphi - \phi) + \hat{T}_y + v_s \cos\varphi \end{aligned} \tag{3.242}$$

where $\hat{T}_x, \hat{T}_y, \hat{T}_z$ are adaptive estimates for the unknown wind and target motion and v_s is a yet-to-be specified signal.

This construction defines a heading given by:

$$\tan\chi = \frac{-u_0 \sin(\varphi - \phi) + \hat{T}_y + v_s \cos\varphi}{-u_0 \cos(\varphi - \phi) + \hat{T}_x - v_s \sin\varphi} \tag{3.243}$$

and an airspeed input given by:

$$u_1^2 = \left(-u_0 \cos(\varphi - \phi) + \hat{T}_x - v_s \sin\varphi\right)^2 + \left(-u_0 \sin(\varphi - \phi) + \hat{T}_y + v_s \cos\varphi\right)^2 \tag{3.244}$$

The heading given by equation (3.243) may be differentiated to obtain the heading rate input. Using relation (3.242) with (3.237) gives:

$$\dot{d} = -u_0 \frac{d^2 - r_d^2}{d^2 + r_d^2} + \tilde{T}_x \cos\varphi + \tilde{T}_y \sin\varphi \tag{3.245}$$

$$d\dot{\varphi} = -u_0 \frac{2dr_d}{d^2 + r_d^2} - \tilde{T}_x \sin\varphi + \tilde{T}_y \cos\varphi + v_s \tag{3.246}$$

where $\tilde{T}_x = \hat{T}_x - T_x$, $\tilde{T}_y = \hat{T}_y - T_y$ are the adaptive estimation errors. The Lyapunov guidance field can be used to define the perfect case relative motion:

$$\dot{d}_p = -u_0 \frac{d_p^2 - r_d^2}{d^2 + r_d^2} \qquad d_p \dot{\varphi}_p = -u_0 \frac{2d_p r_d}{d_p^2 + r_d^2} \tag{3.247}$$

The error signals are defined as :

$$e_d = d - d_p \qquad e_\varphi = \varphi - \varphi_p \tag{3.248}$$

and the corresponding error dynamics are then given by:

$$\dot{e}_d = -u_0 \frac{2r_d^2(d^2 - d_p^2)e_d}{(d^2 + r_d^2)(d^2 + d_p^2)} + \tilde{T}_x \cos\varphi + \tilde{T}_y \sin\varphi \tag{3.249}$$

$$\dot{e}_\varphi = -u_0 \frac{2r_d(d + d_p)e_d}{(d^2 + r_d^2)(d^2 + d_p^2)} + \frac{1}{d}\left(-\tilde{T}_x \sin\varphi + \tilde{T}_y \cos\varphi + v_s\right) \tag{3.250}$$

The actual and perfect guidance trajectories are defined from the same point, which implies that $d_p(0) = d(0)$ and $\varphi_p(0) = \varphi(0)$. Consequently the error signals are zero at the initial time after the heading convergence to the desired Lyapunov guidance field is achieved [86].

3.3.4.6 Suboptimal control

Optimal control of nonlinear dynamics with respect to an index of performance has also been investigated. One of the difficulties in obtaining optimal solutions for a nonlinear system is that optimal feedback control depends on the solution of the **Hamilton–Jacobi–Bellman (HJB)** equation. The Hamilton–Jacobi–Bellman equation is difficult to solve in general. Consequently, a number of papers investigated methods to find suboptimal solutions to nonlinear control problems. One such technique is the power series expansion based method. Another technique that systematically solves the nonlinear regulator problem is the **state-dependent Riccati equation** (SDRE) method. By turning the equations of motion into a linear-like structure, linear optimal control methods such as the LQR methodology and the H_∞ design technique are employed for the synthesis of nonlinear control system. The state-dependent Riccati equation method, however, needs online computation of the algebraic Riccati equation at each sample time [95]. This methodology can be applied to any of the aircraft models presented previously.

3.3.4.6.1 Optimal output transition problem

The minimum time state transition with bounds on the input magnitude leads to the classical bang-bang type input for the fastest state transition. However, the transition time can be reduced further if only the system output needs to be transitioned from one value to another rather than the entire system state. The time-optimal output transition problem is to change the system output from one initial value $Y(t) = \underline{Y}, \forall t \leq 0$ to a final value $Y(t) = \overline{Y}, \forall t \geq T_f$. The output transition problem is posed for the invertible nonlinear system.

If the following invertible affine system is considered:

$$\begin{aligned} \dot{X}(t) &= f(X(t)) + \mathbf{G}(X(t))U(t) \\ Y(t) &= h(X(t)) \end{aligned} \tag{3.251}$$

where $X(t) \in \mathbb{R}^n$ is the state, $Y(t) = (Y_1(t), Y_2(t), ..., Y_p(t))^T$ is the output, with the same number of inputs as outputs, i.e., $U(t), Y(t) \in \mathbb{R}^p$ and the input is bounded as:

$$\|U(t)\|_{\infty} \leq U_{max}, \forall t \tag{3.252}$$

Let $\underline{X}$ and $\overline{X}$ be controlled equilibrium points of the system (3.251) corresponding, respectively, to inputs $\underline{U}$ and $\overline{U}$ and outputs $\underline{Y}$ and $\overline{Y}$.

Definition 3.22. ***Delimiting States for Transitions:***

$$\begin{array}{ll} f(\overline{X}) + \mathbf{G}(\overline{X})\overline{U} = 0 & \overline{U} = h(\overline{X}) \\ f(\underline{X}) + \mathbf{G}(\underline{X})\underline{U} = 0 & \underline{Y} = h(\underline{X}) \end{array} \tag{3.253}$$

The output transition problem is formally stated next.

Definition 3.23. ***Output Transition Problem:*** *Given the delimiting states and a transition*

$$\begin{array}{c} \dot{X}_{ref} = f(X_{ref}(t)) + \mathbf{G}(X_{ref}(t))U_{ref}(t) \\ Y_{ref}(t) = h(X_{ref}(t)) \\ \|U(t)\|_{\infty} \leq U_{max} \end{array} \tag{3.254}$$

and the following two conditions:

1. ***The output transition condition:*** *the output transition in the time interval $I_T = [0, T_f]$ is maintained at the desired value outside the time interval I_T, i.e., from*

$$Y_{ref}(t) = \underline{Y}, \forall t \leq 0 \text{ to } Y_{ref}(t) = \overline{Y}, \forall t \geq T_f \tag{3.255}$$

2. ***The delimiting state condition:*** *the system state approaches the delimiting states as t varies from: $-\infty \rightarrow +\infty$*

$$X(t) \rightarrow \underline{X} \text{ as } t \rightarrow -\infty \qquad X(t) \rightarrow \overline{X} \text{ as } t \rightarrow +\infty \tag{3.256}$$

the time-optimal output transition seeks to minimize the transition time T_f with constraints on the input.

The following output is chosen, for the aircraft:

$$\begin{array}{c} Y_1 = x + b\cos\gamma\cos\chi \\ Y_2 = y + b\cos\gamma\sin\chi \\ Y_3 = z - b\sin\gamma \end{array} \tag{3.257}$$

As long as $b \neq 0$ and $\gamma \neq \pm\frac{\pi}{2}$, the system is invertible and

$$\begin{array}{c} \dot{Y}_1 = U_1 \\ \dot{Y}_2 = U_2 \\ \dot{Y}_3 = U_3 \end{array} \tag{3.258}$$

with the following constraints $U_{i_{min}} \leq U_i \leq U_{i_{max}}$.

Problem 3.2. ***Optimal Output Transition Problem:*** *The minimum time-optimal output transition problem is to find the bounded input-state trajectory U^*, X^* that satisfies the output transition problem and minimizes the transition time T_f with the cost function*

$$J = \int_0^{T_f} 1 \quad dt = T_f \tag{3.259}$$

The solution begins with the standard approach based on time optimal state transition, followed by the solution to the time-optimal output transition problem for invertible systems. The output transition problem can be solved using the time optimal state transition approach defined below:

Problem 3.3. ***Time-optimal State Transition:*** *Given an initial state X_i and a final state X_f, find a bounded input-state trajectory X_{ref}, U_{ref} that satisfies the system equation* (3.251) *and input constraints* (3.252) *in the transition interval I_T as in* (3.254) *and achieves the state transition from $X_{ref}(t) = X_i$ to $X_{ref}(T_f) = X_f$ while minimizing the transition time T_f.*

The optimal output transition approach is presented for invertible systems. As the system (3.251) was assumed to be invertible, it can be rewritten through a coordinate transformation T:

$$X(t) = T\left(\xi(t), \eta(t)\right) \tag{3.260}$$

in the following form

$$\begin{pmatrix} \frac{d^{(r_1)}}{dt^{(r_1)}} Y_1(t) \\ \frac{d^{(r_2)}}{dt^{(r_2)}} Y_2(t) \\ \cdots \\ \frac{d^{(r_p)}}{dt^{(r_p)}} Y_p(t) \end{pmatrix} = A\left(\xi(t), \eta(t)\right) + \mathbf{B}\left(\xi(t), \eta(t)\right) U(t) \tag{3.261}$$

$$\dot{\eta}(t) = \mathbf{S}_1\left(\xi(t), \eta(t)\right) + \mathbf{S}_2\left(\xi(t), \eta(t)\right) U(t) \tag{3.262}$$

where the matrix $\mathbf{B}$ is invertible in the region of interest $X(t) \in \mathbb{X} \subset \mathbb{R}^n$. The state component $\xi(t)$ represents the output and its state derivatives:

$$\xi(t) = \left[Y_1, \dot{Y}_1, \dots, Y^{(r_1)}, Y_2, \dot{Y}_2, \dots, Y^{(r_2)}, \dots, Y_p, \dot{Y}_p, \dots, Y^{(r_p)}\right]^T \tag{3.263}$$

and

$$\overline{\xi} = \left[\overline{Y}_1, 0, \dots, 0, \overline{Y}_2, 0, \dots, 0, \dots, 0, \overline{Y}_p, 0, \dots, 0\right]^T$$

$$\underline{\xi} = \left[\underline{Y}_1, 0, \dots, 0, \underline{Y}_2, 0, \dots, 0, \dots, 0, \underline{Y}_p, 0, \dots, 0\right]^T.$$

Remark 3.16. *The invertibility assumption is satisfied if the system in* (3.251) *has a well defined relative degree* $r = [r_1, r_2, \dots, r_p]$.

During the pre- and post-actuation, the output is constant. Therefore, the state component ξ is known in terms of the desired output, $\overline{\xi}, \underline{\xi}$, as in (3.263) as:

$$\xi(t) = \underline{\xi}, \forall t \leq 0 \qquad \xi(t) = \overline{\xi}, \forall t \geq T_f \tag{3.264}$$

Moreover, the input to maintain the desired constant output, during pre- and post-actuation can be found from equations (3.261) to (3.263) as

$$\begin{aligned} U(t) = \underline{U}(\eta(t)) = -\left(\mathbf{B}\left(\underline{\xi}(t), \eta(t)\right)\right)^{-1} A\left(\underline{\xi}(t), \eta(t)\right) \quad \forall t \leq 0 \\ U(t) = \overline{U}(\eta(t)) = -\left(\mathbf{B}\left(\overline{\xi}(t), \eta(t)\right)\right)^{-1} A\left(\overline{\xi}(t), \eta(t)\right) \quad \forall t \geq T \end{aligned} \tag{3.265}$$

The state dynamics during pre-actuation and post-actuation reduces to the following time-invariant internal dynamics (obtained by rewriting equation (3.262)):

$$\begin{aligned} \dot{\eta} = \mathbf{S}_1\left(\underline{\xi}, \eta\right) - \mathbf{S}_2\left(\underline{\xi}, \eta\right)\left(\mathbf{B}\left(\underline{\xi}, \eta\right)\right)^{-1} A\left(\underline{\xi}, \eta\right) \quad \forall t \leq 0 \\ \dot{\eta} = \mathbf{S}_1\left(\overline{\xi}, \eta\right) - \mathbf{S}_2\left(\overline{\xi}, \eta\right)\left(\mathbf{B}\left(\overline{\xi}, \eta\right)\right)^{-1} A\left(\overline{\xi}, \eta\right) \quad \forall t \geq T_f \end{aligned} \tag{3.266}$$

For specific output transition, the evolution of the internal dynamics (3.266) as well as the inverse input (3.265) needed to maintain a fixed output (during pre- and post-actuation) are both determined by the boundary value of the internal state:

$$\Psi = (\eta(0), \eta(T))^T \tag{3.267}$$

The set Ψ of acceptable boundary values as in (3.267) should satisfy the following three conditions:

1. **Delimiting conditions on the internal state**: the internal states should meet the delimiting state conditions in (3.256). This is satisfied provided the initial boundary condition $\eta(0)$ is on the local manifold $M_u(\underline{X})$ of the internal dynamics at $X = \underline{X}$ and the final boundary condition $\eta(T)$ is on the local stable manifold $M_s(\overline{X})$ of the internal dynamics at $X = \overline{X}$.
2. **Bounded pre- and post-actuation condition**: the inverse input in equation (3.265) needed to maintain a constant output during pre- and post-actuation should be bounded and satisfy the input constraint in equation (3.252).
3. **State transition condition**: there exists an input U that satisfies the input constraint in equation (3.252) and achieves the state transition from $X_{ref}(0) = T\left(\underline{\xi}, \eta(0)\right)$ to $X_{ref}(t) = T\left(\overline{\xi}, \eta(T_f)\right)$, for some transition time T_f while the state trajectories $X(t)$ remain in $\mathbb{R}^n$ during the transition time interval I_T.

The flexibility in the choice of the boundary values of the internal state in the optimal output transition approach can reduce the transition time further compared to the state space transition approach for invertible system [64].

3.3.4.6.2 *Incorporating a class of constraints*

A method is proposed to systematically transform a **constrained optimal control problem** (OCP) into an **unconstrained optimal control problem**, which can be treated in the standard calculus of variations. The considered class of constraints comprises up to m input constraints and m state constraints with well-defined relative degree, where m denotes the number of inputs of the given nonlinear system. Starting from an equivalent normal form representation, the constraints are incorporated into a new system dynamics by means of saturation functions and differentiation along the normal form cascade. This procedure leads to a new unconstrained optimal control problem, where an additional penalty term is introduced [26]. The following nonlinear control affine multiple system is considered:

$$\dot{X} = f(X) + \sum_{i=1}^{m} g_i(X)U_i \tag{3.268}$$

where the state $X = (x, y, z, V, \chi, \gamma, T, \alpha, \sigma)^T \in \mathbb{R}^n$ and the input $U = \left(\dot{T}, \dot{\alpha}, \dot{\sigma}\right)^T \in \mathbb{R}^m$, $f, g_i : \mathbb{R}^n \to \mathbb{R}^n, i = 1, \ldots, m$ are sufficiently smooth vector fields. The following state and input constraints are assumed:

$$C_i(X) \in \left[c_i^-, c_i^+\right], U_i \in \left[U_i^-(X), U_i^+(X)\right], i = 1, \ldots, m \tag{3.269}$$

as $\dot{T}_{min} \leq \dot{T} \leq \dot{T}_{max}$, $\dot{\alpha}_{min} \leq \dot{\alpha} \leq \dot{\alpha}_{max}$, $\dot{\sigma}_{min} \leq \dot{\sigma} \leq \dot{\sigma}_{max}$, $T_{min} \leq T \leq T_{max}$, $\alpha_{min} \leq \alpha \leq \alpha_{max}$, $\sigma_{min} \leq \sigma \leq \sigma_{max}$, $0 < V_{stall} \leq V \leq V_{max}$, $|\gamma| \leq \gamma_{max} < \frac{\pi}{2}$, $C_{L_{min}} \leq C_L \leq C_{L_{max}}$.

The vector relative degree $\{r_1, \ldots, r_m\}$ of the m functions $C_i(X)$ at a point X_0 is defined by:

$$L_{g_j} L_f^k C_i(X) = 0 \tag{3.270}$$

$\forall \quad 1 \leq j \leq m, k < r_i - 1, 1 \leq i \leq m, \forall X \in Neigh(X_0)$.

Moreover, the $m \times m$ matrix

$$\mathbf{A}(X) = \begin{pmatrix} L_{g_1} L_f^{r_1 - 1} C_1(X) & \ldots & L_{g_m} L_f^{r_1 - 1} C_m(X) \\ \vdots & \ldots & \vdots \\ L_{g_1} L_f^{r_m - 1} C_1(X) & \ldots & L_{g_m} L_f^{r_m - 1} C_m(X) \end{pmatrix} \tag{3.271}$$

has to be non singular at $X = X_0$. The m constraints (3.269) have a well-defined relative degree $\{r_1, \ldots, r_m\}$ which means that the condition (3.270) as well as the non singularity of the decoupling matrix (3.271) are satisfied in a sufficiently large neighborhood of X_0.

Problem 3.4. ***Optimal Control Problem:***

$$min \left[J(U) = \varphi(X(T)) + \int_0^T L(X, U, t)dt \right] \tag{3.272}$$

subject to

$$\dot{X} = f(X) + \sum_{i=1}^{m} g_i(X)U_i \tag{3.273}$$

$$X(0) = X_0 \qquad \varsigma(X(T)) = 0 \tag{3.274}$$

$$C_i(X) \in \left[c_i^-, c_i^+\right], U_i \in \left[U_i^-(X), U_i^+(X)\right], i = 1, \dots, m \tag{3.275}$$

Owing to the well-defined relative degree $\{r_1, \dots, r_m\}$, there exists a change of coordinates:

$$\begin{pmatrix} X \\ Z \end{pmatrix} = \begin{pmatrix} \vartheta_Y(X) \\ \vartheta_Z(X) \end{pmatrix} = \vartheta(X) \tag{3.276}$$

with $Y^T = \left(Y_1^T, \dots, Y_m^T\right)$ where $Y_i = \left(Y_{i,1}, \dots, Y_{i,r_i}, \right)^T$ defined by:

$$Y_{i,1} = C_i(X) = \vartheta_{i,1}(X), \dots, Y_{i,j} = L_f^j C_i(X) = \vartheta_{i,j}(X) \tag{3.277}$$

$\forall j = 2, \dots, r_i, i = 1, \dots, m.$

The single functions $\vartheta_{i,j}$ are comprised in $\vartheta = \left(\vartheta_{1,1}, \dots, \vartheta_{m,r_m}\right)^T$, the additional coordinates $Z = \vartheta_Z(X) \in \mathbb{R}^{n-1}$ with $r = \sum_{i=1}^{m} r_i$ if $r < n$ are necessary to complete the transformation. In these coordinates, the original optimal control problem 3.4 can be stated under the following form.

Problem 3.5. ***Optimal Control Problem:***

$$min \left[\overline{J}(U) = \overline{\varphi}(Y(T), Z(T)) + \int_0^T \overline{L}(Y, Z, U, t)dt \right] \tag{3.278}$$

subject to

$$\dot{Y}_{i,j} = Y_{i,j+1} \quad j = 1, \dots, r_{i-1} \tag{3.279}$$

$$\dot{Y}_{i,r_i} = a_{i,0}(Y, Z) + \sum_{i=1}^{n} a_{i,j}(Y, Z)U_j \qquad \forall i = 1, \dots, m \tag{3.280}$$

$$\dot{Z} = b_0(Y, Z) + \mathbf{B}(Y, Z)U \tag{3.281}$$

$$Y(0) = \vartheta_y(X_0), \overline{\chi}(Y(T), Z(T)) = 0 \tag{3.282}$$

$$Y_{i,1} \in \left[c_i^-, c_i^+\right], U_i \in \left[\overline{U}_i^-(Y, Z), \overline{U}_i^+(Y, Z)\right], i = 1, \dots, m \tag{3.283}$$

where

$$\begin{aligned} a_{i,0} &= L_f^{r_i} C_i(X) \circ \vartheta^{-1} \\ a_{i,j} &= L_{g_i} L_f^{r_i - 1} C_i(X) \circ \vartheta^{-1} \\ \overline{\varphi} &= \varphi \circ \vartheta^{-1} \\ \overline{L} &= L \circ \vartheta^{-1} \\ \overline{U_i}^{\pm} &= U_i^{\pm} \circ \vartheta^{-1} \end{aligned} \tag{3.284}$$

The normal form dynamics comprises the input-output dynamics (3.281) with the matrix function $\mathbf{B} : \mathbb{R}^r \times \mathbb{R}^{n-r} \rightarrow \mathbb{R}^{(n-r) \times m}$. The equations (3.280) for $\dot{Y}_{i,r_i}$ can be written in vector notation:

$$\dot{Y}_r = a_0(Y, Z) + \overline{\mathbf{A}}(Y, Z) U \tag{3.285}$$

to determine the input vector U as:

$$U = \overline{\mathbf{A}}^{-1}(Y, Z) \left(\dot{Y}_r - a_0(Y, Z) \right) \tag{3.286}$$

The inverse of the decoupling matrix $\overline{\mathbf{A}}(Y, Z) = \mathbf{A}(X) \circ \vartheta^{-1}$ is well defined due to the full rank condition (3.271).

The state constraints (3.283) can be represented by m saturation functions $Y_{i,1}$. This defines the mapping:

$$Y_{i,1} = h_{i,1}\left(\zeta_{i,1}\right) = \psi\left(\zeta_{i,1}, c_i^{\pm}\right) \qquad i = 1, \dots, m \tag{3.287}$$

$$Y_{i,j} = h_{i,j}\left(\zeta_{i,1}, \dots, \zeta_{i,j}\right) = \gamma_{i,j}\left(\zeta_{i,1}, \dots, \zeta_{i,j-1}\right) + \psi'\left(\zeta_{i,j}\right) \qquad j = 2, \dots, r_i \tag{3.288}$$

The nonlinear terms are determined with respect to the previous equations for $Y_{i,j-1}$. The successive differentiations of $Y_{i,1}$ along the multiple cascades lead to a new set of coordinates. The next step is to introduce the constraints to the objective function via a penalty term. This penalty term is successively reduced during the numerical solution of the unconstrained optimal control problem in order to approach the optimal solution [26].

The drawback of this technique is that the numerical solution can be quite cumbersome.

3.3.4.7 Measurement error input to state stability

Nonlinear output feedback design is an important problem in nonlinear control [81]. One reason for this is that a separated design of a global asymptotic stabilizing state feedback and a global convergent observer does not automatically lead to a global asymptotic stable closed-loop in nonlinear feedback design. Additional effort is necessary in order to guarantee global asymptotic stability, for example, either to redesign the observer or to redesign the state feedback [20, 61]. Any affine model of the aircraft presented before can be used for this analysis. The following theorem can be proposed:

Theorem 3.11

If Assumptions 1 and 2 hold:

Assumption 1: System Class: The nonlinear control system is of the form:

$$\begin{aligned} \dot{X} &= f(X) + \mathbf{G}(X)U \\ Y &= h(X) \end{aligned} \tag{3.289}$$

where $X \in \mathbb{R}^n$ is the state, $U \in \mathbb{R}^m$ is the input and $Y \in \mathbb{R}^p$ the output; the functions $f : \mathbb{R}^n \longrightarrow \mathbb{R}^n$, $\mathbf{G} : \mathbb{R}^n \longrightarrow \mathbb{R}^{n+m}$, $h : \mathbb{R}^n \longrightarrow \mathbb{R}^n$ are assumed to be sufficiently smooth with $f(0) = 0, h(0) = 0$. The function f is given by equation (3.138) while the matrix $\mathbf{G}$ is given by relation (3.139).

Assumption 2: State Feedback: The globally asymptotically state feedback is assumed to be of the form:

$$U = k(X) = -\frac{1}{2}\mathbf{R}^{-1}(X)\mathbf{R}^T(X)\tilde{V}_X^T(X) \tag{3.290}$$

where $\mathbf{R}$ is a positive definite matrix function with $\lambda_{min} I \leq \mathbf{R}(X) \leq \lambda_{max} I$ with $\lambda_{max} > \lambda_{min} > 0$. Moreover, suppose that the state feedback (3.290) is of the form:

$$U = k(X) = m(X) + \tilde{p}(c_i^T X) \tag{3.291}$$

where $m : \mathbb{R}^n \longrightarrow \mathbb{R}^m, m(0) = 0$ is globally **Lipschitz** and $\tilde{p} : \mathbb{R}^n \longrightarrow \mathbb{R}^m$ is of the form

$$\tilde{p}(c_i^T X) = \begin{pmatrix} \tilde{p}_1(c_{i_1}^T X) \\ \cdots \\ \tilde{p}_m(c_{i_m}^T X) \end{pmatrix} \tag{3.292}$$

where the components $\tilde{p}_j$ are polynomial functions in a single state $X_{i_j}, i_j \in \{1, \ldots, n\}, j = 1, \ldots, q$,

Then the control system

$$\dot{X} = f(X) + \mathbf{G}(X)k(X + e) \tag{3.293}$$

is input-to-state stable with respect to the measurement error $e \in \mathbb{R}^n$. ■

Remark 3.17. *In this theorem, it is assumed that the state feedback* (3.290) *is inverse optimal with respect to the performance measure:*

$$\tilde{V}(X(0)) = \int_0^\infty \left[Q(X(t)) + U^T(t)\mathbf{R}(X(t))U(t)\right] dt \tag{3.294}$$

*The **Hamilton-Jacobi-Bellman** (HJB) equation:*

$$\tilde{V}_X(X)f(X) + \tilde{V}_X(X)\mathbf{G}(X)k(X) + Q(X) + k^T(X)\mathbf{R}(X)k(X) = 0 \tag{3.295}$$

is satisfied when $Q(X) \geq c\|X\|^2, c > 0$. Furthermore, $\tilde{V}$ is assumed to be a positive definite radially unbounded C^1 function.

Remark 3.18. *The function $m : \mathbb{R}^n \longrightarrow \mathbb{R}^m, m(0) = 0$ is globally* ***Lipschitz*** *means that $\|m(X+e) - m(X)\| \leq \tilde{\gamma}\,\|e\|$ with $\tilde{\gamma}$ a constant.*

The notation $\tilde{p}(X_i)$ means that $\tilde{p}$ is formed by univariate polynomials and each polynomial depends on a certain state variable, not necessarily different from each other. Proof of this theorem can be found in [20].

Theorem 3.12

If Assumption 3 holds:

Assumption 3: State Observer: It is assumed that a state observer for the estimated state $\hat{X}$ for the control system (3.289) with a globally uniformly asymptotic observer error dynamics

$$\dot{e} = a(e, X) \tag{3.296}$$

$e = X - \hat{X} \in \mathbb{R}^n$ is known. More precisely, it is assumed that there exists a Lyapunov function $\tilde{V}_e$ such that:

$$\tilde{V}_e(e)a(e, X) < -\tilde{\alpha}(\tilde{V}_e(e)) \tag{3.297}$$

for all nonzero e, X where $\tilde{\alpha}$ is a positive-definite function. Moreover suppose that the feedback (3.290) is of the form (3.291) to (3.292),

Then the closed-loop (3.291) is globally asymptotically stable. ■

The following control can also be proposed:

$$U = m(X) + \tilde{p}(c_i^T X) + r(Y) \tag{3.298}$$

where r depends only on the **output measurement**.

3.3.4.8 Control Lyapunov function based adaptive control

Control Lyapunov function based adaptive control designs are proposed for the stabilization of multi-input multi-output affine systems. As opposed to the classical adaptive approaches, where the control law depends on estimates of the system nonlinearities, the controller is based on approximating terms that depend both on the possibly unknown system nonlinearities and on an unknown control Lyapunov function [8].

Definition 3.24. ***Robust Control Lyapunov Function*** *(RCLF): A C^1 function $\tilde{V} : \mathbb{R}^n \rightarrow \mathbb{R}^+$ is a* ***robust control Lyapunov function*** *for system (3.138) to (3.140). If it is positive definite, radially unbounded and moreover there exists a non negative constant such that $L_f\tilde{V}(X) < 0$ if $L_g\tilde{V}(X) = 0, \forall X \in \mathbb{R}^n/\Omega_c(\tilde{V})$ where $\Omega_c(\tilde{V})$ is a compact set defined as $\Omega_c(\tilde{V}) = \left\{X \in \mathbb{R}^n | \tilde{V}(X) \leq c\right\}$.*

The **robust control Lyapunov function** (RCLF) characterizes the resolvability of the problem and at the same time, it raises two important design issues [23]:

1. How does one construct a robust control Lyapunov function for the aircraft considered as an uncertain nonlinear system? Are they significant classes of models for which a systematic construction is possible?
2. Once a robust control Lyapunov function has been found, how does one construct the robust controller?

A known robust control Lyapunov function for an affine model can be used to construct an optimal control law directly and explicitly without recourse to the **Hamilton–Jacobi–Isaacs partial differential equation** (HJI)

$$min_U max_W \left[L(X,U) + \nabla\tilde{V}(X).f(X,U,W) = 0\right] \tag{3.299}$$

for a general system and cost functional

$$\dot{X} = f(X,U,W) \qquad J = \int_0^\infty L(X,U)dt \tag{3.300}$$

This can be accomplished by solving an inverse optimal robust stabilization problem.

Theorem 3.13

If the system represented by equations (3.138) to (3.140) is robustly stabilizable via a continuous control law then there exists a robust control Lyapunov function for (3.119). On the other hand, if there exists a robust control Lyapunov function for system (3.119) then it is robustly stabilizable via a continuous control law. ■

Affine aircraft models of the form (3.138) to (3.140) can be transformed by suitable feedback of the form $U_i = \alpha_i(X) + \sum_j \beta_{ij}(X)V_j$ to a linear system. Conditions for feedback linearizability can be given in terms of a nested sequence of vector fields associated with the problem [8].

3.3.4.9 Noisy system

Path following control aims at driving an aircraft towards a given path without any temporal specifications [59]. It is therefore different from tracking control, where a precise temporal law is prescribed. The path following problem is significant for non holonomic systems [17]. Let the motion of the center of geometry $P(t)$ of a kinematic aircraft moving in a 3D space be given by:

$$\dot{P}(t) = V(t) \begin{pmatrix} \cos\gamma(t)\cos\chi(t) \\ \cos\gamma(t)\sin\chi(t) \\ -\sin\gamma(t) \end{pmatrix} \tag{3.301}$$

where the aircraft velocity $V(t)$ is an assigned bounded smooth positive function defined in $\mathbb{R}^+$.

A point $X(t)$ is chosen rigidly linked to $P(t)$, that is: $\|X(t) - P(t)\| = d > 0$ and the spherical coordinates of the vector $(X(t) - P(t))$ are given by $\chi(t) + \bar{\chi}$ and $\gamma(t) + \bar{\gamma}$ where $\bar{\chi}$ and $\bar{\gamma}$ are two constants such that $-\frac{\pi}{2} \leq \bar{\chi}, \bar{\gamma} \leq \frac{\pi}{2}$. Therefore, $X(t)$ is given by:

$$X(t) = P(t) + d \begin{pmatrix} \cos(\gamma(t) + \bar{\gamma})\cos(\chi(t) + \bar{\chi}) \\ \cos(\gamma(t) + \bar{\gamma})\sin(\chi(t) + \bar{\chi}) \\ -\sin(\gamma(t) + \bar{\gamma}) \end{pmatrix} \tag{3.302}$$

If $Z(t) = \begin{pmatrix} \chi(t) + \bar{\chi} \\ \gamma(t) + \bar{\gamma} \end{pmatrix}$ and $U = \mathbf{A}(Z)Z$ where $\mathbf{A}(Z) = \begin{pmatrix} d\cos Z_2 & 0 \\ 0 & d \end{pmatrix}$ adding process and measurement noise, the motion of $X(t)$ is governed by the following systems:

$$\begin{aligned} \dot{X}(t) = V(t) & \begin{pmatrix} \cos(Z_2(t) + \bar{\gamma})\cos(Z_1(t) - \bar{\chi}) \\ \cos(Z_2(t) + \bar{\gamma})\sin(Z_1(t) - \bar{\chi}) \\ -\sin(Z_2(t) - \bar{\gamma}) \end{pmatrix} + \\ & + \begin{pmatrix} -\sin Z_1(t) & -\cos Z_1(t)\sin Z_2(t) \\ \cos Z_1(t) & -\sin Z_1(t)\sin Z_2(t) \\ 0 & \cos Z_2(t) \end{pmatrix} U(t) + e_X(t) \end{aligned} \tag{3.303}$$

$$\dot{Z} = \mathbf{A}^{-1}(t)U(t) + e_Z(t) \tag{3.304}$$

where the noise terms $e_X : \mathbb{R}^+ \to \mathbb{R}^3, e_Z : \mathbb{R}^+ \to \mathbb{R}^2$ are continuous mappings and X_0, Z_0 are the initial conditions such that:

$$\|e_X\| \leq B_X \qquad \|e_Z\| \leq B_Z \tag{3.305}$$

If $\mathbf{A} = (A_{i,j})$ is an $h \times k$ matrix, then

$$\|\mathbf{A}\| = \left(\sum_{i=1}^{h} \sum_{j=1}^{k} A_{ij}^2 \right)^{1/2} \tag{3.306}$$

$\beta = \arccos(\cos\bar{\chi}\cos\bar{\gamma})$ and

$$\mathbf{H}(z) = \mathbf{A}^{-1}(Z) = \begin{pmatrix} \frac{1}{d\cos Z_2} & 0 \\ 0 & \frac{1}{d} \end{pmatrix} \tag{3.307}$$

$\mathbf{H}(z)$ is a bounded map if $\Omega = \mathbb{R} \times \left(-\frac{\pi}{2} + \epsilon, \frac{\pi}{2} - \epsilon\right)$ where ϵ is any real number such that $0 < \epsilon < \frac{\pi}{2}$. This system is nonholonomic and equivalent to the kinematic constraint:

$$\dot{X} = F(t, Z) + \mathbf{G}(Z)\mathbf{H}^{-1}\dot{Z} \tag{3.308}$$

which, in general, is a non-integrable relation between the velocities.

Problem 3.6. *Let $\tilde{\gamma} : \mathbb{R}^+ \to \Omega$ be a C^2 arc-length parametrized curve i.e., $\left\|\dot{\tilde{\gamma}}(\lambda)\right\| = 1, \forall \lambda \geq 0$ and $\Gamma = \tilde{\gamma}(\mathbb{R}^+)$, find sufficient conditions on $\mathbf{B} = (B_X, B_Z)$ and $\tilde{\gamma}$ such that for any $\epsilon > 0$, any choice of (e_X, e_Z) verifying $\|e_X(t)\| \leq B_X$ and $\|e_Z(t)\| \leq B_Z$, $\forall t \geq 0$ and any $X_0 \in \mathbb{R}^n$, there exists a control U such that the system is solvable on $\mathbb{R}^+$ and $lim_{t \to +\infty} sup\left(d(X(t), \Gamma)\right) < \epsilon$ where $\forall X \in \mathbb{R}^3, d(X, \Gamma) = inf_{t \geq 0}\left(\|X - \tilde{\gamma}(t)\|\right)$.*

If $\left\|\ddot{\tilde{\gamma}}\right\|$ the positive scalar curvature of $\tilde{\gamma}$ at λ is always sufficiently small and if at the initial time:

$$\tilde{\gamma}(0) = X_0 \qquad \dot{\tilde{\gamma}}(0)\mathbf{G}^{\perp}(Z_0) > 0 \tag{3.309}$$

with $\left\|\mathbf{G}^{\perp}(Z)\right\| = 1$ and $\mathbf{G}_i^T(Z_0)\mathbf{G}^{\perp}(Z) = 0, \forall i = 1 \dots n-1, z \in \Omega$ then it is possible to find a parametrization $\mu(t)$ of $\tilde{\gamma}$ and a control $U(t)$ such that the solution X of system (3.308) with noise

$$\begin{aligned} \dot{X} &= F(t, Z) + \mathbf{G}(Z)U \\ \dot{Z} &= \mathbf{H}(Z)U \\ X(0) = X_0 & \qquad Z(0) = Z_0 \end{aligned} \tag{3.310}$$

verifies the property
$X(t) = \tilde{\gamma}(\mu(t)), \forall t \geq 0, \; lim_{t \to +\infty}\mu(t) = +\infty$
and
$\left\|\dot{X}(t)\right\| \leq K_1, \left\|\dot{Z}(t)\right\| \leq K_2, \forall t \geq 0$, where K_1, K_2 are suitable constants; U and μ are given by the following dynamic inversion based controller:

$$\dot{\mu} = F_{\dot{\tilde{\gamma}}}(\mu, Z) \qquad U = -F_G(\mu, Z) \qquad \mu(0) = 0 \tag{3.311}$$

Under the previous hypotheses, it is possible to find a control U such that the solution $X(t)$ of system (3.310) covers all the path Γ with a positive bounded speed; the remaining bounded $\dot{Z}$ may be considered as the internal dynamics of the system. It is shown in [17] that under suitable hypotheses on the curvature of $\tilde{\gamma}$ and the noise amplitude, the distance between the path Γ and the solution $X(t)$ may be estimated in terms of a decreasing function, that is, the system is practically stable.

3.3.4.10 Backstepping control for affine systems with drift

A backstepping control law can be derived for systems of the form [68]:

$$\begin{aligned} \dot{X} &= f(X) + \mathbf{G}(X)U \\ Y &= h(X) \end{aligned} \tag{3.312}$$

for which holds

$$\dot{Y} = \frac{\partial h(X)}{\partial X}\dot{X} = \frac{\partial h(X)}{\partial X}\left(f(X) + \mathbf{G}(X)U\right) = L_f h(X) + L_{\mathbf{G}} h(X)U \tag{3.313}$$

where the Lie derivatives are defined as:

$$L_f h(X) = \frac{\partial h(X)}{\partial X} f(X) \quad L_{\mathbf{G}} h(X) = \frac{\partial h(X)}{\partial X}\mathbf{G}(X) \tag{3.314}$$

3.3.4.10.1 Basic approach for a single-input-single-output system

The single-input-single-output (SISO) form of (3.312) can be written in cascading form:

$$\begin{aligned} \dot{X}_1 &= f_1(X_1) + g_1(X_1)X_2 \\ \dot{X}_2 &= f_2(X_1, X_2) + g_2(X_1, X_2)X_3 \\ \dot{X}_3 &= f_3(X_1, X_2, X_3) + g_3(X_1, X_2, X_3)X_4 \\ &\vdots \\ \dot{X}_{n-1} &= f_{n-1}(X_1, X_2, \dots, X_{n-1}) + g_{n-1}(X_1, X_2, \dots, X_{n-1})X_n \\ \dot{Y}_n &= f_n(X_1, X_2, \dots, X_n) + g_{n-1}(X_1, X_2, \dots, X_n)U \\ Y &= h(X_1) \end{aligned} \tag{3.315}$$

Then, the nth order backstepping SISO controller is given by the recursive relation:

$$\begin{aligned} \dot{\alpha}_1 &= \beta_1 \\ \dot{\alpha}_2 &= \beta_2 \\ &\vdots \\ \dot{\alpha}_i &= \beta_i \\ &\vdots \\ \dot{\alpha}_n &= \beta_n \\ U &= \alpha_n \end{aligned} \tag{3.316}$$

where:

$$\beta_1 = \frac{\left(\dot{Y}_d - L_{f_1} h(X_1) - k_1 Z_1 - n_1(Z_1)\right)}{L_{g_1} h(X_1)}$$

$$\beta_2 = \frac{(\dot{\alpha}_1 - f_2(X_1, X_2) - L_{g_1} h(X_1) Z_1 - k_2 Z_2 - n_2(Z_2) Z_2)}{L_{g_2} h(X_1, X_2)}$$

$$\beta_i = \frac{(\dot{\alpha}_{i-1} - f_i(X_1, \dots, X_i) - L_{g_i} h(X_i) Z_i - k_i Z_i - n_i(Z_i) Z_i)}{L_{g_i} h(X_1, X_2, \dots, X_i)}$$

$$\beta_n = \frac{(\dot{\alpha}_{n-1} - f_i(X_1, \dots, X_n) - L_{g_n} h(X_n) Z_n - k_n Z_n - n_n(Z_n) Z_n)}{L_{g_i} h(X_1, X_2, \dots, X_n)}$$

with

$$\begin{array}{c} Z_1 = h(X_)) - Y_d \\ Z_i = X_i - \alpha_{i-1} \end{array} \tag{3.317}$$

Such a backstepping controller results in closed-loop dynamics given by:

$$\dot{Z} = -\mathbf{K}(Z)Z + \mathbf{S}(X)Z \tag{3.318}$$

with

$$\mathbf{K}(Z) = \begin{pmatrix} k_1 + n_1(Z_1) & 0 & \dots & 0 \\ 0 & k_2 + n_2(Z_2) & \dots & 0 \\ \dots & \dots & \dots & \dots \\ 0 & 0 & \dots & k_n + n_n(Z_n) \end{pmatrix} \tag{3.319}$$

and

$$\mathbf{S}(z) = \begin{pmatrix} 0 & L_{g_1} & 0 & \dots & 0 & 0 & 0 \\ -L_{g_1} & 0 & g_2 & 0 & \dots & 0 & 0 \\ 0 & -g_2 & 0 & \dots & 0 & 0 & 0 \\ \dots & \dots & \dots & \dots & \dots & \dots & \dots \\ 0 & 0 & 0 & \dots & 0 & g_{n-1} & 0 \\ 0 & 0 & 0 & \dots & -g_{n-1} & 0 & 0 \\ 0 & 0 & 0 & \dots & 0 & -g_n & 0 \end{pmatrix} \tag{3.320}$$

3.3.4.10.2 *Basic backstepping for a coordinated flight*

Backstepping is a systematic Lyapunov-based method for nonlinear control design [82, 83]. It can be applied to the aircraft model that can be transformed into lower triangular form such as:

$$\begin{array}{c} \dot{X}_1 = f(X_1) + \mathbf{G}(X_1)X_2 \\ \dot{X}_2 = U \\ Y = X_1 \end{array} \tag{3.321}$$

with the following formulation:
$X_1 = (x - x_r, y - y_r, z - z_r, V - V_r, \chi - \chi_r, \gamma - \gamma_r)^T$,
$X_2 = (T - T_r, \alpha - \alpha_r, \sigma - \sigma_r)^T$
and the control $U = (\dot{T} - \dot{T}_r, \dot{\alpha} - \dot{\alpha_r}, \dot{\sigma} - \dot{\sigma_r})^T$, with

$$f_1(X_1) = \begin{pmatrix} V\cos\gamma\cos\chi - V_r\cos\gamma_r\cos\chi_r \\ V\cos\gamma\sin\chi - V_r\cos\gamma_r\sin\chi_r \\ -V\sin\gamma - V_r\sin\gamma_r \\ -g(\sin\gamma - \sin\gamma_r) \\ 0 \\ -\frac{g}{V}(\cos\gamma - \cos\gamma_r) \end{pmatrix} \tag{3.322}$$

$$\mathbf{G}(X_1) = \begin{pmatrix} 0 & 0 & 0 \\ 0 & 0 & 0 \\ 0 & 0 & 0 \\ \frac{1}{m} & \frac{1}{2m}\rho S V^2 C_D(V) & 0 \\ 0 & 0 & \frac{L}{mV\cos\gamma} \\ 0 & \frac{1}{2m}\rho S V C_L(V) & 0 \end{pmatrix} \tag{3.323}$$

with the following assumptions: $\sin\alpha \approx 0, \cos\alpha \approx 1$, $\sin\sigma \approx \sigma, \cos\sigma \approx 1$, $\alpha\sigma \approx 0$ and the drag and lift are, respectively, approximated by

$$D - D_r \approx \frac{1}{2}\rho S V^2 C_D(V)(\alpha - \alpha_r) \tag{3.324}$$

$$L - L_r \approx \frac{1}{2}\rho S V^2 C_L(V)(\alpha - \alpha_r) \tag{3.325}$$

The aircraft has to track a reference trajectory (x_r, y_r, z_r) with a reference velocity V_r and reference heading χ_r and flight path angle γ_r. The reference controls are calculated accordingly to this reference trajectory.

Using the backstepping procedure, a control law is recursively constructed, along with a **control Lyapunov function** (CLF) to guarantee global stability. For the system (3.321), the aim of the design procedure is to bring the state vector X_1 to the origin. The first step is to consider X_2 as the virtual control of the X_1 subsystem and to find a desired virtual control law $\alpha_1(X_1)$ that stabilizes this subsystem by using the control Lyapunov function

$$\tilde{V}_1(X_1) = \frac{1}{2}X_1^T X_1 \tag{3.326}$$

The time derivative of this control Lyapunov function is negative definite:

$$\dot{\tilde{V}}_1(X_1) = \frac{\partial \tilde{V}_1(X_1)}{\partial X_1}\left(f(X_1) + \mathbf{G}(X_1)\alpha_1(X_1)\right) < 0 \qquad X_1 \neq 0 \tag{3.327}$$

if only the virtual control law:

$$X_2 = \alpha_1(X_1) \tag{3.328}$$

could be satisfied.

The key property of backstepping is that it can be stepped back through the system. If the error between X_2 and its desired value is defined as:

$$Z = X_2 - \alpha_1(X_1) \tag{3.329}$$

then the system (3.321) can be rewritten in terms of this error state:

$$\begin{aligned} \dot{X}_1 &= f(X_1) + \mathbf{G}(X_1)\left(\alpha_1(X_1) + Z\right) \\ \dot{Z} &= U - \frac{\partial \alpha_1(X_1)}{\partial X_1}\left(f(X_1) + \mathbf{G}(X_1)\left(\alpha_1(X_1) + Z\right)\right) \end{aligned} \tag{3.330}$$

The control Lyapunov function (3.326) can now be expanded with a term penalizing the error state Z

$$\tilde{V}_2(X_1, X_2) = \tilde{V}_1(X_1) + \frac{1}{2} Z^T Z \tag{3.331}$$

Differentiating,

$$\begin{array}{c} \dot{\tilde{V}}_2(X_1, X_2) = \frac{\partial \tilde{V}_1(X_1)}{\partial X_1} \left(f(X_1) + \mathbf{G}(X_1) \left(\alpha_1(X_1) + Z \right) \right) + \\ + Z^T \left(U - \frac{\partial \alpha_1(X_1)}{\partial X_1} \left(f(X_1) + \mathbf{G}(X_1) \left(\alpha_1(X_1) + Z \right) \right) \right) \end{array} \tag{3.332}$$

which can be rendered negative definite with the control law:

$$\begin{array}{c} U = -kZ + \frac{\partial \alpha_1(X_1)}{\partial X_1} \left(f(X_1) + g(X_1) \left(\alpha_1(X_1) + Z \right) \right) + \\ \\ - \frac{\partial \tilde{V}_1(X_1)}{\partial X_1} \left(f(X_1) + \mathbf{G}(Z + \alpha_1(X_1)) \right) \end{array} \tag{3.333}$$

with $k > 0$.

This design procedure can also be used for a system with a chain of integrators. The only difference is that there will be more virtual steps to backstep through. Starting with the state farthest from the actual control, each step of the backstepping technique can broken up in three parts:

1. Introduce a virtual control α and an error state Z and rewrite the current state equations in terms of these.
2. Choose a control Lyapunov function for the system, treating it as a final stage.
3. Choose an equation for the virtual control that makes the control Lyapunov stabilizable.

The control Lyapunov function is augmented at subsequent steps to reflect the presence of new virtual states, but the same three stages are followed at each step. Hence backstepping is a recursive design procedure [82].

3.3.4.10.3 Adaptive backstepping

For systems with parametric uncertainties, there exists a method called adaptive backstepping which achieves boundedness of the closed-loop states and convergence of the tracking error to zero. The following parametric strict-feedback system is considered as given by equations (3.138) to (3.140):

$$\begin{array}{c} \dot{X}_1 = f(X_1) + \mathbf{G}(X_1)X_2 + W \\ \dot{X}_2 = U \\ Y = X_1 \end{array} \tag{3.334}$$

where W is a vector of unknown constant parameters representing the wind. The control objective is to make the aircraft asymptotically track a given reference $Y_r(t)$. All derivatives of $Y_r(t)$ are assumed to be known.

The adaptive backstepping design procedure is similar to the normal backstepping procedure, only this time a control law (static part) and a parameter update law (dynamic part) are designed along with a control Lyapunov function to guarantee global stability. The control law makes use of a parameter estimate $\hat{W}$, which is constantly adapted by the dynamic parameter update law. Furthermore, the control Lyapunov function now contains an extra term that penalizes the parameter estimation error $\tilde{W} = W - \hat{W}$ [82, 83].

Theorem 3.14

To stabilize the system (3.334), an error variable is introduced for each state:

$$Z = X_2 - Y_r^{i-1} - \alpha_1(X_1) \tag{3.335}$$

along with a virtual control law:

$$\begin{gathered}\alpha_i(\bar{X}_i, \bar{\vartheta}, \bar{Y}_r^{i-1}) = -c_i Z_i - Z_{i-1} - \omega_i^T \hat{\vartheta} + \tfrac{\partial \alpha_{i-1}}{\partial \hat{\vartheta}} \Gamma \tau_i \\ + \sum_{k=1}^{i-1} \left(\tfrac{\partial \alpha_{i-1}}{\partial X_k} X_{k+1} + \tfrac{\partial \alpha_{i-1}}{\partial Y_r^{k-1}} Y_r^k \right) + \sum_{k=2}^{i-1} \tfrac{\partial \alpha_{i-1}}{\partial \hat{\vartheta}} \Gamma \omega_i Z_k\end{gathered} \tag{3.336}$$

for $i = 1, 2 \dots n$ where the tuning function τ_i and the regressor vectors ω_i are defined as:

$$\tau_i(\bar{X}_i, \hat{\vartheta}, \bar{Y}_r^{i-1}) = \tau_{i-1} + \omega_i Z_i \tag{3.337}$$

and

$$\omega_i(\bar{X}_i, \hat{\vartheta}, \bar{Y}_r^{i-1}) = \varphi - \sum_{k=1}^{i-1} \frac{\partial \alpha_{i-1}}{\partial X_k} \varphi_k \tag{3.338}$$

where $\bar{X}_i = (X_1, \dots, X_i)$ and $\bar{Y}_r^i = (Y_r, \dot{Y}_r, \dots, Y_r^i), c_i > 0$ are design constants. With these new variables, the control and adaptation laws can be defined as:

$$U = \frac{1}{\varsigma(X)} \left(\alpha_n \left(X, \hat{\vartheta}, \bar{Y}_r^{n-1} \right) + \bar{Y}_r^n \right) \tag{3.339}$$

and

$$\dot{\hat{\vartheta}} = \Gamma \tau_n \left(X, \hat{\vartheta}, \bar{Y}_r^{n-1} \right) = \Gamma W_z \tag{3.340}$$

where $\Gamma = \Gamma^T > 0$ is the adaptation gain matrix and W the regressor matrix

$$W(Z, \hat{\vartheta}) = (\omega_1, \dots, \omega_i) \tag{3.341}$$

The control law (3.339) with (3.341) renders the derivative of the Lyapunov function:

$$\tilde{V} = \frac{1}{2} \sum_{i=1}^{n} Z_i^2 + \frac{1}{2} \tilde{\vartheta}^T \Gamma^{-1} \tilde{\vartheta} \tag{3.342}$$

negative definite and thus this adaptive controller guarantees global boundedness of $X(t)$ and asymptotically tracking of a given reference $Y_r(t)$ with X_1. ■

3.3.4.11 Sliding mode control

The **variable structure based sliding mode** is a robust control technique for control of nonlinear systems. The basic design of sliding mode control involves the construction of a switching surface with desirable dynamics in the first step, followed by the derivation of a discontinuous control law in the second step. Moreover, the sliding mode control law derived in the second step has to ensure the attractiveness and finite-time reachability of the constructed switching surface.

3.3.4.11.1 Coordinated flight

The approach is applied on the dynamic model of the aircraft as represented by equations (3.174) to (3.185) and $Y = X_1$. The basic idea of sliding mode control defines a sliding surface $S(X) = 0$ for the controlled system where the system evolves according to a desired behavior on the surface. The following sliding surface is chosen

$$S = \ddot{e} + K_V \dot{e} + K_P e + K_I \int e dt \tag{3.343}$$

The candidate control Lyapunov function is:

$$\tilde{V} = \frac{1}{2} S^T S \tag{3.344}$$

After selecting the sliding surface, the control law is designed such that the sliding surface becomes an attractive surface. This is achieved by enforcing the sliding condition $S\dot{S} < 0$ and thus turning the sliding surface to be invariant. The time derivative is given by:

$$\dot{\tilde{V}} = S^T \dot{S} \tag{3.345}$$

where

$$\dot{S} = f_4 + \mathbf{B}_2 U \tag{3.346}$$

with

$$\begin{aligned} e &= Y - Y_r \\ \dot{e} &= f_0(X_2) - \dot{Y}_r \\ \ddot{e} &= \mathbf{B}_1 \left(f_1(X_2) + f_2(X_2, X_3) \right) - \ddot{Y}_r \\ e^{(3)} &= f_3(X_2, X_3) + \mathbf{B}_2 U - Y_r^{(3)} \end{aligned} \tag{3.347}$$

and

$$\begin{aligned} f_4 = &-Y_r^{(3)} - K_V \ddot{Y}_r - K_P \dot{Y}_r - K_I Y_r + \\ &+ K_V \mathbf{B}_1 (f_1 + f_2) + K_P f_0 + K_I Y + f_3 \end{aligned} \tag{3.348}$$

The equivalent control input is computed for $\dot{S} = 0$ giving:

$$U_{eq} = \mathbf{B}_2^{-1} f_4 \tag{3.349}$$

and the sliding mode control law is given by:

$$U = U_{eq} - S - Sgn(S) \tag{3.350}$$

where the $Sgn(S)$ is the sign function.

Sliding mode control is a control strategy that uses high frequency switching to provide control. The controller switches from one control law to the next, sliding along the boundaries of the control strategies.

3.3.4.11.2 Non-coordinated flight

A robust control design using the variable structure control approach is utilized in this section, similar to the one presented in [80], which uses the coordinated flight model.

The non-coordinated flight model presented in the second chapter is used for this derivation. First, an appropriate sliding surface on which the trajectory has desirable property is selected. The sliding surface is chosen as

$$S = \begin{pmatrix} \ddot{e} + K_v \dot{e} + K_p e + K_i \int_0^t e d\tau \\ \beta + K_{b_0} \int_0^t \beta d\tau \end{pmatrix} \tag{3.351}$$

where

$$e = \begin{pmatrix} x \\ y \\ z \end{pmatrix} - \begin{pmatrix} x_r \\ y_r \\ z_r \end{pmatrix} = Y - Y_r \tag{3.352}$$

These gains are chosen so that $S = 0$ yields exponentially stable response for the error e and the slide-slip angle β. Integral feedback in (3.351) provides additional flexibility for a robust design. The motion of the closed-loop including the variable structure control law evolves in two phases:

1. The trajectory beginning from arbitrary initial state is attracted towards $S = 0$,
2. On the sliding phase, the trajectory slides on $S = 0$.

The nonlinear aircraft model has uncertain aerodynamic derivatives and can be written under the following form:

$$\dot{X}_1 = \begin{pmatrix} V\cos\chi\cos\gamma \\ V\sin\chi\cos\gamma \\ -V\sin\gamma \end{pmatrix} = f_0(X_2) \tag{3.353}$$

$$\dot{X}_2 = f_1(X_2) + f_2(T, X_2, X_3) \tag{3.354}$$

where

$$X_1 = \begin{pmatrix} x \\ y \\ z \end{pmatrix} \quad X_2 = \begin{pmatrix} V \\ \chi \\ \gamma \end{pmatrix} \quad X_3 = \begin{pmatrix} \sigma \\ \alpha \\ \beta \end{pmatrix} \quad \omega = \begin{pmatrix} p \\ q \\ r \end{pmatrix} \tag{3.355}$$

with

$$U = \begin{pmatrix} \dot{T} \\ p \\ q \\ r \end{pmatrix} \qquad f_1(X_2) = \begin{pmatrix} -g \sin\gamma \\ 0 \\ -\frac{g}{V}\cos\gamma \end{pmatrix} \tag{3.356}$$

and

$$f_2(X) = \begin{pmatrix} \frac{1}{m}\left(T\cos\alpha\cos\beta - D\right) \\ \frac{1}{mV\cos\gamma}\left(T\left(\sin\alpha\sin\sigma - \cos\alpha\sin\beta\cos\sigma\right) - C\cos\sigma + L\sin\sigma\right) \\ \frac{1}{mV}\left(T\left(\cos\alpha\sin\beta\sin\sigma + \sin\alpha\cos\sigma\right) + C\sin\sigma + L\cos\sigma\right) \end{pmatrix} \tag{3.357}$$

T is the thrust and L, D, C are, respectively, the lift, drag and side forces. The control system is decomposed into a variable structure outer loop and an adaptive inner loop. The outer loop feedback control system accomplishes (x, y, z) position trajectory following Y_r and side-slip angle control using the derivative of thrust and three angular velocity components (p, q, r) as virtual control inputs. The set of equations are:

$$\dot{X}_3 = \mathbf{f}_3(T, X_2, X_3) + \mathbf{B}_3(X_3)\omega \tag{3.358}$$

with

$$f_3(T, X_2, X_3) = \begin{pmatrix} 0 & \sin\gamma + \cos\gamma\sin\sigma\tan\beta & \cos\sigma\tan\beta \\ 0 & -\cos\gamma\sin\sigma\sec\beta & -\cos\sigma\sec\beta \\ 0 & \cos\gamma\cos\sigma & -\sin\sigma \end{pmatrix}(f_1 + f_2) \tag{3.359}$$

and

$$\mathbf{B}_3(X_3) = \begin{pmatrix} \cos\alpha & 0 & \sin\alpha\sec\beta \\ -\cos\alpha\tan\beta & 1 & -\sin\alpha\tan\beta \\ \sin\alpha & 0 & -\cos\alpha \end{pmatrix} \tag{3.360}$$

and

$$\dot{X}_4 = \mathbf{f}_\omega(X_2, X_3, \omega) + \mathbf{B}_\omega\delta \tag{3.361}$$

with

$$f_\omega(X_2, X_3, \omega) = \begin{pmatrix} -i_1 qr \\ i_2 pr \\ -i_3 pq \end{pmatrix} + \left(\frac{V}{V_0}\right)^2 f_V$$

$$f_V = \begin{pmatrix} l_\beta\beta + l_q q + \left(l_{\beta\alpha}\beta + l_{r\alpha}\right)\Delta\alpha + l_p p \\ m_\alpha\Delta\alpha + m_q q - m_{\dot{\alpha}} p\beta + m_V\Delta V + m_{\dot{\alpha}}\frac{g}{V}\left(\cos\theta\cos\phi - \cos\theta\right) \\ n\beta\beta + n_r r + n_p p + n_{p\alpha}\Delta\alpha + n_q q \end{pmatrix}$$

and

$$\mathbf{B}_{\omega} = \left(\frac{V}{V_0}\right)^2 \begin{pmatrix} l_{\delta_a} & l_{\delta_r} & 0 \\ 0 & 0 & m_{\delta_e} \\ n_{\delta_a} & n_{\delta_r} & 0 \end{pmatrix} \quad (3.362)$$

V_0 being the value of the trim velocity.

In the presence of uncertainty, a discontinuous control law is used for accomplishing sliding motion. In the sliding phase, e and β converge to zero because $S = 0$. Once the choice of sliding surface has been made, a controller must be designed such that $S = 0$ becomes an attractive surface. Differentiating $X_1 = (x, y, z)^T$ successively gives

$$\ddot{X}_1 = \begin{pmatrix} \cos\chi\cos\gamma & -V\sin\chi\cos\gamma & -V\cos\chi\sin\gamma \\ \sin\chi\cos\gamma & V\cos\chi\cos\gamma & -V\sin\chi\sin\gamma \\ -\sin\gamma & 0 & -V\cos\gamma \end{pmatrix} \begin{pmatrix} \dot{V} \\ \dot{\chi} \\ \dot{\gamma} \end{pmatrix} \quad (3.363)$$

$$\begin{aligned} X_1^{(3)} &= \dot{\mathbf{B}}_1(f_0 + f_1) + \mathbf{B}_1(\dot{f}_0 + \dot{f}_1) = \\ &= f_3(X_2, X_3, T) + B_1(X_2)\mathbf{B}_2(T, X_2, X_3)\left(\dot{T}, \dot{X}_3\right)^T \end{aligned} \quad (3.364)$$

where

$$X_1^{(3)} = \frac{d^3 X_1}{dt^3} \qquad f_3 = \dot{B}_1(f_0 + f_1) + \mathbf{B}_1\dot{f}_0 + \mathbf{B}_1\frac{\partial f_1}{\partial V} \quad (3.365)$$

$\mathbf{B}_2$ is a 3×4 matrix given by

$$\mathbf{B}_2 = \begin{pmatrix} \frac{\partial f_1}{\partial T} & \frac{\partial f_1}{\partial \sigma} & \frac{\partial f_1}{\partial \alpha} & \frac{\partial f_1}{\partial \beta} \end{pmatrix} \quad (3.366)$$

Using (3.358), the following relation can be written:

$$\begin{pmatrix} \dot{T} \\ \dot{X}_3 \end{pmatrix} = f_4(T, X_2, X3) + \mathbf{B}_4(\alpha, \beta)U_0 \quad (3.367)$$

where $\mathbf{B}_4$ is a 4×4 matrix given by

$$\mathbf{B}_4 = \begin{pmatrix} 1 & 0 \\ 0 & 1 \end{pmatrix} \quad (3.368)$$

,

$$f_4 = \begin{pmatrix} 0 \\ f_2 \end{pmatrix} \quad (3.369)$$

and

$$U_0 = \begin{pmatrix} T \\ \omega \end{pmatrix} \quad (3.370)$$

Differentiating S given in (3.351) and (3.365) gives

$$\dot{S} = f_5(T, X_2, X_3) + \mathbf{B}_1\dot{T} + \mathbf{B}_2\dot{X}_3 = f_5(T, X_2, X_3) + \mathbf{B}_5\left(\dot{T} \quad \dot{X}_3\right) \quad (3.371)$$

where

$$f_5(T, X_2, X_3) = \left(f_3 + k_v \left(\mathbf{B}_1(f_0 + f_1) - \ddot{Y}_r\right) - Y_r^{(3)} + k_p\dot{e} + k_i e\right) \quad (3.372)$$

This gives

$$\dot{S} = f_5 + \mathbf{B}_5 f_4 + \mathbf{B}_5\mathbf{B}_4 U_0 = f_5^* + \Delta f_5 + (\mathbf{B}_5 + \Delta\mathbf{B}_5)U_0 \quad (3.373)$$

where starred functions denote nominal values of functions and $\Delta f, \Delta B_5$ denote uncertain functions. For the existence of a variable structure control, invertibility of matrix $\mathbf{B}_5\mathbf{B}_4$ is required. In a neighborhood of the trim value, a variable structure control law can be designed. For trajectory control, maneuvering through the region in which singularity lies must be avoided by a proper trajectory planning [10].

For the derivation of the control law, the Lyapunov approach is used. The following shows the Lyapunov function

$$\tilde{V} = \frac{1}{2}S^T S \quad (3.374)$$

where $S = (S_1, S_2, S_3, S_4)^T$. The derivative of $\tilde{V}$ is given by:

$$\dot{\tilde{V}} = S^T\left(f_5^* + \Delta f_5 + (\mathbf{B}_5 + \Delta\mathbf{B}_5)U_eq\right) \quad (3.375)$$

For making $\dot{\tilde{V}}$ negative, the control law is chosen of the form:

$$U_0 = \left(\mathbf{B}_5^*\right)^{-1}\left(-f_5^* - \mathbf{K}_1 S - k_2 sign(S)\right) \quad (3.376)$$

where $sign(S) = (sign(S_1), sign(S_2), sign(S_3), sign(S_4))^T$, $\mathbf{K}_1$ is a diagonal matrix and $k_2 > 0$ is yet to be chosen. This control system includes discontinuous functions [80].

Remark 3.19. *Classical sliding mode control is known to generate high-frequency control signals to enforce the sliding condition while under disturbances. This phenomenon is usually called chattering. It can be avoided by smoothing the discontinuous functions replacing sign functions by saturation functions.*

3.3.4.11.3 Integral sliding mode control

The nonlinear aircraft model is considered:

$$\dot{X} = f(X, t) + \mathbf{B}U(X, t) + D_w(X, t) \quad (3.377)$$

where $X \in \mathbb{R}^n$ is the state, $U(X, t)$ is the control and $D_w(X, t)$ is a perturbation due to external disturbances.

The following assumptions are made:

1. **Assumption 1**: $Rank(\mathbf{B}) = m$
2. **Assumption 2**: The unknown actual value of D_w is bounded by a known function $\|D_w(X,t)\| \leq \bar{D_w}(X,t), \forall X$ and t.

In the **integral sliding mode control** (ISMC) approach, a law of the form:

$$U(X,t) = U_0(X,t) + U_1(X,t) \tag{3.378}$$

is proposed. The nominal control $U_0(X,t)$ is responsible for the performance of the nominal system; $U_1(X,t)$ is a discontinuous control action that rejects the perturbations by ensuring the sliding motion. The sliding manifold is defined by the set

$$\{X | S(X,t) = 0\}$$

with

$$S(X,t) = \mathbf{G}\left(X(t) - X(t_0) - \int_{t_0}^{t} \left(f(X,t) + BU_0(X,\tau)\right) d\tau\right) \tag{3.379}$$

$\mathbf{G} \in \mathbb{R}^{m \times n}$ is a projection matrix such that the matrix $\mathbf{GB}$ is invertible.

The discontinuous control U_1 is usually selected as:

$$U_1(X,t) = -\Xi(X,t)\frac{(\mathbf{GB})^T S(X,t)}{\left\|(\mathbf{GB})^T S(X,t)\right\|} \tag{3.380}$$

where $\Xi(X,t)$ is a gain high enough to enforce the sliding motion [15].

Remark 3.20. *If the following assumption is made:*

$$f(X,t) = \mathbf{A}X \qquad D_w = \mathbf{B}_w W \qquad Y = \mathbf{C}X \tag{3.381}$$

with the pair $(\mathbf{A}, \mathbf{B})$ *being stabilizable and the pair* $(\mathbf{A}, \mathbf{C})$ *being detectable, then the proposed methodology is summarized in the following algorithm:*

1. *Solve the Riccati equation:*

$$\mathbf{PA} + \mathbf{A}^T\mathbf{P} - \mathbf{P}\left(\mathbf{BB}^T - \Xi^{-2}\bar{\mathbf{B}}_w\bar{\mathbf{B}}_w^T\right)\mathbf{P} + \mathbf{C}^T\mathbf{C} = 0 \tag{3.382}$$

where $\bar{\mathbf{B}}_w = \mathbf{B}^{\perp}\mathbf{B}^{\perp^+}\mathbf{B}_w$ *and the left inverse of* $\mathbf{B}$ *is* $\mathbf{B}^+ = \left(\mathbf{B}^T\mathbf{B}\right)^{-1}\mathbf{B}^T$. *The columns of* $\mathbf{B}^{\perp} \in \mathbb{R}^{n \times (n-m)}$ *span the null space of* $\mathbf{B}^T$.

2. *Set the sliding manifold as:*

$$S = \mathbf{B}^+\left(X(t) - X(t_0) - \int_{t_0}^{t} \left(\mathbf{A} - \mathbf{BB}^T\mathbf{P}\right) X(\tau) d\tau\right) \tag{3.383}$$

3. *Set the control as:*

$$U = -\mathbf{B}^T\mathbf{P}X - \Xi\frac{S}{\|S\|} \qquad \Xi > \left\|\mathbf{B}^+\mathbf{B}_w W\right\| \tag{3.384}$$

3.3.4.11.4 Short and fast modes approach

The general formulation of the six degrees of freedom aircraft model presented in the second chapter is used for this derivation.

Aircraft flight dynamics can generally be segregated into **slow and fast modes**: $X_1 = (\alpha, \beta, \phi)^T \in \mathbb{R}^3$ correspond to the slow modes of the system, and the body angular rates $X_2 = (p, q, r)^T \in \mathbb{R}^3$ represent the fast mode. This time-scale separation property of aircraft flight dynamics can be exploited to design a two-loop sliding mode control. The rigid-body aircraft flight dynamics equations can be expressed in the following square cascade structure:

$$\begin{aligned} \dot{X}_1 &= f_1(X_1, X_3) + \Delta f_1(X_1, X_3, U) + \mathbf{B}_1(X_1, X_3)X_2 \\ \dot{X}_2 &= f_2(X_1, X_2, X_3) + \mathbf{B}_2(X_1, X_3)U \\ \dot{X}_3 &= f_3(X_1, X_2, X_3, U) \end{aligned} \tag{3.385}$$

where $X_3 = (V, \theta)^T \in \mathbb{R}^2$ refers to the internal dynamics of the system, $U = (\delta_a, \delta_e, \delta_r) \in \mathbb{U} \subset \mathbb{R}^3$; the engine thrust T is varied in an open loop manner and hence is not included in U. Δf_1 is the bounded disturbance acting on the system; $\mathbf{B}_1, \mathbf{B}_2$ are the control input matrices for the outer and inner loops of the sliding mode control algorithm, respectively. Further system (3.385) is assumed to be of minimum phase. It is also required that all the states of a given system be available for feedback to synthesize the sliding mode control law.

3.3.4.11.4.1 Outer-loop sliding mode control design

The outer-loop sliding mode control is formulated by considering the dynamics of the slow mode:

$$\dot{X}_1 = f_1(X_1, X_3) + \Delta f_1(X_1, X_3, U) + \mathbf{B}_1(X_1, X_3)X_{2d} \tag{3.386}$$

where $X_{2d} = (p_d, q_d, r_d)^T$ is regarded as the virtual control input for the outer loop. Aerodynamic control forces are considered as a disturbance term, which is ultimately rejected by the sliding mode control law. The vector relative degree of (3.386) is $(1, 1, 1)^T$. The reference signal X_R is defined as:

$$X_R = (\alpha_d, \beta_d, \phi_d) \tag{3.387}$$

and a vector of error variables:

$$e = X_1 - X_R = (\alpha_d - \alpha, \beta_d - \beta, \phi_d - \phi)^T \tag{3.388}$$

A sliding surface can be chosen as: $S = (S_1, S_2, S_3)^T$ such that:

$$S_i = e_i + k_i \int_0^t e_i d\tau \tag{3.389}$$

where e_i denotes the ith component of the error vector. The coefficients $k_i, i = 1, 2, 3$ in equation (3.389) are selected such that the error dynamics on a sliding surface is asymptotically stable. Differentiating equation (3.389), $i = 1, 2, 3$ with respect to time gives:

$$\dot{S}_i = f_1 + \Delta f_1 + \mathbf{B}_1 X_{2d} + \mathbf{P}e - \dot{X}_R \tag{3.390}$$

where $\mathbf{P} = diag(k_i)$ represents a diagonal matrix. Then the following control input is chosen:

$$X_{2d} = \mathbf{B}_1^{-1}\left(-f_1 - \mathbf{P}e + \dot{X}_R - \mathbf{K}_1(S)sgn(S)\right) \tag{3.391}$$

where $sgn(S) = (sgn(S_1), sgn(S_2), sgn(S_3))$. The matrix $\mathbf{B}_1$ is assumed to be non-singular so that the nominal system can be decoupled using state feedback. The gain matrix $\mathbf{K}_1(S)$ in equation (3.391) is chosen as a sum of power law and a constant term:

$$\mathbf{K}_1(S) = diag(X_{3i}|S_i|^{\alpha} + b_1) \tag{3.392}$$

where $X_{3i} > 0, i = 1, 2, 3$ and $\alpha \in (0, 1)$. The constant b_1 in equation (3.392) is determined by the bounds of the disturbance term

$$\|\Delta f_1\| \leq b_1 \tag{3.393}$$

which can be estimated through numerical simulations. Substituting equations (3.391) to (3.392) in (3.390), the following relation is obtained:

$$\dot{S} = \Delta f_1 - \mathbf{K}_1(S)sgn(S) \tag{3.394}$$

To prove that the switching surface defined in equation (3.389) is attractive and finite time reachable, the following candidate Lyapunov function is considered:

$$\tilde{V} = \frac{1}{2}S^T S > 0 \tag{3.395}$$

Taking the time derivative of equation (3.395), one obtains:

$$\dot{\tilde{V}} = S^T\left(\Delta f_1 + \mathbf{K}_1(S)sgn(S)\right) \tag{3.396}$$

Substituting equations (3.392) to (3.393) in equation (3.396), the following inequality is obtained:

$$\dot{\tilde{V}} \leq X_{3i}\sum_{i=1}^{3}|S_i|^{\alpha+1} < 0, \forall X \neq 0 \tag{3.397}$$

From equations (3.395) to (3.397), it can be inferred that the selected switching surface is both attractive and finite time reachable. This finishes the design of the outer-loop SMC [24, 63].

3.3.4.11.4.2 Inner-loop sliding mode control design The governing equation for the fast mode is:

$$\dot{X}_2 = f_2(X_1, X_2, X_3) + \mathbf{B}_2(X_1, X_3)U \quad (3.398)$$

The vector relative degree of equation (3.398) is $(1,1,1)^T$. The following error variable is considered:

$$e = X_2 - X_{2d} = (p - p_d, q - q_d, r - r_d)^T \quad (3.399)$$

Choose the switching surface $S = (S_1, S_2, S_3)^T = 0$ such that the switching variable is:

$$S_i = e_i \Rightarrow \dot{S} = f_2 + \mathbf{B}_2 U \quad (3.400)$$

The term $\dot{X}_{2d}$ being a function of slow variables has been neglected. Choose the following control input as

$$U = \mathbf{B}_2^{-1}(-f_2 - \mathbf{K}_2(s)sgn(S)) \quad (3.401)$$

The gain matrix $\mathbf{K}_2(s)$ is chosen as a power law term:

$$\mathbf{K}_2(S) = diag\left(X_{3i}|S_i|^{\alpha''}\right) \quad X_{3i} > 0 \quad \alpha'' \in (0,1) \quad (3.402)$$

The attractiveness and finite-time reachability of the chosen sliding surface in equation (3.400) can be established in a similar manner as demonstrated for the outer-loop SMC.

The sliding mode control law is inherently discontinuous and hence results in control chattering, which can excite the unmodeled high frequency plant dynamics. To reduce chattering, several approaches have been suggested such as continuation approximation of the discontinuous sliding mode control laws, power rate reaching law method and higher-order sliding modes. The power rate reaching law method is used to mitigate chattering [24].

Remark 3.21. ***Higher-order sliding mode control*** *(HOSMC) enforces higher-order derivative constraints on the sliding surface, while keeping the advantages of the classical sliding mode control. HOSMC removes chattering completely and provides better control accuracy and robustness. The* r^{th} *order sliding mode can be defined by:*

$$\sigma = \dot{\sigma} = \ddot{\sigma} = \cdots = \sigma^{r-1} \quad (3.403)$$

which forms an r-dimensional constraint set on the dynamics of the system. Two of the most common HOSMC design approaches are twisting and super-twisting algorithms [15].

A dynamic sliding mode control approach is presented in [91] for the longitudinal loop mode.

3.3.5 MODEL PREDICTIVE CONTROL

Model predictive control is among the techniques that have been used for real time optimization. It is essentially a feedback control scheme in which the optimal control problem is solved over a finite horizon $[t, t+T_f]$, where at each time step t, the future states are predicted over the horizon length T_f based on the current measurements. The first control input of the optimal sequence is applied to the aircraft and the optimization is repeated. The closed-loop implementation provides robustness against modeling uncertainties and disturbances. However, because of the finite horizon, the closed-loop stability cannot be guaranteed if no special precautions are taken in the design and implementation. One way to address this issue is to use terminal constraints or cost-to-go functions combined with a control Lyapunov function.

Model predictive control (MPC) is a strategy that explicitly uses the model of the aircraft to predict its behavior [94]. The model is used to find the best control signal possible by minimizing an objective function:

1. The future outputs for a determined horizon N, called the **prediction horizon**, are predicted at each instant t, using the system model. These predicted outputs $Y(k+j|k)$ for $j = 1 \dots N$ depend on $X(k|k)$ and the future control signals $U(k+j|k), j = 0 \dots N-1$.
2. The set of future control signals is calculated by optimizing a criterion in order to keep the process as close as possible to a reference trajectory $X_r(k+j|l)$. The **objective function** usually takes the form of a quadratic function of errors between the predicted output signal and the predicted reference trajectory. The control effort is also included in the objective function in most cases. An explicit solution can be obtained if the **objective function** is quadratic, the model is linear and there are no constraints. Otherwise, a general optimization method must be used.
3. The control signal $U(k|k)$ is sent to the system, while the control signal $U(k+j|k), j = 1 \dots N-1$ are rejected and step 1 is repeated with all the states brought up to date. Thus $U(k+1|k+1)$ is calculated.

In aircraft flight control problems, only noisy outputs or partial state measurements are available. In such a case, the standard approach is to design an observer to reconstruct the partially unknown state and its estimate is exploited for regulation purposes [22]. As a consequence, an unknown estimation error, acting as an additional uncertainty source on the system, has to be taken into account. Along these lines, contributions on output feedback control MPC share as common denominators the stability of the augmented system: observer and moving horizon controllers.

The aim is to develop a memoryless output MPC strategy by avoiding the design of an observer/controller pair that gives rise to nonconvex conditions, when uncertain model plants are taken into consideration. An output receding horizon controller is designed by imposing pre-definite matrix structures to relevant optimization variables so that the state reconstruction is no longer necessary. Hence, the design is formulated as a semi-definite programming problem in terms of the linear matrix inequalities condition.

The norm bounded uncertain linear description of the linear description for the aircraft dynamics is:

$$X(t+1) = f(X(t), U(t)) = \Phi X(t) + \mathbf{G}U(t) + \mathbf{B}_\nu \nu \tag{3.404}$$

$$Y(t) = h(X(t), U(t)) = \mathbf{C}X(t) \tag{3.405}$$

$$Z(t) = \mathbf{C}_z X(t) + \mathbf{D}_z U(t) \tag{3.406}$$

$$\nu(t) = \Delta(t) Z(t) \tag{3.407}$$

with $X \in \mathbf{R}^{n_x}$ denoting the state, $U \in \mathbf{R}^{n_u}$ denoting the control input, $Y \in \mathbf{R}^{n_y}$ denoting the output and $\nu, Z \in \mathbf{R}^{n_z}$ denoting additional variables which account for the uncertainty. The state information is not fully available at each time instant. The aircraft is subject to the following component-wise input and state evolution constraint $U(t) \in \mathbb{O}(U)$ and $X(t) \in \mathbb{O}(X)$ where:

$$\mathbb{O}(U) = \{U \in \mathbb{R}^{n_u} : |U_i(t+k|t)| \leq U_{i,max}\} \tag{3.408}$$

and

$$\mathbb{O}(X) = \{X \in \mathbb{R}^{n_x} : |X_i(t+k|t)| \leq X_{i,max}\} \tag{3.409}$$

The following matrices can be defined:

$$\tilde{\mathbf{A}} = \begin{pmatrix} \Phi & \mathbf{G} \\ \mathbf{C} & 0 \end{pmatrix} \tag{3.410}$$

$$\tilde{\mathbf{B}} = \begin{pmatrix} \mathbf{B}_\nu \\ 0 \end{pmatrix} \tag{3.411}$$

$$\tilde{\mathbf{C}} = \begin{pmatrix} \mathbf{C}_z & \mathbf{D}_z \end{pmatrix} \tag{3.412}$$

The following set

$$\mathbb{O}(\Delta) = \left\{ \tilde{\mathbf{A}} + \tilde{\mathbf{B}}\Delta\tilde{\mathbf{C}} \text{ where } \|\Delta\| \leq 1 \right\} \tag{3.413}$$

is the image of the matrix norm unit ball under a matrix linear fractional mapping.

The following closed-loop approach is used

$$U(t) = -\mathbf{K}(t)X(t) \tag{3.414}$$

The key idea is to determine at each instant time t, on the basis of the current state $X(t)$, the pair $(\mathbf{P}, \mathbf{K})$ by minimizing the cost index $J\left(X(t), \mathbf{K}(t)\right)$ and by ensuring the constraints' fulfillment from t onward:

$$J\left(X(0), \mathbf{K}X(t)\right) = \max_{\nu \in \mathbf{O}_\nu} \sum_{t=0}^{\infty} \|X(t)\|^2_{\mathbf{R}_X} + \|\mathbf{K}X(t)\|^2_{\mathbf{R}_U} \tag{3.415}$$

where $\mathbf{R}_X, \mathbf{R}_U$ are symmetric positive definite weighting matrices.

$$\mathbb{O}(\nu) = \left\{ \nu \in \mathbb{R}^{n_z} : \|\nu\|_2^2 \leq \|\mathbf{C}_K X(t)\|_2^2 \right\} \tag{3.416}$$

represents aircraft uncertainty regions at each time instant.

The symmetrical definite positive matrix satisfies the following linear matrix inequalities:

$$\begin{pmatrix} \Phi_K^T \mathbf{P} \Phi_K - \mathbf{P} + \mathbf{K}^T \mathbf{R}_U \mathbf{K} + \mathbf{R}_X + \lambda \mathbf{C}_K^T \mathbf{C}_K & \Phi_K^T \mathbf{P} \mathbf{B}_\nu \\ \mathbf{B}_\nu^T \mathbf{P} \Phi_K & \mathbf{B}_\nu^T \mathbf{P} \mathbf{B}_\nu \end{pmatrix} \leq 0 \tag{3.417}$$

where $\lambda > 0$ and

$$\begin{aligned} \Phi_K &= \Phi + \mathbf{G}\mathbf{K} \\ \mathbf{C}_K &= \mathbf{C}_z + \mathbf{D}_z \mathbf{K} \end{aligned}$$

$\mathbf{R}_U, \mathbf{R}_X$ being symmetric weighting matrices.

Remark 3.22. *The various aircraft outputs tend to react over different periods because they are affected differently by each input and are often coupled. Controlling one output with an input at a specified rate to obtain a desirable response in another coupled output is difficult. Naturally, certain outputs may need to be managed at higher rates than others to maintain adequate levels of performance [54]. The use of multiple prediction horizons becomes particularly useful when attempting to control systems with high degrees of cross couplings between outputs with multiple degrees of freedom such as an aircraft. In such cases, prediction horizons may be specific to certain tracked outputs with multiple degrees of freedom. It may allow for a certain level of decoupling to be achieved, where the response of a particular output can be adjusted without greatly affecting the response of other tracked outputs [11].*

3.4 FUZZY FLIGHT CONTROL

Fuzzy flight control is part of intelligent control, using tools of computational intelligence [75]. Evolutionary computation is a kind of optimization methodology inspired by the mechanisms of biological evolution and behaviors of living organisms [104]. It includes also genetic algorithms (GA), advanced neural networks (ANN), evolutionary programming (EP), swarm intelligence (SI), ant colony optimization (ACO), particle swarm optimization (PSO). Variants machine learning (ML) techniques have been used in evolutionary computation algorithms to enhance the algorithm performance. These machine learning techniques can include: statistical methods (e.g., mean and variance), interpolation and regression, clustering analysis, artificial neural networks (ANN), bayesian network (BN), reinforcement learning (RL). These machine learning techniques can be incorporated into different evolutionary computation algorithms in various ways and they affect evolutionary computation also on various aspects, namely: population initialization, fitness evaluation and selection, population reproduction and variation, algorithm adaptation and local search.

A number of authors [21, 68, 76] have reported the development and application of neural network based or adaptive/intelligent control algorithms. A fuzzy logic controller is based on fuzzy logic, an approach that uses logic variables with continuous values, as opposed to classical logic which operates on discrete values of either 0 (false) or 1 (true). In the input stage, the data are mapped to the values obtained from a membership function (triangular, trapezoidal or bell shaped function). The processing stage consists of a number of rules.

The majority of the fuzzy control system applied to aircraft has been a heuristic logic system blended through fuzzy rules. These systems are designed as intelligent systems for navigation tracking control. Fuzzy gain scheduling is the most used technique for fuzzy flight control [35, 93].

3.4.1 FUZZY APPROACH FOR TRACKING MOVING TARGETS

The approach proposed in this section is fuzzy-logic based. Three fuzzy modules are designed; one module is used for adjusting the bank angle value to control the latitude and the longitude coordinates and the other two are used for adjusting the elevator and the throttle controls to obtain the desired altitude value [45].

Basically, a fuzzy logic system consists of three main parts: the fuzzifier, the fuzzy-inference engine and the defuzzifier. The fuzzifier maps a crisp input into some fuzzy sets. The fuzzy inference engine uses fuzzy *if-then* rules from a rule base to reason for the fuzzy output. The output in fuzzy terms is converted back to a crisp value by the defuzzifier.

Mamdani-type fuzzy rules are used to synthesize the fuzzy logic, controllers which adopt the following fuzzy *if-then* rules:

$$\begin{array}{c} R^{(l)}: \text{ IF } (X_1 \text{ is } X_1^{(l)} \text{ AND } \dots \text{ AND } (X_n \text{ is } X_n^{(l)}) \\ \text{THEN } (Y_1 \text{ is } Y_1^{(l)}) \text{ AND } \dots \text{ AND } (Y_k \text{ is } Y_k^{(l)}) \end{array} \tag{3.418}$$

where R^l is the lth rule, X, Y are the input and the output state linguistic variables of the controller, respectively, and $\mathbb{U}, \mathbb{V} \subset \mathbb{R}^n$ are the universe of input and output variables, respectively.

A **multi-input-single-output (MISO)** fuzzy logic controller with a singleton fuzzifier is considered: k=1.Using triangular membership functions, algebraic product for logical AND operation, product-sum inference and centroid defuzzification method, the output of the fuzzy controller has the following form:

$$Y = \frac{\sum_{l=1}^{M} \left(\prod_{i=1}^{N} \mu_{X_i^l}(X_i) \right) Y_l}{\left(\prod_{i=1}^{N} \mu_{X_i^l}(X_i) \right)} \tag{3.419}$$

where N and M represent the number of input variables and the total number of rules, respectively. $\mu_{X_i^l}$ denote the membership function of the lth input fuzzy set for the ith input variable:

1. **FLC1: Altitude Controller**: Inputs: altitude error and derivative of altitude error, output throttle;
2. **FLC2: Altitude Controller**: Inputs: air speed error and derivative of air speed error, output elevator;
3. **FLC3: Longitude and Latitude Controller**: Inputs: bank angle error and derivative of bank angle altitude error, output bank angle.

There are two main classes of fuzzy controllers:

1. Position type fuzzy logic controller which generates the control input U from the error e and the error rate Δe: PD fuzzy logic control.
2. Velocity type fuzzy logic controller which generates incremental control input ΔU from the error and the error rate: PI fuzzy logic control.

In this section, PI type fuzzy logic controllers are preferred for the bank angle and the altitude controller, because of the nonlinearities of the model and the inference between the controlled parameters. It is easier to derive the required change in the control input instead of predicting its exact value:

$$e(t) = Y_{ref} - Y \qquad \Delta e(t) = e(t) - e(t-1) \tag{3.420}$$

where Y_{ref} and Y denote the applied set point input and plant output, respectively. The output of the controller is the incremental change in the control signal ΔU

$$U(t) = U(t-1) + \Delta U(t) \tag{3.421}$$

The membership functions used for each input of the fuzzy logic controllers are of triangular types. As the membership functions of the altitude error and its derivative, five triangular functions were chosen. As the output membership functions, the bank angle and the throttle control outputs were represented with seven membership functions [46].

For terrain following/terrain avoidance (TF/TA) [71], to obtain optimal terrain following/terrain avoidance trajectories, costs such as the mission time, fuel consumption and height of the aircraft are minimized with different relative weights. Optimal trajectories can be generated using a global differential evolution optimization algorithm, then a m multi-output nonlinear model predictive controller is established to enable the aircraft to track the optimal paths in real time. The controller uses a neuro-fuzzy predictor model that is trained using the local linear model tree algorithm. A robustness analysis shows that the designed controller can effectively reject wind disturbances while maintaining stability in the presence of uncertainties in the physical parameters [41]. Since the fuzzy controllers depend on simple rules, they are much easier to understand and to implement [43].

3.4.2 STABILIZATION OF TAKAGI–SUGENO SYSTEMS UNDER THE IMPERFECT PREMISE MATCHING

Based on the Takagi–Sugeno system, the Lyapunov stability theorem is the main approach for solving stabilization problems. If there is a solution for the Lyapunov inequalities, the equilibrium point of the closed-loop system is guaranteed to be asymptotically stable. The common Lyapunov function is a usual way for obtaining Lyapunov inequalities but it has very narrow feasible regions that give solutions for the Lyapunov inequalities. In order to resolve this problem, various techniques have been proposed such as piece-wise Lyapunov function, fuzzy Lyapunov function, polynomial fuzzy systems, membership function dependent approach. Among them, the **fuzzy Lyapunov function** method is defined as a fuzzy blending of common Lyapunov functions and therefore only one matrix does not need to meet all of the Lyapunov inequalities. However, the fuzzy Lyapunov function has a restriction that the maximum norm value of the first derivative of the Takagi–Sugeno model membership function should be known, but usually, it is very hard or even impossible to derive it using the mathematical ways [42].

On the other hand, the general method for designing fuzzy controllers is the **parallel distributed compensation** (PDC) scheme. There is, however, a considerable disadvantage that the fuzzy controller should be designed based on the premise membership function of the fuzzy model. To solve this problem, the imperfect premise matching method is proposed while the Takagi–Sugeno model and fuzzy controller do not share the same premise membership functions. Under the imperfect premise matching conditions, the design flexibility and robustness property of the fuzzy controller can be enhanced compared with the PDC one. The following nonlinear model is considered:

$$\dot{X}(t) = f\left(X(t), U(t)\right) \tag{3.422}$$

$X \in \mathbb{R}^n$ is the state vector, $U \in \mathbb{R}^m$ is the control input vector and $f : \mathbb{R}^{n+m} \rightarrow \mathbb{R}^n$ is a nonlinear function belonging to class $C^\ell, \ell \in I_\sigma = \{1, 2, \dots, \sigma\}$. The nonlinear function (3.422) can be modeled as the Takagi–Sugeno system within the following compact sets for $X(t), U(t)$:

$$C_1 = \left\{X(t) \in \mathbb{R}^n, \|X(t)\| \leq \bar{X}\right\} \quad \bar{X} \in \mathbb{R}^{n+} \tag{3.423}$$

$$C_2 = \left\{U(t) \in \mathbb{R}^m, \|U(t)\| \leq \bar{U}\right\} \quad \bar{U} \in \mathbb{R}^{m+} \tag{3.424}$$

The wide range of nonlinear system (3.422) within the sets (3.423) to (3.424), as shown in the second chapter, can be represented as the Takagi–Sugeno system.

Rule

$$R_i \text{ IF } \zeta_1(t) \text{ is } \Gamma_1^i \text{ AND } \dots \text{ AND } \zeta_p(t) \text{ is } \Gamma_p^i \text{ THEN } \dot{X}(t) = \mathbf{A}_i X(t) + \mathbf{B}_i U(t) \tag{3.425}$$

where R_i is the ith fuzzy rule, $\zeta_k(t), k \in I_p$ is the kth premise variable and $\Gamma_k^i(i,k) \in \mathbb{I}_r \times \mathbb{I}_p$ is the fuzzy set of the kth premise variable in the ith fuzzy rule.

Using the singleton fuzzifier, product inference engine and center-average defuzzification, the Takagi–Sugeno system (3.425) is described as the following equation:

$$\dot{X}(t) = \sum_{i=1}^{r} \varpi_i(\zeta(t)) \left(\mathbf{A}_i X(t) + \mathbf{B}_i U(t)\right) \tag{3.426}$$

in which

$$\varpi_i(\zeta(t)) = \frac{\mu_i(\zeta(t))}{\sum_{i=1}^{r} \mu_i(\zeta(t))} \tag{3.427}$$

$$\mu_i(\zeta(t)) = \prod_{k=1}^{p} \Gamma_k^i(\zeta_k(t)) \tag{3.428}$$

$\Gamma_k^i(\zeta_k(t))$ is the membership value of the kth premise variable $\zeta_k(t)$ in Γ_k^i. The $\varpi_i(\zeta(t))$ satisfies the following properties:

$$0 \leq \varpi_i(\zeta(t)) \leq 1 \qquad \sum_{i=1}^{r} \varpi_i(\zeta(t)) = 1, i \in \mathbb{I}_r \tag{3.429}$$

A fuzzy controller for system (3.422) is based on the imperfect premise matching method with r rules.

Rule

$$R_j \text{ IF } S_1(t) \text{ is } \Phi_1^i \text{ AND } \dots \text{ AND } S_q(t) \text{ is } \Phi_q^i \text{ THEN } U(t) = -K_j X(t) \tag{3.430}$$

where R_j is the jth fuzzy rule, $S_l(t), l \in I_q$ is the lth premise variable and $\Phi_l^j(i,k) \in I_r \times I_q$ is the fuzzy set of the lth premise variable in the jth fuzzy rule.

The defuzzified output of the fuzzy controller (3.430) is described as the following equation:

$$U(t) = -\sum_{j=1}^{r} k_j(S(t)) \mathbf{K}_j X(t) \tag{3.431}$$

in which

$$k_j(S(t)) = \frac{m_j(S(t))}{\sum_{j=1}^{r} m_j(S(t))} \tag{3.432}$$

$$m_j(S(t)) = \prod_{k=1}^{p} \Phi_l^j(S_l(t)) \tag{3.433}$$

$\Phi_l^j(S_l(t))$ is the membership value of the lth premise variable $S_l(t)$ in Φ_l^j. The function $k_j(S(t))$ satisfies the following properties:

$$0 \leq k_j(S(t)) \leq 1 \qquad \sum_{i=1}^{r} k_i(S(t)) = 1, i \in I_r \tag{3.434}$$

Definition 3.25. ***Fuzzy Lyapunov Function:*** *The fuzzy Lyapunov function (FLF) is defined as:*

$$\tilde{V}(X) = \sum_{i=1}^{r} \varpi_i(\zeta(t)) X^T(t) \mathbf{P}_i X(t) \tag{3.435}$$

The main problem with the fuzzy Lyapunov function is that the stabilization conditions cannot be handled via LMI because of the time derivative of the membership function. To overcome this problem, the upper bounds of the membership functions are considered:

$$\|\dot{\varpi}_k(\zeta(t))\| \leq \Phi_{hk}, \varpi_k(\zeta(t)) \in C^1, k \in \mathbb{I}_r \tag{3.436}$$

where $\Phi_{hk} \in \mathbb{R}^{n+}$.

Theorem 3.15

Consider a scalar $\mu > 0$ and assumption (3.436). The continuous fuzzy system (3.426) is stabilized by the fuzzy set controller:

$$U(t) = -\sum_{j=1}^{r} \varpi_j(\zeta(t)) \mathbf{K}_j X(t) \tag{3.437}$$

if there exists symmetric matrices $\mathbf{T}_i, \mathbf{Y}$ and any matrices $\mathbf{R}_i, \mathbf{S}_i$ satisfying the following LMI:

$$\mathbf{T}_i > 0, i \in \mathbb{I}_r \tag{3.438}$$

$$\mathbf{T}_i + \mathbf{Y} > 0, i \in \mathbb{I}_r \tag{3.439}$$

$$\mathbf{O}_{ii} < 0, i \in I_r \tag{3.440}$$

$$\mathbf{O}_{ij} + \mathbf{O}_{ji} < 0, i, j \in \mathbb{I}_r, i < j \tag{3.441}$$

where

$$\mathbf{O}_{ij} = \begin{pmatrix} \mathbf{O}_{ij11} & \mathbf{O}_{ij12} \\ \mathbf{O}_{ij12}^T & \mu(\mathbf{R} + \mathbf{R}^T) \end{pmatrix} \tag{3.442}$$

where:

$$\mathbf{O}_{ij_{11}} = \mathbf{T}_\phi - \left(\mathbf{A}_i \mathbf{R}^T - \mathbf{B}_i \mathbf{S}_j^T\right) - \left(\mathbf{R}\mathbf{A}_i^T - \mathbf{S}_j \mathbf{B}_i^T\right)$$

$$\mathbf{O}_{ij_{12}} = \mathbf{O}_{ij_{11}} \mathbf{T}_j^T - \mu\left(\mathbf{A}_i \mathbf{R}^T - \mathbf{B}_i \mathbf{S}_j^T\right)^T + \mathbf{R}^T$$

and

$$\mathbf{T}_\phi = \sum_{p=1}^{r} \Phi_{gp}(\mathbf{T}_p + \mathbf{Y}) \qquad \mathbf{Y} = \mathbf{R}\mathbf{M}_2\mathbf{R}^T \tag{3.443}$$

■

3.4.3 FUZZY MODEL PREDICTIVE CONTROL FOR WIND DISTURBANCES REJECTION

The application of a hybrid approach for the problem of control, combining neural and fuzzy technologies, is called the adaptive neuro fuzzy inference system (ANFIS). The ANFIS model is trained with the back-propagation gradient descent method. This is a hybrid neuro-fuzzy technique that brings learning capabilities of neural networks to the fuzzy inference system. The learning algorithm tunes the membership function of a Sugeno-type fuzzy inference system using the training input/output data. ANFIS has a five-layered structure given below [101]:

1. Layer 1 contains membership functions of inputs and all inputs are applied to these functions:
$$L1X_i = \mu_{A_i}(X), i = 1 \dots n \qquad L1Y_i = \mu_{B_i}(Y), i = 1 \dots n \quad (3.444)$$
There are different types and shapes of the membership functions.
2. In Layer 2, each function value is multiplied by other values coming from other inputs due to defined rules, and rule base and result values are named as firing strengths of each rule:
$$L2_i = W_i = \mu_{A_i}(X), i = 1 \dots n \quad \mu_{B_i}(Y), i = 1 \dots n \quad (3.445)$$
3. In Layer 3, firing strengths are normalized:
$$L3_i = \tilde{W}_i = \frac{W_i}{\sum_{j=1}^{n} W_j}, i = 1 \dots n \quad (3.446)$$
4. In Layer 4, normalized firing strengths are multiplied by a first order function of inputs:
$$L4_i = \tilde{W}_i f_i = \tilde{W}_i \left(P_i X + Q_i Y + R_i\right) \quad (3.447)$$
(P_i, Q_i, R_i) are parameters of the first order function and these parameters are consequent parameters.
5. In Layer 5, values coming from all Layer 4 outputs are summed and output value is obtained:
$$L5 = \sum_{i=1}^{n} \tilde{W}_i f_i \quad (3.448)$$

The limitation of ANFIS is that it cannot be applied to fuzzy systems when the membership functions are overlapped by pairs.

ANFIS is the Takagi–Sugeno model put in the framework to facilitate learning and adaptation procedures. Such a network makes fuzzy logic more systematic and less reliant on expert knowledge. The objective of ANFIS is to adjust the parameters of a fuzzy system by applying a learning procedure using input-output training data. A combination technique of least square algorithm and back propagation are used for training fuzzy inference systems [76].

Basic architecture of ANFIS has two inputs, X and Y, and one output, f. Assume that the rule base contains two Takagi–Sugeno *if-then* rules as follows:

$$\text{Rule 1: If } X \text{ is } A_1 \text{ AND } Y \text{ is } B_1 \text{ THEN} f_1 = P_1 X + Q_1 Y + R_1 \tag{3.449}$$

$$\text{Rule 2: If } X \text{ is } A_2 \text{ AND } Y \text{ is } B_2 \text{ THEN} f_2 = P_2 X + Q_2 Y + R_2 \tag{3.450}$$

ANFIS has five layers. The parameters in the adaptive nodes are adjustable. The rules of the neuro-fuzzy inference subsystems are formulated so that possible potential situations that may occur to the aircraft are taken into consideration.

A MIMO nonlinear model predictive controller is established to enable the aircraft to track the paths in real time. The controller uses a **neuro-fuzzy** predictor model that is trained using the local linear model tree algorithm. A robustness analysis shows that the designed controller can effectively reject wind disturbance while maintaining stability in the presence of uncertainties in the physical parameters. To reduce the complexity of the plant for prediction purposes, the input-output data used for training the prediction model are taken from a simulation of the aircraft with no aerodynamic modeling. Then the trained predictor is used in a MIMO nonlinear model predictive control to retrieve the closed-loop control forces and moments. As a result a set of closed-loop thrust and aerodynamic control surface commands are obtained with full consideration of the aerodynamic model [41].

The local linear neuro-fuzzy predictor model consists of a set of L neurons where for $i = 1, \dots, L$, the ith neuron represents a **local linear model** (LLM) with input vector $U \in \mathbb{R}^p$ and output $\hat{Y}_i \in \mathbb{R}$ defined as

$$\hat{Y}_i = \alpha_{i_0} + \alpha_{i_1} U_1 + \cdots + \alpha_{i_p} U_p \tag{3.451}$$

where α_{i_j} are the parameters of the ith neuron. The output of the predictor model is determined as:

$$\hat{Y} = \sum_{i=1}^{L} \hat{Y}_i \eta_i(\underline{U}) = \sum_{i=1}^{L} \left(\alpha_{i_0} + \alpha_{i_1} U_1 + \cdots + \alpha_{i_p} U_p \right) \eta_i(\underline{U}) \tag{3.452}$$

where $\underline{U}$ is the normalized input and η_i is the validity function that specifies the activity of the ith local linear model and is defined as

$$\eta_i(\underline{U}) = \frac{\lambda_i(\underline{U})}{\sum_{i=1}^{L} \lambda_j(\underline{U})} \tag{3.453}$$

where for $i = 1, \dots, L$, the membership functions λ_i are defined as

$$\lambda_i(\underline{U}) = \exp\left(-\frac{(U_1 - m_{1i})^2}{2\tilde{\sigma}_{i1}^2} \right) \dots \exp\left(-\frac{(U_p - m_{pi})^2}{2\tilde{\sigma}_{ip}^2} \right) \tag{3.454}$$

where m_{ij} and $\tilde{\sigma}_{ij}$ are the centers and the standard deviations of the Gaussian distribution functions.

At sample time k, the optimizer calculates the optimal control $\bar{U}^*(k) \in \mathbb{R}^{m \times N_u}$ by minimizing the performance index:

$$J = \sum_{i=1}^{n} \left[\tilde{\alpha}_i(k) \sum_{j=1}^{N_y} \left(R_i(k+j) - \left[\hat{Y}_c\right]_{i,j}(k) \right)^2 \right] + \\ + \sum_{i=1}^{m} \left[\tilde{\beta}_i(k) \sum_{j=1}^{N_u} \left([\Delta U]_{i,j}(k) \right)^2 \right] \tag{3.455}$$

where R_i are the reference signal maps and $\tilde{\alpha}_i(k), \tilde{\beta}_i(k)$ are the penalty factors of the future tracking errors and the changes in control inputs, respectively. N_y is the prediction horizon and N_u is the control horizon. The change in the control input from the current sample time k to $k + N_u - 1$ is:

$$\Delta U(k) = \begin{pmatrix} \Delta U_1(k) & \dots & \Delta U_1(k+N_u-1) \\ \vdots & \dots & \vdots \\ \Delta U_m(k) & \dots & \Delta U_m(k+N_u-1) \end{pmatrix} \tag{3.456}$$

Once the optimizer computes $\bar{U}^*(k), U^*(k) = U^*(k-1) + \bar{U}^*(k)$ is fed into the neuro-fuzzy predictor. Initially set $U^*(0) = 0$, the predictor computes the current outputs prediction $\hat{Y}(k)$ and the future outputs prediction

$$\hat{Y}(k) = \begin{pmatrix} \hat{Y}_1(k+1) & \dots & \hat{Y}_1(k+N_y) \\ \vdots & \dots & \vdots \\ \hat{Y}_m(k+1) & \dots & \hat{Y}_m(k+N_y) \end{pmatrix} \tag{3.457}$$

for the entire prediction horizon N_y. The current error is:

$$e(k) = \begin{pmatrix} \hat{Y}_1(k) - Y_1(k+1) \\ \dots \\ \hat{Y}_n(k) - Y_n(k+1) \end{pmatrix} \tag{3.458}$$

which is produced by both predictive model error and the existence of the unmeasurable disturbances $v(k)$. The same error for all sample times is considered within the prediction horizon, that is:

$$E(k) = \begin{pmatrix} 1 & \dots & 1 \\ \vdots & \dots & \vdots \\ 1 & \dots & 1 \end{pmatrix} e(k) \tag{3.459}$$

Next, $E(k)$ is used to correct the future outputs prediction $\hat{Y}(k)$ and define the corrected outputs prediction as:

$$\hat{Y}_c(k) = \hat{Y}(k) + E(k) \tag{3.460}$$

Finally, $\hat{Y}_c(k)$ and $R_i(k+1), \ldots, R_i(k+N_y)$ are used to calculate the performance measure J defined in relation (3.455). The thrust region least square algorithm is used as the optimizer to minimize J and calculate $U^*(k+1)$ to use in $(k+1)^{th}$ sample time.

Type 1 and type 2 fuzzy logic systems in airplane control were compared in [102]. It was found that under high uncertainty levels the type 2 fuzzy logic system outperformed the type 1 fuzzy logic system. Specifically, the type 1 fuzzy logic system showed oscillatory behavior around the reference altitude set points. The interval type 2 fuzzy logic system $\tilde{A}$ is described by its **footprint of uncertainty** $FOU(\tilde{A})$ which can be thought of as the blurring of a type 1 **membership function** (MF). The footprint of uncertainty is completely described by its two bounding functions, the lower membership function and the upper membership function, both of which are type 1 fuzzy logic systems. Consequently, it is possible to use type 1 mathematics to characterize and work with **interval type 2 fuzzy system** (IT2FS).

Adaptive fuzzy controllers (AFC) are usually classified into two categories, namely indirect adaptive fuzzy controllers and direct adaptive fuzzy controllers. In the indirect adaptive fuzzy controllers, two fuzzy logic systems (FLS) are used as estimation models to approximate the aircraft dynamics. In the direct scheme, only one fuzzy logic system is applied as a controller to approximate an ideal control law. An adaptive fuzzy controller system includes uncertainties caused by unmodeled dynamics, fuzzy approximation errors, external disturbances, which cannot be effectively handled by the fuzzy logic system and may degrade the tracking performance of the closed-loop [62]. The interval type 2 fuzzy systems can deal with linguistic and numerical uncertainties simultaneously. Interval type 2 fuzzy systems can handle the uncertainties and give performance that outperform their type 1 counterparts. The third dimension of the interval type 2 fuzzy system and its footprint of uncertainty gives them more degrees of freedom sufficient for better modeling the uncertainties [55].

3.5 CONCLUSION

Some well-known techniques of classical linear and nonlinear control methods as well as some fuzzy approaches were presented in this chapter. Other techniques have been used for flight control of UAV, but mostly for some particular missions. While the nonlinear and robust nature of fuzzy control complements the requirements of flight control well, shortcomings have prevented more pronounced adoption as a flight control technique. Thus, it is important to conduct a systematic analysis needed for flight approval and certification.

REFERENCES

1. Alwi, H.; Edwards, C.; Hamayun, M.T. (2013): *Fault tolerant control of a large transport aircraft using an LPV based integral sliding mode controller*, IEEE Conference on Control and Fault Tolerant System, Nice, France, pp. 637–642.
2. Anderson, R.; Bakolas, E.; Milutinovic, D.; Tsiotras, P. (2013): *Optimal feedback guidance of a small aerial vehicle in a stochastic wind*, AIAA Journal of Guidance, Control and Dynamics, vol. **36**, pp. 975–985.
3. Aouf, N.; Boulet, B.; Boetz, R. (2000): *H_2 and H_∞ optimal gust load alleviation for a flexible aircraft*, American Control Conference, Chicago, Il, pp. 1872–1876.
4. Arino, C.; Perez, E.; Sala, R. (2010): *Guaranteed cost control analysis and iterative design for constrained Takagi–Sugeno systems*, Journal of Engineering Applications of Artificial Intelligence, vol. **23**, pp. 1420–1427, DOI: 10.1016/j.engappai.2010.03.004.
5. Astrom, K. J.; Murray, R. M (2008): *Feedback Systems: an Introduction for Scientists and Engineers*, Princeton University Press.
6. Athans, M. (1986): *A tutorial on the LQG/LTR method*, American Control Conference, pp. 1289–1296.
7. Beard, R. W.; Ferrin, J.; Umpherys, J. H. (2014): *Fixed wing UAV path following in wind with input constraints*, IEEE Transactions on Control System Technology, vol. **22**, pp. 2103–2117, DOI 10.1109/TCST.2014.2303787.
8. Bechlioulis, C. P.; Rovithakis, G. A. (2010): *Prescribed performance adaptive control for multi-input multi-output affine in the control nonlinear systems*, IEEE Transactions on Automatic Control, vol. **55**, pp. 1220–1226.
9. Ben Asher, J. Z. (2010): *Optimal Control Theory with Aerospace Applications*, AIAA Press.
10. Biggs, J.; Holderbaum, W.; Jurdjevic, V. (2007): *Singularities of optimal control problems on some 6D Lie groups*, IEEE Transactions on Automatic Control, vol. **52**, pp. 1027–1038.
11. Blackmore, L.; Ono, M.; Bektassov, A.; Williams, B. (2010): *A probabilistic particle control approximation of chance constrained stochastic predictive control*, IEEE Transactions on Robotics, vol. **26**, pp. 502–517.
12. Blakelock, J. (1991): *Automatic Control of Aircraft and Missiles*, Wiley.
13. Brockett R. W. (1983): *Asymptotic stability and feedback stabilization*, In **Differential Geometric Control Theory**, eds: Brockett, R. W.; Millman, R. S.; Sussmann, H. J.; Birkhauser, Basel-Boston, pp. 181–191.
14. Calvo, O.; Sousa, A.; Rozenfeld, A.; Acosta, G. (2009): *Smooth path planning for autonomous pipeline inspections*, IEEE Multi-conference on Systems, Signals and Devices, SSD'09. pp. 1–9, IEEE, DOI 978-1-4244-4346-8/09/
15. Castanos, F.; Fridman, L. (2006): *Analysis and design of integral sliding manifolds for systems with unmatched perturbations*, IEEE Transactions on Automatic Control, vol. **51**, pp. 853–858.
16. Chaturvedi, N.; Sanyal A. K.; McClamroch N. H. (2011): *Rigid body attitude control*, IEEE Control Systems Magazine, vol. **31**, pp. 30–51.
17. Consolini, L.; Tosques, M. (2005): *A path following problem for a class of non holonomic control system with noise*, Automatica, vol. **41**, pp. 1009–1016.
18. Cook, M. V. (1997): *Flight Dynamics Principle: A Linear Systems Approach to Aircraft Stability and Control*, Elsevier Aerospace Series.

19. Coron, J.-M. (1998): *On the stabilization of some nonlinear control systems: results, tools, and applications*, NATO Advanced Study Institute, Montreal.
20. Ebenbauer, C.; Raff, T.; Allgower, F. (2007): *Certainty equivalence feedback design with polynomial type feedback which guarantee ISS*, IEEE Transactions on Automatic Control, vol. **52**, pp. 716–720.
21. Erginer, B.; Altug, E. (2012): *Design and implementation of a hybrid fuzzy logic controller for a quad-rotor VTOL vehicle*, Int. Journal of Control, Automation and Systems, vol.**11**, pp. 61–70.
22. Franze, G.; Mattei, M., Ollio, L.; Scardamaglia, V. (2013): *A receding horizon control scheme with partial state measurements: control augmentation of a flexible UAV*, IEEE Conference on Control and Fault-tolerant Systems (SYSTOL), Nice, France, pp. 158–163.
23. Freeman, R.A; Kokotovic P. V (1996): *Robust Nonlinear Control Design: State Space and Lyapunov Techniques*, Birkhauser.
24. Friedman, L.M. (2001): *An averaging approach to chattering*, IEEE Transactions on Automatic Control, vol. **46**, pp. 1260–1265.
25. Godhavn, J. M.; Balluchi, A.; Crawford, L. S.; Sastry, S. (1999): *Steering for a class of nonholonomic systems with drift term*, Automatica, vol. **35**, pp. 837–847.
26. Graichen, K.; Petit, N. (2009): *Incorporating a class of constraints into the dynamics of optimal control problems*, Optimal Control Applications and Methods, vol. **30**, pp. 537–561.
27. Grochowski, M. (2010): *Some properties of reachable sets for control affine systems*, Analysis and Mathematical Physics, vol. **1**, pp. 3–13, DOI 10.1007/s13324-010-0001-y.
28. Guo, J.; Tao, G.; Liu, Y. (2011): *A multivariable MRAC scheme with application to a nonlinear aircraft model*, Automatica, vol. **47**, pp. 804–812.
29. Heo, S.; Georgis, D.; Daoutidis, P. (2013): *Control structure design for complex energy integrated networks using graph theoretic methods*, IEEE MED conference on Control and Automation, Crete, Greece, pp. 477–482.
30. Hou, Y.; Wang, Q.; Dong, C. (2011): *Gain scheduled control: switched polytopic system approach*, AIAA Journal of Guidance, Control and Dynamics, vol. **34**, pp. 623–628.
31. Ibrir, S.; Su, C. Y. (2014): *Robust nonlinear feedback design for wing rock stabilization*, AIAA Journal of Guidance, Control and Dynamics, **37**, pp. 321–324.
32. Ichalal, D.; Marx, B.; Ragot, J.; Maquin, D. (2009): *An approach for the state estimation of Takagi–Sugeno models and application to sensor fault diagnosis*, 48th IEEE Conference on Decision and Control, Shanghai, China, pp. 7787–7794.
33. Innocenti, M.; Pollini, L.; Marullo, A. (2003): *Gain Scheduling stability issues using differential inclusion and fuzzy systems*, AIAA Journal of Guidance, Control and Dynamics, vol. **27**, pp. 720–723.
34. Innocenti, M.; Pollini L.; Giuletti F. (2004): *Management of communication failures in formation flight*, AIAA Journal of Aerospace Computing, Information and Communication, vol. **1**, pp. 19–35.
35. Innocenti, M.; Pollini, L.; Turra, D. (2008): *A Fuzzy approach to the guidance of unmanned air vehicles tracking moving targets*, IEEE Transactions on

Control Systems Technology, vol. **16**, pp. 1125–1137.
36. Ioannou, P.; Kokotovic, P. (1984): *Instability analysis and improvement of robustness of adaptive control*, Automatica, vol. **20**, pp. 583–594.
37. Isidori, A. (1995): *Nonlinear Control*, Springer-Verlag.
38. Jayawardhana, B.; Lagemann, H.; Ryan, E. P. (2011): *The circle criterion and input to state stability*, IEEE Control Systems Magazine, vol. **31**, pp. 32–67.
39. Joshi, S. M.; Kelkar, A. (1998): *Inner loop control of supersonic aircraft in the presence of aeroelastic modes*, IEEE Transactions on Control System Technology, vol. **6**, pp. 730–739.
40. Jurdjevic, V. (2008): *Geometric Control Theory*, Cambridge University Press.
41. Kamyar, R., Taheri, E. (2014): *Aircraft optimal terrain threat based trajectory planning and control*, AIAA Journal of Guidance, Control and Dynamics, vol. **37**, pp. 466–483.
42. Kim, H. J., Park, J. B., Joo, Y. H. (2013): *Stabilization conditions of Takagi–Sugeno fuzzy systems based on the fuzzy Lyapunov function under the imperfect premise matching*, American Control Conference, Washington, DC, pp. 5643–5647.
43. Kim, K.; Hwang, K.; Kin, H. (2013): *Study of an adaptive fuzzy algorithm to control a rectangural shaped unmanned surveillance flying car*, Journal of Mechanical Science and Technology, vol. **27**, p. 2477–2486.
44. Kladis, G.; Economou, J.; Knowles, K.; Lauber, J.; Guerra, T. M. (2011): *Energy conservation based fuzzy tracking for unmanned aerial vehicle missions under a priori known wind information*, Engineering Applications of Artificial Intelligence, vol. **24**, pp. 278–294.
45. Kurnaz, S.; Eroglu, E.; Kaynak, O.; Malkoc, U. (2005): *A frugal fuzzy logic based approach for autonomous flight control of UAV*, MICAI 2005, LNAI 3789, Gelbukh, A.; de Albornoz, A.; Terashima, H. (eds), Springer-Verlag, Berlin, pp. 1155–1163.
46. Kurnaz, S.; Cetin, O.; Kaynak, O. (2008): *Fuzzy logic based approach to design of flight control and navigation tasks for an autonomous UAV*, Journal of Intelligent and Robotic Systems, vol. **54**, pp. 229–244.
47. Kuwata, Y.; Schouwenaars, T.; Richards, A.; How, J. (2005): *Robust constrained receding horizon control for trajectory planning*, AIAA Conference on Guidance, Navigation and Control, paper AIAA 2005–6073, DOI 10.2514/6.2005-6079.
48. Levine, W. S. (2011): *Control System Advanced Methods*, CRC Press.
49. Low, C.B. (2010): *A trajectory tracking design for fixed wing unmanned aerial vehicle*, IEEE. Int. Conference on Control Applications, pp. 2118–2123.
50. Lozano, R. (ed) (2010): *Unmanned Aerial Vehicles - Embedded Control*, Wiley.
51. Manikonda, V.; Krishnaprasad, P.S. (2002): *Controllability of a class of underactuated mechanical system with symmetry*, Automatica, vol. **38**, pp. 1837–1850.
52. Markdahl, J.; Hoppe, J.; Wang L.; Hu, X. (2012): *Exact solution to the closed-loop kinematics of an almost globally stabilizing feedback law on S0(3)*, 51st IEEE conference on Decision and Control, Maui, HI, USA, pp. 2274–2279.
53. Marconi, L.; Naldi, R. (2012): *Control of aerial robots*, IEEE Control System Magazine, vol. **32**, pp. 43–65.
54. Medagoda, E.; Gibbens, P. (2014): *Multiple horizon model predictive flight*

control, AIAA Journal of Guidance, Control and Dynamics, vol. **37**, pp. 949–951.

55. Melingui, A.; Chettibi, T.; Merzouki, R.; Mbede, J. B. (2013): *Adaptive navigation of an omni-drive autonomous mobile robot in unstructured dynamic environment*, IEEE Int. Conference on Robotics and Biomimetics, pp. 1924–1929.
56. Muller, J.F. (1967): *Systematic determination of simplified gain scheduling programs*, AIAA Journal of Aircraft, vol. **4**, pp. 529–533.
57. Nakamura, N.; Nakamura, H.; Nishitani, H. (2011): *Global inverse optimal control with guaranteed convergence rates of input affine nonlinear systems*, IEEE Transactions on Automatic Control, vol. **56**, pp. 358–369.
58. Nejjari, F.; Rotondo, D.; Puig, V.; Innocenti, M. (2012): *Quasi-LPV modeling and nonlinear identification of a twin rotor system*, IEEE Mediterranean Conference on Control and Automation, Barcelona, Spain, pp. 229–234.
59. Nelson, R.; Barber, B.; McLain, T.; Beard, R. (2007): *Vector field path following for miniature air vehicle.* IEEE Transactions on Robotics, vol. **23**, pp. 519–529.
60. Nonami, K.; Kendoul, F.; Suzuki, S.; Wang, W.; Nakazawa, D. (2010):*Autonomous Flying Robots: Unmanned Aerial Vehicles and Micro-aerial Vehicles*, Springer.
61. Olfati-Saber, R. (2001): *Nonlinear control of under-actuated mechanical systems with application to robotics and aerospace vehicles*, PhD Thesis, MIT, Cambridge, MA, USA.
62. Pan, Y.; Joo Er, M.; Hwang, D.; Wang, Q. (2011): *Fire-rule based direct adaptive type 2 fuzzy H_∞ tracking control*, Engineering Applications of Artificial Intelligence, vol. **24**, pp. 1174–1185.
63. Papageorgiou, C.; Glover, K. (2004): *Robustness analysis of nonlinear dynamic inversion control laws with application to flight control*, 43th IEEE conference on Decision and Control, Bahamas, pp. 3485–3490.
64. Pappas, G.; Simic, S. (2002): *Consistent abstractions of affine control systems*, IEEE Transactions on Automatic Control, vol. **47**, pp. 745–756.
65. Peng, H.; Gao, Q.; Wu, Z.; Zhang, W. (2013): *Efficient sparse approach for solving receding horizon control problems*, AIAA Journal of Guidance, Control and Dynamics, vol. **36**, pp. 1864–1872.
66. Ren, W.; Beard, R. (2004): *Trajectory tracking for unmanned air vehicles with velocity and heading rate constraints*, IEEE Transactions on Control Systems Technology, vol. **12**, pp. 706–716.
67. Richardson, T.; Lowenberg, M.; Dibernardo, M.; Charles, G. (2006): *Design of a gain-scheduled flight control system using bifurcation analysis*, AIAA Journal of Guidance, Control and Dynamics, vol. **29**, pp. 444–453.
68. Rigatos, G. G. (2011): *Adaptive fuzzy control for field-oriented induction motor drives*, Neural Computing and Applications, vol. **21**, pp. 9–23, DOI 10.1007/s00521-011-0645-z.
69. Rotondo, D.; Nejjari, F.; Torren, A.; Puig, V. (2013): *Fault tolerant control design for polytopic uncertain LPV system: application to a quadrotor*, IEEE Conference on Control and Fault-tolerant Systems, Nice, France, pp. 643–648.
70. Rugh, W. J. (1991): *Analytical framework for gain scheduling*, IEEE Transactions on Control Systems, vol. **11**, pp. 79–84.

71. Samar, R.; Rehman, A. (2011): *Autonomous terrain following for unmanned air vehicles*, Mechatronics, vol. **21**, pp. 844–860.
72. Sastry, S. (1999): *Nonlinear Systems, Analysis, Stability and Control*, Springer, Berlin.
73. Sato, M. (2013): *Robust gain scheduling flight controller using inexact scheduling parameters*, American Control Conference, pp. 6829–6834.
74. Schirrer, A.; Westermayer, C.; Hemedi, M.; Kozek, M. (2010): *LQ based design of the inner loop lateral control for a large flexible BWB-type aircraft*, IEEE Int. Conference on Control Applications, Yokohama, Japan, pp. 1850–1855.
75. Schwefel, H. P.; Wegener, I.; Weinert, K. (2002): *Advances in Computational Intelligence*, Springer.
76. Selma, B.; Chouraqui, S. (2013): *Neuro fuzzy controller to navigate an unmanned vehicle*, Springer Plus Journal, vol. **2**, pp. 1–8.
77. Shaked, U. (2002): *A LPV approach to robust H_2 and H_∞ static output feedback design*, American Control Conference, pp. 3476–3481, DOI 0-7803-7516-5.
78. Shin, J.; Balas, G. J. (2002): *Optimal blending functions in linear parameter varying control synthesis for F16 aircraft*, American Control Conference, pp. 41–46, DOI 0-7803-7298-0.
79. Shue, S.; Agarwal, R. (1999): *Design of automatic landing system using mixed $H2$ / H_∞ control*, AIAA Journal of Guidance, Control and Dynamics, vol. **22**, pp. 103–114.
80. Singh, S.; Steinberg, A. (2003): *Nonlinear adaptive and sliding mode flight path control of F/A 18 model*, IEEE Transactions on Aerospace and Electronic Systems, vol. **39**, pp. 1250–1262.
81. Slegers,N., Kyle, J. , Costello, M. (2006): *Nonlinear Model Predictive Control Technique for Unmanned Air Vehicles*, AIAA Journal of Guidance, Control and Dynamics, Vol. **29**, pp. 1179–1188.
82. Sonneveld, L.; Chu, Q. P.; Mulder, J. A. (2007): *Nonlinear flight control design using constrained adaptive backstepping*, AIAA Journal of Guidance, Control and Dynamics, vol. **30**, pp. 322–336.
83. Sonneveld, L.; van Oort, E. R.; Chu, Q. P.; Mulder, J. A. (2009): *Nonlinear adaptive trajectory control applied to an F-16 model*, AIAA Journal of Guidance, Control and Dynamics, vol. **32**, pp. 25–39.
84. Sontag, E. D. (1998): *Mathematical Control Theory*, Springer.
85. Stevens, B. L.; Lewis, F. L. (2007): *Aircraft Control and Simulation*, Wiley.
86. Summers, T. H.; Akella, M. A.; Mears, M. J. (2009): *Coordinated standoff tracking of moving targets: control laws and information architectures*, AIAA Journal of Guidance, Control and Dynamics, vol. **32**, pp. 56–69.
87. Sun, Y. (2007): *Necessary and sufficient conditions for global controllability of planar affine control systems*, IEEE Transactions on Automatic Control, vol. **52**, pp. 1454–1460.
88. Sussmann, H. J. (1987): *A general theorem on local controllability*, SIAM Journal on Control and Optimization, vol. **25**, pp. 158–195.
89. Tewari, A. (2010): *Automatic Control of Atmospheric and Space Flight Vehicles*, Birkhauser.
90. Turcio, W.; Yoneyama, T.; Moreira, F. (2013): *Quasi-LPV gain scheduling control of a nonlinear aircraft pneumatic system*, IEEE Mediterranean Conference on Control and Automation, Crete, Greece, pp. 341–350.

91. Ure, N. K.; Inalhan, G. (2012): *Autonomous control of unmanned combat air vehicle*, IEEE Control System Magazine, vol. **32**, pp. 74–95.
92. UrRehman, O.; Petersen, I. R.; Fidan, B. (2012): *Minimax linear quadratic Gaussian control of nonlinear MIMO system with time-varying uncertainties*, Australian Control Conference, Sydney, Australia, pp. 138–143.
93. Wang, Q. (2005): *Robust nonlinear flight control of a high-performance aircraft*, IEEE Transactions on Control Systems Technology, vol. **13**, pp. 15–26.
94. Xiaowei, G.; Xiaoguang, G. (2014): *Effective real time unmanned air vehicle path planning in presence of threat netting*, AIAA Journal of Aerospace Information System, vol. **11**, pp. 170–177.
95. Xin, M.; Balakrishnan, S. N. (2005): *A new method for suboptimal control of a class of nonlinear systems*, Optimal Control Applications and Methods, vol. **26**, pp. 55–83.
96. Yang, C.; Chang, C.; Sun, Y. (1996): *Synthesis of H_∞ controller via LQG-based loop shaping design*, IEEE Conference on Decision and Control, pp. 739–744.
97. Yang, W.; Hammoudi, M. N.; Hermann, G.; Lowenberg, M.; Chen, X. (2012): *Two state dynamic gain scheduling control applied to an aircraft model*, International Journal of Nonlinear Mechanics, vol. **47**, pp. 1116–1123.
98. Yavrucuk, I.; Prasad, J. V.; Unnikrishnan, S. U. (2009) :*Envelope protection for autonomous unmanned aerial vehicles*, AIAA Journal of Guidance, Control and Dynamics, vol. **32**, pp. 248–261.
99. Yedavalli, R. K.; Banda, S.; Ridgely, D. B. (1985): *Time-domain stability robustness measures for linear regulators*, AIAA Journal of Guidance, Control, and Dynamics, vol. **8**, pp. 520–524.
100. Yucelen, T.; Calise, A. J. (2011): *Derivative free model reference adaptive control*, AIAA Journal of Guidance, Control and Dynamics, vol. **34**, pp. 933-950.
101. Yuksel, T.; Sezgin, A. (2010): *Two fault detection and isolation schemes for robot manipulator using soft computing techniques*, Applied Soft Computing, vol. **10**, pp. 125–134.
102. Zaheer, S.; Kim, J. (2011): *Type 2 fuzzy airplane altitude control: a comparative study*, IEEE Int. Conference on Fuzzy Systems, Taipei, pp. 2170–2176.
103. Zarafshan, P.; Moosavian, S. A.; Bahrami, M. (2010): *Comparative controller design of an aerial robot*, Aerospace Science and Technology, vol. **14**, pp. 276–282.
104. Zhang, J.; Zhan, Z.; Liu, Y.; Gong, Y. (2011): *Evolutionary computation meets machine learning: a survey*, IEEE Computational Intelligence Magazine, vol. **6**, pp.68–75.
105. Zou, Y.; Pagilla, P. R.; Ratliff, R. T. (2009): *Distributed formation flight control using constraint forces*, AIAA Journal of Guidance, Control and Dynamics, vol. **32**, pp. 112–120.

4 Flight Planning

ABSTRACT

Flight planning is defined as finding a sequence of actions that transforms some initial state into some desired goal state. This chapter begins with path and trajectory planning: trim trajectories, time optimal trajectories and nonholonomic motion planning. Trajectory generation refers to determining a path in free configuration space between an initial configuration of the aircraft and a final configuration consistent with its kinematic and dynamic constraints. The optimal approach can be used to realize the minimum time trajectory or minimum energy to increase the aircraft's endurance. Zermelo's problem is then considered; it allows the study of aircraft's trajectories in the wind. In the middle of the chapter, guidance and collision/obstacle avoidance are considered. Planning trajectories is a fundamental aspect of autonomous aircraft guidance. It can be considered as a draft of the future guidance law. The guidance system can be said to fly the aircraft on an invisible highway in the sky by using the attitude control system to twist and turn the aircraft. Guidance is the logic that issues the autopilot commands to accomplish certain flight objectives. Algorithms are designed and implemented such that the motion constraints are respected while following the given command signal. Flight planning is also the process of automatically generating alternate paths for an autonomous aircraft, based on a set of predefined criteria, when obstacles are detected in the way of the original path. Aircraft operate in a three-dimensional environment where there are static and dynamic obstacles as well as other aircraft and they must avoid turbulence and storms. As obstacles may be detected while the aircraft moves through the environment or their locations may change over time, the trajectory needs to be updated and satisfy the boundary conditions and motion constraints. Then, mission planning is introduced by route optimization and fuzzy planning.

4.1 INTRODUCTION

Flight planning generates paths that are consistent with the physical constraints of the autonomous aircraft, the obstacle and collision avoidance and weighed regions. Weighed regions are regions with abnormally low or high pressure, wind speeds or any other factor affecting flight. 3D mission planning involves creating a path generation system which helps the aircraft to reach the mission goal but also creates a path to satisfy different constraints during the mission. This path generation system generates the path from the initial point to the mission goal and navigates the aircraft. Flight planning requires an awareness of the environment in which it is operating [23]. The position, orientation and velocity of the aircraft are known from the sensors and the flight management system has information about the meteorological conditions and probable obstacles to avoid. In this chapter, the assumption is made that the information required will be available. More information about situation awareness can be found in the following chapter.

The human approach to navigation is to make maps and erect sign posts or use landmarks. Robust navigation in natural environments is an essential capability of smart autonomous aircraft. In general, they need a map of their surroundings and the ability to locate themselves within that map, in order to plan their motion and successfully navigate [208, 210]. Map based navigation requires that the aircraft's position is always known.

Smart autonomous aircraft should be able to make decisions for performing tasks or additional actions or for changing current tasks. They should have the capacity to perceive their environment and consequently update their activity. The autonomy of these systems is increased as high level decisions, such as aircraft way point assignment and collision avoidance, are incorporated in the embedded software [184].

Within the autonomy area, automated guidance and trajectory design play an important role. onboard maneuver planning and execution monitoring increase the aircraft maneuverability, enabling new mission capabilities and reducing costs [31, 127].

The main tasks of the flight planning system are:

1. Given a mission in terms of waypoints, generate a series of paths while providing a minimum clearance.
2. Generate the reference trajectories while satisfying aircraft constraints.

The route planning problem is about finding an optimum path between a start point and a destination point considering the dynamics of the aircraft, the environment and specific constraints implied by operational reasons. The calculation of a flight plan involves the consideration of multiple elements. They can be classified as either continuous or discrete, and they can include nonlinear aircraft performance, atmospheric conditions, wind forecasts, airspace structure, amount of departure fuel and operational constraints [181]. Moreover, multiple differently characterized flight phases must be considered in flight planning. The multi-phase motion of an aircraft can be modeled by a set of differential algebraic dynamic subsystems:

$$\Upsilon = \{\Upsilon_0, \Upsilon_1, \dots, \Upsilon_{N-1}\}$$

so that for $k \in \{0, \dots, N-1\}$,

$$\Upsilon_k = \{f_k : \mathbb{X}_k \times \mathbb{U}_k \times \mathbb{R}^{n_{l_k}} \longrightarrow \mathbb{R}^{n_{X_k}}, g_k : \mathbb{X}_k \times \mathbb{U}_k \times \mathbb{R}^{n_{l_k}} \longrightarrow \mathbb{R}^{n_{Z_k}}\}$$

where f_k represents the differential equation

$$\dot{X} = f_k(X, U, p)$$

for the kth subsystem, g_k describes the algebraic constraints and k represents the index for phases. The state set has the following property $\mathbb{X}_k \subset \mathbb{R}^{nX_k} \subseteq \mathbb{R}^{nX}$ and the control set is such that $\mathbb{U}_k \subset \mathbb{R}^{nU_k} \subseteq \mathbb{R}^{nU}$. A vector of parameter is $p \in \mathbb{R}^{n_p}$. Let the switch times between phases be defined as:

$$t_I = t_0 \leq t_1 \leq \cdots \leq t_N = t_f$$

That is at time t_k, the dynamics subsystem changes from Υ_{k-1} to Υ_k. As a consequence, in the time subinterval $[t_k, t_{k+1}]$, the system evolution is governed by the dynamic subsystem Υ_k. In the sub-interval $[t_{N-1}, t_N]$, the active dynamics subsystem is Υ_{N-1}. The switch is triggered by a sequence of switch conditions in the set $\mathbb{S} = \{\mathbb{S}_1, \mathbb{S}_2, \ldots, \mathbb{S}_{N-1}\}$; $\mathbb{S} = \mathbb{S}_A \cup \mathbb{S}_c$ provides logic constraints that relate the continuous state and mode switch. $\mathbb{S}_A$ corresponds to the set of autonomous switch and $\mathbb{S}_c$ to the set of controlled switch. For instance, for an autonomous switch, when the state trajectory intersects a certain set of the state space at subsystem $k-1$ the system is forced to switch to subsystem k. For a controlled switch, only when the state belongs to a certain set, the transition from $k-1$ to k is possible. This controlled switch might take place in response to the control law. Key parameters depend on the mission. There is no universal way of picking them.

The configuration space may be altered if the aircraft properties or the characteristics of the environment change. The scenarios in which the UAS missions are executed are dynamic and can change continuously. Part of the mission or in the worst case the initial mission objectives could thus be modified. Based on the particular situation, the aircraft could fly toward a destination point, to monitor a set of objectives or a specific rectangular area or to survey a target of interest. When the aircraft must monitor an area or survey a target, the payload parameters and the altitude of the maneuver must be set. The planning algorithm is composed of three subalgorithms:

1. **Path planning**: An integral part of UAV operation is the design of a flight path or mission path that achieves the objectives of the mission. If a monitor or a survey task is commanded, it runs assuming subsequent couples of **primary mission waypoints** (PMW) as input data. If a *fly-to* task is commanded, input data consist of the current aircraft position as the starting point and the commanded position as the destination point. The fundamentals of flight are in general: straight and level flight (maintenance of selected altitude), ascents and descents, level turns and wind drift correction. The algorithm calculates the path between each couple of primary mission waypoints, called a **macroleg**, generating an appropriate sequence of route waypoints and corresponding speed data in a 3D space. Each macroleg can be composed of climb, cruise and descent phases depending on the relative altitude and position of the primary points and the obstacles. The global route is the connected sequence of the macrolegs. Such a route is safe and efficient and is provided as a set

of waypoints to pass through at a specified speed. The safe algorithm calculates a set of geometric safe paths from the first to the second primary waypoints for each couple of waypoints. The cost algorithm manipulates the safe paths and generates the path that fulfills the aircraft performance and mission priority. It generates also the reference trajectories while satisfying aircraft constraints.

2. **Mission planning**: Mission planning ensures that the UAV operates in a safe and efficient manner. It identifies a series of ordered point locations called primary mission waypoints, which will be included in the route to accomplish the mission objectives. The mission planning system needs to find motion plans that avoid collision scenarios and maximize mission efficiency and goal attainment. Furthermore, the plans cannot exceed the performance limitations of the aircraft. There are complex trade-offs that must be made with regard to mission goals, mission efficiency and safety objectives. As autonomous aircraft operate in an unpredictable and dynamic outdoor environment, it is necessary to combine pre-flight strategic planning with in-flight tactical re-planning. There is significant time pressure on tactical re-planning due to aircraft velocities and the additional constraint for fixed wing UAV motion, maintaining a minimum stall speed [209].
3. **Mission management**: It provides the translation of mission objectives into quantifiable, scientific descriptions giving a measure to judge the performance of the platform, i.e., the system in which the mission objectives are transformed into system parameters. The mission management functions can be split into two different functional elements:
    a. The payload functions are mission specific and directly relate to the mission.
    b. The aircraft management system is defined as the set of functions that are required for the onboard embedded software to understand, plan, control and monitor the aircraft operations. They usually represent the safety critical functionality required for the safe employment of the platform; hence, they include all the flight critical and safety related functions.

Remark 4.1. *Online re-planning onboard the aircraft ensures continued conformance with the* ***National Airspace System*** *(NAS) requirements in the event of an outage in the communication link.*

4.2 PATH AND TRAJECTORY PLANNING

The general problem of path planning is to determine a motion for an aircraft allowing it to move between two configurations while respecting a number of constraints and criteria. These arise from several factors of various nature and generally depend on the characteristics of the system, environment and type of task.

Problem 4.1. ***Planning:*** *Given a mechanical system (S), whose motion is governed by a system of differential equations: find aircraft path and trajectory such that its movement is constrained limiting the configuration space and the constraints on the controls.*

The planning problem implies the calculation of a trajectory (X (t), U(t)) satisfying the differential equation such that $X(t_0) = X_0$ and $X(t_f) = X_f$.

1. A **path** is a set of configurations reached by the aircraft to go from one configuration to another. Path planning (finding a path connecting two configurations without collision) is a kinematical/geometrical problem. A path is defined as the interpolation of position coordinates. A path does not specify completely the motion of the system in question.
2. A **trajectory** is a path over a law describing the time instants of passage of each system configuration. The path planning is not only a kinematical/geometrical problem but also a dynamical problem. A trajectory refers to timely annotated paths; it is aircraft specific.

In this case, the constraints relate to aircraft geometry, kinematics and dynamics. A solution must optimize a cost function expressed in terms of distance traveled by the aircraft between two extremal configurations, time or energy necessary to the implementation of its motion. Optimization problems are divided in two categories, those with continuous variables and those with discrete variables.

Trajectory planning can find a path expressed in terms of the degrees of freedom of the aircraft and velocity/angle rates. A 4D motion planning comprises a referenced sequence of 3D waypoints and the desired track velocities between them. Such tracks are also referred to as trajectory segments. It is necessary to incorporate an approximation of aircraft dynamics to ensure that the generated paths are physically realizable [85].

A motion plan consists of two classes of motion primitives [95]:

1. the first class is a special class of trajectories: trim trajectories. A trim is a steady state or quasi-steady flight trajectory.
2. The second class consists of transitions between trims: maneuvers.

Each flight segment is defined by two end flight constraints which together with the dynamic model form a system of **differential algebraic equations** (DAE). The resolution of the differential algebraic equations for the different flight segments is often based on the reduction of the aircraft equations of motion to a system of ordinary differential equations through the explicit utilization of the flight constraints. A continuously differentiable path should be preferred to enable smooth transitions. Typically waypoints and paths are planned when UAS are operated autonomously. After planning waypoints, paths are then typically planned joining these waypoints. As there are dynamic constraints, the paths are planned by using various geometric curves instead of straight lines. After planning paths, guidance laws are designed for path following.

4.2.1 TRIM TRAJECTORIES

A trimmed flight, condition is defined as one in which the rate of change (of magnitude) of the state vector is zero (in the body-fixed frame) and the resultant of the applied forces and moments is zero. In a trimmed trajectory, the autonomous aircraft will be accelerated under the action of non-zero resultant aerodynamic and gravitational forces and moments; these effects will be balanced by effects such as centrifugal and gyroscopic inertial forces and moments.

$$\dot{u} = \dot{v} = \dot{w} = 0 \qquad \dot{p} = \dot{q} = \dot{r} = 0 \tag{4.1}$$

Under the trim condition, the aircraft motion is uniform in the body-fixed frame. The aerodynamic coefficients which are variable in time and space become stationary under this condition and their identification becomes easier [21, 30]. Their geometry depends on the body-fixed linear velocity vector V_e, the roll angle ϕ_e, pitch angle θ_e and the rate of yaw angle $\dot{\psi}_e$. The choice of these quantities should satisfy the dynamic equations, the control saturation and envelope protection constraints.

For trim trajectories, the flight path angle γ is a constant γ_0 while the angle χ is linearly varying versus time t.

$$\chi(t) = \chi_0 + t\chi_1 \tag{4.2}$$

The parameters γ_0, χ_0, χ_1 being constants, the following relations can be proposed:

$$x(t) = x_0 + \frac{\cos\gamma_0}{\chi_1}\left(\cos(\chi_0 + \chi_1 t) - \cos\chi_0\right) \tag{4.3}$$

$$y(t) = y_0 - \frac{\cos\gamma_0}{\chi_1}\left(\sin(\chi_0 + \chi_1 t) - \sin\chi_0\right) \tag{4.4}$$

$$z(t) = z_0 + \sin(\gamma_0)t \tag{4.5}$$

Trim trajectories are represented in general by helices, with particular cases such as straight motion or circle arcs. For this kind of helix, curvature κ and torsion τ are constant:

$$\begin{aligned} \kappa &= \chi_1 \cos(\gamma_0) \\ \tau(s) &= \chi_1 \sin(\gamma_0) \end{aligned} \tag{4.6}$$

The dynamic model allows to compute the following relations

$$T = \frac{D + mg\sin\gamma_0}{\cos\alpha} \tag{4.7}$$

$$\sigma = 0 \tag{4.8}$$

and

$$(L + T\sin\alpha) - mg\cos\gamma_0 = 0 \tag{4.9}$$

A part of a helix can be used to join two configurations, respectively, $X_0 = (x_0, y_0, z_0, \chi_0, \gamma_0)$ and $X_f = (x_f, y_f, z_f, \chi_f, \gamma_f)$. This particular case occurs when $\gamma_0 = \gamma_f$ and the following relationships are verified

$$\chi_1 = \sin(\gamma_0)\frac{\chi_f - \chi_0}{z_f - z_0} \tag{4.10}$$

with the constraint between the initial and final positions:

$$[\chi_1(x_f - x_0) + \cos\gamma_0 \sin\chi_0]^2 + [\chi_1(y_f - y_0) - \cos\gamma_0\cos\chi_0]^2 = \cos^2\gamma_0 \tag{4.11}$$

L the length of the path being given by

$$L = \frac{\chi_f - \chi_0}{\chi_1} \tag{4.12}$$

The trim trajectories have the advantage of facilitating the planning and control problems. The role of the trajectory generator is to generate a feasible time trim trajectory for the aircraft.

4.2.2 TRAJECTORY PLANNING

Aircraft trajectory planning goes from the mathematical optimization of the aircraft trajectory to the automated parsing and understanding of desired trajectory goals, followed by their formulation in terms of a mathematical optimization programming [53, 80, 156, 157].

4.2.2.1 Time optimal trajectories

The subject of this section is to formulate the trajectory generation problem in minimum time as this system has bounds on the magnitudes of the inputs and the states [55]. The velocity is assumed to be linearly variable. As the set of allowable inputs is convex, the time optimal paths result from saturating the inputs at all times (or zero for singular control). For a linear time-invariant controllable system with bounded control inputs, the time-optimal control solution to a typical two point boundary value problem is a bang-bang function with a finite number of switches [16, 43, 154].

Problem 4.2. *The Dubins problem is the problem of describing the minimum time trajectories for differential system defined as:*

$$\begin{aligned} \dot{x} &= \cos\chi \\ \dot{y} &= \sin\chi \\ \dot{\chi} &= U \\ |U| &\leq 1 \end{aligned} \tag{4.13}$$

It has been proved by Dubins [58] that the optimal arcs of this problem (4.13) are a concatenation of at most three pieces: S (straight), R (turn to the right) and L (turn to the left). The shortest path for a Dubins vehicle consists of three consecutive path segments, each of which is a circle of minimum turn radius C or a straight line L. So the Dubins set $\mathbb{D}$ includes six paths: $\mathbb{D} = \{LSL, RSR, RSL, LSR, RLR, LRL\}$. If the paths are sufficiently far apart, the shortest path will always be of CSC type. In the case of an airplane, this is always the case.

In [124], the problem of finding a fastest path between an initial configuration $(0, 0, \chi_s)$ and a final configuration (x_t, y_t, χ_t) is considered. The direction dependent model represents an extension of the original Dubins aircraft model which assumes isotropic speed and minimum turning radius $\frac{V}{R(\theta)}U$.

In [190], the optimal $3D$ curves are helicoidal arcs. This direction dependent framework generalizes some of the previous work, in particular Dubins-like vehicles moving in constant and uniform wind [215].

Problem 4.3. *The Markov–Dubins problem is the problem of describing the minimum time trajectories for the system:*

$$\begin{aligned} \dot{x} &= \cos\chi \\ \dot{y} &= \sin\chi \\ \dot{\chi} &= \omega \\ \dot{\omega} &= U \\ |U| &\leq 1 \end{aligned} \tag{4.14}$$

This system is a dynamic extension of the Dubins system (4.13).

Flight planning's aim is leg-based navigation. A leg specifies the flight path to get to a given waypoint. The crossing points define the rough paths of aircraft. Thus, the goal is to refine these paths to generate trajectories parametrized function of time, satisfying the kinematic constraints of the aircraft.

Remark 4.2. *The principle of maximum of Pontryagin (PMP) provides necessary conditions for excluding certain types of trajectories [33] . Most often, the conclusions derived by applying the PMP characterize a family of sufficient controls containing the optimal control between two points. An application of PMP combined with the tools of Lie algebra helped, in the case of mobile robots to refine the set of optimal trajectories, initially obtained by Reed and Sheep [39, 109].*

4.2.2.2 Nonholonomic motion planning

Nonholonomic motion planning relies on finding a trajectory in the state space between given initial and final configurations subject to nonholonomic constraints [28, 29, 160]. The Lie algebraic method relies on a series of local planning around consecutive current states [125]. Global trajectory results from joining local trajectories. At a current state, a direction of motion towards the goal state is established. Then, a rich enough space of controls is taken [165]. As the system is controllable, via some vector fields, the controls are able to generate the fields [61, 171]. The steering method for affine driftless systems exploits different properties of such a system, namely nilpotence, chained form and differential flatness [5, 23].

The methods involving sinusoid at integrally related frequencies can be modified using some elementary **Fourier analysis** to steer the system (3.131).

The kinematic equations of motion of an aircraft in 3D are given by:

$$\begin{aligned} \dot{x} &= V\cos\gamma\cos\chi \\ \dot{y} &= V\cos\gamma\sin\chi \\ \dot{z} &= -V\sin\gamma \\ \dot{\chi} &= \omega_1 \\ \dot{\gamma} &= \omega_2 \end{aligned} \tag{4.15}$$

If the angles are assumed to be small, an approximation to this system is obtained by setting $\cos\gamma \approx 1, \sin\gamma \approx \gamma, \cos\chi \approx 1, \sin\chi \approx \chi$. Relabeling the variables, the preceding system can be written under a chained form:

$$\begin{aligned} \dot{X}_1 &= U_1 \\ \dot{X}_2 &= U_2 \\ \dot{X}_3 &= U_3 \\ \dot{X}_4 &= X_2 U_1 \\ \dot{X}_5 &= -X_3 U_1 \end{aligned} \tag{4.16}$$

where $X = (x, \chi, \gamma, y, z)^T$, $U = (\dot{x}, \dot{\chi}, \dot{\gamma})^T$. This system can also be written under the following form:

$$X = \begin{pmatrix} 1 \\ 0 \\ 0 \\ X_2 \\ -X_3 \end{pmatrix} U_1 + \begin{pmatrix} 0 \\ 1 \\ 0 \\ 0 \\ 0 \end{pmatrix} U_2 + \begin{pmatrix} 0 \\ 0 \\ 1 \\ 0 \\ 0 \end{pmatrix} U_3 = g_1 U_1 + g_2 U_2 + g_3 U_3 \tag{4.17}$$

Using Lie brackets, the following vectors are calculated:

$$g_4 = [g_1, g_2] = \begin{pmatrix} 0 \\ 0 \\ 0 \\ 1 \\ 0 \end{pmatrix} \qquad g_5 = [g_1, g_3] = \begin{pmatrix} 0 \\ 0 \\ 0 \\ 0 \\ -1 \end{pmatrix} \tag{4.18}$$

The determinant of the matrix $(g_1, g_2, g_3, g_4, g_5)$ being different from zero, the controllability rank condition is satisfied. However, it should be noticed that for a fixed-wing aircraft, U_1 is not symmetric as $0 < V_{stall} \leq U_1 \leq V_{max}$, while U_2, U_3 are symmetric. Hence, small time local controllability cannot be ensured.

This multi-chained form system can be steered using sinusoid at integrally related frequencies. To steer this system, first, the controls U_1, U_2, U_3 are used to steer x, χ, γ to their desired locations:

$$U_1 = \delta_1 \cos(\omega t) \qquad U_2 = \delta_2 \cos(k_2 \omega t) \qquad U_3 = \delta_3 \cos(k_3 \omega t) \tag{4.19}$$

where k_2, k_3 are positive integers.

By integration, x, χ, γ are all periodic and return to their initial values

$$x = \frac{\delta_1}{\omega} \sin(\omega t) + X_{10} \tag{4.20}$$

$$\chi = \frac{\delta_2}{k_2 \omega} \sin(k_2 \omega t) + X_{20} \tag{4.21}$$

$$\gamma = \frac{\delta_3}{k_3 \omega} \sin(k_3 \omega t) + X_{30} \tag{4.22}$$

$$y = -\frac{\delta_1 \delta_2}{2k_2 \omega} \left(\frac{\cos((k_2+1)\omega t)}{(k_2+1)\omega} + \frac{\cos((k_2-1)\omega t)}{(k_2-1)\omega} \right) + X_{20} \frac{\delta_1}{\omega} \sin(\omega t) + X_{40} \tag{4.23}$$

and

$$z = \frac{\delta_1 \delta_3}{2k_3 \omega} \left(\frac{\cos((k_3+1)\omega t)}{(k_3+1)\omega} + \frac{\cos((k_3-1)\omega t)}{(k_3-1)\omega} \right) - X_{30} \frac{\delta_1}{\omega} \sin(\omega t) + X_{50} \tag{4.24}$$

where $X_{10}, X_{20}, X_{30}, X_{40}, X_{50}$ are integration constants.

The problem of steering the approximate model from $X_0 \in \mathbb{R}^5$ at $t = 0$ to $X_f \in \mathbb{R}^5$ at $t = 1$ is considered. The initial conditions allow the calculation of the integration constants:

$$X_{10} = x_0 \quad X_{20} = \chi_0 \quad X_{30} = \gamma_0 \tag{4.25}$$

$$X_{40} = y_0 + \frac{\delta_1 \delta_2}{2k_2 \omega} \left(\frac{1}{(k_2+1)\omega} + \frac{1}{(k_2-1)\omega} \right) \tag{4.26}$$

and

$$X_{50} = z_0 - \frac{\delta_1 \delta_3}{2k_3 \omega} \left(\frac{1}{(k_3+1)\omega} + \frac{1}{(k_3-1)\omega} \right) \tag{4.27}$$

The final conditions allow to write:

$$\delta_1 = \frac{\omega}{\sin \omega} (x_f - x_0) \tag{4.28}$$

$$\delta_2 = \frac{k_2 \omega}{\sin(k_2 \omega)} (\chi_f - \chi_0) \tag{4.29}$$

$$\delta_3 = \frac{k_3\omega}{\sin(k_3\omega)}(\gamma_f - \gamma_0) \tag{4.30}$$

while the following nonlinear equations must be solved in k_2, k_3, ω to characterize entirely the system:

$$y_f - y_0 - \chi_0(x_f - x_0) = -\frac{\delta_1\delta_2}{2k_2\omega}\left(\frac{\cos\left((k_2+1)\omega\right)-1}{(k_2+1)\omega} + \frac{\cos\left((k_2-1)\omega\right)-1}{(k_2-1)\omega}\right) \tag{4.31}$$

$$z_f - z_0 - \gamma_0(x_f - x_0) = -\frac{\delta_1\delta_3}{2k_3\omega}\left(\frac{\cos\left((k_3+1)\omega\right)-1}{(k_3+1)\omega} + \frac{\cos\left((k_3-1)\omega\right)-1}{(k_3-1)\omega}\right) \tag{4.32}$$

Once this has been done, all the reference trajectories are characterized.

In the general case, motion planning may be facilitated by a preliminary change of state coordinates which transforms the kinematics equations of the aircraft into a simpler canonical form.

4.2.3 PATH PLANNING

The path planning problem can be formulated as follows:

Problem 4.4. *Given a C-space Ω, the path planning problem is to find a curve:*

$$C : [0,1] \longrightarrow C_{free} \qquad s \rightarrow C(s)$$

where s is the arc-length parameter of C.

C_{free} represents the set of configurations free of obstacles. An optimal path is a curve C that minimizes a set of internal and external constraints (time, fuel consumption or risk). The complete set of constraints is described in a cost function τ, which can be isotropic or anisotropic [150]:

1. **Isotropic case**: The cost function τ depends only on the configuration X.
2. **Anisotropic case**: The cost function τ depends on the configuration X and a vector of field force.

The aircraft needs to move smoothly along a path through one or more waypoints. This might be to avoid obstacles in the environment or to perform a task that involves following a piecewise continuous trajectory. Several path models, such as straight lines segments, Lagrange interpolation, Hermite interpolation, piecewise linear (quadratic, cubic) interpolation, spline interpolation (cubic, Bezier) can be used [53, 152, 203]. Other techniques can also be used such as wavefront algorithm, Pythagorean hodograph [52, 87, 90].

The unmanned aircraft minimum **curvature radius** and pitch angle constraints should be satisfied because the aircraft's curvature radius is highly related to the geometry, kinematics and dynamics of the aircraft [17, 18]. In addition, the pitch angle is generally limited in a certain range for the sake of the aircraft safety in 3D space; the pitch angle at all points on the trajectory must be constrained between the assigned lower and upper bounds [40].

The optimal path is essential for path planning. Path planning should produce not only a feasible path, but also the optimal one connecting the initial configuration to the final configuration. In [204], a real time dynamic **Dubins helix** method for trajectory smoothing is presented. The projection of 3D trajectory on the horizontal plane is partially generated by the Dubins path planner such that the curvature radius constraint is satisfied. The helix curve is constructed to satisfy the pitch angle constraint, even in the case where the initial and final configurations are close.

4.2.3.1 B-spline formulation

Splines are a set of special parametric curves with certain desirable properties. They are piecewise polynomial functions, expressed by a set of control points. There are many different forms of splines, each with their own attributes. However, there are two desirable properties:

1. Continuity: the generated curve smoothly connects its points.
2. Locality of the control points: the influence of a control point is limited to a neighborhood region.

Different alternate paths can be represented by B-spline curves to minimize computation, because a simple curve can be easily defined by three control points. **A parametric B-spline curve** $p(s)$ of the order k or degree $k-1$ is defined by $(n+1)$ control points p_i, knot vector X and by the relationship

$$p(s) = \sum_{i=0}^{n} p_i N_{i,k}(s) \tag{4.33}$$

where $N_{i,k}(s)$ are the **Bernstein basis functions** and are generated recursively using

$$N_{i,k}(s) = \frac{(s - X_i)N_{i,k-1}(s)}{X_{i+k-1} - X_i} + \frac{(X_{i+k} - s)N_{i+1,k-1}(s)}{X_{i+k-1} - X_{i+1}} \tag{4.34}$$

and

$$N_{i,1} = \left\{ \begin{array}{cc} 1 & \text{If } X_i \leq s \leq X_{i+1} \\ 0 & \text{Otherwise} \end{array} \right\} \tag{4.35}$$

The control points define the shape of the curve. By definition, a low degree B-spline will be closer and more similar to the control polyline (the line formed by connecting the control points in order). The B-splines used can be third degree B-splines to ensure that the generated curves stay as close to the control points as possible [64].

4.2.3.2 Cubic Hermite spline

The Hermite spline is a special spline with the unique property that the curve generated from the spline passes through the control points that define the spline. Thus a set of pre-determined points can be smoothly interpolated by simply setting these points as control points for the Hermite spline [75].

Cubic Hermite spline interpolation passes through all the waypoints and it is possible to assign the derivative values at the control points and also obtain local control over the paths. A solution to the path generation problem is to use a cubic Hermite spline for each pair of successive waypoints [115].

Given a non negative integer n, P_n denotes the set of all real-valued polynomials. The partition of the interval $I = [a, b]$ is given as $a = X_1 < X_2 < \cdots < X_n = b$ and $f_i, i = 1, \dots, n$, the corresponding set of monotone data at the partition points:

$$p(s) = f_i H_1(s) + f_{i+1} H_2(s) + h_i H_3(s) + h_{i+1} H_4(s) \tag{4.36}$$

where $H_k(s)$ are the cubic Hermite basis functions for the interval I_i:

$$\begin{array}{ll} H_1(s) = \varphi \frac{X_{i+1}-s}{h_i} & H_2(s) = \varphi \frac{s-X_i}{h_i} \\ H_3(s) = -h_i \eta \frac{X_{i+1}-s}{h_i} & H_4(s) = h_i \eta \frac{s-X_i}{h_i} \end{array} \tag{4.37}$$

where $h_i = X_{i+1} - X_i$, $\varphi = 3t^2 - 2t^3$, $\eta = t^3 - t^2$.

This methodology can be extended to parametric splines. This entails the introduction of the independent variable ϑ and the formulation of one separate equation for each one of the data variable:

$$x_d(\vartheta) = C_{x_3}(\vartheta - \vartheta_i)^3 + C_{x_2}(\vartheta - \vartheta_i)^2 + C_{x_1}(\vartheta - \vartheta_i) + C_{x_0} \tag{4.38}$$

where

$$\begin{array}{ll} C_{x0} = X_i & C_{x_1} = X_i' \\ C_{x_2} = \frac{3S_i^x - x_{i+1}' - 2x_i'}{\Delta\vartheta_i} & C_{x3} = \frac{-2S_i^x + x_{i+1}' + x_i'}{\Delta\vartheta_i} \end{array} \tag{4.39}$$

where $(.)'$ denotes differentiation with respect to parameter ϑ, $\Delta\vartheta_i = \vartheta_{i+1} - \vartheta_i$ is the local mesh spacing and $S_i^x = \frac{x_{i+1}+x_i}{\Delta\vartheta_i}$ is the slope of the linear interpolant.

4.2.3.3 Quintic Hermite spline

The task is to find a trajectory, a parametrized curve, $\eta(t) = \begin{pmatrix} x(t) \\ y(t) \\ z(t) \end{pmatrix}$ with $t \in [0, T_f]$ from a start point $\eta(0)$ with specified velocity $\dot{\eta}(0)$ and acceleration $\ddot{\eta}(0) = 0$ to a destination point $\eta(T_f)$ with specified velocity $\dot{\eta}(T_f)$ and acceleration $\ddot{\eta}(T_f) = 0$, taking into account the kinematics of the aircraft, the operational requirements of the mission and the no-fly areas.

The objective function to be minimized can be the flight time T_f, the length of the trajectory $\int_0^{T_f} |\dot{\eta}(t)|dt$, the mean flight height above ground $\frac{1}{T_f}\int_0^{T_f} (z(t) - h_{terrain}(x(t), y(t)))$ with $h_{terrain}(x(t), y(t))$ denoting the terrain elevation of point $(x(t), y(t))$.

The approach is based on a discretization of the airspace by a 3D network [2]. Its topology depends on the kinematic properties of the aircraft, the operational requirements of the mission and the relief of the terrain. Each directed path in the network corresponds to a trajectory which is both flyable and feasible [65]. Generation of the flight path segment is split into two subsystems:

1. A twice continuously differentiable parametrized curve with appropriate conditions is determined using quintic Hermite interpolation

$$\begin{array}{l} x(t) = a_{51}t^5 + a_{41}t^4 + a_{31}t^3 + a_{21}t^2 + a_{11}t + a_{01} \\ y(t) = a_{52}t^5 + a_{42}t^4 + a_{32}t^3 + a_{22}t^2 + a_{12}t + a_{02} \\ z(t) = a_{53}t^5 + a_{43}t^4 + a_{33}t^3 + a_{23}t^2 + a_{13}t + a_{03} \end{array} \tag{4.40}$$

   The parameters a_{ij} can be determined by the endpoint conditions.
2. The flyability and feasibility of the trajectories are checked using a simplified model of the aircraft.

4.2.3.4 Pythagorean hodographs

The Pythagorean hodograph condition in $\mathbb{R}^3$ is given by:

$$x'^2(t) + y'^2(t) + z'^2(t) = \tilde{\sigma}^2(t) \tag{4.41}$$

where $\tilde{\sigma}(t)$ represents the parametric speed. The problem lies in finding an appropriate characterization for polynomial solutions [62].

Theorem 4.1

If relatively prime real polynomials $a(t), b(t), c(t), d(t)$ satisfy the Pythagorean condition:

$$a^2(t) + b^2(t) + c^2(t) = d^2(t) \tag{4.42}$$

they must be expressible in terms of other real polynomials $\tilde{u}(t), \tilde{v}(t), \tilde{p}(t), \tilde{q}(t)$ in the form:

$$\begin{array}{c} a(t) = \tilde{u}^2(t) + \tilde{v}^2(t) - \tilde{p}^2(t) - \tilde{q}^2(t) = x'(t) \\ b(t) = 2\left(\tilde{u}(t)\tilde{q}(t) + \tilde{v}(t)\tilde{p}(t)\right) = y'(t) \\ c(t) = 2\left(\tilde{v}(t)\tilde{q}(t) - \tilde{u}(t)\tilde{p}(t)\right) = z'(t) \\ d(t) = \tilde{u}^2(t) + \tilde{v}^2(t) + \tilde{p}^2(t) + \tilde{q}^2(t) = \sigma(t) \end{array} \tag{4.43}$$

■

This form can be written in several different ways corresponding to permutations of $a(t), b(t), c(t)$ and $\tilde{u}(t), \tilde{v}(t), \tilde{p}(t), \tilde{q}(t)$.

If the polynomials $\tilde{u}(t), \tilde{v}(t), \tilde{p}(t), \tilde{q}(t)$ are specified in terms of Bernstein coefficients on $t \in [0,1]$, the Bezier control points of the spatial PH curves they define can be expressed in terms of these coefficients. For example for $\tilde{u}(t) = u_0(1-t) + u_1 t$ and similarly for $\tilde{v}(t), \tilde{p}(t), \tilde{q}(t)$, the control points of the spatial Pythagorean hodograph cubic are found to be of the form:

$$P_1 = P_0 + \frac{1}{3}\begin{pmatrix} u_0^2 + v_0^2 - p_0^2 - q_0^2 \\ 2\left(u_0 q_0 + v_0 p_0\right) \\ 2\left(v_0 q_0 - u_0 p_0\right) \end{pmatrix} \tag{4.44}$$

$$P_2 = P_1 + \frac{1}{3}\begin{pmatrix} u_0 u_1 + v_0 v_1 - p_0 p_1 - q_0 q_1 \\ \left(u_0 q_1 + u_1 q_0 + v_0 p_1 + v_1 p_0\right) \\ \left(v_0 q_1 + v_1 q_0 - u_0 p_1 - u_1 p_0\right) \end{pmatrix} \tag{4.45}$$

$$P_3 = P_2 + \frac{1}{3}\begin{pmatrix} u_1^2 + v_1^2 - p_1^2 - q_1^2 \\ 2\left(u_1 q_1 + v_1 p_1\right) \\ 2\left(v_1 q_1 - u_1 p_1\right) \end{pmatrix} \tag{4.46}$$

The point P_0 corresponds to the integration constants.

4.2.4 ZERMELO'S PROBLEM

Zermelo's problem corresponds to the problem of aircraft in the wind [102].

4.2.4.1 Initial Zermelo's problem

Zermelo's problem was originally formulated to find the quickest nautical path for a ship at sea in the presence of currents, from a given departure point in $\mathbb{R}^2$ to a given destination point [98]. It can also be applied to the particular case of an aircraft with a constant altitude and a zero flight path angle and the wind velocity represented by $W = (W_N, W_E)$ [176].

A UAV has to travel through a region of strong constant wind at a constant altitude. The wind is assumed to have a constant wind velocity W in the y direction. The autopilot modulates the aircraft's heading χ to minimize travel time to the origin.

4.2.4.1.1 First case study

In the first case study, the UAV is assumed to have a constant velocity V and its heading χ is chosen as an input. The goal is to minimize time with the following boundary conditions: $x_0 = y_0 = 0$ and $x_f = 1; y_f = 0$. The control is assumed to be unconstrained. The minimal time problem can be formulated as follows:

$$\text{Min} \int_0^{T_f} dt \tag{4.47}$$

subject to

$$\begin{aligned} \dot{x} &= V\cos\chi \\ \dot{y} &= V\sin\chi + W_E \\ x(0) &= y(0) = 0 \\ x(T_f) &= 1 \quad y(T_f) = 0 \end{aligned} \tag{4.48}$$

Using the Pontryagin minimum principle, the following optimal control can be calculated:

$$\chi^* = -\arcsin\left(\frac{W_E}{V}\right) \tag{4.49}$$

while the optimal trajectories are:

$$\begin{aligned} x^*(t) &= tV\cos\chi \\ y^*(t) &= t(V\sin\chi + W_E) \end{aligned} \tag{4.50}$$

The final time is:

$$T_f = \frac{1}{\sqrt{V^2 - W_E^2}}$$

This resolution is only possible if $|W_E| \leq V$.

4.2.4.1.2 *Second case study*

The second case study is an attempt to be more realistic. The input now is the rate of the heading $\dot{\chi}$, constrained to belong to the interval $[-U_{max}, U_{max}]$. The boundary conditions are slightly different from the first case: $x_0 = y_0 = \chi_0 = 0$ and $\chi_f = 0; y_f = 0$. The minimal time problem can be formulated as follows:

$$\text{Min} \int_0^{T_f} dt \tag{4.51}$$

subject to

$$\begin{aligned} \dot{x} &= V\cos\chi \\ \dot{y} &= V\sin\chi + W_E \\ \dot{\chi} &= U \\ x(0) &= y(0) = \chi(0) = 0 \\ \chi(T_f) &= 0 \quad y(T_f) = 0 \end{aligned} \tag{4.52}$$

Using the Pontryagin minimum principle, the following optimal control can be calculated:

$$U^* = \left\{ \begin{array}{ll} U_{max} & 0 \leq t \leq t_1 \\ -U_{max} & t_1 \leq t \leq T_f \end{array} \right\} \tag{4.53}$$

with $t_1 = \frac{T_f}{2}$ while the optimal trajectories are:

$$\chi^* = \left\{ \begin{array}{ll} U_{max}t & 0 \leq t \leq t_1 \\ U_{max}(T_f - t) & t_1 \leq t \leq T_f \end{array} \right\} \tag{4.54}$$

$$x^* = \left\{ \begin{array}{lr} \frac{V}{U_{max}} \sin\left(U_{max} t\right) & 0 \leq t \leq t_1 \\ \frac{V}{U_{max}} \sin\left(U_{max}(t - T_f)\right) + 2\frac{V}{U_{max}} \sin\left(\frac{T_f U_{max}}{2}\right) & t_1 \leq t \leq T_f \end{array} \right\} \tag{4.55}$$

$$y^* = \left\{ \begin{array}{lr} -\frac{V}{U_{max}} \cos\left(U_{max} t\right) + W_E t + \frac{V}{U_{max}} & 0 \leq t \leq t_1 \\ \frac{V}{U_{max}} \cos\left(U_{max}(t - T_f)\right) + W_E(t - T_f) - \frac{V}{U_{max}} & t_1 \leq t \leq T_f \end{array} \right\} \tag{4.56}$$

The final time can be found from the resolution of the following equation:

$$\frac{U_{max}}{2} \frac{W_E}{V} T_f - \cos\left(\frac{U_{max} T_f}{2}\right) = 1 \tag{4.57}$$

Depending on the values of U_{max}, V, W, this equation may have or not a real positive solution.

Another case study is treated next. The equations describing the optimal path for the case of linearly varying wind velocity are:

$$\begin{aligned} \dot{x} &= V \cos\chi + W_N(y) \\ \dot{y} &= V \sin\chi \end{aligned} \tag{4.58}$$

where (x, y) are its coordinates and $W_N = \frac{Vy}{h}$ is the velocity of the wind. The initial value of χ is chosen so that the path passes through the origin. For the linearly varying wind strength considered here, the optimal steering angle can be related to the aircraft position through a system of implicit feedback equation [60]

$$\dot{\chi} = -\cos^2\chi \frac{dW_N}{dy} \tag{4.59}$$

If $W_N = \frac{W}{a} y$, a being a constant then

$$\chi = \arctan\left(\frac{W}{a} t + \tan\chi_0\right) \tag{4.60}$$

The optimal trajectory is then given by:

$$y = a\left(\frac{1}{\cos\chi_f} - \frac{1}{\cos\chi}\right) \tag{4.61}$$

and

$$x = \frac{a}{2}\left(\frac{1}{\cos\chi_f}\left(\tan\chi_f - \tan\chi\right) - \tan\chi\left(\frac{1}{\cos\chi_f} - \frac{1}{\cos\chi}\right)\right) + \\ + \frac{a}{2} \ln\left(\frac{\tan\chi_f + \frac{1}{\cos\chi_f}}{\tan\chi + \frac{1}{\cos\chi}}\right) \tag{4.62}$$

If $W_N(x, y) = -Wy$, it has been proved in [91] that the time to go is given by

$$T_f = \frac{1}{W}(\tan \chi_f - \tan \chi_0) \tag{4.63}$$

Remark 4.3. *The following approximate relations can be implemented, for N waypoints:*

$$\dot{\chi}_k = \frac{y_k - y_{k-1}}{x_k - x_{k-1}} \cos^2 \chi_k \tag{4.64}$$

$$\dot{\gamma}_k = -\frac{(z_k - z_{k-1}) \cos^2 \gamma_k}{\cos \chi_k (x_k - x_{k-1}) + \sin \chi_k (y_k - y_{k-1})} \tag{4.65}$$

for $k = 1 \dots N$.

4.2.4.2 2D Zermelo's problem on a flat Earth

The following problem can be solved in the same way using the Pontryagin maximum principle:

Problem 4.5. ***2D Zermelo's Problem on a Flat Earth:*** *Time optimal trajectory generation can be formulated as follows:*

$$min \int_0^{T_f} dt \tag{4.66}$$

subject to

$$\begin{aligned} \dot{x} &= U_1(t) + W_N(x, y) \\ \dot{y} &= U_2(t) + W_E(x, y) \end{aligned} \tag{4.67}$$

with

$$U_1^2(t) + U_2^2(t) \leq V_{max}^2 \tag{4.68}$$

The heading angle is the control available for achieving the minimum time objective [91].

Zermelo's navigation formula consists of a differential equation for $U^*(t)$ expressed in terms of only the drift vector and its derivatives. The derivation can be explained as follows. Let the angle $\chi(t)$ be given by $U_1(t) = V_{max} \cos \chi(t)$ and $U_2(t) = V_{max} \sin \chi(t)$; the following ordinary differential equation must be solved:

$$\frac{d\chi}{dt} = -\cos^2 \chi \frac{\partial W_N}{\partial y} + \sin \chi \cos \chi \left(\frac{\partial W_N}{\partial x} - \frac{\partial W_E}{\partial y} \right) + \sin^2 \chi \frac{\partial W_E}{\partial x} \tag{4.69}$$

Remark 4.4. *Flying exact trajectories relative to the ground may be good for optimizing operations (fixed approach trajectories) but the autopilot has to control the aircraft that drifts with the air mass. Relative to the air mass to adhere to the ground trajectory, the autopilot must continuously vary the bank angle during turns.*

The problem is generating an optimal path from an initial position and orientation to a final position and orientation in the 2D plane for an aircraft with bounded turning radius in the presence of a constant known wind. Some researchers have addressed the problem of optimal path planning of an aircraft at a constant altitude and velocity in the presence of wind with known magnitude and direction [144, 205]. A dynamic programming method to find the minimum time waypoint path for an aircraft flying in known wind was proposed by [94, 95]. A target trajectory in the horizontal plane can be described by connecting straight lines and arcs. Clearly, this type of description forces the aircraft to attain a bank angle instantaneously at the point of transition between the straight and curved segments. In [68], the curved part of a trajectory is designed assuming a certain steady bank angle and initial speed in a no-wind condition. If a strong wind exists, the autopilot has to change the bank angle depending on the relative direction of the wind during the curved flight phase or continuously fly with a steeper than nominal bank angle. A steeper bank angle and changing relative wind direction both affect thrust control.

In the absence of wind, this is the Dubins problem [66]. The original problem of finding the optimal path with wind to a final orientation is transformed over a moving virtual target whose velocity is equal and opposite to the velocity of the wind. An UAV approaching a straight line under a steady wind can be considered as a virtual target straight line moving with an equal and opposite velocity to the wind acting on it in a situation where the wind is absent. So in the moving frame, the path will be a Dubins path. The ground path is indeed the time optimal path [80].

Time optimal navigation for aircraft in a planar time varying flow field has also been considered in [193]. The objective is to find the fastest trajectory between initial and final points. It has been shown that in a point symmetric time varying flow field, the optimal steering policy necessarily has to be such that the rate of the steering angle equals the angular rotation rate of the fluid particles.

In many real scenarios, the direction of wind is not known a priori or it changes from time to time [131]. An approach based on overlaying a vector field of desired headings and then commanding the aircraft to follow the vector field was proposed by [138]. A receding horizon controller was used in [107] to generate trajectories for an aircraft operating in an environment with disturbances. The proposed algorithm modifies the online receding horizon optimization constraints (such as turn radius and speed limits) to ensure that it remains feasible even when the aircraft is acted upon by unknown but bounded disturbances [130].

4.2.4.3 3D Zermelo's problem on a flat Earth

Now, the wind optimal time trajectory planning problem for an aircraft in a 3D space is considered [91, 92]. An aircraft must travel through a windy region. The magnitude and the direction of the winds are known to be functions of position, i.e., $W_N = W_N(x, y, z)$, $W_E = W_E(x, y, z)$ and $W_D = W_D(x, y, z)$ where (x, y, z) are 3D coordinates and (W_N, W_E, W_D) are the velocity components of the wind. The aircraft velocity relative to the air V is constant. The minimum-time path from point A to point B is sought. The kinematic model of the aircraft is

$$\begin{aligned} \dot{x} &= V \cos\chi \cos\gamma + W_N \\ \dot{y} &= V \sin\chi \cos\gamma + W_E \\ \dot{z} &= -V \sin\gamma + W_D \end{aligned} \tag{4.70}$$

where χ is the heading angle of the aircraft relative to the inertial frame and γ is the flight path angle.

Using the principle of the maximum of Pontryagin, the evolution of the heading is obtained as a nonlinear ordinary differential equation:

$$\begin{aligned} \dot{\chi} = \sin^2\chi \tfrac{\partial W_E}{\partial x} + \sin\chi\cos\chi \left(\tfrac{\partial W_N}{\partial x} - \tfrac{\partial W_E}{\partial y} \right) \\ - \tan\gamma \left(\sin\chi \tfrac{\partial W_D}{\partial x} - \cos\chi \tfrac{\partial W_D}{\partial y} \right) - \cos^2\chi \tfrac{\partial W_N}{\partial y} \end{aligned} \tag{4.71}$$

while the evolution of the flight path angle is given by a nonlinear ordinary differential equation:

$$\begin{aligned} \dot{\gamma} = \cos^2\gamma \cos\chi \tfrac{\partial W_N}{\partial z} + \cos^2\gamma \tfrac{\partial W_E}{\partial z} + \sin\gamma\cos\gamma \left(\tfrac{\partial W_N}{\partial x} - \tfrac{\partial W_D}{\partial z} \right) + \\ + \sin^2\gamma \tfrac{\partial W_E}{\partial x} - \sin^2\gamma \sec\chi \tfrac{\partial W_D}{\partial x} - \sin\gamma\cos\gamma\tan\chi\sin^2\chi \tfrac{\partial W_E}{\partial x} + \\ - \sin\gamma\cos\gamma\sin^2\chi \left(\tfrac{\partial W_N}{\partial x} - \tfrac{\partial W_E}{\partial y} \right) + \tan\chi \sin^2\gamma \left(\sin\chi \tfrac{\partial W_D}{\partial x} - \cos\chi \tfrac{\partial W_D}{\partial y} \right) + \\ + \sin\gamma\cos\gamma\sin\chi\cos\chi \tfrac{\partial W_N}{\partial y} \end{aligned} \tag{4.72}$$

More information about the derivation of these equations can be found in [23, 74].

Remark 4.5. *When it is assumed that there is no wind, the velocity of the aircraft is constant and the available control inputs are the flight path angle and the heading angle. The following optimal trajectory can be calculated*

$$\begin{aligned} x &= Vt \cos\gamma \cos\chi + x_0 \\ y &= Vt \cos\gamma \sin\chi + y_0 \\ z &= -Vt \sin\gamma + z_0 \end{aligned} \tag{4.73}$$

where (x_0, y_0, z_0) is the initial position of the aircraft. If the final position is given by (x_f, y_f, z_f) then the predicted arrival time is:

$$T = \frac{1}{V} \sqrt{(x_f - x_0)^2 + (y_f - y_0)^2 + (z_f - z_0)^2} \tag{4.74}$$

while the heading angle is given by:

$$\chi = \arctan\left(\frac{y_f - y_0}{x_f - x_0}\right) \tag{4.75}$$

and the flight path angle by:

$$\gamma = \arctan\left(\frac{z_f - z_0}{\cos\chi(x_f - x_0) + \sin\chi(y_f - y_0)}\right) \tag{4.76}$$

When the considered control inputs are the flight path angle rate $\dot{\gamma}$, the heading angle rate $\dot{\chi}$ and the derivative of the velocity $\dot{V}$, the idea is to use the structure and to apply simple bang-bang controls in the planning. The amount of control available is a concern in the planning for this system due to the drift term. The class of bang-bang controls is often a sufficiently rich class of controls for analysis of nonlinear systems. This simple class of controls makes it possible to integrate the equations forward.

4.2.4.4 3D Zermelo's problem on a spherical Earth

This paragraph presents the practical trajectory optimization algorithm that approximates the minimization of the total cost of travel time and fuel consumption for aircraft on a spherical Earth. A typical aircraft trajectory consists of an initial climb, a steady-state cruise and a final descent. Here, aircraft performance is optimized for the cruise phase only. The cruise trajectory is divided into segments on several altitudes as the optimal cruise altitude increases due to the reduction in aircraft weight as fuel is used. The aircraft optimal heading during cruise is the solution of the **Zermelo's problem** derived on a spherical Earth surface in the absence of constraints. The horizontal trajectory segments are optimized based on the cost-to-go associated with extremal trajectories generated by forward-backward integrating the dynamical equations for optimal heading and aircraft motion from various points in the airspace [27, 139].

The direct operating cost for a cruising aircraft can be written as:

$$J = \int_{t_0}^{T_f} \left(C_t + C_f \tilde{f}(m, z, V)\right) dt \tag{4.77}$$

where C_t and C_f are the cost coefficient of time and fuel. The **fuel flow rate** $\tilde{f}$ can be approximated by a function of aircraft mass m, altitude z and airspeed V. The fuel burn for aircraft during cruise $\tilde{F}$ is calculated as:

$$\tilde{F} = t\tilde{f} \tag{4.78}$$

where t is the elapsed time. The fuel burn rate f for jets and turboprops is determined by the specific fuel consumption (SFC) and thrust T:

$$\tilde{f} = \frac{C_{fcr}}{1000}.SFC.T_f \qquad SFC = C_{f_1}\left(1 + \frac{V_{TAS}}{C_{f_2}}\right) \tag{4.79}$$

where $C_{fcr}, C_{f_1}, C_{f_2}$ are the thrust **specific fuel consumption** coefficients and V_{TAS} is the true airspeed.

During cruise, thrust equals the aerodynamic drag forces and lift equals the weight:

$$\begin{aligned} T = D = \tfrac{1}{2}C_D(V,\alpha)\rho S V^2 \\ C_D(V,\alpha) = C_{D_0} + KC_L^2 \\ C_L(V,\alpha) = \tfrac{2mg}{\rho S V^2} \end{aligned} \tag{4.80}$$

Under the **international standard atmosphere** (ISA), the tropopause altitude is at $11000m$ and the optimal cruise altitude z_{opt} at or below the tropopause can be calculated as:

$$z_{opt} = \left(1 - \exp\left(-f(m,V)K_T R_{gas}/2(g + K_T R_{gas})\rho_{0ISA}^2\right)\right)\left(\frac{1000 T_{0ISA}}{6.5}\right) \tag{4.81}$$

Above the tropopause, it is:

$$z_{opt} = \frac{-f(m,V) R_{gas} T_{propISA}}{2g\rho_{tropISA}} + 11000 \text{ m} \tag{4.82}$$

where

$$f(m,V) = \ln\left(\frac{4m^2g^2K}{S^2V^4C_{D_0}}\right) \tag{4.83}$$

R_{gas} is the real gas constant for air, the temperature gradient K_T, the sea level density ρ_{0ISA} and the sea level temperature T_{0ISA} considered constant under the ISA. The air density $\rho_{tropISA}$ and temperature $T_{tropISA}$ are all constant at the troposphere.

Optimal cruise altitudes are computed from relations (4.81) to (4.82) based on the atmospheric constants and aerodynamic drag coefficients that are aircraft-type dependent. They vary also with aircraft mass and airspeed.

The aircraft equations of motion at a constant altitude above the spherical Earth's surface are

$$\dot{\ell} = \frac{V\cos\chi + W_E(\ell,\lambda,z)}{R\cos\lambda} \tag{4.84}$$

$$\dot{\lambda} = \frac{V\sin\chi + W_N(\ell,\lambda,z)}{R} \tag{4.85}$$

$$\dot{m} = -f \tag{4.86}$$

subject to the conditions that the thrust equals the drag, ℓ is the longitude and λ is the latitude, χ is the heading angle and R is the Earth radius $R >> z$. The dynamical equation for the optimal aircraft heading is:

$$\dot{\chi} = -\frac{F_{wind}(\chi,\ell,\lambda,W_E,W_N,V)}{R\cos\lambda} \tag{4.87}$$

where $F_{wind}(\chi, \ell, \lambda, W_E, W_N, V)$ is aircraft heading dynamics in response to winds and is expressed as:

$$F_{wind} = -\sin\chi\cos\chi\frac{\partial W_E}{\partial \ell} + \cos^2\chi\sin\lambda W_E + \cos^2\chi\cos\lambda\left(\frac{\partial W_E}{\partial \lambda} - \frac{\partial W_N}{\partial \ell}\right) +$$
$$+\sin\chi\cos\chi\sin\lambda W_N + \cos\chi\sin\chi\cos\lambda\frac{\partial W_N}{\partial \lambda} + V\cos\chi\sin\lambda + \cos^2\chi\frac{\partial W_N}{\partial \ell} \tag{4.88}$$

The minimum-time trajectory is the combination of wind optimal extremals on several different altitudes, each solved using conditions on that altitude. The optimal virtual profile provides the initial and subsequent optimal cruise altitudes as well as the transition time between the altitudes.

4.2.4.5 Virtual goal

The problem is to determine the optimal path in 3D space between the initial configuration, position X_1 and orientation e_1 and the final configuration, position X_2 and orientation e_2, for a constant speed aircraft and with turn rate constraints. The unit orientation vectors at the initial and final points are $e_1 = (\cos\gamma_1\cos\chi_1, \cos\gamma_1\sin\chi_1, -\sin\gamma_1)^T$ and $e_2 = (\cos\gamma_2\cos\chi_2, \cos\gamma_2\sin\chi_2, -\sin\gamma_2)^T$. The proposed path planning algorithm is based on the following kinematic equations of motion:

$$\begin{aligned}\dot{x} &= V\cos\gamma\cos\chi + W_N \\ \dot{y} &= V\cos\gamma\sin\chi + W_E \\ \dot{z} &= -V\sin\gamma + W_D \\ \dot{\chi} &= \omega_1 \\ \dot{\gamma} &= \omega_2\end{aligned} \tag{4.89}$$

where the state vector is defined as $X = (x, y, z, \chi, \gamma)$, the velocity V is assumed to be constant and ω_1, ω_2 are the control inputs. The trajectory must satisfy a maximum turn rate constraint or the curvature bound $\pm\kappa_{max}$ and torsion $\pm\tau_{max}$ for the aircraft.

The original problem of computing an optimal path in the presence of wind can be expressed as one of computing the optimal path from an initial position X_1 and orientation e_1 with no wind to a final orientation e_2 and a virtual goal that moves with a velocity equal and opposite to the velocity of the wind [163]. The air path defined as the path traveled by the aircraft with respect to the moving frame can be of CSC type or of helicoidal type. Let the minimum time required to reach the final point be T_f. At T_f, the virtual final point moves from the given final position $X_2 = (x_2, y_2, z_2)$ to a new position $X_{2v} = (x_{2v}, y_{2v}, z_{2v})$ which can be expressed as a function of T_f.

$$X_{2v} = X_2 - (W\cos\gamma_1\cos\chi_1, W\cos\gamma_1\sin\chi_1, -W\sin\gamma_1)^T T_f \tag{4.90}$$

or equivalently

$$X_{2v} = X_2 - \int_0^{T_f} W(t)dt \tag{4.91}$$

with $W = \sqrt{W_N^2 + W_E^2 + W_D^2}$.

Finally, the position to be reached is X_{2v}. The heading angle and the flight path angle at the virtual goal are χ_2, γ_2. The reformulated problem (4.90) is similar to the kinematic model of a 3D Dubins vehicle in the absence of wind except that the final point X_{2v} is also dependent on T_f. Thus a CSC or helicoidal path can be computed and the variables X_{2v}, T_f can be obtained using nonlinear equations solving the algorithm and the ground path can be computed in the inertial frame using the state equations in (4.89).

4.3 GUIDANCE AND COLLISION/OBSTACLE AVOIDANCE

Aircraft collision is a serious concern as the number of aircraft in operation increases. In the future, they will be expected to carry sophisticated avoidance systems when flying together with conventional aircraft. **Onboard sensor systems** combined with self-operating algorithms will ensure collision avoidance with little intervention from ground stations. Onboard sensors can detect other aircraft nearby. Information related to the other aircraft such as position, velocity and heading angle can be used to build an avoidance command. In order for an aircraft to maneuver successfully in such a dynamic environment, a feasible and collision-free trajectory needs to be planned in the physical configuration space. The avoidance law should be generated in real time and simple to implement. The ability to sense and avoid natural and man made obstacles and to rebuild its flight path is an important feature that a smart autonomous aircraft must possess [195]. Guidance, trajectory generation and flight and mission planning are the core of the flight management system of a smart autonomous aircraft [19, 34, 108, 136]. The computational abilities provide the strongest constraints on the autonomous aircraft, although advances in the hardware mechanisms are to be expected. Improvements in software are essential [48, 93, 97].

Collision avoidance is of vital importance for small UAV flying on low heights which usually encounter a large number of mostly static obstacles [42]. Due to a limited knowledge of the environment and small computing power, collision avoidance needs to rely on a little information while being computationally efficient. Guidance laws are one way of tackling such difficulties [173]. For autonomous missions, sense and avoid capability is a critical requirement [77].

Collision avoidance can be broadly classified into global and local path planning algorithms, to be addressed in a successful mission. Whereas global path planning broadly lays out a path that reaches the goal point, local collision avoidance algorithms which are usually fast, reactive and carried out online ensure safety of the aircraft from unexpected and unforeseen obstacles and collisions. The algorithm in [135] first plans a path to the goal avoiding the obstacles known a priori. If a collision is predicted to occur, the path is replanned so that the obstacle is avoided. However, the objective is to always move toward the goal point after collision is avoided. Hence, these algorithms are considered to be global path planning algorithms with local collision avoidance features embedded into them.

The problem of trajectory prediction is encountered whenever it is necessary to provide the control system with a projection of the future position of the aircraft, given the current flight conditions, together with an envelope of feasible trajectory. The prediction can be integrated in a ground proximity warning system, to reveal a possible ground collision with a time margin sufficient for undertaking an appropriate control action for avoiding obstacles on the prescribed flight path. In this case, the knowledge of an envelope of feasible future position is sought. In [8], an algorithm is presented for determining future positions of the aircraft center of gravity inside a given prediction horizon from measurements of acceleration and angular velocity. The technique is based on the weighted combination of two estimates:

1. The projection in the **Frenet frame** of a helix with vertical axis for long term prediction in steady and quasi-steady maneuver segments.
2. A third order accurate power series expansion of the trajectory in the Frenet frame, useful for short term trajectory prediction during transient maneuvering phases.

4.3.1 GUIDANCE

Guidance is a dynamic process of directing an object toward a given point that may be stationary or moving [79, 213]. Inertia of the aircraft is ignored in most approaches and their dynamics are ignored [106]. In this section, three approaches are presented, two conventional and the last one based on fuzzy techniques.

Definition 4.1. ***Guidance*** *is the logic that issues steering commands to the aircraft to accomplish certain flight objectives. The guidance algorithm generates the autopilot commands that steer the autonomous aircraft. A guidance system is defined as a group of components that measures the position of the guided aircraft with respect to the target and changes its flight path in accordance with a guidance law to achieve the flight mission goal.*

A guidance law is defined as an algorithm that determines the required commanded aircraft accelerations. In guidance studies, only local information on the wind flow field is assumed to be available and a near optimal trajectory, namely a trajectory that approximates the behavior of the optimal trajectory, is determined[132].

There are various approaches to both static and moving obstacle detection that are mostly based on the collision cone approach. Often collision avoidance is achieved by tracking an aim-point at a safe distance from the obstacle, using a homing guidance law. Explicit avoidance laws can be derived by limiting the consideration to a plane, defined by the relative geometry between the aircraft and the obstacle. The main issue is that primary mission objectives are not fully considered which may cause trajectories to be far from optimal [178].

Geometric techniques including pure pursuit and variants of pursuit and **line of sight** guidance laws are mainly found in [62]. The path-following algorithms based on pure pursuit and line of sight based guidance laws use a **virtual target point** (VTP) on the path. The guidance laws direct the aircraft to chase the virtual target point which eventually drives the aircraft onto the path. The distance between the virtual target point and the UAV position projected on the path is often called virtual distance. The stability of line of sight guidance laws for path following depends significantly on the selection of the virtual distance parameter. Pure pursuit and line of sight guidance laws can be combined to create a new guidance law for path following [186].

3D guidance refers to following the mission path in both the horizontal and vertical planes. It covers both 2D ground track following, as well as altitude profile following of the desired trajectory. In the 3D case, guidance commands are generated for the lateral directional control system in terms of reference roll or heading angles, and for the longitudinal control system in terms of pitch or altitude commands [178]. A subset of the general 3D problem is the 2D lateral guidance problem in which the guidance objective is to ensure accurate ground track following of the aircraft. Thus it must exactly fly over lines and arcs joining mission waypoints, with minimum cross-track or lateral deviation.

4.3.1.1 Proportional navigation

Line of sight (LOS) and its variations are still the simplest and most popular guidance laws in use today [37]. The **proportional navigation** (PN) guidance law is a strategy that can be applied to any situation where the nominal trajectory is well known. According to this law, the maneuver of the aircraft is proportional to the line of sight rate. It is based on the fact that if the target and the aircraft have constant velocities, then on collision course, the line of sight rate is zero. There are two basic disturbances that may influence the guidance loop: the target and the initial conditions. The guidance law is designed to steer the aircraft in a flight path toward the boundary of a safety bound. The safety bound can be a minimal radius circle and/or cylinder to prevent collision.

The center of mass of the aircraft is instantaneously located at $R(t)$ and its desired value $R_T(t)$ with respective velocities $V(t)$ and $V_T(t)$ relative to a stationary frame of reference. The instantaneous position of the target relative to the aircraft is given by:

$$e(t) = R_T(t) - R(t) \tag{4.92}$$

with

$$V_e(t) = \frac{de}{dt} = V_T(t) - V(t) \tag{4.93}$$

The following control law for guidance is used:

$$V_e(t) = \mathbf{K}(t)e(t) \tag{4.94}$$

where $\mathbf{K}(t)$ is a time-varying gain matrix.
The required acceleration control $U(t)$ to be applied to the aircraft is:

$$U(t) = \frac{dV_e}{dt} = \mathbf{K}(t)V(t) + \dot{\mathbf{K}}(t)e(t) \tag{4.95}$$

A choice of state vector for the guidance problem is:

$$X(t) = (e(t), \dot{e}(t))^T \tag{4.96}$$

which yields the linear feedback law:

$$U(t) = \left(\dot{\mathbf{K}}(t), \mathbf{K}(t)\right) X(t) \tag{4.97}$$

This proportional navigation guidance can be easily implemented.

A suitable navigation strategy for a rendezvous would be to achieve simultaneously a zero miss distance and a zero relative speed.

A simple approach is to have the instantaneous velocity error $V(t)$ become aligned with the acceleration error $a = \dot{V}_T(t) - \dot{V}(t)$. This implies that the cross product of velocity and acceleration errors must vanish:

$$V \times a = 0 \tag{4.98}$$

Such a navigation law is termed cross product steering [213].

4.3.1.2 Method of adjoints

A **terminal guidance** is considered where two objects, having constant speeds, move in a plane toward a collision point. Small perturbations are assumed with respect to the nominal collision triangle. That is the line of sight (LOS) deviation λ is small. Let R be the range and $V_c = -\dot{R}$ the closing speed. The usual small perturbation is assumed with the reference line; in that case the closing speed is assumed to be constant. Then along the line of sight, the final time T_f and time to go t_{go} are defined as follows:

$$T_f = \frac{R_0}{-\dot{R}} = \frac{R_0}{V_c} \tag{4.99}$$

$$t_{go} = \frac{R}{-\dot{R}} \tag{4.100}$$

The miss distance m is defined as

$$m = y(T_f) \tag{4.101}$$

where y is the relative separation perpendicular to the initial line of sight. Proportional navigation guidance law states that the command acceleration of the aircraft normal to the line of sight n_c is proportional to the line of sight rate:

$$n_c = N'V_c\dot{\lambda} \tag{4.102}$$

where N' is the navigation constant [213].

4.3.1.3 Fuzzy guidance scheme

The overall guidance scheme has two components: a waypoint generator and a fuzzy guidance system. The desired trajectory is specified in terms of a sequence of waypoints without any requirements on the path between two successive waypoints [88]. The waypoint generator holds a list of waypoints in $5D$, checks aircraft position and updates the desired waypoint when the previous one has been reached with a given tolerance. The waypoint generator's only task is to present the actual waypoint to the **fuzzy guidance system** (FGS) [216, 217].

A tolerance ball is included around the waypoint, defining that as actual target reached. Between the waypoint generator and the fuzzy guidance system, a coordinate transformation is performed to convert Earth-fixed frame position errors into waypoint frame components. Each waypoint defines a coordinate frame centered at the waypoint position (x_w, y_w, z_w) and rotated by its heading angle χ_w around the z axis. γ_w is the flight path angle of the waypoint. The coordinate transformation allows the synthesis of a fuzzy ruleset valid in the waypoint fixed coordinate frame, which is invariant with respect to the desired approach direction χ_w. When a waypoint is reached, the next one is selected, the actual reference value is changed and the rotation matrix is updated to transform position and orientation errors into the new waypoint coordinate frame.

The aircraft autopilots are designed to track desired airspeed, heading and flight path angles V_d, γ_d, χ_d using decoupled closed-loop inertial dynamics and so three independent Takagi–Sugeno controllers were synthesized to constitute the fuzzy guidance system.

1. The first controller generates the desired flight path angle γ_d for the autopilot using altitude error:

$$e_z = z_w - z_A \qquad \gamma_d = f_\gamma(e_z) \tag{4.103}$$

The state vector $(V_A, \gamma_A, \chi_A, z_A)^T$ represents aircraft speed, flight path angle, heading and altitude, respectively.

2. The second controller computes desired aircraft velocity:

$$e_V = V_w - V_A \qquad V_d = V_w + f_V(e_V) \tag{4.104}$$

3. The third one is responsible for the generation of the desired heading angle χ_d using the position error along the X, Y axes and the heading error e_χ.

A fuzzy rule set designed at a specific trim airspeed value could yield insufficient tracking performance when the desired waypoint crossing-speed V_w differs significantly from V [88]. To accommodate large values of e_V and to investigate the effect of disturbance, modeled as aircraft's speed differential with respect to crossing-speed V_w, a speed-correlated scale coefficient of position error is introduced. The rotation matrix is defined as:

$$\mathbf{R}(\chi_w) = \begin{pmatrix} \cos(\chi_w + \frac{\pi}{2}) & \sin(\chi_w + \frac{\pi}{2}) \\ -\sin(\chi_w + \frac{\pi}{2}) & \cos(\chi_w + \frac{\pi}{2}) \end{pmatrix} \tag{4.105}$$

The position errors in the fixed waypoint coordinate frame are given by:

$$\begin{pmatrix} e_x^w \\ e_y^w \end{pmatrix} = \mathbf{R}(\chi_w) \begin{pmatrix} x - x_w \\ y - y_w \end{pmatrix} \tag{4.106}$$

The velocity compensated position errors $e_{x_c}^w, e_{y_c}^w$ are defined by

$$\begin{pmatrix} e_{x_c}^w \\ e_{y_c}^w \end{pmatrix} = \frac{V^*}{V^w} \begin{pmatrix} x - x_w \\ y - y_w \end{pmatrix} \tag{4.107}$$

where V^* represents the airspeed value used during fuzzy guidance system membership rules design. In this way, position errors used by the fuzzy guidance system to guide the aircraft toward the waypoint with desired approach direction are magnified when V^w requested waypoint crossing-speed is larger than V^* or reduced otherwise.

Remark 4.6. *Equation* (4.107) *may diverge if* $V^* \to 0$. *However, this is not an operationally relevant condition because the requested waypoint crossing speed should be defined according to aircraft flight parameters and the stall velocity must be avoided.*

Finally, the desired heading angle produced by the fuzzy controller is:

$$\chi_d = \chi_w + f_\chi(e_{X_c}^w, e_{Y_c}^w) \tag{4.108}$$

The fuzzy guidance system is based on a Takagi–Sugeno system model described by a blending of IF-THEN rules. Using a weighted average defuzzifier layer, each fuzzy controller output is defined as follows:

$$Y = \frac{\sum_{k=1}^{m} \mu_k(X) U_k}{\sum_{k=1}^{m} \mu_k(X)} \tag{4.109}$$

where $\mu_i(X)U_i$ is the ith membership function of input X to the ith zone. The membership functions are a combination of Gaussian curves of the form

$$f(X, \sigma, c) = \exp\left(-\frac{(X-c)^2}{\sigma^2}\right) \tag{4.110}$$

The fuzzy rules are defined according to the desired approach direction and the angular rate limitations of the aircraft. The fuzzy knowledge base is designed to generate flyable trajectories using the max linear and angular velocities and accelerations. The fuzzy guidance system provides different desired flight path and heading angle commands for different waypoints. The altitude and velocity controllers are implemented using a Takagi–Sugeno model directly. For the altitude, the input is the altitude error e_z and the output is the desired flight path angle γ_d. Inputs and outputs are mapped with four fuzzy sets each:

1. If e_z is N_∞ then γ_d is P_∞, for big negative errors
2. If e_z is N_s then γ_d is P_s, for small negative errors
3. If e_z is P_s then γ_d is N_s, for small positive errors
4. If e_z is P_∞ then γ_d is N_∞, for big positive errors

Here, the generic output constant P_s represents the output value s and the constant N_s represents the output value $-s$.

The velocity controller is similar to the altitude controller. Three input fuzzy sets are used for the velocity error e_V and three for the resulting ΔV_d output:

1. If e_V is N_∞ then ΔV_d is P_s, for negative errors,
2. If e_V is ZE then ΔV_d is P_0, for near to zero errors,
3. If e_V is P_∞ then ΔV_d is N_s, for positive errors.

Guidance in the horizontal (x, y) plane is more complex. The horizontal plane fuzzy controller takes its input from scaled position errors $e^w_{x_c}, e^w_{y_c}$ and heading error e_χ. The error along the X axis is coded into five fuzzy sets:

1. N_∞ for big negative lateral errors
2. N_s for small negative lateral errors
3. ZE for near exact alignment
4. P_s for small positive lateral errors
5. P_∞ for big positive lateral errors

Three sets are also defined over the Y^w axis error

1. ZE for aircraft over the waypoint
2. N_s for waypoint behind the aircraft
3. P_s for waypoint in front of the aircraft

Finally, the heading error may be coded in seven fuzzy sets. In the application of equation (4.109), the m fuzzy rules are grouped into S groups, each with K rules: $m = SK$. The S groups correspond to S areas on the xy plane. From the preceding:

$$Y = \frac{1}{C(X)} \sum_{i=1}^{s} \sum_{j=1}^{K} \mu_i^{xy}\left(e^w_{X_c}, e^w_{Y_c}\right) \mu^x_{ij}(e_x) U_{ij} \tag{4.111}$$

or

$$Y = \frac{1}{C(X)} \sum_{i=1}^{s} \mu_i^{xy} \left(e_{X_c}^w, e_{Y_c}^w\right) \delta_i^x (e_x) \tag{4.112}$$

where

$$Y = \sum_{i=1}^{s} \frac{\mu_i^{xy} \left(e_{X_c}^w, e_{Y_c}^w\right)}{C(x)} \delta_i^x (e_x) = \sum_{i=1}^{s} \bar{\mu}_i^{xy} \left(e_{X_c}^w, e_{Y_c}^w\right) \delta_i^x (e_x) \tag{4.113}$$

Fixing $\left(e_{X_c}^w, e_{Y_c}^w\right)$ in the middle of the pth zone under the assumption that the contribution from the other zones is near zero yields:

$$Y\left(e_{X_c}^{wP}, e_{Y_c}^{wP}\right) = \bar{\mu}_p^{xy} \left(e_{X_c}^{wP}, e_{Y_c}^{wP}\right) \delta_p^x (e_x) + \sum_{i=1; i \neq p}^{s} \bar{\mu}_i^{xy} \left(e_{X_c}^w, e_{Y_c}^w\right) \delta_i^x (e_x) \tag{4.114}$$

or

$$Y\left(e_{X_c}^{wP}, e_{Y_c}^{wP}\right) \approx \bar{\mu}_p^{xy} \left(e_{X_c}^{wP}, e_{Y_c}^{wP}\right) \delta_p^x (e_x) \tag{4.115}$$

Equation (4.115) shows that once the fuzzy sets for the position errors $\left(e_{X_c}^{wP}, e_{Y_c}^{wP}\right)$ are fixed, the definition of fuzzy sets for e_x should be computed by looking first at each area on the XY plane and then adding the cumulative results. Under this assumption, seven fuzzy sets are defined for the heading error $e_x \in \{N_b, N_m, N_s, ZE, P_s, P_m, P_b\}$, b is for big, m for medium and s for small.

The design goals of this fuzzy guidance system are thus:

1. Capability of reaching a set of waypoints in a prescribed order,
2. Possibility of specifying the speed and heading of the aircraft at a waypoint crossing,
3. Capability of quickly reconfiguring the waypoint set in response to changes in the mission scenario,
4. Reaching fixed waypoints as well as tracking and reaching moving waypoints.

In an operational scenario, the waypoint generator may be interfaced with a mission management system that updates the waypoints when needed.

4.3.2 STATIC OBSTACLES AVOIDANCE

The basic representations of an environment are configuration space and occupancy grid. In configuration space, the dimensions of the environment plus the coordinates of all obstacles are given. In an occupancy grid, the environment is specified at a certain resolution with individual voxels either representing free space or obstacles [31]. There are many ways to represent a map and the position of the aircraft within the map. The free regions and obstacles may be represented as polyhedral, each comprising a list of vertices or edges. This is potentially a very compact form but determining potential collisions between the aircraft and obstacles may involve testing against long lists of edges. A simpler representation is the occupancy grid. The environment is treated as a grid of cells and each cell is marked as occupied or unoccupied.

Motion planning is realized by an integrated set of algorithms such as collision checking [78], configuration sampling [103, 128] and path planning [123]. It can be categorized as static, in which all obstacle configuration is known prior to planning or dynamic in which environment (obstacles) information becomes known to the planner only through real time sensing of its local environment.

Given the nature of these aspects as diverse and difficult, most of the proposed work in the field of motion planning has focused on the consideration of some versions of the general problem. The ability to perform autonomous mission planning is considered one of the key enabling technologies for UAV. The development of algorithms capable of computing safe and efficient routes in terms of distance, time and fuel is very important [148].

The uniform framework to study the path planning problem among static obstacles is the **configuration space** (C-space). The main idea of the C-space is to represent the aircraft as a point called a configuration. The C-space is the set of all possible configurations; C-free are the regions of C-space which are free of obstacles. Obstacles in the environment become C-obstacles in the C-space.

The main idea in the sampling-based path planning method is to avoid an exhaustive construction of C-obstacles by sampling the C-space. The sampling scheme may be deterministic [151, 153] or probabilistic [133, 149]. The key issue is then to use an efficient grid-search algorithm to find an optimal path in the sense of a metric.

Definition 4.2. *A metric defines the distance between two configurations in the C-space, which becomes a metric space. This metric can be seen as the* cost-to-go *for a specific aircraft to reach a configuration from another one.*

A grid search algorithm is an optimization technique that successively performs an exploration and an exploitation process:

1. The **exploration process** builds a minimum cost-to-go map, called distance function, from the start to the goal configuration.
2. The **exploitation process** is a backtracking from the goal to the start configuration.

Accurate environmental mapping is essential to the path planning process [37]:

1. **Qualitative or topological mapping** represents features without reference to numerical data and is therefore not geometrically exact. It consists of nodes and arcs, with vertices representing features or landmarks.

2. **Quantitative or metric mapping** adopts a data structure which is feasible for path planning based on waypoints or subgoals: meadow maps, Voronoi diagrams, regular occupancy grid, quad-tree mapping.

Planning can be formulated as either a continuous or a discrete task [199, 162]. There are continuous methods such as potential fields, vector field histogram and bug algorithms and discrete methods such as visibility graph planning or Voronoi diagram or A^* algorithm [7, 9, 23, 41, 117].

4.3.2.1 Discrete methods

In discrete planning, the system is typically modeled as a graph and is called a transition system. **Nodes** represent states and **edges** represent transitions between states and are labeled by **actions** that enable these transitions. An important feature of sampling-based approaches is that the required controllers for feasible trajectories are automatically constructed as a result of the exploration process.

While many navigation algorithms work directly on the environment description, some algorithms such as Dijkstra or A^* require a **distance graph** as an input. A distance graph is an environment description at a higher level. It does not contain the full environment information but it allows for an efficient path planning step. While the **quadtree method** identifies points with some free space around them, the **visibility graph** method uses corner points of obstacles instead.

Other approaches can abstract a geometric environment into a topological map based on landmarks. Planning is carried out on this topological map. A planned route has to be converted back to the geometric space for continuous motion control. After obtaining discrete routing information, an admissible and safe path has to be generated for an aircraft to travel from one node to another, compliant with all kinematic and dynamic constraints. Many planning algorithms address obstacle avoidance while planning a path to reach a destination point using A^*, D^*, **Voronoi diagrams, probabilistic roadmap** (PRM) or rapidly exploring random trees (RRT) methods. Goerzen in [72] reviewed deterministic motion planning algorithms in the literature from the autonomous aircraft guidance point of view.

Definition 4.3. *A **directed graph** (N, E) is a structure where N is a set of nodes or vertices and E is a set of edges connecting the nodes ($E \subseteq N \times N$). In path planning, nodes usually stand for positions in the space and edges determine whether it is possible to transit directly between these positions.*

UAV have strict payloads and power constraints which limit the number and variety of sensors available to gather information about knowledge of its surroundings. In [105], the paths are planned to maximize collected amounts of information from desired regions while avoiding forbidden regions.

Graph search algorithms look for a feasible path in a given environment. Examples of graph search are deterministic graph search techniques such as the A^* search algorithm, Voronoi graph search and probabilistic sampling-based planners. Although these algorithms are mainly used for global path planning, reactive planners have been achieved by modifying the algorithms in order to reduce the computation time.

Differential constraints naturally arise from the kinematics and dynamics of the smart autonomous aircraft. When incorporated into the planning process, a path is produced that already satisfies the constraints [112]. Because of the difficulty of planning under differential constraints, nearly all planning algorithms are sampling-based as opposed to combinatorial. From an initial state X_0, a reachability tree can be formed by applying all sequences of discretized actions. Sampling-based algorithms proceed by exploring one or more reachability trees that are derived from discretization. In some cases, it is possible to trap the trees onto a regular lattice structure. In this case, planning becomes similar to grid search. A solution trajectory can be found by applying standard graph search algorithms to the lattice. If a solution is not found, then the resolution may need to be increased.

A planning algorithm that allows the aircraft to autonomously and rapidly calculate 3D routes can also follow a heuristic approach [211]. This algorithm is developed and verified in a virtual world that replicates the real world in which the aircraft is flying. The elements of the real world such as the terrain and the obstacles are represented as mesh-based models. Such an algorithm is also based on a graph but aims at increasing the portion of space explored, identifying several nodes in the 3D space with an iterative approach.

4.3.2.1.1 Deterministic methods

Several methods have been proposed and are mainly based on the construction of **visibility graphs**. Following this approach, the nodes of the graph are candidate path points that the aircraft will fly through and each arc represents an approximate path between them. To build the visibility graph, the set of nodes and the arcs that join them without intersecting obstacles have to be determined.

4.3.2.1.1.1 Visibility graph The **visibility graph** uses corner points of obstacles. If the environment is represented as a configuration space, the polygon description of all obstacles is already available. The list of all start and end points of obstacle border lines plus the autonomous aircraft's start and goal position is available. A complete graph is then constructed by linking every node position of every other one. Finally, all the lines that intersect an obstacle are deleted, leaving only the lines that allow the flight from one node to another in a direct line. The characteristic of algorithm 1 is as follows. Let $V = \{v_1, ..., v_n\}$ be the set of vertices of the polygons in the configuration space as well as the start and goal configurations. To construct the visibility graph, other vertices visible to $v \in V$ must be determined. The most obvious way to make this determination is to test all line segments $vv_i, v \neq v_i$, to see if they intersect an edge of any polygon. A more efficient way is the rotational sweep algorithm. For the problem of computing the set of vertices visible from v, the sweep line I is a half-line emanating from v and a rotational sweep rotating I from 0 to 2π is used. .

Algorithm 1 Visibility Algorithm

1. **Input**: A set of vertices $\{v_i\}$ (whose edges do not interset) and a vertex v
2. **Output**: A subset of vertices from $\{v_i\}$ that are within line of sight of v
3. For each vertex v_i, calculate α_i, the angle from the horizontal axis to the line segment vv_i
4. Create the vertex list ϵ, containing the α_i sorted in increasing order
5. Create the active list S, containing the sorted list of edges that intersect the horizontal half line emanating from v
6. for all α_i do
7. if v_i is visible to v then
8. add the edge (v, v_i) to the visibility graph
9. endif
10. if v_i is the beginning of an edge E, not in S, then
11. insert the edge E into S
12. endif
13. if v_i is the end of an edge in S then
14. delete the edge from S
15. endif
16. endfor

This algorithm incrementally maintains the set of edges that intersect I, sorted in order of increasing distance from v. If a vertex v_i is visible to v, then it should be added to the visibility graph. It is straightforward to determine if v_i is visible to v. Let S be the sorted list of edges that intersects the half line emanating from v. The set is incrementally constructed as the algorithm runs. If the line segment vv_i does not intersect the closed edge in S and if I does not lie between the two edges incident on v (the sweep line does not intersect the interior of the obstacles at v) then v_i is visible from v [17].

4.3.2.1.1.2 Voronoi algorithm The **Voronoi tessellation** of a set of planar points, known as sites, is a set of Voronoi cells. Each cell corresponds to a side and consists of all points that are closer to its site than to any other site. The edges of the cells are the points that are equidistant to the two nearest sites. A generalized Voronoi diagram comprises cells defined by measuring distances to objects rather than points.

Definition 4.4. ***Planar Ordinary Voronoi Diagram:*** *Given a set of a finite number of distinct points in the Euclidean plane, all locations in that space are associated with the closest members of the point set with respect to the Euclidean distance. The result is a tessellation of the plane into a set of the regions associated with members of the point set. This tessellation is called a* ***planar ordinary Voronoi diagram*** *generated by the point set, and the regions constituting the Voronoi diagram, ordinary Voronoi polygons [143].*

In hostile environments, a Voronoi diagram can decompose a space defined by random scattered points into separate cells in such a manner that each cell will include one point that is closer to all elements in this cell than any other points [49]. The graph is constructed by using Delaunay triangulation and its dual graph Voronoi diagrams. The procedure of this cell decomposition starts with a priori knowledge of the location and of the number of the scattered points. Formation of triangles formed by three of the Voronoi sites without including any other sites in this circumcircle is called Delaunay triangulation. By connecting all the edges together, polygons are formed and these polygons construct the Voronoi graph.

Considering the formulation of 3D networks, **3D Delaunay method** is used in [159] to partition the space because of uniqueness: regional, nearest and convex polyhedron.

As it is difficult to control aircraft precisely enough to follow the minimum-distance path without risk of colliding with obstacles, many skeleton-based road map approaches have been taken. The Voronoi approach builds a skeleton that is maximally distant from the obstacles, and finds the minimum distance path that follows this skeleton. This algorithm is a 2D algorithm, complete but not optimal. Voronoi diagram is a special kind of decomposition of a metric space determined by distances to a specified discrete set of objects in the space [143]. Given a set of points S, the corresponding Voronoi diagram is generated; each point P has its own Voronoi cell which consists of all points closer to P than any other points. The border points between polygons are the collection of the points with the distance to shared generators [7].

The Voronoi algorithm works by constructing a skeleton of points with minimal distances to obstacles. A free space F in environment (white voxels) is defined as well an occupied space F' (black voxels). The point $b' \in F'$ is a basis point for $p \in F$ if and only if b has minimal distance to p, compared with all other points in F'. Voronoi diagram is defined as $\{p \in F | p$ has at least two basis points $\}$. Typical computational units are given in algorithm 2 on page 235. Line 1 set ϵ to maximum precision while line 3 compute some data. Line 4 checks topological conditions while line 6 relaxes the ϵ threshold and line 7 makes sure to reset ϵ. Line 10 checks locally for soundness of input and line 13 fixes the problem in the input data. Finally line 14 replaces *correct* by *best possible.* If the relaxation of the ϵ threshold and the heuristics of the multi-level recovery process have not helped to compute data that meets the topological conditions then the code finally enters

Algorithm 2 Voronoi Algorithm

1. ϵ = lower-bound
2. Repeat
3. x = ComputeData(ϵ)
4. success = CheckConditions (x, ϵ)
5. If (not success) then
6. $\epsilon = 10\epsilon$
7. reset data structure appropriately
8. until (success OR $\epsilon >$ upper-bound
9. ϵ = lower-bound
10. If (not success) then
11. illegal = CheckInput()
12. If (illegal) then
13. clean data locally
14. restart computation from scratch
15. else
16. x = DesperateMode()

desperate mode, because the *optimum* is replaced by *best possible.*

One problem of this approach is that it allows lines to pass very closely to an obstacle, and so this would only work for a theoretical point robot. However, this problem can be easily solved by virtually enlarging each obstacle by at least half of the autonomous aircraft's largest dimension before applying the algorithm.

4.3.2.1.1.3 Dijkstra's algorithm **Dijkstra's algorithm** is a method for computing all shortest paths from a given starting node in a fully connected graph. Relative distance information between all nodes is required. In the loop, nodes are selected with the shortest distance in every step. Then distance is computed to all of its neighbors and path predecessors are stored. The cost associated with each edge is often the distance but other criteria such as safety, clearance can be incorporated into the cost function as well.

The algorithm works by maintaining for each vertex the shortest path distance from the start vertex. Further, a back-pointer is maintained for each vertex indicating from which neighboring vertex the shortest path from the start comes. Hence, a shortest path to some vertex can be read out by following the back-pointers back to the start vertex. From the start vertex, the shortest path distances are propagated through the graph until all vertices have received their actual shortest path distance.

Dijkstra's algorithm solves the single source shortest paths problem on a weighted directed graph $G = (V, E)$ for the case in which all edge weights are nonnegative, $w(u, v) \geq 0$ for each edge $(u, v) \in E$. This algorithm maintains a set S of vertices whose final shortest-path weights from the source s have already been determined. The algorithm repeatedly selects the vertex $u \in V - S$, with the minimum shortest path estimate, adds u to S and relaxes all edges leaving u.

Algorithm 3 Dijkstra's Algorithm

1. INITIALIZE-SINGLE-SOURCE(G, s)
2. S = $\emptyset$
3. Q= G.V
4. while $Q \neq \emptyset$
5. u = EXTRACT-MIN(Q)
6. S = S $\cup \{u\}$
7. for each vertex $v \in G.Adj\,[u]$
8. RELAX(u, v, w)

The Dijkstra's algorithm 3 works as follows. Line 1 initializes the d and π values and line 2 initializes the set S to the empty set. The algorithm maintains the invariant that $Q = V - S$ at the start of each iteration of the while loop of lines 4 to 8. Line 3 initializes the min-priority queue Q to contain all the vertices in V. Since $S = \emptyset$ at that time, the invariant is true after line 3. Each time through the while loop of lines 4 to 8, line 5 extracts a vertex u from $Q = V - S$ and line 6 adds it to set S, thereby maintaining the invariant. Vertex u, therefore, has the smallest shortest path estimate of any vertex in $V - S$. Then, lines 7 to 8 relax each edge (u, v) leaving u, thus updating the estimate $v.d$ and the predecessor $v.\pi$ if the shortest path to v can be improved.

Because Dijkstra's algorithm always chooses the lightest or closest vertex in $V - S$ to add to set S, it is said to be a greedy strategy. Greedy algorithms do not always yield optimal results in general, but Dijkstra's algorithm computes shortest paths.

Some hybrid methods can be proposed. One method consists of building the visibility graph using Dijkstra's algorithm to find the approximate shortest paths from each node to the goal and finally implementing a mixed integer linear programming receding horizon control to calculate the final trajectory.

4.3.2.1.1.4 A^* algorithm The A^* **algorithm** is based on Dijkstra's algorithm. It focuses the search in the graph towards the goal. A heuristic value is added to the objective function. It must be a lower bound estimate of the distance from the current vertex to the goal vertex. A^* heuristic algorithm computes the shortest path from one given start node to one given goal node. Distance graphs with relative distance information between all nodes plus lower bounds of distance to goal from each node are required. In every step, it expands only the currently shortest path by adding the adjacent node with the shortest distance including an estimate of remaining distance to goal. The algorithm stops when the goal vertex has been treated in the priority queue.

Algorithm 4 A* Algorithm

1. Input: A graph
2. Output: A path between start and goal nodes
3. Repeat
   a. Pick n_{best} from 0 such that $f(n_{best}) < f(n)$
   b. Remove n_{best} from O and add to C
   c. If $n_{best} = q_{goal}$, EXIT
   d. expand n_{best}: for all $x \in Star(n_{best})$ that are not in C
   e. if $x \notin O$ then
   f. add x to O
   g. else if $g(n_{best}) + C(n_{best}, x) < g(x)$ then
   h. update x's back pointer to point to n_{best}
   i. end if
4. Until O is empty

The pseudocode for this approach can be formulated as algorithm 4. If the heuristic is admissible, i.e., if $h(s) \leq c^*(s, s_{goal})$ for all s, A^* is guaranteed to find the optimal solution in optimal running time. If the heuristic is also consistent, i.e., if $h(s) \leq c^*(s, s') + h(s')$, it can be proven that no state is expanded more than once by the A^* algorithm. It has a priority queue which contains a list of nodes sorted by priority, determined by the sum of the distance traveled in the graph thus far from the start node and the heuristic. The first node to be put into the priority queue is naturally the start node. Next, the start node is expanded by putting all adjacent nodes to the start node into the priority queue sorted by their corresponding priorities. These nodes can naturally be embedded into the autonomous aircraft free space and thus have values corresponding to the cost required to traverse between the adjacent nodes. The output of the A^* algorithm is a back-pointer path, which is a sequence of nodes starting from the goal and going back to the start. The difference from Dijkstra is that A^* expands the state s in OPEN with a minimal value of $g(s)+h(s)$ where $h(s)$ is the heuristic that estimates the cost of moving from s to s_{goal}. Let $c^*(s, s')$ denote the cost of the optimal solution between s and s'. The open list saves the information about the parental nodes found as a candidate solution. The 3D cells in the grid not only have elements in the neighbourhood on the same height level used but also have cell nodes with locations above and below. Two additional structures are used, an open set O and a closed set C. The open set O is the priority queue and the closed set C contains all processed nodes [17]. The Euclidean distance between the current point and the destination goal, divided by the maximum possible nominal speed, can be employed as a heuristic function. This choice ensures that the heuristic cost will always be lower than the actual cost to reach the goal from a given node and thus the optimum solution is guaranteed.

Trajectory primitives provide a useful local solution method within the sampling-based path planning algorithm to produce feasible but suboptimal trajectories through complicated environments with relatively low computational cost. Extensions to the traditional trajectories' primitives allow 3D maneuvers with continuous heading and flight path angles through the entire path. These extensions, which include simple transformations as well as additional maneuvers, can maintain closed form solutions to the local planning problem and therefore maintain low computational cost [96].

4.3.2.1.2 Probabilistic methods

Sampling-based motion planning algorithms have been designed and successfully used to compute a probabilistic complete solution for a variety of environments. Two of the most successful algorithms include the **probabilistic roadmap method** (PRM) and **rapidly exploring random trees** (RRT). At a high level, as more samples are generated randomly, these techniques provide collision-free paths by capturing a larger portion of the connectivity of the free space [114].

4.3.2.1.2.1 Probabilistic roadmap method The **probabilistic roadmap method** (PRM) sparsely samples the word map, creating the path in two phase process: planning and query. The **query** phase uses the result of the planning phase to find a path from the initial configuration to the final one. The planning phase finds N random points that lie in free space. Each point is connected to its nearest neighbors by a straight line path that does not cross any obstacles, so as to create a network or graph, with a minimal number of disjoint components and no cycles. Each edge of the graph has an associated cost which is the distance between its nodes. An advantage of this planner is that once the roadmap is created by the planning phase, the goal and starting points can be changed easily. Only the query phase needs to be repeated.

The PRM algorithm combines an off-line construction of the roadmap with a randomized online selection of an appropriate path from the roadmap. However, this algorithm cannot be applied in a rapidly changing environment due to off-line construction of the roadmap.

The roadmap methods described above are able to find a shortest path on a given path. The issue most path planning methods are dealing with is how to create such a graph. To be useful for path planning applications, the roadmap should represent the connectivity of the free configuration space well and cover the space such that any query configuration can be easily connected to the roadmap. Probabilistic roadmap approach (PRM) is a probabilistic complete method that is able to solve complicated path planning problems in arbitrarily high dimension configuration spaces. The basic concept in PRM is that rather than attempt to sample all of C-space, one instead samples it probabilistically. This algorithm operates in two phases, a roadmap construction phase in which a roadmap is constructed within the C-space and a query phase, in which probabilistic searches are conducted using the roadmap to speed the search:

1. **Roadmap Construction Phase**: tries to capture the connectivity of free configuration space. An undirected, acyclic graph is constructed in the autonomous aircraft C-space in which edges connect nodes if and only if a path can be found between the nodes corresponding to waypoints. The graph is grown by randomly choosing new locations in C-space and attempting to find a path from the new location to one of the nodes already in the graph while maintaining the acyclic nature of the graph. This relies on a local path planner to identify possible paths from the randomly chosen location and one or more of the nodes in the graph. The choice of when to stop building the graph and the design of the local path planner are application specific, although performance guarantees are sometimes possible. **Local planning** (m milestones, e edges) connect nearby milestones using a local planner and form a roadmap. Local planning checks whether there is a local path between two milestones, which corresponds to an edge on the roadmap. Many methods are available for local planning. The most common way is to discretize the path between two milestones into n_i steps and the local path exists when all the intermediate samples are collision free, performing discrete collision queries at those steps. It is the most expensive part of the PRM algorithm.
2. **Query Phase**: When a path is required between two configurations s and g, paths are first found from s to node $\bar{s}$ in the roadmap and from g to some node $\bar{g}$ in the roadmap. The roadmap is then used to navigate between $\bar{g}$ and $\bar{s}$. After every query, the nodes s and g and the edges connecting them to the graph can be added to the roadmap. As in the learning phase, the query phase relies on a heuristic path planner to find local paths in the configuration space.

Algorithm 5 Roadmap Algorithm

1. nodes $\leftarrow$ sample N nodes random configuration
2. for all nodes
3. find $k_{nearest}$ nearest neighbors
4. if collision check and $\gamma \leq \gamma_{max}$ then roadmap $\leftarrow$ edge
5. end

The local planner should be able to find a path between two configurations in simple cases in a small amount of time. Given a configuration and a local planner, one can define the set of configurations to which a local planning attempt will succeed. This set is called the visibility region of a node under a certain local planner. The larger the visibility region is, the more powerful the local planner. The most straightforward sampling scheme shown in algorithm 5 in page 239 is to sample configurations uniformly randomly over the configuration space.

For every node, a nearest neighbor search is conducted. Several constraints have to be satisfied during the construction phase before an edge connection between two nodes is possible.

4.3.2.1.2.2 Rapidly-exploring random tree method The method **rapidly exploring random tree** (RRT) is able to take into account the motion model of the aircraft. A graph of aircraft configurations is maintained and each node is a configuration. The first node in the graph is the initial configuration of the aircraft. A random configuration is chosen and the node with the closest configuration is found. This point is near in terms of a cost function that includes distance and orientation. A control is computed that moves the aircraft from the *near* configuration to the random configuration over a fixed period of time. The point that it reaches is a new point and this is added to the graph. The distance measure must account for a difference in position and orientation and requires appropriate weighting of these quantities. The random point is discarded if it lies within an obstacle. The result is a set of paths or roadmap. The trees consist of feasible trajectories that are built online by extending branches towards randomly generated target states.

The rapidly expanding random tree approach is suited for quickly searching high-dimensional spaces that have both algebraic and differential constraints. The key idea is to bias the exploration toward unexplored portions of the space by sampling points in the state space and incrementally pulling the search tree toward them, leading to quick and uniform exploration of even high-dimensional state spaces. A graph structure must be built with nodes at explored positions and with edges describing the control inputs needed to move from node to node. Since a vertex with a larger Voronoi region has a higher probability to be chosen as (x_{near}) and it is pulled to the randomly chosen state as close as possible, the size of larger Voronoi regions is reduced as the tree grows. Therefore, the graph explores the state space uniformly and quickly. The basic RRT algorithm 6 in page 241 operates as follows: the overall strategy is to incrementally grow a tree from the initial state to the goal state. The root of this tree is the initial state; at each iteration, a random sample is taken and its nearest neighbor in the tree computed. A new node is then created by growing the nearest neighbor toward the random sample.

For each step, a random state (x_{rand}) is chosen in the state space. Then (x_{near}) in the tree that is the closest to the (x_{rand}) in metric ρ is selected. Inputs $u \in \mathbf{U}$, the input set, are applied for Δt, making motions toward (x_{rand}) from (x_{near}). Among the potential new states, the state that is as close as possible to (x_{rand}) is selected as a new state (x_{new}). The new state is added to the tree as a new vertex. This process is continued until (x_{new}) reaches (x_{goal}).

To improve the performance of the RRT, several techniques have been proposed such as biased sampling and reducing metric sensitivity. **Hybrid systems** models sometimes help by switching controllers over cells during a

Algorithm 6 RRT Basic Algorithm

1. Build RRT (x_{init})
2. G_{sub}, init(x_{init})
3. for k = 1 to maxIterations do
4. $x_{rand} \leftarrow$ RANDOM-STATE()
5. $x_{near} \leftarrow$ NEAREST-NEIGHBOR(x_{rand}, G_{sub})
6. u_{best}, x_{new}, success $\leftarrow$ CONTROL($x_{near}, x_{rand}, G_{sub}$);
7. if success
8. G_{sub}.add-vertex x_{new}
9. G_{sub}.add-edge $x_{near}, x_{new}, u_{best}$
10. end
11. Return G_{sub}
12. RRT-EXTEND
13. $V \leftarrow \{x_{init}\}, E \leftarrow \emptyset, i \leftarrow 0$;
14. While $i < N$, do
15. $G \leftarrow (V, E)$
16. $x_{rand} \leftarrow$ Sample(i); $i \leftarrow i + 1$)
17. (V, E) $\leftarrow$ Extend(G, x_{rand})
18. end

decomposition. Another possibility is to track space-filling trees, grown backwards from the goal.

4.3.2.2 Continuous methods

4.3.2.2.1 Receding horizon control

Receding horizon control, a variant of model predictive control, repeatedly solves online a constrained optimization problem over a finite planning horizon. At each iteration, a segment of the total path is computed using a dynamic model of the aircraft that predicts its future behavior. A sequence of control inputs and resulting states is generated that meet the kino-dynamic and environmental constraints and that optimize some performance objective. Only a subset of these inputs is actually implemented, however, and the optimization is repeated as the aircraft maneuvers and new measurements are available. The approach is specially useful when the environment is explored online.

4.3.2.2.2 Mixed integer linear programming

Mixed integer linear programming (MILP) approaches the problem of collision avoidance as an optimization problem with a series of constraints. The goal or objective function is to minimize the time needed to traverse several waypoints. The constraints are derived from the problem constraints (flight speed, turning radius) and the fact that the aircraft must maintain a safer distance from obstacles and other aircraft.

Autonomous aircraft trajectory optimization including collision avoidance can be expressed as a list of linear constraints, involving integer and continuous variables known as the mixed integer linear program. The mixed integer linear programming (MILP) approach in [161] uses indirect branch-and-bound optimization, reformulating the problem in a linearized form and using commercial software to solve the MILP problem. A single aircraft collision avoidance application was demonstrated. Then this approach was generalized to allow for visiting a set of waypoints in a given order. Mixed integer linear programming can extend continuous linear programming to include binary or integer decision variables to encode logical constraints and discrete decisions together with the continuous autonomous aircraft dynamics. The approach to optimal path planning based on MILP was introduced in [24, 122, 174]. The autonomous aircraft trajectory generation is formulated as a 3D optimization problem under certain conditions in the Euclidean space, characterized by a set of decision variables, a set of constraints and the objective function. The decision variables are the autonomous aircraft state variables, i.e., position and speed. The constraints are derived from a simplified model of the autonomous aircraft and its environment. These constraints include:

1. Dynamics constraints, such as a maximum turning force which causes a minimum turning radius, as well as a maximum climbing rate.
2. Obstacle avoidance constraints like no-flight zones.
3. Target reaching constraints of a specific way point or target.

The objective function includes different measures of the quality in the solution of this problem, although the most important criterion is the minimization of the total flying time to reach the target. As MILP can be considered as a geometric optimization approach, there is usually a protected airspace set up around the autonomous aircraft in the MILP formulation. The stochasticity that stems from uncertainties in observations and unexpected aircraft dynamics could be handled by increasing the size of protected airspaces. An advantage of the MILP formulation is its ability to plan with non-uniform time steps between waypoints. A disadvantage of this approach is that it requires all aspects of the problem (dynamics, ordering of all waypoints in time and collision avoidance geometry) to be specified as a carefully designed and a usually long list of many linear constraints, and then the solver's task is basically to find a solution that satisfies all of those constraints simultaneously [194].

Then a MILP solver takes the objective and constraints and attempts to find the optimal path by manipulating the force effecting how much a single aircraft turns at each time step. Although mixed integer linear programming is an elegant method, it suffers from exponential growth of the computations [161, 174].

4.3.2.2.3 Classical potential field method

The **artificial potential field method** is a collision-avoidance algorithm based on electrical fields. Obstacles are modeled as repulsive charges and destinations (waypoints) are modeled as attractive charges. The summation of these charges is then used to determine the safest direction to travel.

Let $X = (x, y, z)^T$ denote the UAV current position in airspace. The usual choice for the **attractive potential** is the standard parabolic that grows quadratically with the distance to the goal such that:

$$U_{att} = \frac{1}{2} k_a d_{goal}^2(X) \tag{4.116}$$

where $d_{goal} = \|X - X_{goal}\|$ is the Euclidean distance of the UAV current position X to the goal X_{goal} and k_a is a scaling factor. The attractive force considered is the negative gradient of the attractive potential:

$$F_{att}(X) = -k_a \left(X - X_{goal}\right) \tag{4.117}$$

By setting the aircraft velocity vector proportional to the vector field force, the force $F_{att}(X)$ drives the aircraft to the goal with a velocity that decreases when the UAV approaches the goal.

The **repulsive potential** keeps the aircraft away from obstacles. This repulsive potential is stronger when the UAV is closer to the obstacles and has a decreasing influence when the UAV is far away. A possible repulsive potential generated by obstacle i is:

$$U_{rep_i}(X) = \begin{pmatrix} \frac{1}{2} k_{rep} \left(\frac{1}{d_{obs_i}(X)} - \frac{1}{d_0} \right)^2 & d_{obs_i}(X) \leq d_0 \\ 0 & \text{otherwise} \end{pmatrix} \tag{4.118}$$

where i is the number of the obstacle close to the UAV, $d_{obs_i}(X)$ is the closest distance to the obstacle i, k_{rep} is a scaling constant and d_0 is the obstacle influence threshold.

$$F_{rep_i}(X) = \begin{pmatrix} k_{rep} \left(\frac{1}{d_{obs_i}(X)} - \frac{1}{d_0} \frac{1}{d_{obs_i}(X)} \right) \hat{e}_i & d_{obs_i}(X) \leq d_0 \\ 0 & \text{otherwise} \end{pmatrix} \tag{4.119}$$

where $\hat{e}_i = \frac{\partial d_{obs_i}(X)}{\partial X}$ is a unit vector that indicates the direction of the repulsive force; therefore:

$$\begin{pmatrix} \dot{x}_d \\ \dot{y}_d \\ \dot{z}_d \end{pmatrix} = -\left(F_{att}(X) + F_{rep_i}(X)\right) \tag{4.120}$$

After the desired global velocity is calculated by the potential field method, the corresponding desired linear velocity V_d and attitude χ_d, γ_d can also be obtained:

$$V_d = k_u \sqrt{\dot{x}_d^2 + \dot{y}_d^2 + \dot{z}_d^2} \tag{4.121}$$

$$\gamma_d = atan2\left(-\dot{z}_d, \sqrt{\dot{x}_d^2 + \dot{y}_d^2}\right) \tag{4.122}$$

$$\chi_d = atan2(\dot{y}_d, \dot{z}_d) \tag{4.123}$$

where the gain k_u is introduced to allow for additional freedom in weighting the velocity commands. The pitch and yaw angle guidance laws are designed so that the aircraft's longitudinal axis steers to align with the gradient of the potential field. The roll angle guidance law is designed to maintain the level flight.

Harmonic field approach is useful in avoiding local minima of the classical potential field methods [23].

4.3.3 MOVING OBSTACLES AVOIDANCE

Smart autonomous aircraft require a collision-avoidance algorithm, also known as **sense and avoid**, to monitor the flight and alert the aircraft to necessary avoidance maneuvers. The challenge is now autonomous navigation in open and dynamic environments, i.e., environments containing moving objects or other aircraft as potential obstacles; their future behavior is unknown. Taking into account these characteristics requires to solve three main categories of problems [110]:

1. Simultaneous localization and mapping in dynamic environments. This topic will be discussed in the next chapter.
2. Detection, tracking, identification and future behavior prediction of the moving obstacles.
3. Online motion planning and safe navigation.

In such a framework, the smart autonomous aircraft has to continuously characterize with onboard sensors and other means. As far as the moving objects are concerned, the system has to deal with problems such as interpreting appearances, disappearances and temporary occlusions of rapidly maneuvering vehicles. It has to reason about their future behavior and consequently make predictions. The smart autonomous aircraft has to face a double constraint: constraint on the response time available to compute a safe motion which is a function of the dynamicity of the environment and a constraint on the temporal validity of the motion planned which is a function of the validity duration of the predictions.

Path planning in a priori unknown environments cluttered with dynamic objects and other aircraft is a field of active research. It can be addressed by using explicit time representation to turn the problem into the equivalent static problem, which can then be solved with an existing static planner. However, this increases the dimensionality of the representation and requires exact motion models for surrounding objects. The dimensionality increase raises the computational effort (time and memory) to produce a plan and motion modeling raises difficult prediction issues.

There are various approaches to both static and moving obstacles that are mostly based on the collision cone approach. Often, collision avoidance is achieved by tracking a waypoint at a safe distance from the obstacles using a homing guidance law [20]. Atmosphere can be very dynamic. The **cloud behavior** is very complex. Typically, in aircraft navigation, clouds and turbulence should be avoided [214]. They are considered as moving obstacles. There are several ways to model their behavior in a complex environment. Physical modeling of the cloud/turbulence can be done using **Gaussian dispersion methods**, which predict the cloud behavior using statistical dispersion techniques. Another modeling approach is to define the points picked up by the UAV as the vertices $\{v_i, i = 1, \dots, N\}$. The vertices are connected by line segments of constant curvature $\{\kappa_{ij}\}$ with C^2 contact at the vertices. The **splinegon** representation assumes some reasonably uniform distribution of vertices [179]. Each vertex has a curvature and length and these can be used to determine matrices for each segment.

The work presented in [173] is concerned with developing an algorithm that keeps a certain safety distance while passing an arbitrary number of possibly moving but non maneuvering obstacles. Starting from a 3D collision cone condition, input-output linearization is used to design a nonlinear guidance law [38]. The remaining design parameters are determined considering convergence and performance properties of the closed-loop guidance loop. Then, the guidance framework is developed in terms of a constrained optimization problem that can avoid multiple obstacles simultaneously while incorporating other mission objectives.

4.3.3.1 D^* algorithm

The algorithm D^* has a number of features that are useful for real world applications. It is an extension of the A^* algorithm for finding minimum cost paths through a graph for environments where the environment changes at a much slower speed than the aircraft. It generalizes the occupancy grid to a cost map $c \in \mathbb{R}$ of traversing each cell in the horizontal and vertical directions. The cost of traversing the cell diagonally is $c\sqrt{2}$. For cells corresponding to obstacles $c = \infty$. The key features of D^* is that it supports incremental replanning. If a route has a higher than expected cost, the algorithm D^* can incrementally replan to find a better path. The incremental replanning has a lower computational cost than completely replanning. Even though D^* allows the path to be recomputed as the cost map changes, it does not support a changing goal. It repairs the graph allowing for an efficient updated searching in dynamic environments.

Notation

1. X represents a state
2. O is the priority queue
3. L is the list of all states

4. S is the start state
5. t(x) is the value of state with respect to priority queue
   a. t(x): New if x has never been in O
   b. t(x): Open if x is currently in O
   c. t(x): Closed if x was in O but currently is not

Algorithm 7 D^* Algorithm

---

1. **Input**: List of all states L
2. **Output**: The goal state, if it is reachable, and the list of states L are updated so that back-pointer list describes a path from the start to the goal. If the goal state is not reachable, **return NULL**
3. For each $X \in L$ do
4. $t(X) = New$
5. $endfor$
6. $h(G) = 0$; $0 = \{G\}$; $X_c = S$
7. The following loop is Dijkstra's search for an initial path.
8. repeat
9. k_{min} = process-state (0, L)
10. until ($k_{min} > h(x_c)$) or ($k_{min} = -1$)
11. P=Get-Pointer-list (L, X_c, G)
12. If P = Null then
13. return (Null)
14. end if
15. end repeat
16. $endfor$
17. X_c is the second element of P Move to the next state in P
18. P= Get-Back-Pointer-List(L, X_c, G)
19. until X_c=G
20. return (X_c)

The D^* algorithm 7 is devised to locally repair the graph allowing efficient updated searching in dynamic environments, hence the term D^*. D^* initially determines a path starting with the goal and working back to the start using a slightly modified Dijkstra's search. The modification involves updating a heuristic function. Each cell contains a heuristic cost h which for D^* is an estimate of path length from the particular cell to the goal, not necessarily the shortest path length to the goal as it was for A^*. These h values will be updated during the initial Dijkstra search to reflect the existence of obstacles. The minimum heuristic values h are the estimate of the shortest path length to the goal. Both the h and the heuristic values will vary as the D* search runs, but they are equal upon initialization [17].

Field D^*, like D^*, is an incremental search algorithm which is suitable for navigation in an unknown environment. It makes an assumption about the unknown space and finds a path with the least cost from its current location to the goal. When a new area is explored, the map information is updated and a new route is replanned, if necessary. This process is repeated until the goal is reached or it turns out that that goal cannot be reached (due to obstacles for instance). The algorithm A^* can also be used in a similar way [56].

4.3.3.2 Artificial potential fields

The harmonic potential field approach works by converting the goal, representation of the environment and constraints on behavior into a reference velocity vector field [126]. A basic setting is

Problem 4.6. *Solve*

$$\nabla^2 V(P) = 0 \qquad P \in \mathbf{\Omega} \tag{4.124}$$

subject to $V(P) = 1$ *at* $P = \Gamma$ *and* $V(P_r) = 0$

A provably correct path may be generated using the gradient dynamical system

$$\dot{P} = -\nabla V(P) \tag{4.125}$$

The harmonic potential field approach can incorporate directional constraints along with regional avoidance constraints in a provably correct manner to plan a path to a target point. The navigation potential may be generated using the **boundary value problem** (BVP).

Problem 4.7. *Solve*

$$\nabla^2 V(P) = 0 \qquad P \in \mathbf{\Omega} - \mathbf{\Omega}' \tag{4.126}$$

and

$$\nabla\left(\Sigma(P)V(P)\right) = 0, P \in \mathbf{\Omega}' \tag{4.127}$$

subject to $V(P) = 1$ *at* $P = \Gamma$ *and* $V(P_r) = 0$.

$$\Sigma(P) = \begin{pmatrix} \sigma(P) & 0 & \dots & 0 \\ 0 & \sigma(P) & 0 & 0 \\ 0 & \dots & 0 & \sigma(P) \end{pmatrix} \tag{4.128}$$

A provably correct trajectory to the target that enforces both the regional avoidance and directional constraints may be simply obtained using the gradient dynamical system in (4.125). The approach can be modified to take into account ambiguity that prevents the partitioning of an environment into admissible and forbidden regions [126].

Dynamic force fields can be calculated based on aircraft position and velocity. The force field uses scalar modifications to generate a larger and stronger field in front of the aircraft. Therefore, the force exerted by one aircraft on another can be calculated using the difference between the bearing of the aircraft exerting the force and the one feeling that force. Furthermore, a secondary calculation occurs that scales the force exerted into a force belt. This scaling is computed by determining the difference between the exerted force and the bearing of the aircraft feeling the force. Finally, the repulsive forces are summed, scaled and added to an attractive force of constant magnitude. This resultant vector is bound by the maximum turning rate and then used to inform the aircraft of its next maneuvers: a new target waypoint is sent to the aircraft.

When an aircraft is close to its destination and because it begins to ignore the other aircraft, it is necessary to extend its force field such that the forces are exerted on non priority aircraft at a larger distance. This combination allows the prioritized aircraft to go directly to its goal while providing other aircraft with an early alert through the expanded force field [164].

The artificial potential method is modified in [164] to handle special cases by including a **priority system** and techniques to prevent an aircraft from circling its final destination. 2D constraints can be imposed because many UAS must operate in a limited range of altitude. Each UAS has a maximum operating altitude and the minimum altitude may be imposed by stealth or the application requirements.

The design and implementation of a potential field obstacle algorithm based on fluid mechanics panel methods is presented in [47]. Obstacles and the UAV goal positions are modeled by harmonic functions, thus avoiding the presence of local minima. Adaptations are made to apply the method to the automatic control of a fixed wing aircraft, relying only on a local map of the environment that is updated from sensors onboard the aircraft. To avoid the possibility of collision due to the dimension of the UAV, the detected obstacles are expanded. Considering that the detected obstacles are approximated by rectangular prisms, the expansion is carried out by moving the faces outwards by an amount equal to the wingspan. The minimum value that assured clearance to the obstacle is a half-span. Then obstacles are further extended creating prisms.

4.3.3.3 Online motion planner

The virtual net comprises a finite set of points $X_e(R)$ corresponding to a finite set of prescribed relative positions [206]:

$$R \in \mathbb{M} = \{R_1, R_2, \dots, R_n\} \subset \mathbb{R}^3 \tag{4.129}$$

$$X_e(R_k) = (R_k, 0)^T = (R_{x,k}, R_{y,k}, R_{z,k}, 0, 0, 0)^T \qquad k = 1 \dots n \tag{4.130}$$

where velocity states are zero and n is the number of points in the virtual net. The obstacle position and uncertainty is represented by an ellipsoid. The set $O(q, Q)$ centered around the position $q \in \mathbb{R}^3$ is used to over-bound the position of the obstacle i.e.:

$$\mathbb{O}(q, Q) = \left\{ X \in \mathbb{R}^6, (\mathbf{S}X - q)^T \mathbf{Q} (\mathbf{S}X - q) \leq 1 \right\} \tag{4.131}$$

where $\mathbf{Q} = \mathbf{Q}^T$ and $\mathbf{S} = \begin{bmatrix} 1 & 0 & 0 & 0 & 0 & 0 \\ 0 & 1 & 0 & 0 & 0 & 0 \\ 0 & 0 & 1 & 0 & 0 & 0 \end{bmatrix}$

The online motion planning with obstacle avoidance is performed according to the following algorithm 8.

Algorithm 8 Online Motion Planner

1. Determine the obstacle location and shape (i.e., q and Q).
2. Determine the growth distance.
3. Construct a graph connectivity matrix between all $R_i, R_j \in \mathbb{M}$. In the graph connectivity matrix, if two vertices are not connected, the corresponding matrix element is $+\infty$. If they are connected, the corresponding matrix element is 1. The graph connectivity matrix is multiplied element-wise to produce a constrained cost of transition matrix.
4. Perform graph search using any standard graph search algorithm to determine a sequence of connected vertices, $R(k) \in \mathbb{M}$ such that $R[1]$ satisfies the initial constraints, $R[l_p]$ satisfies the final constraints and the cumulative transition cost computed from the constrained cost of the transition matrix is minimized.
5. After the path has been determined as a sequence of the waypoints, the execution of the path proceeds by checking if the current state $X(t)$ is in the safe positively invariant set corresponding to the next reference R^+.

The set $\mathbb{O}(q, Q)$ can account for the obstacle and aircraft physical sizes and for the uncertainties in the estimation of the obstacle/aircraft positions. The set $\mathbb{O}(q, Q)$ has an ellipsoidal shape in the position directions and is unbounded in the velocity directions. Ellipsoidal sets, rather than polyhedral sets, can be used to over-bound the obstacle because ellipsoidal bounds are typically produced by position estimation algorithms, such as the extended Kalman filter. This filter will be presented in the next chapter.

4.3.3.4 Zermelo–Voronoi diagram

In many applications of autonomous aircraft, ranging from surveillance, optimal pursuit of multiple targets, environmental monitoring and aircraft routing problems, significant insight can be gleaned from data structures associated with **Voronoi-like partitioning** [11, 12]. A typical application can be the following: given a number of landing sites, divide the area into distinct non-overlapping cells (one for each landing site) such that the corresponding site in the cell is the closest one (in terms of time) to land for any aircraft flying over this cell in the presence of winds. A similar application that fits in the same framework is the task of subdividing the plane into **guard/safety** zones such that a **guard/rescue** aircraft residing within each particular zone can reach all points in its assigned zone faster than any other guard/rescuer outside its zone. This is the generalized minimum distance problems where the relevant metric is the minimum intercept or arrival time. Area surveillance missions can also be addressed using a frequency based approach where the objective implies to optimize the elapsed time between two consecutive visits to any position known as the refresh time [1].

Recent work in patrolling can be classified as [185]:

1. **offline versus online**: offline computes patrols before sensors are deployed, while online algorithm controls the sensor's motion during operation and is able to revise patrols after the environment has changed.
2. **finite versus infinite**: finite planning horizon algorithm computes patrols that maximizes reward over a finite horizon while an infinite horizon maximize an expected sum of rewards over an infinite horizon.
3. **controlling patrolling versus single traversal**: it is dynamic environment monitoring versus one snapshot of an environment.
4. **strategic versus non strategic patrolling**.
5. **spatial or spatio-temporal dynamics**.

The construction of **generalized Voronoi diagrams** with time as the distance metric is in general a difficult task for two reasons: First, the distance metric is not symmetric and it may not be expressible in closed form. Second, such problems fall under the general case of partition problems for which the aircraft's dynamics must be taken into account. The topology of the agent's configuration space may be non-Euclidean; for example, it may be a manifold embedded in a Euclidean space. These problems may not be reducible to **generalized Voronoi diagram** problems for which efficient construction schemes exist in the literature [143].

The following discussion deals with the construction of Voronoi-like partitions that do not belong to the available classes of generalized Voronoi diagrams. In particular, Voronoi-like partitions exist in the plane for a given finite set of generators, such that each element in the partition is uniquely associated with a particular generator in the following sense: an aircraft that resides in a particular set of the partition at a given instant of time can arrive at the generator associated with this set faster than any other aircraft that may be located anywhere outside this set at the same instant of time. It is assumed that the aircraft's motion is affected by the presence of temporally varying winds.

Since the generalized distance of this Voronoi-like partition problem is the minimum time to go of the Zermelo' problem, this partition of the configuration space is known as the **Zermelo–Voronoi diagram** (ZVD). This problem deals with a special partition of the Euclidean plane with respect to a generalized distance function. The characterization of this Voronoi-like partition takes into account the proximity relations between an aircraft that travels in the presence of winds and the set of Voronoi generators. The question of determining the generator from a given set which is the closest in terms of arrival time, to the agent at a particular instant of time, reduces the problem of determining the set of the Zermelo–Voronoi partition that the aircraft resides in at the given instant of time. This is the **point location problem**.

The **dynamic Voronoi diagram problem** associates the standard Voronoi diagram with a time-varying transformation as in the case of time-varying winds. The **dual Zermelo–Voronoi diagram** problem leads to a partition problem similar to the Zermelo–Voronoi Diagram with the difference that the generalized distance of the dual Zermelo–Voronoi diagram is the minimum time of the Zermelo's problem from a Voronoi generator to a point in the plane. The minimum time of the Zermelo's navigation problem is not a symmetric function with respect to the initial and final configurations. The case of non stationary spatially varying winds is much more complex.

The problem formulation deals with the motion of an autonomous aircraft. It is assumed that the aircraft's motion is described by the following equation:

$$\dot{X} = U + W(t) \tag{4.132}$$

$X = (x, y, z)^T \in \mathbb{R}^3, U \in \mathbb{R}^3$ and $W = (W_N, W_E, W_D)^T \in \mathbb{R}^3$ is the wind which is assumed to vary uniformly with time, known a priori. In addition, it is assumed that $|W(t)| < 1, \forall t \geq 0$ which implies that system (4.132) is controllable. Furthermore, the set of admissible control inputs is given by $\mathbb{U} = \{U \in \mathbf{U}, \forall t \in [0, T], T > 0\}$ where $\mathbf{U} = \left\{(U_1, U_2, U_3) \in \mathbb{U} | U_1^2 + U_2^2 + U_3^2 = 1\right\}$ the closed unit ball and U is a measurable function on $[0, T]$.

The Zermelo's problem solution when $W = 0$ is the control $U^*(\chi^*, \gamma^*) = (\cos\gamma^* \cos\chi^*, \cos\gamma^* \sin\chi^*, -\sin(\gamma^*))$ where χ^*, γ^* are constants, as shown in the previous section. Furthermore, the Zermelo's problem is reduced to the shortest path problem in 3D.

Next, the Zermelo–Voronoi diagram problem is formulated:

Problem 4.8. ***Zermelo–Voronoi Diagram Problem:*** *Given the system described by (4.132), a collection of goal destination* $\mathbb{P} = \left\{p_i \in \mathbb{R}^3, i \in \ell\right\}$ *where* ℓ *is a finite index set and a transition cost*

$$C(X_0, p_i) = T_f(X_0, p_i) \tag{4.133}$$

determine a partition $\mathbb{B} = \{\mathbb{B}_i : i \in \ell\}$ *such that:*

1. $\mathbb{R}^3 = \cup_{i \in \ell} \mathbb{B}_i$
2. $\bar{\mathbb{B}}_i = \mathbb{B}_i, \forall i \in \ell$
3. *for each* $X \in int(\mathbb{B}_i), C(X, p_i) < C(X, P_j), \forall j \neq i$

It is assumed that the wind $W(t)$ induced by the winds is known in advance over a sufficiently long (but finite) time horizon. Henceforth, $\mathbb{P}$ is the set of Voronoi generators or sites, $\mathbb{B}_i$ is the Dirichlet domain and $\mathbb{B}$ the Zermelo–Voronoi diagram of $\mathbb{R}^3$, respectively. In addition, two Dirichlet domains $\mathbb{B}_i$ and

$\mathbb{B}_j$ are characterized as neighboring if they have a non-empty and non-trivial (i.e., single point) intersection.

The Zermelo's problem can be formulated alternatively as a moving target problem as follows:

Problem 4.9. ***Moving Target Problem:*** *Given the system described by*

$$\dot{\mathbf{X}} = \dot{X} - W(t) = U(t) \quad \mathbf{X}(0) = X_0 \tag{4.134}$$

determine the control input $U^* \in \mathbb{U}$ such that:

1. *The control U^* minimizes the cost functional $J(U) = T_f$ where T_f is the free final time*
2. *The trajectory $\mathbf{X}^* : [0, T_f] \to \mathbb{R}^3$ generated by the control U^* satisfies the boundary conditions*

$$\mathbf{X}^*(0) = X_0 \quad \mathbf{X}^*(T_f) = X_f - \int_0^{T_f} W(\tau)d\tau \tag{4.135}$$

The Zermelo's problem and problem 4.9 are equivalent in the sense that a solution of the Zermelo's problem is also a solution of problem 4.9 and vice versa. Furthermore, an optimal trajectory $\mathbf{X}^*$ of problem 4.9 is related to an optimal trajectory X^* of the Zermelo's problem by means of the time-varying transformation

$$\mathbf{X}^*(t) = X(t) - \int_0^t W(\tau)d\tau \tag{4.136}$$

The Zermelo's minimum time problem can be interpreted in turn as an optimal pursuit problem as follows:

Problem 4.10. ***Optimal Pursuit Problem:*** *Given a pursuer and the moving target obeying the following kinematic equations*

$$\dot{X}_p = \dot{X} = U \quad X_p(0) = X_0 = \mathbf{X}_0 \tag{4.137}$$

$$\dot{X}_T = -W(t) \quad X_T(0) = X_f \tag{4.138}$$

where X_p and X_T are the coordinates of the pursuer and the moving target, respectively, find the optimal pursuit control law U^* such that the pursuer intercepts the moving target in minimum time T_f:

$$X_p(T_f) = \mathbf{X}(T_f) = X_T(T_f) = X_f - \int_0^{T_f} W(\tau)d\tau \tag{4.139}$$

The optimal control of Zermelo's problem is given by

$$U^*(\chi^*, \gamma^*) = (\cos\gamma^* \cos\chi^*, \cos\gamma^* \sin\chi^*, -\sin\gamma^*)$$

The same control is also the optimal control for the moving target in problem 4.9. Because the angles χ^*, γ^* are necessarily constant, the pursuer is constrained to travel along a ray emanating from X_0 with constant unit speed (constant bearing angle pursuit strategy) whereas the target moves along the time parametrized curve $X_T : [0, \infty] \to \mathbb{R}^3$ where $X_T(t) = X_f - \int_0^t W(\tau)d\tau$.

4.3.4 TIME OPTIMAL NAVIGATION PROBLEM WITH MOVING AND FIXED OBSTACLES

A fast autonomous aircraft wishes to go through a windy area so as to reach a goal area while needing to avoid n slow moving aircraft and some very turbulent areas. The fast aircraft is regarded as a fast autonomous agent and the n slow aircraft are regarded as n slow agents. It is assumed that the trajectories of the n slow agents are known to the fast aircraft in advance. The objective is to find a control such that the mission is accomplished within a minimum time. A time optimal navigation problem with fixed and moving obstacles is considered in this section. This problem can be formulated as an optimal control problem with continuous inequality constraints and terminal state constraints. By using the control parametrization technique together with the time scaling transform, the problem is transformed into a sequence of optimal parameter selection problems with continuous inequality constraints and terminal state constraints. For each problem, an exact penalty function method is used to append all the constraints to the objective function yielding a new unconstrained optimal parameter selection problem. It is solved as a nonlinear optimization problem [22].

An exact penalty function method is applied to construct a constraint violation function for the continuous inequality constraints and the terminal state constraints. It is then appended to the control function, forming a new cost function. In this way, each of the optimal parameter selections is further approximated as an optimal parameter selections subject to a simple non-negativity constraint or a decision parameter. This problem can be solved as a nonlinear optimization by any effective gradient based optimization technique, such as the sequential quadratic programming (SQP) method [118].

4.3.4.1 Problem formulation

Given $n+1$ agents in a 3D flow field, where n slow aircraft follow navigated trajectories, while the fastest aircraft is autonomously controllable, let the trajectories of the n slow agents be denoted as:

$$\eta_i = \begin{pmatrix} x_i(t) \\ y_i(t) \\ z_i(t) \end{pmatrix}, i = 1, \dots, n, \qquad t \geq 0.$$

The flow velocity components at any point (x, y, z) in the 3D flow field can be denoted by $W_N(x, y, z, t), W_E(x, y, z, t), W_D(x, y, z, t)$, respectively. Then the motion of the autonomous fast aircraft can be modeled as:

$$\begin{aligned} \dot{x} &= V\cos\chi\cos\gamma + W_N(x,y,z,t) \\ \dot{y} &= V\sin\chi\cos\gamma + W_E(x,y,z,t) \\ \dot{z} &= -V\sin\gamma + W_D(x,y,z,t) \end{aligned} \tag{4.140}$$

where V is the velocity of the controlled agent and the angles $\chi(t), \gamma(t)$ are considered as the control variables, subject to limitation constraints:
$|\chi(t)| \leq \chi_{max}$ $\quad |\gamma(t)| \leq \gamma_{max}$
The relations (4.140) are equivalent to:

$$\dot{\eta}(t) = f(\eta(t), \chi(t), \gamma(t), t) \quad \eta(0) = \eta_0 \quad t \geq 0 \tag{4.141}$$

where η_0 is the initial position of the fast autonomous aircraft. The objective of the Zermelo's problem is to find an optimal trajectory for the fast agent A_{n+1} such as the shortest route, the fastest route or the least fuel consumption to arrive at its goal area without colliding with the fixed obstacles and the other n slow agents.

Time optimal control problem 4.11 is formulated as given next:

Problem 4.11. ***Optimal Control Problem***

$$min_{\chi,\gamma} T_f$$

subject to

$$\dot{\eta}(t) = f(\eta(t), \chi(t), \gamma(t), t) \quad \eta(0) = \eta_0 \quad t \geq 0$$

$$\sqrt{(x(t) - x_i(t))^2 + (y(t) - y_i(t))^2 + (z(t) - z_i(t))^2} \geq max\left\{R, R_i\right\} \tag{4.142}$$

$$\eta(t) \in \aleph = \{x_{min} \leq x \leq x_{max}, y = 2h_y, z = 2h_z\}$$

where T_f represents the time instant at which the fast agent reaches the goal area. The terminal time T_f depends implicitly on the control function, which is defined at the first time when the fast autonomous aircraft enters the target set $\aleph$. For each $i = 1, \dots, n$, R_i is the safety radius of the ith slow aircraft and R is the safety radius of the fast autonomous aircraft.

4.3.4.2 Control parametrization and time scaling transform

Problem 4.11 is a nonlinear optimal control problem subject to continuous inequality constraints. Control parametrization and time scaling transform are applied to transform this problem into a nonlinear semi-infinite optimization problem to be solved by an exact penalty function method. The control parametrization is achieved as follows:

$$\chi_p(t) = \sum_{k=1}^{p} \vartheta_k^{\chi} \chi^c_{\tau^{\chi}_{k-1}, \tau^{\chi}_k}(t) \qquad \gamma_p(t) = \sum_{k=1}^{p} \vartheta_k^{\gamma} \gamma^c_{\tau^{\gamma}_{k-1}, \tau^{\gamma}_k}(t) \tag{4.143}$$

where $\tau_{k-1} \leq \tau_k, k = 1, \ldots, p$ and χ^c, γ^c are the characteristic functions defined by

$$\chi^c(t) = \left\{ \begin{array}{cc} 1 & t \in I \\ 0 & \text{otherwise} \end{array} \right\} \quad \gamma^c(t) = \left\{ \begin{array}{cc} 1 & t \in [\tau_{k-1}, \tau_k] \\ 0 & \text{otherwise} \end{array} \right\} \tag{4.144}$$

The switching times $\tau_i^{\chi}, \tau_i^{\gamma}, 1 \leq i \leq p-1$ are also regarded as decision variables. The time scaling transforms these switching times into fixed times $k/p, k = 1, \ldots, p-1$ on a new time horizon $[0, 1]$. This is achieved by the following differential equation:

$$\dot{t}(s) = \vartheta^t(s) \sum_{k=1}^{p} \vartheta_k^t \chi^t_{\tau^{\chi}_{k-1}, \tau^{\chi}_k}(t) \tag{4.145}$$

Observations of weather avoidance maneuvering typically reveal reactive (tactical) deviations around hazardous weather cells. Safety constraints dictate that aircraft must remain separated from one another as from hazardous weather. Because of weather forecasting errors, weather constraints are not usually known with certainty. The uncertainty is smaller for short range forecasts but the uncertainty increases and becomes substantial for long range forecasts. Model weather constraints can be modeled as deterministic constraints varying with time according to a piece-wise constant function that is based on a weather forecast model; the most recent short range weather forecast is made. Aircraft are modeled as points in motion. Their dynamics are specified in terms of bounds on the aircraft velocity and magnitude of acceleration. Whereas acceleration bounds give rise to bounds on the radius of curvature of flight, the scale of the solution is assumed to be large enough that aircraft dynamics can be approximated with a single representation of piece-wise linear flight legs connected at way points.

4.3.4.3 RRT variation

In [120], a path planning algorithm based on the 3D Dubins curve for UAV to avoid both static and moving obstacles is presented. A variation of RRT is used as the planner. In tree expansion, branches of the trees are generated by propagating along the 3D Dubins curve. The node sequence of shortest length together with the Dubins curves connecting them is selected as the path. When the UAV executes the path, the path is checked for collision with an updated obstacles state. A new path is generated if the previous one is predicted to collide with obstacles. Such checking and re-planning loop repeats until the UAV reaches the goal. The 3D Dubins curve is used in mode connection because:

1. It allows to assign initial and final heading of the UAV as well as position.

2. It is the shortest curve that connects two points with a constraint on the curvature determined by UAV turning radius of the path and prescribed initial and final headings.

To connect node (x_0, y_0, z_0, χ_0) to (x_1, y_1, z_1, χ_1), a 2D Dubins curve C is first created from (x_0, y_0, χ_0) to (x_1, y_1, χ_1); it is then extended to 3D by assigning:

$$z = z_0 + \frac{\ell(x,y)}{\ell(x_1,y_1)}(z - z_0) \tag{4.146}$$

to each (x, y) in C where $\ell(x, y)$ stands for the length along C from (x_0, y_0) to (x, y).

The next step is to propagate the UAV model along the 3D Dubins curve until reaching the end or being blocked by obstacles. If the end is reached, the propagatory trajectory is the connection between the two nodes; otherwise, a connection does not exist due to collision with obstacles.

4.4 MISSION PLANNING

A mission describes the operation of an aircraft in a given region, during a certain period of time while pursuing a specific objective. waypoints are locations to which the autonomous aircraft is required to fly. A flight plan is defined as the ordered set of waypoints executed by the aircraft during a mission. Along the way, there may be a set of areas to visit and a set of areas to avoid. In addition, the mission planning strategy should be dynamic as the mission planning problem is to create a path in a dynamic environment. The aim is to replace the human expert with a synthetic one that can be deployed onboard the smart autonomous aircraft [83, 84].

A mission is carried out by performing different actions: actions of movements, actions on the environment, information gathering. The resources used for the implementation of actions are available in limited quantities [99]. For the autonomous aircraft, resources are consumable as fuel and electricity levels decline gradually as the mission proceeds. Mission planning adapts flight to mission needs. The mission planning problem is to select and order the best subset among the set of objectives to be achieved and to determine the dates of start and end of each objective, maximizing the rewards obtained during the objectives and criteria for minimizing the consumption of resources while respecting the constraints on resources and mission.

Mission planning can be considered as a selection problem. The objectives are linked to rewards whose values vary depending on the importance of each objective. Planning must choose a subset of objectives to be achieved in time and limited resources. The existing planning systems are mostly unsuitable for solving smart autonomous aircraft problems: they address a problem where the goal is a conjunction of goals and fail if the goal is not reached. Moreover, the selection of targets does not entirely solve the problem of mission planning. Indeed, the selection is often based on a simplified model for the resources of the problem and ignores the various ways to achieve the same goal. In most cases, a practical solution is obtained by combining the selection of objectives, planning and task scheduling in multi-levels: each level defines the problem to the scheduling algorithm of the lower level. Unlike the planning carried out during the mission preparation, planning online is characterized by the fact that the time taken to find a plan is one of the main criteria for judging the quality of a method. Models to formalize a planning problem and associated methods can be classified into three types:

1. Representations of **logical type**: The dynamics of the aircraft translates into a succession of consecutive states indexed by a time parameter. The states are described by a set of logical propositions and action is an operator to move from one state to another. The purpose of planning is to synthesize a trajectory in state space, predict rewards earned for the course, select and organize different types of actions to achieve a goal or optimize the reward functions.
2. Representations of **graph type**: they offer a more structured representation. Among the approaches using graphs include Petri nets and Bayesian networks.
3. Representations of **object type**: they have spread because of their relevance to the object-oriented programming languages.

The aim is to formulate the art of flying an aircraft into logical tasks with a series of events to maintain control of specific functions of the aircraft. This concept is a key component with systems designed for future use. The initial concept of using an expert system to control an aircraft seems simple but proves difficult to apply. An expert pilot's decision-making processes are difficult to initiate with computers [147]. The dynamic environment and conditions affecting an aircraft are areas that have to be adapted to such an expert system. The many tasks involved in the control of flight must be divided into manageable steps.

The purpose of the **mission planning** is to select the objectives to achieve and find a way to achieve them, taking into account the environment. Among the possible solutions, the planner must choose the one that optimizes a criterion taking into account the rewards for each goal and cost to achieve them, and respects the constraints of time and resources. Rewards and constraints are nonlinear functions of time and resources to the different times when the aircraft performs the actions that lead to achieving the objectives. For achieving a goal, there is a beginning of treatment, end of treatment and when the reward associated with the target is obtained. The mission planner must choose and order a subset of targets, to achieve among all mission objectives. It should optimize the choice of its actions, knowing its resources, the environment, the maximum reward associated with each objective and the time constraints associated with them. Techniques were first designed to solve classical problems from combinatorial optimization such as the traveling salesman problem or the Chinese postman problem [202], the maximum flow problem and the independent set point [140, 142, 146]. Most of these problems are closely related to graph theory.

Formalism could be based on the decomposition of the problem into two levels:

1. The top level corresponds to the objectives of the mission,
2. The lowest level describes this achievement as a function of time and resources.

It formalizes a problem with uncertainties where the number of objectives of the plan is not fixed a priori.

4.4.1 TRAVELING SALESMAN PROBLEM

A salesman has to visit several cities (or road junctions). Starting at a certain city, he wants to find a route of minimum length which traverses each of the destination cities exactly once and leads him back to his starting point. Modeling the problem as a complete graph with n vertices, the salesman wishes to make a tour or Hamiltonian cycle, visiting each cycle exactly once and finishing at the city he starts from [50].

A traveling salesman problem (TSP) instance is given by a complete graph G on a node set $V = \{1, 2, .., m\}$ for some integer m and by a cost function assigning a cost c_{ij} to the arc (i, j) for any i, j in V. The salesman wishes to make the tour whose total cost is minimum where the total cost is the sum of the individual costs along the edges of the tour [44].

Problem 4.12. *The formal language for the corresponding decision problem is:*

$$TSP = \left\{ \begin{array}{c} (G, c, k) : G = (V, E) \textit{ is a complete graph },\\ c \textit{ is a function from } V \times V \rightarrow N, k \in \mathbb{N},\\ G \textit{ has a traveling salesman tour with cost at most } k \end{array} \right\}$$

The data consist of weights assigned to the edges of a finite complete graph, and the objective is to find a Hamiltonian cycle, a cycle passing through all the vertices, of the graph while having the minimum total weight. $c(A)$ denotes the total cost of the edges in the subset $A \subseteq E$:

$$c(A) = \sum_{(u,v) \in A} c(u, v) \tag{4.147}$$

In many practical situations the least costly way to go from a place u to a place w is to go directly, with no intermediate steps. The cost function c satisfies the triangle inequality if for all the vertices, $u, v, w \in V$

$$c(u, w) \leq c(u, v) + c(v, w) \tag{4.148}$$

This triangle inequality is satisfied in many applications but not in all. It depends on the chosen cost. In this case, the minimum spanning tree can be used to create a tour whose cost is no more than twice that of the minimum tree weight, as long as the cost function satifies the triangle inequality. Thus, the pseudocode of the TSP approach can be presented in algorithm 9 below.

Algorithm 9 TSP with Triangle Inequality

1. Select a vertex $r \in G, V$ to be a root vertex
2. Compute a minimum spanning tree T for G from root r using MST-PRIM(G, c, r)
3. Let H be a list of vertices, ordered according to when they are first visited in a preorder tree walk of T.
4. A state transition rule is applied to incrementally build a solution.
5. return the Hamiltonian cycle H
6. **Minimum Spanning Trees**: Procedure MST-PRIM(G, c, r)
7. For each $u \in G, V$
8. u.key $= \infty$
9. u.π = NULL
10. r.key = 0
11. Q = G. V
12. While Q $\neq$ 0
13. u= EXTRACT-MIN(Q)
14. for each $v \in G.Adj\,[u]$
15. if $v \in Q$ and $w(u, v) < v.key$
16. v.π = u
17. v.key = w(u,v)

Lines 7 to 11 of algorithm 9 set the key of each vertex to ∞ (except for the root r, whose key is set to 0 so that it will be the first vertex processed), set the parent of each vertex to NULL and initialize the min-priority queue Q to contain all the vertices.

The algorithm maintains the following three-part loop invariant prior to each iteration of the while loop of lines 12 to 17.

1. $A = \{(\nu, v, \pi) : \nu \in V - \{r\} - Q\}$.
2. The vertices already placed into the minimum spanning tree are those in $V - Q$.
3. For all vertices $v \in Q$, if $v.\pi \neq NULL$, then $v.key < \infty$ and $v.key$ is the weight of a light edge (ν, v, π) connecting v to some vertex, already placed into the minimum spanning tree.

Line 13 identifies a vertex $u \in Q$ incident on a light edge that crosses the cut $(V - Q, Q)$ (with the exception of the first iteration, in which $u = r$ due to line 4). Removing u from the set Q adds it to the set $V - Q$ of vertices in the tree, thus adding $(u, u.\pi)$ to A. The for loop of lines 14 to 17 updates the *key* and π attributes of every vertex v adjacent to u but not in the tree, thereby maintaining the third part of the loop invariant.

There are other approximation algorithms that typically perform better in practice. If the cost c does not satisfy the triangle inequality, then good approximate tours cannot be found in polynomial time. There are different approaches for solving the TSP. Classical methods consist of heuristic and exact methods. Heuristic methods like cutting planes and branch and bound can only optimally solve small problems whereas the heuristic methods, such as Markov chains, simulated annealing and tabu search, are good for large problems [113]. Besides, some algorithms based on greedy principles such as nearest neighbor and spanning tree can be introduced as efficient solving methods. Nevertheless, classical methods for solving the TSP usually result in exponential computational complexities. New methods such as nature based optimization algorithms, evolutionary computation, neural networks, time adaptive self-organizing maps, ant systems, particle swarm optimization, simulated annealing and bee colony optimization are among solving techniques inspired by observing nature. Other algorithms are intelligent water drops algorithms and artificial immune systems [46].

In an instance of the traveling salesman problem, the distances between any pair of n points are given. The problem is to find the shortest closed path (tour) visiting every point exactly once. This problem has been traditionally been solved in two steps with the layered controller architecture for mobile robots. The following discussion is mainly based on [116, 172].

Problem 4.13. ***Dubins Traveling Salesman Problem (DTSP):*** *Given a set of n points in the plane and a number $L > 0$, DTSP asks if there exists a tour for the Dubins vehicle that visits all these points exactly once, at length at most L.*

At the higher decision-making level, the dynamics of the autonomous aircraft are usually not taken into account and the mission planner might typically choose to solve the TSP for the Euclidean metric (ETSP) i.e., using the Euclidean distances between waypoints. For this purpose, one can directly exploit many existing results on the ETSP on graphs. The first step determines the order in which the waypoints should be visited by the autonomous aircraft. At the lower level, a path planner takes as an input this waypoint ordering and designs feasible trajectory between the waypoints respecting the dynamics of the aircraft. In this section, the aircraft at a constant altitude is assumed to have a limited turning radius and can be modeled as Dubins vehicles. Consequently, the path planner could solve a sequence of Dubins shortest path problems (DSPP) between the successive waypoints. Even if each problem is solved optimally, however, the separation into two successive steps can be inefficient since the sequence of points chosen by the TSP algorithm is often hard to follow for the physical system.

In order to improve the performance of the autonomous aircraft system, mission planning and path planning steps are integrated. The Dubins vehicle can be considered as an acceptable approximation of a fixed wing aircraft at a constant altitude. Motivated by the autonomous aircraft applications, the traveling salesman problem is considered for Dubins vehicle DTSP:

Problem 4.14. *Given n points on a plane, what is the shortest Dubins tour through these points and what is its length?*

The worst case length of such a tour grows linearly with n and an algorithm can be proposed with performance within a constant factor of the optimum for the worst case point sets. An upper bound on the optimal length is also obtained [172]. A practical motivation to study the Dubins traveling salesman problem arises naturally for autonomous aircraft monitoring a collection of spatially distributed points of interest. In one scenario, the location of the points of interest might be known and static. Additionally, autonomous aircraft applications motivate the study of the dynamic traveling repairman problem (DTRP) in which the autonomous aircraft is required to visit a dynamically changing set of targets [14]. Such problems are examples of distributed task allocation problems and are currently generating much interest: complexity issues related to autonomous aircraft assignment problems, Dubins vehicles keeping under surveillance multiple mobile targets, missions with dynamic threats ...[67]. Exact algorithms, heuristics as well as polynomial time constant factor approximation algorithms, are available for the Euclidean traveling salesman problem. A variation of the TSP is the angular metric problem. Unlike other variations of the TSP, there are no known reductions of the Dubins TSP to a problem on a finite dimensional graph, thus preventing the use of well-established tools in combinatorial optimization.

Definition 4.5. ***Feasible Curve:*** *A feasible curve is defined for the Dubins vehicle or a Dubins path as a curve $\gamma : [0,T] \to \mathbb{R}^2$ that is twice differentiable almost everywhere and such that the magnitude of its curvature is bounded above by $1/\rho$ where $\rho > 0$ is the minimum turning radius.*

The autonomous aircraft configuration is represented by the triplet $(x, y, \psi) \in SE(2)$ where (x, y) are the Cartesian coordinates of the vehicle and ψ its heading. Let $P = p_1, \dots, p_n$ be a set of n points in a compact region $Q \subseteq \mathbb{R}^2$ and P_n be the collection of all point sets $P \subset Q$ with the cardinality n. Let ETSP(P) denote the cost of the Euclidean TSP over P, i.e., the length of the shortest closed path through all points in P. Correspondingly, let DTSP(P) denote the cost of the Dubins path through all points in P, with minimum turning radius ρ. The initial configuration is assumed to be $(x_{init}, y_{init}, \psi_{init}) = (0,0,0)$. Let $C_\rho : SE(2) \to \bar{\mathbb{R}}$ associate to a configuration (x, y, ψ) the length of the shortest Dubins path and define $F_0 :]0,\pi[\times]0,\pi[\to]0,\pi[$, $F_1 :]0,\pi[\to \mathbb{R}$ and $F_2 :]0,\pi[\to \mathbb{R}$

$$F_0(\psi, \theta) = 2 \arctan\left(\frac{\sin(\psi/2) - 2\sin(\psi/2 - \theta)}{\cos(\psi/2) + 2\cos(\psi/2 - \theta)}\right) \tag{4.149}$$

$$F_1(\psi) = \psi + \sin\left(\frac{F_0(\psi,\psi/2-\alpha(\psi))}{2}\right) \\ +4\arccos\left(\sin\left(\frac{0.5(\psi - F_0(\psi,\psi/2-\alpha(\psi)))}{2}\right)\right) \tag{4.150}$$

$$F_2(\psi) = 2\pi - \psi + 4\arccos\left(\frac{\sin(\psi/2)}{2}\right) \tag{4.151}$$

where

$$\alpha(\psi) = \frac{\pi}{2} - \arccos\left(\frac{\sin(0.5\psi)}{2}\right) \tag{4.152}$$

The objective is the design of an algorithm that provides a provably good approximation to the optimal solution of the Dubins TSP. The alternating algorithm works as follows: Compute an optimal ETSP tour of P and label the edges on the tour in order with consecutive integers. A DTSP tour can be constructed by retaining all odd-numbered (except the nth) edges and replacing all even-numbered edges with minimum length Dubins paths preserving the point ordering. The pseudocode for this approach is presented in algorithm 10 below.

Algorithm 10 Dubins Traveling Salesman Problem

1. Set $(a_1, .., a_n)$ = optimal ETSP ordering of P
2. Set ψ_1: orientation of segment from a_1 to a_2
3. For $i \in 2, .., n-1$ do
   if i is even then set $\psi_i = psi_{i-1}$
   else set ψ_i = orientation of segment from a_i to a_{i+1}
4. If n is even then set $\psi_n = \psi_{n-1}$ else set ψ_n =orientation of segment from a_n to a_1
5. Return the sequence of configurations $(a_i, \psi_i)_{i \in 1...n}$

The nearest neighbor heuristic produces a complete solution for the Dubins traveling salesman problem, including a waypoint ordering and a heading for each point. The heuristic starts with an arbitrary point and chooses its heading arbitrarily, fixing an initial configuration. Then at each step, a point is found which is not yet on the path but close to the last added configuration according to the Dubins metric. This closest point is added to the path with the associated optimal arrival heading. When all nodes have been added to the path, a Dubins path connecting the last obtained configuration and the initial configuration is added. If K headings are chosen for each point, then an a priori finite set of possible headings is chosen for each point and a graph is constructed with n clusters corresponding to the n waypoints, and each cluster containing K nodes corresponding to the choice of the headings. Then the Dubins distances between configurations corresponding to a pair of nodes in distinct clusters are computed. Finally, a tour through the n clusters is computed which contains exactly one point in each cluster. This problem is called the generalized asymmetric traveling salesman problem over nK nodes. A path planning problem for a single fixed wing aircraft performing a reconnaissance mission using one or more cameras is considered in [141]. The aircraft visual reconnaissance problem for static ground targets in terrain is formulated as a polygon-visiting Dubins traveling salesman problem.

4.4.2 REPLANNING OR TACTICAL AND STRATEGIC PLANNING

The mission parameters are provided by a higher level automated scheduling system. **Strategic planning**, which occurs before takeoff, takes a priori information about the operating environment and the mission goals and constructs a path that optimizes for the given decision objectives. **Tactical planning** involves re-evaluation and re-generation of a flight plan during flight based on updated information about the goals and operating environment. The generated plan should be as close as possible to the optimal plan given available planning information.

An autonomous aircraft must choose and order a subset of targets, to achieve among all mission objectives. It should optimize the choice of its actions, knowing its resources, the environment, the maximum reward associated with each objective and the time constraints associated with them. Formalism could be based on the decomposition of the problem into two levels: the top level corresponds to the objectives of the mission, the lowest level describes this achievement as a function of time and resources. It formalizes a problem with uncertainties where the number of objectives of the plan is not fixed a priori. A static algorithm is used offline to produce one or more feasible plans. A dynamic algorithm is then used online to gradually build the right solution to the risks that arise. The terms static and dynamic characterize the environment in which the plan is carried out. Classical planning assumes that the environment is static, meaning that there is no uncertainty. A predictive algorithm is then used offline to produce a single plan which can then be executed on line without being questioned. In the case of a dynamic environment, several techniques are possible [70].

1. Keep a predictive offline, supplemented by one or reactive algorithms that are executed when a hazard line makes incoherent the initial plan, calling it into question and forced most often to re-plan.
2. Take into account the uncertainties from the construction phase offline: this is called proactive approaches.
3. Plan always predictively but this time online, short term, in a process of moving horizon, in which case the execution will gradually resolve uncertainties and allow for further planning steps.

The level of decision-making autonomy is referred to the planning board. It requires a calculation of plan online, called re-planning. Updating the plan online involves the development of a hybrid architecture, incorporating the outbreak of calculations of new plans in case of hazard and the inclusion of the results of this calculation. The proposed architecture allows to set up an online event planning, with many hierarchical levels of management of the mission. The mission planning problem is to select and order the best subset among the set of objectives to be achieved and to determine the dates of start and end of each objective, maximizing the rewards obtained during the objectives and criteria for minimizing the consumption of resources, while respecting the constraints on resources and mission.

Mission planning can be considered as a selection problem. The objectives are linked to rewards whose values vary depending on the importance of each objective. Planning must choose a subset of objectives to be achieved in time and limited resources. The existing planning systems are mostly unsuitable for solving such problems: they address a problem where the goal is a conjunction of goals and fail if the goal is not reached. The selection of targets is often based on a simplified model for the resources of the problem and ignores the various ways to achieve the same goal. In most cases, a practical solution is obtained by combining the selection of objectives, planning and task scheduling in a multi-levels architecture: each level defines the problem to the scheduling algorithm of the lower level. Unlike the planning carried out during the mission preparation, planning online is characterized by the fact that the time taken to find a plan is one of the main criteria for judging the quality of a method.

The onboard intelligence allows the aircraft to achieve the objectives of the mission and ensuring its survival, taking into account the uncertainties that occur during a mission. The objectives of the planning function are in general: order the passage of the various mission areas, calculate a path between each element of the route; order the realization of the task. The use of deterministic/random hybrid techniques is intended to provide solutions to this problem. A mission planning system must be able to:

1. evaluate multiple objectives,
2. handle uncertainty,
3. be computationally efficient.

The mission planning task is non trivial due to the need to optimize for multiple decision objectives such as safety and mission objectives. For example, the safety objective might be evaluated according to a midair collision risk criterion and a risk presented to third parties' criterion. The degree of satisfaction of the safety objective is obtained by aggregating the constituent criteria. A constraint refers to limits imposed on individual decision criteria (a decision variable) such as the maximum allowable risk.

For some applications, the mission tasks, for example, spray crops or perform surveillance, are conducted at the destination point. Another important consideration is online or in-flight re-planning. A plan that is optimal when it is generated can become invalidated or suboptimal by changes to assumptions in the flight plan. For example, the unanticipated wind conditions can increase fuel consumption, it may take an unexpectedly long time to reach a waypoint and there may be changes to mission goals as new information becomes available.

As with manned aircraft, the dependability and integrity of a UAV platform can be influenced by the occurrence of endogenous and exogenous events. There are a number of safety related technical challenges which must be addressed including provision of a safe landing zone detection algorithm which would be executed in the event of a UAV emergency. In the event of such an emergency, a key consideration of any safety algorithm is remaining flight time which can be influenced by battery life or fuel availability and weather conditions. This estimate of the UAV remaining flight time can be used to assist in autonomous decision-making upon occurrence of a safety critical event [145].

4.4.3 ROUTE OPTIMIZATION

In general, the routing approach consists in reducing a general routing problem to the shortest path problem. The specification of routes is a problem in its own right. If the route network is modeled as a directed graph, then the routing problem is the discrete problem of finding paths in a graph, which must be solved before the speed profile is sought [168].

4.4.3.1 Classic approach

An automated mission planning can enable a high level of autonomy for a variety of operating scenarios [74, 116]. Fully autonomous operation requires the mission planner to be situated onboard the smart autonomous aircraft. The calculation of a flight plan involves the consideration of multiple elements. They can be classified as either continuous or discrete, and they can include nonlinear aircraft performance, atmospheric conditions, wind forecasts, aircraft structure, amount of departure fuel and operational constraints. Moreover, multiple differently characterized flight phases must be considered in flight planning. The flight planning problem can be regarded as a trajectory optimization problem [181]. The mission planning has to define a series of steps to define a flight route. In the context of mission planning for an autonomous aircraft, the plan is necessarily relative to displacement. The plan then contains a sequence of waypoints in geometric space considered. A possible path of research of the current aircraft position to destinations in geometric space, avoiding obstacles in the best way, is sought. Scheduling algorithms must be integrated into an embedded architecture to allow the system to adapt its behavior to its state and dynamic environment [73].

The approach reduces the uncertainty inherent in a dynamic environment through online re-planning and incorporation of tolerances in the planning process [121]. The motion plan is constrained by aircraft dynamics and environmental/operational constraints. In addition, the planned path must satisfy multiple, possibly conflicting objectives such as fuel efficiency and flight time.

It is not computationally feasible to plan in a high dimensional search space consisting of all the aforementioned decision variables. It is common instead to plan the path in the world space (x, y, z, t) by aggregating the decision variables into a single, non-binary cost term.

Integration must take into account that the activation calculations plan is triggered by events. A random event may occur during a mission whose date of occurrence is unpredictable. An autonomous system has two main goals:

1. make out its mission while remaining operational,
2. and react to the uncertainties of mission, environment or system.

The embedded architecture must meet these two objectives by organizing the physical tasks of the mission and the tasks of reasoning. This reaction is conditioned by the inclusion of planning during execution in a control architecture. Each controller is composed of a set of algorithms for planning, monitoring, diagnosing and execution [45] . This architecture can be applied to complex problem solving, using hierarchical decomposition. Hierarchical decomposition is employed to break down the problem into smaller parts both in time and function. This approach provides a viable solution to real time, closed-loop planning and execution problems:

1. The higher levels create plans with greatest temporal scope, but low level of detail in planned activities.
2. The lower levels' temporal scope decreases, but they have an increase in detail of the planned activities.

Remark 4.7. ***Situation awareness*** *includes both monitoring and diagnosis. Plan generation and execution are grouped. A hierarchical planning approach is chosen because it enables a rapid and effective response to dynamic mission events.*

A **functional and temporal analysis** of the mission has to be performed, identifying the activities that need to be performed. As the mission problem is progressively broken down into smaller subproblems, functional activities emerge. At the lowest level of decomposition, the functional activities are operating on timescales of seconds. These activities are related to each other in a tree structure, with the lowest level (leaf) nodes providing the output commands to the guidance, navigation and control systems.

For the autonomous aircraft, the state is given, at least, by three position coordinates, three velocity coordinates, three orientation angles and three orientation rate angles, for a total of twelve variables. The dynamic characteristics of the aircraft determine the dimension of the system, and many systems may use a reduced set of variables that adequately describe the physical state of the aircraft. It is common to consider smaller state space with coupled states, or to extend the state space to include higher order derivatives.

Planning schemes may be classified as explicit or implicit.

1. An **implicit method** is one in which the dynamic behavior of the aircraft is specified. Then the trajectory and the actuator inputs required to go from the start configuration to the goal configuration are derived from the interaction between the aircraft and the environment. The best known example of this method is the potential field method [104] and its extensions. Some other examples include the methods that apply randomized approaches [111] or graph theory [41].
2. **Explicit methods** attempt to find solutions for the trajectories and for the inputs of actuators explicitly during the motion. Explicit methods can be discrete or continuous. Discrete approaches focus primarily on the geometric constraints and the problem of finding a set of discrete configurations between the end states that are free from collisions.

Mission planning problems have been considered from the point of view of artificial intelligence, control theory, formal methods and hybrid systems for solving such problems [25]. A class of complex goals impose **temporal constraints** on the trajectories for a given system, referred also as temporal goals. They can be described using a formal framework such as **linear temporal logic** (LTL), **computation tree logic** and **μ-calculus**. The specification language, the discrete abstraction of the aircraft model and the planning framework depend on the particular problem being solved and the kind of guarantees required. Unfortunately, only linear approximations of the aircraft dynamics can be incorporated. Multi-layered planning is used for safety analysis of hybrid systems with reachability specifications and motion planning involving complex models and environments. The framework introduces a discrete component to the search procedure by utilizing the discrete structure present in the problem.

The framework consists of the following steps:

1. Construction of a discrete abstraction for the system,
2. High level planning for the abstraction using the specifications and exploration information from a low level planner,
3. Low level sampling-based planning using the physical model and the suggested high level plans.

There is a two way exchange of information between the high level and low level planning layers. The constraints arising due to temporal goals are systematically conveyed to the low level layer from the high level layer using synergy. The construction of the discrete abstraction and two way interaction between the layers are critical issues that affect the overall performance of the approach.

4.4.3.2 Dynamic multi-resolution route optimization

An approach is described in this section for **dynamic route optimization** for an autonomous aircraft [169]. A multi-resolution representation scheme is presented that uses **B-spline basis functions** of different support and at different locations along the trajectory, parametrized by a dimensionless parameter. A multi-rate receding horizon problem is formulated as an example of online multi-resolution optimization under feedback. The underlying optimization problem is solved with an anytime evolutionary computing algorithm. By selecting particular basis function coefficients as the optimization variables, computing resources can flexibly be devoted to those regions of the trajectory requiring the most attention. Representations that can allow a UAV to dynamically re-optimize its route while in flight are of interest. A popular technique for route optimization is dynamic programming, often in combination with other methods. waypoints generated by dynamic programming serve as input to either an optimal control or a virtual potential field approach. The potential field method models the route with point masses connected by springs and dampers. Threats and targets are modeled by virtual force fields of repelling and attracting forces. The optimized route then corresponds to the lowest energy state of this mechanical equivalent.

The general requirements are wind optimal routes, avoiding regions for several reasons, minimizing fuel costs and making allowances for a required time of arrival. The **dynamic programming method** can be used for the intended application. Dynamic programming is, however, a global optimizer and more flexible methods are preferred.

4.4.3.2.1 Route optimization problem formulation

The route is represented by a sequence of waypoints: $(x_k, y_k, z_k, t_k)_{k=1}^{K}$ where (x_k, y_k, z_k) are Cartesian coordinates of the waypoints and t_k is the scheduled time of arrival to the waypoints.

Problem 4.15. *The route optimization problem can be expressed as:*

$$X = arg_{X \in D_X} min J(X) \tag{4.153}$$

where X is a list containing the waypoint parameters, J is the route optimality index and D_x is the domain of the allowed routes. The route optimization problems of interest can involve thousands of waypoints.

Therefore, direct solution of (4.153) is not possible given onboard processing constraints. The optimization set is limited to a parametric family of trajectories represented by spline functions. The trajectory is parametrized by a dimensionless parameter u and is represented by samples of $x(u)$, $y(u)$ and $z(u)$.

Remark 4.8. *The span of u for the entire route has to be chosen carefully, taking into account the maneuverability and velocity of the aircraft.*

Assuming constant velocity, the distance spanned by a fixed Δu should be approximately the same along the route. Since the optimization will change the position of the waypoints when the distance spanned by $\Delta u = 1$ is small there will be a higher likelihood of generating routes that are not flyable; the turns required at waypoints may exceed the aircraft's maneuvering capability. On the other hand, a large distance spanned by $\Delta u = 1$ will not allow much flexibility to find the best route. The route parametrization uses the following B-spline expansion:

$$x(u) = \sum_{n=0}^{N_{max}} a_n \tilde{\psi}(u-n) + x_0(u) \tag{4.154}$$

$$y(u) = \sum_{n=0}^{N_{max}} b_n \tilde{\psi}(u-n) + y_0(u) \tag{4.155}$$

$$z(u) = \sum_{n=0}^{N_{max}} c_n \tilde{\psi}(u-n) + z_0(u) \tag{4.156}$$

where $\tilde{\psi}(u)$ is a basis function and $(x_0(u), y_0(u), z_0(u))$ is the initial approximation of the route (from offline mission planning). The following second-order B-spline basis function is used

$$\tilde{\psi}(w) = \left\{ \begin{array}{cc} 0 & w < 0 \\ w^2 & 0 \leq w \leq 1 \\ -2w^2 + 6w - 3 & 1 \leq w \leq 2 \\ (3-w)^2 & 2 \leq w \leq 3 \\ 0 & w \geq 3 \end{array} \right\} \tag{4.157}$$

This basis function has support on the interval $w \in [0, 3]$. The representation (4.154) and (4.155) defines a 3D trajectory. This has to be complemented by the time dependence of the aircraft position (for evaluating turn constraints). Assuming a constant speed V,

$$\dot{s} = V \tag{4.158}$$

the path from the trajectory start at $u = 0$ to a point parametrized by $u = w$ on the route is given by:

$$s(w) = \int_{u=0}^{w} \sqrt{\left(\left(\frac{dx}{du}\right)^2 + \left(\frac{dy}{du}\right)^2 + \left(\frac{dz}{du}\right)^2 \right)} du \tag{4.159}$$

By equations (4.158), (4.159) and solving for w, it is possible to generate a route, represented by K time stamped waypoints $(x_k, y_k, z_k, t_k)_{k=1}^{K}$.

4.4.3.2.2 Online optimization under feedback

Most existing aircraft route optimization is used off-line. Criteria for the route optimization are given ahead of the mission and the route is pre-computed. An algorithm that can re-optimize the route inflight within available computational time limits is developed. A receding horizon control problem can be formulated and addressed in [169]. At every sample, an online optimization algorithm is used to update the route taking into account situation changes and disturbances. Consider the path point $u = k$. The planning of the path for $u \geq k$ is done as:

$$x_{k+1}(u) = \sum_{n=0}^{N_{max}} a_n \tilde{\psi}(u-n-k) + x_k(u) + \Delta x_k(u) \tag{4.160}$$

$$y_{k+1}(u) = \sum_{n=0}^{N_{max}} b_n \tilde{\psi}(u-n-k) + y_k(u) + \Delta y_k(u) \tag{4.161}$$

$$z_{k+1}(u) = \sum_{n=0}^{N_{max}} c_n \tilde{\psi}(u-n-k) + z_k(u) + \Delta z_k(u) \tag{4.162}$$

where $x_{k+1}(u), y_{k+1}(u), z_{k+1}(u)$ is the trajectory as computed before the update and $\Delta x_k(u), \Delta y_k(u), \Delta z_k(u)$ are corrections to deal with disturbances. The expansion weights a_n, b_n, c_n are recomputed at each step by solving an optimization problem similar to (4.153). The corrections $\Delta x_k(u), \Delta y_k(u), \Delta z_k(u)$ are introduced in (4.160) to (4.162) because disturbances such as the wind cause the position and velocity of the aircraft at time $u = k$ to be different from those given by the nominal trajectory $x_k(k), y_k(k), z_k(k)$. The corrections allow the route optimizer to generate a new route from the actual position of the aircraft. Assume that at the path point coordinate $u = k$, the guidance and navigation system of the aircraft determines a position deviation $\Delta x_k(u), \Delta y_k(u), \Delta z_k(u)$ and a velocity deviation $\Delta V_x(k), \Delta V_y(k), \Delta V_z(k)$ from the previously planned route. The route is then adjusted around the aircraft position. Since the B-spline approximation (4.154) to (4.156) and (4.157) gives a trajectory that is a piece-wise second order polynomial, the most natural way of computing the correction is also a piece-wise second order polynomial spline. By matching the trajectory coordinates and derivatives at $u = k$, the correction can be computed as

$$\Delta x_k(u) = \Delta x(k)\tilde{\alpha}(u-k) + \Delta V_x(k)\tilde{\beta}(u-k) \tag{4.163}$$

$$\Delta y_k(u) = \Delta y(k)\tilde{\alpha}(u-k) + \Delta V_y(k)\tilde{\beta}(u-k) \tag{4.164}$$

$$\Delta z_k(u) = \Delta z(k)\tilde{\alpha}(u-k) + \Delta V_z(k)\tilde{\beta}(u-k) \tag{4.165}$$

where

$$\tilde{\alpha}(w) = \left\{ \begin{array}{cc} 1-0.5w^2 & 0 < w \leq 1 \\ 0.5(w-2)^2 & 1 < w \leq 2 \end{array} \right\} \tag{4.166}$$

$$\tilde{\beta}(w) = \left\{ \begin{array}{cc} w - 0.75w^2 & 0 < w \leq 1 \\ 0.25(w-2)^2 & 1 < w \leq 2 \end{array} \right\} \quad (4.167)$$

The receding horizon update at point $u = k$ computes a trajectory of the form (4.160) to (4.162) where the expansion weights $a_n, b_n, c_n (n = 1 \dots N_{max})$ are such as to minimize a modified optimality index of the form (4.153). The modified optimality index takes into account only the *future* part of the trajectory for $u \geq k$.

4.4.3.2.3 Multi-scale route representation

The route has to be dynamically updated in flight to compensate for both disturbances and changes in the overall mission structure such as the emergence or disappearance of threats and targets. Route planning in the vicinity of the current location can usually be done in detail because the current local information can be expected to be reliable. It must also be done quickly before the aircraft moves too far. The longer term planning is also less critical at the current time. An optimal trade-off between different optimization horizons can be effected with limited computational resources using a multiscale representation. A technique similar to wavelet expansion is used as it provides an efficient way to represent a signal, such as a trajectory over time at multiple temporal and frequency scales.

4.4.4 FUZZY PLANNING

Many traditional tools for formal modeling, reasoning and computing are of crisp, deterministic and precise character. However, most practical problems involve data that contain uncertainties. There has been a great amount of research in **probability theory**, **fuzzy set theory**, **rough set theory**, **vague set theory**, **gray set theory**, **intuitionistic fuzzy set theory** and **interval math** [187]. Soft set and its various extensions have been applied with dealing with decision-making problems. They involve the evaluation of all the objects which are decision alternatives.

The imposition of constraints, such as aircraft dynamic constraints and risk limits, corresponds to skill level decision-making [10]. The evaluation function used to calculate path costs is rule based, reflecting the rules level. Finally, the selection of an evaluation function and scheduling of planning activities, such as in an anytime framework, mimics the level of knowledge.

Heuristics primary role is to reduce the search space and thus guide the decision maker onto a satisfying or possibly optimal solution in a short space of time. It is applicable to mission planning as flight plans tend to follow the standard flight profile. Additionally, due to the time pressure of online re-planning, the satisfying heuristic could be used to quickly find a negotiable path rather than deliver an optimal solution that is late.

The planner can consider as **decision criteria**: obstacles, roads, ground slope, wind and rainfall. **Decision variables** are aggregated into a single cost value which is used in a heuristic search algorithm. The rain criterion, for example, is represented by the membership function: light rain, moderate rain and heavy rain. *If-then* rules are then used to implicate the degree of membership to the output membership function on the mobility (i.e., difficulty of traversal) universe. A vector neighborhood can be identified as a suitable method for overcoming the limited track angle resolution to ensure path optimality. Uncertainty is often mitigated with online re-planning, multi-objective decision-making and incorporation of tolerance into the planned path [48].

In the context of mission planning, each decision criterion has associated with it a constraint and ideal value. For example, there is an ideal cruise velocity and a maximum and minimum airspeed limit. Constraints can be enforced with simple boundary checking and correspond to the skills' level of cognition. Optimization of the selected paths, however, requires evaluation of multiple decision rules as per the rules' level. The safety objective may be decomposed into a risk criterion, no-fly zones criterion, wind and maximum velocity constraints. Additionally, an approximation of the dynamic constraints of the aircraft is required to ensure that the final problem is traversable. Each membership function objective can be decomposed into individual decision criteria [170, 175].

4.4.4.1 Fuzzy decision tree cloning of flight trajectories

A graph based mission design approach is developed in [197], involving maneuvers between a large set of trim trajectories. In this approach, the graph is not an approximation of the dynamics in general, but rather each node consists of one of the available maneuvers and a connection exists if there is a low cost transition between trim trajectories. The method can be divided into two primary sections: the off-line mission planning phase and the onboard phases. The mission planning steps are relatively expensive computationally and involve creating, cataloging and efficiently storing the maneuvers and boundary conditions, as well as describing the relationship between the discretized dynamics and fuel estimates. On the other hand, the onboard process quickly leverages the work done on the ground and stored in memory to quickly generate a transfer using search and correction methods. The onboard and off-line portions are complementary images of each other: the mission planning stages translate and compress the continuous system dynamics into a discrete representation whereas the onboard portion searches and selects discrete information and reconstructs continuous trajectories [26, 36, 59].

Trajectory optimization aims at defining optimal flight procedures that lead to time/energy efficient flights. A **decision tree** algorithm is used to infer a set of **linguistic decision rules** from a set of 2D obstacle avoidance trajectories optimized using mixed integer linear programming in [198]. A method to predict a discontinuous function with a **fuzzy decision tree** is proposed and shown to make a good approximation to the optimization behavior with significantly reduced computational expense. Decision trees are shown to generalize to new scenarios of greater complexity than those represented in the training data and to make decisions on a timescale that would enable implementation in a real-time system. The transparency of the rule based approach is useful in understanding the behavior exhibited by the controller. Therefore, the decision trees are shown to have the potential to be effective online controllers for obstacle avoidance when trained on data generated by a suitable optimization technique such as mixed integer linear programming.

Adaptive dynamic programming (ADP) and **reinforcement learning** (RL) are two methods for solving decision-making problems where a performance index must be optimized over time. They are able to deal with complex problems, including features such as uncertainty, stochastic effects, and nonlinearity. Adaptive dynamic programming tackles these challenges by developing optimal control methods that adapt to uncertain systems over time. Reinforcement learning takes the perspective of an agent that optimizes its behavior by interacting with an initially unknown environment and learning from the feedback received.

The problem is to control an autonomous aircraft to reach a target obscured by one or more threats by following a near optimal trajectory, minimized with respect to path length and subject to constraints representing the aircraft dynamics. The path must remain outside known threat regions at all times. The **threat size** is assumed to be constant and the threats and target are assumed to be stationary. Furthermore, it is assumed that the aircraft flies at constant altitude and velocity and is equipped with an autopilot to follow a reference heading input. The output from the controller shall be a change in demanded heading $\Delta\chi$ and the inputs are of the form $\{R_{target}, \theta_{target}, R_{threat}, \theta_{threat}\}$ where $R_{target}, \theta_{target}$ are range and angle to target and $R_{threat}, \theta_{threat}$ are the ranges and angles to any threats present. All angles are relative to the autonomous aircraft current position and heading.

The system has two modes: learn and run. When in the learning mode, the mixed integer linear programming is used to generate heading deviation decisions. The heading deviations are summed with the current heading and used as a reference input to the aircraft and recorded with the optimization inputs in a training set. Once sufficient training runs (approximately 100 in [198]) have been performed, the run mode is engaged where the decision tree is used to generate the heading deviation commands and performance evaluated.

Predicting a heading deviation and choosing to fix the frame of reference relative to the aircraft heading results in a data representation that is invariant under global translation and rotation. The independence of these basic transformations reduces the problem space and improves generalization by allowing many different scenarios to be mapped on to a single representation.

The optimization solves for the minimum time path to the target using a linear approximation to the aircraft dynamics. Variables are time to target N and acceleration $a(k)$ for predicted steps $k = 0, \ldots, N-1$.

Problem 4.16. *The optimization problem can be formulated as follows:*

$$min_{N,a(k)} \left[N + \tilde{\gamma} \sum_{k=0}^{N} \|a(k)\|^2 \right] \tag{4.168}$$

subject to $\forall k \in \{0, \ldots, (N-1)\}$

$$R(0) = R_0 \quad V(0) = V_0 \tag{4.169}$$

$$\begin{aligned} V(k+1) &= V(k) + a(k)\delta z \\ R(k+1) &= R(k) + V(k)\delta z + \tfrac{1}{2}a(k)\delta z^2 \end{aligned} \tag{4.170}$$

$$\|a(k)\|_2 \leq a_{max} \qquad \|V(k)\|_2 \leq a_{max} \tag{4.171}$$

$$\|R(N) - R_{target}\|_\infty \leq D_T \qquad \|R(k) - R_{threat}\|_2 \geq R_0 \tag{4.172}$$

The cost (4.168) primarily minimizes time to target N in steps of δz, with a small weight $\tilde{\gamma}$ on acceleration magnitudes to ensure a unique solution. Equations (4.169) are initial condition constraints defining position, heading and velocity. Equations (4.170) are forward Euler approximations to the aircraft dynamics and kinematics. Equations (4.171) are constraints on maximal accelerations and velocities. Equation (4.172) ensures that the aircraft is within a given tolerance of the target at time N and that at all times the distance from the aircraft to the obstacle, r_{threat}, is outside the obstacle radius R_0. These are non convex constraints, features that makes the problem resolution difficult. The 2-norm constraint of (4.172) is implemented by approximating it with one of a choice of linear constraints and using binary variables to selectively relax all but one of these. The resulting optimization is mixed linear integer programming.

The heading deviation command $\Delta\chi$ is found by taking the difference of the current heading and the heading predicted by the optimization. The heading predicted by the optimization is found from the velocity vectors $V(k+1)$ that are part of its output. Heading deviation is calculated in place of the required heading to obtain the translation and rotation invariance. The optimization is integrated with the model using the algorithm 11 shown next.

Algorithm 11 Receding Horizon MILP Controller

1. Convert $\{R_{target}, \theta_{target}, R_{threat}, \theta_{threat}\}$ to $\{r_{target}, r_{threat}\}$
2. Solve the optimization problem
3. Derive initial heading deviation $\Delta\chi$ from optimization output
4. Set new desired heading χ_{k+1}
5. Run simulation for δz, i.e., one time step of the optimization
6. Return to step 1

A linguistic decision tree is a tree structured set of linguistic rules formed from a set of attributes that describe a system. The leaf nodes of the tree represent a **conditional probability distribution** on a set, $\mathbb{L}_t$, of descriptive labels covering a target variable given the branch is true. The algorithm constructs a decision tree on the basis of minimizing the **entropy** in a training database across random set partitions on each variable represented in the training data. The rules are formulated from sets of labels, LA, describing the universe of each attribute that provides information about the current state. The **degree of appropriateness** of a particular label, L, for a given instance quantifies the belief that L can be appropriately used to describe X. An automatic algorithm is used to define the appropriateness measures that best partition the data. The set of focal elements $\mathbb{F}$ for the same universe are defined as all the sets of labels that might simultaneously be used to describe a given X.

Definition 4.6. ***Entropy*** *is a scalar value measure of the compactness of a distribution. When a probability distribution is used to represent the knowledge, the smaller the entropy of the distribution, the more probability mass is assigned to a smaller area of the state space and thus the more informative the distribution is about the state. The entropy $h(X)$ of a multivariate Gaussian probability distribution over the variable X can be calculated from its covariance matrix as follows:*

$$h(X) = \frac{1}{2}\log\left((2\pi e)^n |\mathbf{P}|\right) \tag{4.173}$$

In order to try to mimic the data most readily available to an aircraft and to achieve the rotational and translational invariance, ranges and bearings are used to relate threat locations to the aircraft position. Each attribute that describes the current state, i.e., $\{R_{target}, \theta_{target}, R_{threat}, \theta_{threat}\}$, is covered by a set of labels or focal elements.

In practice, up to nearly 20 focal elements can be used to capture final details; the target attribute in this application is the aircraft heading deviation, $\Delta\chi$, which is covered by a similar set of labels. The **membership** of the **focal** elements can be thought of as the belief that a particular label is an appropriate description for a given value of the attribute. The membership is used to enable the probability that a particular branch is a good description of the current simulation state.

Remark 4.9. *The characteristic of bearing data is that it is circular. This can be achieved by merging the two boundary fuzzy sets from a linear domain to give a coverage in a polar plot where angle denotes bearing and radius denotes attribute membership of each focal element.*

4.4.4.2 Fuzzy logic for fire fighting aircraft

The focus of this section is a single fire fighting aircraft that plans paths using fuzzy logic. In this scenario, the system uses basic heuristics to travel to a continuously updated target while avoiding various, stationary or moving obstacles [167]. In a fire fighting scenario it is important that the aircraft is capable of performing evasive maneuvering in real time. Effective collaboration between all the relevant agents can be difficult in a situation where communication can be minimal and, therefore, obstacle avoidance is crucial [137].

This method performs a multi-solution analysis supervised by a fuzzy decision function that incorporates the knowledge of the fire fighting problem in the algorithm. This research explores the use of fuzzy logic in obtaining a rule base that can represent a fire fighting heuristic. In the fire fighting scenario, using information from ground troops and the incident commander on the target drop location $(x_t, y_t, z_t)^T$ and the system's current location $(x_0, y_0, z_0)^T$, the system drives the difference between its heading angle and the angle to the target to zero. The system has a sensing range that is considered able to sense obstacles of $\pm\frac{\pi}{2} rad$ within a certain radius. When obstacles are detected within this area, the system alters its velocity and heading angle using information about obstacles' distances and angles and the location of the target to avoid and then recover from the obstruction. Similarly, when the system reaches its target location, the target alters its velocity to slow down and apply the fire retardant.

For this setup, four inputs can be used for the fuzzification interface and two outputs are given after defuzzification. Inputs into the system are distance from obstacle, angle from the obstacles, heading angle error and distance to the target. With these inputs and a rule base, the control input is obtained, that is, the percentage of the maximum velocity and the heading angle is outputted from the fuzzy inference system and used as inputs into the system. The main objective of the controller when there are no obstacles within its sensing range is to plan a direct path to the target. With an inertial frame as a reference, the heading angle θ is measured as the angle of the agent's current heading from the horizontal and the target angle χ is measured as the angle to the target from the horizontal. Therefore with information about the agent location (x_0, y_0, z_0) and the target location (x_t, y_t, z_t), the corresponding angles are determined by:

$$\chi = \arctan\left(\frac{y_0}{x_0}\right) \qquad \phi = \arctan\left(\frac{y_t}{x_t}\right) \qquad e = \phi - \chi \tag{4.174}$$

The agent tries to drive the heading angle error e to zero by making small heading angle adjustments according to simple *if-then* rules. Since the target location is considered to be known, the target distance D_t is easily determined within the distance formula and given information. When no obstacles are within the sensing range, the control objective is simple. Once the agent reaches the target, it will slow down to apply its fire retardant to the location. Once this is done, the new target location is the aircraft base to allow the agent to refuel and reload. If the agent senses an obstacle along its path to the target, it must slow down and change its heading to avoid collision. The obstacle distance D_0 and angle β are obtained from information from the agent's sensors.

Rules relating the inputs and outputs for the fuzzy logic controller are setup in the form of IF-THEN statements and are based on heuristics and human experience. There is a total of 40 rules in [167] for this setup that can be broken up into two situations: if there is an obstacle within the sensing range or not. If there is no obstacle detected, the default is set to *very far* away and moving toward the target is the agent's primary objective. However, when an obstacle is within range, the agent must slow down and change course to avoid it. Again, once it is clear of this obstacle, it can continue its path toward the target. The obstacle angle can be described by NB (negative big), NM (negative medium), NS (negative small), PS (positive small), PM (positive medium), PB (positive big). The distance to the obstacle is described by either close, medium, far or very far away (out of range). Similarly, the target angle is NB (negative big), NM (negative medium), NS (negative small), ZE (zero), PS (positive small), PM (positive medium) or PB (positive big). The output velocity is slow, fast or very fast and the output angle change is parallel to the target angle. The sensing radius used is considered the safe distance from an obstacle. The first two sets of rules are for an input of obstacle distance very far while the target distance and heading angle vary. The second three sets describe the change in heading speed and angle when an obstacle is detected. The last set case is for when the obstacles are at extreme angles and pose no threat of collision.

4.4.5 COVERAGE PROBLEM

The **coverage** of an unknown environment is also known as the **sweeping problem**, or mapping of an unknown environment. Basically, the problem can either be solved by providing ability for localization and map building first, or by directly deriving an algorithm that performs sweeping without explicit mapping of the area. Instead of a **measure of coverage**, an average event detection time can be used for evaluating the algorithm.

4.4.5.1 Patrolling problem

In [100, 101], the following base perimeter **patrol problem** is addressed: a UAV and a remotely located operator cooperatively perform the task of perimeter patrol. **Alert stations** consisting of **unattended ground sensors** (UGS) are located at key locations along the perimeter. Upon detection of an incursion in its sector, an alert is flagged by the unattended ground sensors. The statistics of the alerts' arrival process are assumed known. A camera equipped UAV is on continuous patrol along the perimeter and is tasked with inspecting unattended ground sensors with alerts. Once the UAV reaches a triggered unattended ground sensor, it captures imagery of the vicinity until the controller dictates it to move on. The objective is to maximize the information gained and at the same time reduce the expected response time to alerts elsewhere [129]. The problem is simplified by considering discrete time evolution equations for a finite fixed number m of unattended ground sensors' locations. It is assumed that the UAV has access to real time information about the status of alerts at each alert station. Because the UAV is constantly on patrol and is servicing a triggered unattended ground sensor, the problem is a **cyclic polling system** in the domain of a discrete time controlled queuing system. The patrolled perimeter is a simple closed curve with $N \geq m$ nodes that are spatially uniformly separated of which m are the alert stations (unattended ground sensors locations).

The objective is to find a suitable policy that simultaneously minimizes the service delay and maximizes the information gained upon loitering. A stochastic optimal control problem is thus considered [177]. A Markov decision process is solved in order to determine the optimal control policy [32, 89]. However, its large size renders exact dynamic programming methods intractable. Therefore, a state aggregation based approximate linear programming method is used instead, to construct provably good suboptimal patrol policies [183]. The state space is partitioned and the optimal cost to go or value function is restricted to be a constant over each partition. The resulting restricted system of linear inequalities embeds a family of Markov chains of lower dimension, one of which can be used to construct a lower bound on the optimal value function. The perimeter patrol problem exhibits a special structure that enables tractable linear programming formulation for the lower bound [6].

Definition 4.7. ***Markov Chain:*** *A discrete time Markov chain (MC) is a tuple $\mathcal{M} = \langle S, P, s_{init}, \Pi, L \rangle$ where S is a countable set of states, $P : S \times S \rightarrow [0,1]$ is the transition probability function such that for any state $s \in S, \sum_{s' \in S} P(s,s') = 1$, $s_{init} \in S$ is the initial state, Π is a set of atomic propositions and $L : S \rightarrow 2^{\Pi}$ is a labeling function.*

An observable first-order discrete Markov chain is encoded as the matrix of state transition properties. Its rows sum to one but the columns do not necessarily do so. A state S_i in a **Markov chain** is said to be absorbing if $a_{ii} = 1$. Otherwise such a state is said to be transient.

Definition 4.8. ***Hidden Markov Model:*** *A hidden Markov model (HMM) is a Markov model in which the states are hidden. Rather than have access to the internal structure of the hidden Markov model, all that is available are observations that are described by the underlying Markov model. A hidden Markov model λ is described in terms of:*

1. *N, the number of states in the model.*
2. *M, the number of distinct observation symbols per state. The individual symbol is $V = \{V_1, ..., V_M\}$*

3. A, the state transition probability distribution. As for Markov chains $A = \{a_{ij}\} = Prob\left(q_t = S_j | q_{t-1} = S_i\right)$
4. B, the observation symbol probability distribution for state $B = \{b_{ij}\} = Prob\left(V_k(t) | q_j = S_i\right)$. *B is known as the emission matrix*
5. π *the initial state distribution* $\pi = \{\pi_i\} = Prob(q_1 | S_i)$

Definition 4.9. ***Markov Decision Process:*** *A Markov decision process (MDP) is defined in terms of the tuple* $\langle S, A, T, R \rangle$*, where:*

1. *S is a finite set of environmental states*
2. *A is a finite set of actions*
3. *$T : S \times A \rightarrow S$ is the state transition function. Each transition is associated with a transition probability $T(s, a, s')$, the probability of ending in state s', given that the agent starts in s and executes action a.*
4. *$R : S \times A \rightarrow S$ is the immediate reward function, received after taking a specific action from a specific state.*

Definition 4.10. ***Finite Markov Decision Process:*** *A particular finite Markov decision process is defined by its state and action sets and by the one-step dynamics of the environment. Given any state s and action a, the probability of each possible nest state s' is:*

$$P^a_{ss'} = Prob\left\{s_{t+1} = s' | s_t = s, a_t = a\right\} \tag{4.175}$$

where $P^a_{ss'}$ represents transition probabilities and t denotes a finite time step.

In the Markov decision process, the value of $P^a_{ss'}$ does not depend on the past state transition history. The agent receives a reward r every time it carries out the one-step action. Given any current state s and action a, together with any next state s', the expected value of the next reward is:

$$R^a_{ss'} = E\left[r_{t+1} | s_r = s, a_t = a, s_{t+1} = s'\right] \tag{4.176}$$

where $P^a_{ss'}$ and $R^a_{ss'}$ completely specify the dynamics of the finite Markov Decision Process. In the finite Markov Decision Process, the agent follows the policy Π. The policy Π is a mapping from each state s and action a to the probability $\Pi(s, a)$ of taking action a when in state s. In the stochastic planning calculation, based on the Markov Decision Process, the policy Π is decided so as to maximize the value function $V^{\Pi}(s)$. The $V^{\Pi}(s)$ denotes the expected return when starting in S and following Π thereafter. The definition of $V^{\Pi}(s)$ is:

$$V^{\Pi}(s) = E_{\Pi}\left[\sum_{k=0}^{\infty} \gamma^k r_{t+k+1} | s_t = s\right] \tag{4.177}$$

where E_{Π} denotes the expected value given when the agent follows the policy Π and γ is the discount rate $0 < \gamma < 1$. If the values of $P^a_{ss'}$ and

$R^a_{ss'}$ are known, dynamic programming is used to calculate the best policy Π that maximizes the value function $V^\Pi(s)$. When the values of $P^a_{ss'}$ and $R^a_{ss'}$ are unknown, a method such as online reinforcement learning is useful in obtaining the best policy Π in the learning environment [57]. After the planning calculation has finished, a greedy policy that selects action value a that maximizes $V^\Pi(s)$ is optimal.

Definition 4.11. ***Path:*** *A path through a Markov Decision process is a sequence of states, i.e.,*

$$\omega = q_0 \xrightarrow{(a_0,\sigma_{a_0}^{q_0})(q_1)} q_1 \longrightarrow \dots q_i \xrightarrow{(a_i,\sigma_{a_i}^{q_i})(q_{i+1})} \longrightarrow \dots \tag{4.178}$$

where each transition is induced by a choice of action at the current step $i \geq 0$. The ith state of a path ω is denoted by $\omega(i)$ and the set of all finite and infinite paths by $Path^{fin}$ and $Path$, respectively.

A control policy defines a choice of actions at each state of a Markov decision process. Control policies are also known as **schedules** or **adversaries** and are formally defined as follows:

Definition 4.12. ***Control Policy:*** *A control policy μ of an MDP model $\mathbb{M}$ is a function mapping a finite path, i.e., $\omega^{fin} = q_0, q_1, \dots, q_n$ of $\mathbb{M}$, onto an action in $\mathbb{A}(q_n)$. A policy is a function: $\mu : Path^{fin} \rightarrow Act$ that specifies for every finite path, the next action to be applied. If a control policy depends only on the last state of ω^{fin}, it is called a* ***stationary policy.***

For each policy μ, a probability measure $Prob_\mu$ over the set of all paths under μ, $Path_\mu$, is induced. It is constructed through an infinite state Markov chain as follows: under a policy μ, a Markov Decision Process becomes a Markov chain that is denoted D_μ whose states are the finite paths of the Markov Decision Process. There is a one-to-one correspondence between the paths of D_μ and the set of paths $Path_\mu$ in the Markov Decision Process. Hence, a probability measure $Prob_\mu$ over the set of paths $Path_{\mu^{fin}}$ can be defined by setting the probability of $\omega^{fin} \in Path^{fin}$ equal to the product of the corresponding transition probabilities in D_μ [200].

With fuzzy logic, the cost elements are expressed as fuzzy membership functions reflecting the inherent uncertainty associated with the planned trajectory, the obstacles along the path and the maneuvers the aircraft is required to perform as it navigates through the terrain. If employed, the algorithm A^* can use heuristic knowledge about the closeness of the goal state from the current state to guide the search. The cost of every searched cell, n, is composed of two components: the cost of the least-cost route (found in the search so far) from the start cell to cell n, and the heuristic (i.e., estimated) cost of the minimum-cost route from cell n to the destination cell. Given a search state space, an initial state (start node) and final state (goal node), the algorithm A^* will find the optimal (least cost) path from the start node to the goal node, if such a path exists. The generated cell route is further optimized and smoothed by a filtering algorithm.

The filtered route is a series of consecutive waypoints that the autonomous aircraft can navigate through. The supervisory module reads the objectives and the status of the mission and based on that it configures the search engine and assigns weights to the route's three cost components. Furthermore, the supervisory module chooses the start and the destination cells for the search engine depending on the current status of the aircraft, i.e., whether it is stationary or already navigating towards a destination and needs to be redirected to another destination. The learning-support module acquires route cost data from the search engine at certain map landmarks and updates a cost database that is used later to provide better heuristics to guide the search engine. Thus a **two point boundary value problem** (TPBVP) has to be solved for creating a reference path to be followed by the tracking system.

4.4.5.2 Routing problem

The **UAV sensor selection and routing problem** is a generalization of the **orienteering problem**. In this problem, a single aircraft begins at a starting location and must reach a designated destination prior to time T. Along with the starting and ending points, a set of locations exists with an associated benefit that may be collected by the aircraft. Mission planning can be viewed as a complex version of path planning where the objective is to visit a sequence of targets to achieve the objectives of the mission [50, 51]. The integrated sensor selection and routing model can be defined as a mixed integer linear programming formulation [134].

In [82], a path planning method for sensing a group of closely spaced targets is developed that utilizes the planning flexibility provided by the sensor footprint, while operating within dynamic constraints of the aircraft. The path planning objective is to minimize the path length required to view all of the targets. In addressing problems of this nature, three technical challenges must be addressed: coupling between path segments, utilization of the sensor footprint and determination of the viewing order of the targets. A successful path planning algorithm should produce a path that is not constrained by end points or a heading that utilizes the full capability of the aircraft's sensors and that satisfies the dynamic constraints on the aircraft. These capabilities can be provided by discrete time paths which are built by assembling primitive turn and straight segments to form a flyable path. For this work, each primitive segment in a discrete step path is of specified length and is either a turn or a straight line. Assembling the left turn, right turn and straight primitives creates a tree of flyable paths. Thus the objective for the path planner is to search the path tree for the branch that accomplishes the desired objectives in the shortest distance. The learning real time A^* algorithm can be used to learn which branch of a defined path tree best accomplishes the desired path planning objectives.

Field operations should be done in a manner that minimizes time and travels over the field surface. A coverage path planning in 3D has a great potential to further optimize field operations and provide more precise navigation [76].

Another example follows. Given a set of stationary ground targets in a terrain (natural, urban or mixed), the objective is to compute a path for the reconnaissance aircraft so that it can photograph all targets in minimum time, because terrain features can occlude visibility. As a result, in order for a target to be photographed, the aircraft must be located where both the target is in close enough range to satisfy the photograph's resolution and the target is not blocked by terrain. For a given target, the set of all such aircraft positions is called the target's visibility region. The aircraft path planning can be complicated by wind, airspace constraints, aircraft dynamic constraints and the aircraft body itself occluding visibility. However, under simplifying assumptions, if the aircraft is modeled as a Dubins vehicle, the target's visibility regions can be approximated by polygons and the path is a closed tour [141]. Then the 2D reconnaissance path planning can be reduced to the following: for a Dubins vehicle, find a shortest planar closed tour that visits at least one point in each of a set of polygons. This is referenced to as the **polygon visiting Dubins traveling salesman problem** (PVDTSP). Sampling-based roadmap methods operate by sampling finite discrete sets of poses (positions and configurations) in the target visibility regions in order to approximate a polygon visiting Dubins traveling salesman problem instance by a **finite-one in set traveling salesman problem** (FOTSP). The finite-one in set traveling salesman problem is the problem of finding a minimum cost closed path that passes through at least one vertex in each of a finite collection of clusters, the clusters being mutually exclusive finite vertex sets. Once a road-map has been constructed, the algorithm converts the finite-one in set traveling salesman problem instance into an **asymmetric traveling salesman problem** (ATSP) instance to solve a standard solver can be applied.

Another example is the case of a UAV that has to track, protect or provide surveillance of a ground based target. If the target trajectory is known, a deterministic optimization or control problem can be solved to give a feasible UAV trajectory. The goal in [4] is to develop a feedback control policy that allows a UAV to optimally maintain a nominal standoff distance from the target without full knowledge of the current target position or its future trajectory. The target motion is assumed to be random and described by a 2D stochastic process. An optimal feedback control is developed for fixed speed, fixed altitude UAV to maintain a nominal distance from a ground target in a way that it anticipates its unknown future trajectory. Stochasticity is introduced in the problem by assuming that the target motion can be modeled as Brownian motion, which accounts for possible realizations of the unknown target kinematics. The tracking aircraft should achieve and maintain a nominal distance d to the target.To this end, the expectation of an infinite-horizon cost function is minimized, with a discounting factor and with penalty for control. Moreover the possibility for the interruption of observations is included by assuming that the duration of observation times of the target is exponentially distributed, giving rise to two discrete states of operation. A Bellman equation based on an approximating Markov chain that is consistent with the stochastic kinematics is used to compute an optimal control policy that minimizes the expected value of a cost function based on a nominal UAV target distance.

4.4.5.3 Discrete stochastic process for aircraft networks

This section considers a network of autonomous aircraft. Each one is equipped with a certain kind of onboard sensors, for example, a camera or a different sensor, taking snapshots of the ground area [212]. The general aim of this network is to explore a given area, i.e., to cover this area using several applications: target or even detection and tracking in an unknown area, monitoring geographically inaccessible or dangerous areas (e.g., wildfire or volcano) or assisting emergency personnel in case of disasters.

The objective is to provide a simple analytical method to evaluate the performance of different mobility patterns in terms of their coverage distribution. To this end, a stochastic model can be proposed using a **Markov chain**. The states are the location of the aircraft and the transitions are determined by the mobility model of interest. Such a model can be created for independent mobility models such as the random walk and random direction.

However, for a cooperative network, in which each autonomous aircraft decides where to move based on the information received from other aircraft in its communication range, creating a simple Markov model is not straightforward. Therefore, in addition to providing the necessary transition probabilities for random walk and random direction, an approximation to these probabilities is also proposed for a cooperative network. While intuitive rules for the movement paths are chosen when one or more autonomous aircraft meet each other, the proposed model can be extended such that other rules can be incorporated.

Several **mobility models** for autonomous agents have been proposed recently. Some of these are synthetic like the random walk and random direction, others are realistic and all of them are used mainly to describe the movement of the users in a given environment. In the autonomous aircraft domain, such models are good for comparison of different approaches but can give incorrect results when the autonomous aircraft are performing cooperative tasks. Mobility can increase throughput energy efficiency, coverage and other network parameters. Therefore, the analysis of mobility models has become a highlight to design the mobility of the nodes in a way to improve the network importance.

4.4.5.3.1 Markov chain

A discrete-time discrete value stochastic process is introduced that can be used to analyze the coverage performance of the autonomous aircraft network. Nodes can operate independently or in a cooperative manner. The system area is modeled as a 2D lattice where autonomous aircraft move from one grid point to another in each time step. It is assumed that an autonomous aircraft can only move to the four nearest neighboring grid points: the **Von Neumann neighborhood** of radius 1. The probability of moving to a neighboring grid point is determined by the mobility model of interest. In the following, two main components of the proposed Markov chain are presented: state probabilities and transition probabilities.

In this model, the states are defined as [current location, previous location]. Depending on the location, the number of associated states is different. If the current location is at a corner, boundary or middle grid point, there are 2, 3 and 4 associated states, respectively: P_f, P_b, P_l, P_r are, respectively, the probabilities to move forward, backward, left and right. The steady state probabilities of this Markov chain are denoted by $\pi = [\pi(i,j,k,l)]$ and the transition probability matrix by T, where the entities of the matrix are the transition, probabilities between the states $[(i,j);(k,l)]$. Accordingly, the transient state probabilities are denoted by $\pi^{(n)} = \left[\pi^{(n)}_{i,j,k,l}\right]$ at time step n. The following relations for the steady state and transient state probabilities can thus be written:

$$\begin{array}{rl} \vec{\pi} = \pi.T & \text{For steady state} \\ \pi^{(n)} = \pi^{(0)}T^n & \text{For transient state} \\ lim_{n\to\infty}\pi^{(n)} = \vec{\pi} & \end{array} \tag{4.179}$$

where $\sum \pi_{i,j,k,l} = 1$. The initial state $\pi^{(0)}$ can be chosen to be $[1,0,\ldots,0]$, since the solution for $\vec{\pi}$ is independent of the initial condition. From these linear equations, the steady and transient state probabilities can be obtained. This is used to determine the coverage of a given mobility pattern.

4.4.5.3.2 Coverage metrics

The steady state coverage probability distribution for an $a \times a$ area is denoted by $\mathbf{P} = [P(i,j)], 1 \leq i \leq a, 1 \leq j \leq a$. The probability matrix represents the percentage of time a given location (i,j) is occupied and can be computed by adding the corresponding steady state probabilities obtained from (4.179):

$$P(i,j) = \sum_{k,l} \pi(i,j;k,l) \tag{4.180}$$

where $(k,l) = \{(i-1,j),(i,j-1),(i+1,j),(i,j+1)\}$ for the non-boundary states. The (k,l) pairs for boundary states can be determined in a straightforward manner. The transient coverage probability distribution $\mathbf{P}^{(n)} =$

$[P(i,j)], 1 \leq i \leq a, 1 \leq j \leq a$:

$$P^{(n)} = \sum_{k,l} \pi^{(n)}(i,j;k,l) \tag{4.181}$$

Using the obtained $P^{(n)}$, the probability that location (i,j) is covered by a time step can be computed as follows:

$$C^{(n)}(i,j) = 1 - \prod_{\nu=0}^{n} \left(1 - P^{(\nu)}(i,j)\right) \tag{4.182}$$

In the case of multiple autonomous aircraft, the state probabilities can be computed. Given the steady state coverage distribution matrix of the autonomous aircraft k is P_k entities obtained using relation (4.180) and assuming independent/decoupled mobility, the steady state coverage distribution of an m-autonomous aircraft network can be obtained as:

$$p^{multi}(i,j) = 1 - \prod_{k=1}^{m} \left(1 - P_k(i,j)\right) \tag{4.183}$$

The transient behavior of the $m-$aircraft network can be computed similarly, by substituting the (i,j) entry of the transient coverage probability matrix $\mathbf{P}_n^{(k)}$ from relations (4.181) to (4.183). Some potential **metrics** of interest are now defined besides the coverage distribution of a mobility model in a grid, at time step n for a grid of size $a \times a$:

1. **Average Coverage**:

$$E\left[C^{(n)}\right] = \frac{1}{a^2} \sum_{i,j} C^{(n)}(i,j) \tag{4.184}$$

2. **Full coverage**:

$$\epsilon^{(n)} = \Pr\left(C^{(n)} = \vec{1}_{a \times a}\right) = \prod_{i,j} C^{(n)}(i,j) \tag{4.185}$$

where $\vec{1}_{a \times a}$ is an $a \times a$ matrix of ones. These metrics carry some valuable information regarding the coverage performance, e.g., how well a given point is covered, how well the whole area is covered and how much time would be necessary to cover the whole area.

4.4.5.3.3 Transition probabilities: independent mobility

The state transition probabilities for the random walk and direction mobility models are first discussed where the transition probabilities are very intuitive. For a random walk, the knowledge of the previous location is not necessary.

Therefore, the states of the analytical tool (i, j, k, l) can be further simplified to (i, j). For a random walk, it is assumed that at each time step, the autonomous aircraft can go to any one of the neighboring grid points with equal probability. Clearly, the number of neighboring points change depending on the location. On the other hand, for a random direction model, the direction is changed only when the autonomous aircraft reaches the boundary of the grid. Therefore, the previous location, which is also equivalent to direction for the lattice, needs to be taken into account. For both of these schemes, as well as for the cooperative scheme at the boundaries and corners, the next location is chosen randomly among the available neighboring points with equal probability.

4.4.5.3.4 Transition probabilities: cooperative mobility

A method to approximate the coverage performance of a cooperative mobile network is proposed in this section. In such a network, the nodes interact with each other: exchange information, whenever they meet. The objective is to come up with an appropriate transition probability matrix that can be used by the proposed stochastic tool. For independent mobility, the proposed Markov chain can be easily extended to multiple autonomous aircraft. However, for cooperative mobility, this Markov chain is not sufficient to model the interactions. The states of a Markov chain that exactly models all the interactions will grow exponentially with the number of autonomous aircraft [158]. Therefore, an approximation method can be proposed to model the behavior of the aircraft in a way that would allow to treat the cooperative mobility as independent mobility [201].

To decouple the actions for the aircraft from each other, for an m aircraft network the following probabilities are defined:

$$P_X = \sum_{k=0}^{m-1} P_{X|k} \Pr\left(k + 1 \text{nodes meet}\right), X \in \{B, F, L, R\} \tag{4.186}$$

where the backward, forward, left-turn and right-turn probabilities are given by the decision metric $P_{X|k}$ of the cooperative mobility as well as the number of aircraft that will meet. With the assumption that any node can be anywhere in the grid with equal probability, probability that exactly k aircraft out of a total of m aircraft will also be at (i, j) is given by the binomial distribution:

$$\Pr\left(k + 1 \text{ nodes meet }\right) = \binom{m-1}{k} \left(\frac{1}{a^2}\right)^k \left(1 - \frac{1}{a^2}\right)^{m-1-k} \tag{4.187}$$

The entries of the corresponding transition probability matrix can be computed using relations (4.186) to (4.187), given the decision metric $P_{X|k}$.

4.4.5.4 Sensor tasking in multi-target search and tracking applications

The problem of managing uncertainty and complexity of planning and executing an **intelligence surveillance reconnaissance** (ISR) mission is addressed in this section. This intelligence surveillance reconnaissance mission uses a network of UAV sensor resources [86]. In such applications, it is important to design uniform coverage dynamics such that there is little overlap of the sensor footprints and little space left between the sensor footprints. The **sensor footprints** must be uniformly distributed so that it becomes difficult for a target to evade detection. For the search of a stationary target, the uncertainty in the position of the target can be specified in terms of a fixed probability distribution. The spectral multiscale coverage algorithm makes the sensors move so that points on the sensor trajectories uniformly sample this stationary probability distribution. Uniform coverage dynamics coupled with sensor observations help to reduce the uncertainty in the position of the target [13].

Coverage path planning determines the path that ensures a complete coverage in a free workspace. Since the aircraft has to fly over all points in the free workspace, the coverage problem is related to the covering salesman problem. Coverage can be a static concept, i.e., it is a measure of how a static configuration of agents covers a domain or samples a probability distribution. Coverage can also be a dynamic concept, i.e., it is a measure of how well the points on the trajectories of the sensor cover a domain. Coverage gets better and better as every point in the domain is visited or is close to being visited by an agent. Uniform coverage uses a metric inspired by the ergodic theory of dynamical system. The behavior of an algorithm that attempts to achieve uniform coverage is going to be inherently multi-scale. Features of large size are guaranteed to be detected first, followed by features of smaller and smaller size [192, 194, 196].

Definition 4.13. ***Ergodic Dynamics:*** *A system is said to exhibit* ***ergodic dynamics*** *if it visits every subset of the phase space with a probability equal to the measure of that subset. For a good coverage of a stationary target, this translates to requiring that the amount of time spent by the mobile sensors in an arbitrary set be proportional to the probability of finding the target in that set. For good coverage of a moving target, this translates to requiring that the amount of time spent in certain tube sets be proportional to the probability of finding the target in the tube sets.*

A model is assured for the motion of the targets to construct these tube sets and define appropriate metrics for coverage. The model for the target motion can be approximate and the dynamics of targets for which precise knowledge is not available can be captured using stochastic models. Using these metrics for uniform coverage, centralized feedback control laws are derived for the motion of the mobile sensors.

For applications in environmental monitoring with a mobile sensor network, it is often important to generate accurate spatio-temporal maps of scalar fields such as temperature or pollutant concentration. Sometimes, it is important to map the boundary of a region. In [191], a multi-vehicle sampling algorithm generates trajectories for non uniform coverage of a non stationary spatio-temporal field characterized by spatial and temporal decorrelation scales that vary in space and time, respectively. The sampling algorithm uses a nonlinear coordinate transformation that renders the field locally stationary so that the existing multi-vehicle control algorithm can be used to provide uniform coverage. When transformed back to original coordinates, the sampling trajectories are concentrated in regions of short spatial and temporal decorrelation scales.

For applications of multi-agent persistent monitoring, the goal can be to patrol the whole mission domain while driving the uncertainty of all targets in the mission domain to zero [182]. The uncertainty at each target point is assumed to evolve nonlinearly in time. Given a closed path, multi-agent persistent monitoring with the minimum patrol period can be achieved by optimizing the agent's moving speed and initial locations on the path [63, 207].

Remark 4.10. *The main difference between multi-agent persistent monitoring and dynamic coverage lies in that dynamic coverage task is completed when all points attain satisfactory coverage level while the persistent monitoring would last forever.*

4.4.5.4.1 Coverage dynamics for stationary targets

There are N mobile agents assumed to move either by first order or second order dynamics. An appropriate metric is needed to quantify how well the trajectories are sampling a given probability distribution μ. It is assumed that μ is zero outside a rectangular domain $\mathbb{U} \in \mathbb{R}^n$ and that the agent trajectories are confined to the domain $\mathbb{U}$. For a dynamical system to be ergodic, the fraction of the time spent by a trajectory must be equal to the measure of the set. Let $B(X,R) = \{R : \|Y - X\| \leq R\}$ be a spherical set and $\chi(X,R)$ be the indicator function corresponding to the set $B(X,R)$. Given trajectory $X_j : [0,t] \longrightarrow \mathbb{R}^n, j = 1 \dots N$ the fraction of the time spent by the agents in the set $B(X,R)$ is given as:

$$d^t(X,R) = \frac{1}{Nt}\sum_{j=1}^{N}\int_0^t \chi(X,R)(X_j)(\tau)d\tau \tag{4.188}$$

The measure of the set $B(X,R)$ is given as

$$\bar{\mu}(X,R) = \int_{\mathbb{U}} \mu(Y)\chi(X,R)(Y)dY \tag{4.189}$$

For ergodic dynamics, the following relation must be verified:

$$\underbrace{lim}_{t\to\infty} d^t(X,R) = \bar{\mu}(X,R) \tag{4.190}$$

Since the equation above must be true for almost all points X and all radii R, this is the basis for defining the metric

$$E^2(t) = \int_0^R \int_{\mathbb{U}} \left(d^t(X,R) - \bar{\mu}(X,R)\right)^2 dXdR \tag{4.191}$$

$E(t)$ is a metric that quantifies how far the fraction of the time spent by the agents in the spherical sets is from being equal to the measure of the spherical sets. Let the distribution C^t be defined as

$$C^t(X) = \frac{1}{Nt}\sum_{j=1}^{N}\int_0^t \delta\left(X - X_j(\tau)\right) d\tau \tag{4.192}$$

Let $\phi(t)$ be the distance between C^t and μ as given by the Sobolev space norm of negative index H^{-1} for $s = \frac{n+1}{2}$, i.e.,

$$\phi^2(t) = \left\|C^t - \mu\right\|^2_{H^{-s}} = \sum_K \Lambda_k |s_k(t)|^2 \tag{4.193}$$

where

$$s_k(t) = C_k(t) - \mu_k \qquad \Lambda_k = \frac{1}{\left(1 + \|k\|^2\right)^s} \tag{4.194}$$

$$C_k(t) = \left\langle C^t, f_k \right\rangle \qquad \mu_k = \langle \mu, f_k \rangle \tag{4.195}$$

Here, f_k are Fourier basis functions with wave number vector k. The metric $\phi^2(t)$ quantifies how much the time averages of the Fourier basis functions deviate from their spatial averages, but with more importance given to large scale modes than the small scale modes. The case is considered where the sensors are moving by first order dynamics described by:

$$\dot{X}_j(t) = U_j(t) \tag{4.196}$$

The objective is to design feedback laws:

$$U_j(t) = F_j(X) \tag{4.197}$$

so that the agents have ergodic dynamics. A model predictive control problem is formulated to maximize the rate of decay of the coverage metric $\phi^2(t)$ at the end of a short time horizon and the feedback law is derived in the limit as the size of the receding horizon goes to zero. The cost function is taken to be the first time derivative of the $\phi^2(t)$ at the end of the horizon $[t, t + \Delta t]$, i.e.,

$$C(t, \Delta t) = \sum_K \Lambda_k s_k(t + \Delta t) \dot{s}_k(t + \Delta t) \tag{4.198}$$

The feedback law in the limit as $\Delta t \to 0$ is given as:

$$U_j(t) = -U_{max} \frac{B_j}{\|B_j(t)\|_2} \tag{4.199}$$

where

$$B_j(t) = \sum_K \Lambda_k s_k(t) \nabla f_k(X_j(t)) \tag{4.200}$$

and $\nabla f_k(t)$ is the gradient field of the Fourier basis functions f_k.

4.4.5.4.2 Coverage dynamics for moving targets

Let the target motion be described by a deterministic set of ordinary differential equations:

$$\dot{Z}(t) = V(Z(t), t) \tag{4.201}$$

with $Z(t) \in \mathbb{U}, \mathbb{U} \subset \mathbb{R}^3$ being the zone in which the target motion is confined over a period $[0, T_f]$. Let T be the corresponding mapping that describes the evolution of the target position, i.e., if the target is at point $Z(t_0)$ at time $t = t_0$, its position $Z(t_f) = T(Z(t_0), t_0, t_f)$.

Given a set $\mathbb{A} \subset \mathbb{U}$, its inverse image under the transformation $T(., t_0, t_f)$ is given as:

$$T^{-1}(., t_0, t_f)(\mathbb{A}) = \{Y : T(Y, t_0, t_f) \in \mathbb{A}\} \tag{4.202}$$

The initial uncertainty in the position of the target is specified by the probability distribution $\mu(0, X) = \mu_0(X)$.

Let $[P^{t_0,t_f}]$ be the family of Perron–Frobenius operators corresponding to the transformations $T(., t_0, t_f)$ i.e.,

$$\int_{\mathbb{A}} [P^{t_0,t_f}] \mu(t_0, Y) dY = \int_{\mathbb{A}} \mu(t_f, Y) dY = \int_{T^{-1}(., t_0, t_f)(\mathbb{A})} \mu(t_0, Y) dY \tag{4.203}$$

At time t, the spherical set $B(X, R)$ with radius R and center X is considered as well as the corresponding tube set:

$$H^t(B(X, R)) = \{(Y, \tau) : \tau \in [0, t] \text{ and } T(Y, \tau, t) \in B(X, R)\} \tag{4.204}$$

The tube set $H^t(B(X, R))$ is a subset of the extended space-time domain and is the union of the sets

$$T^{-1}(., \tau, t)(B(X, R)) \times \{\tau\}, \forall \tau \in [0, t] \tag{4.205}$$

This tube set can be thought of as the set of all points in the extended space-time domain that end up in the spherical set $B(X,R)$ at time t when evolved forward in time according to the target dynamics.

The probability of finding a target within any time slice of the tube set is the same i.e.:

$$\mu\left(\tau_1, T^{-1}(.,\tau_1,t)(B(X,R))\right) = \mu\left(\tau_2, T^{-1}(.,\tau_2,t)(B(X,R))\right) = \mu(t, B(X,R)) \tag{4.206}$$

$\forall \tau_1, \tau_2 \leq t$.

This is because none of the possible target trajectories either leave or enter the tube set $H^t(B(X,R))$. Let the sensor trajectories be $X_j : [0,t] \to \mathbb{R}^2, \forall j = 1..N$. The fraction of the time spent by the sensor trajectories $(X_j(t), t)$ in the tube set $H^t(B(X,R))$ is given as:

$$d^t(X,R) = \frac{1}{Nt}\sum_{j=1}^{N}\int_0^t \chi T^{-1}\left(.,\tau,t\right)\left(B(X,R)\right)\left(X_j(\tau)\right)d\tau \tag{4.207}$$

or

$$d^t(X,R) = \frac{1}{Nt}\sum_{j=1}^{N}\int_0^t \chi B(X,R)\left(T\left(X_j(\tau),\tau,t\right)\right)d\tau \tag{4.208}$$

$\chi B(X,R)$ is the indicator function on the set $B(X,R)$. $d^t(X,R)$ can be computed as the spherical integral

$$d^t(X,R) = \int_{B(X,R)} C^t(Y)dY \tag{4.209}$$

of a distribution

$$C^t(X) = \frac{1}{Nt}\sum_{j=1}^{N}\int_0^t P^{\tau,t}\delta X_j(\tau)(x)d\tau \tag{4.210}$$

referred to as the coverage distribution. $\delta X_j(\tau)$ is the delta distribution with mass at the point $X_j(\tau)$. The coverage distribution C^t can be thought of as the distribution of points visited by the mobile sensors when evolved forward in time according to the target dynamics.

For uniform sampling of the target trajectories, it is desirable that the fraction of time spent by the sensor trajectories in the tube must be close to the probability of finding a target trajectory in the tube which is given as:

$$\mu(t, B(X,R)) = \int_{B(X,R)} \mu(t,Y)dY = \int_{T^{-1}(.,0,t)(B(X,R))} \mu_0(Y)dY \tag{4.211}$$

This is the basis for defining the metric:

$$\Psi^2(t) = \left\|C^t - \mu(t,.)\right\|^2_{H^{-s}} \tag{4.212}$$

Using the same receding horizon approach as described before for stationary targets, feedback laws similar to that in equation (4.199) are obtained.

During search missions, efficient use for flight time requires flight paths that maximize the probability of finding the desired subject. The probability of detecting the desired subject based on UAV sensor information can vary in different search areas due to environment elements like varying vegetation density or lighting conditions, making it likely that the UAV can only partially detect the subject. In [119], an algorithm that accounts for partial detection in the form of a difficulty map is presented. It produces paths that approximate the payoff of optimal solutions, the path planning being considered as a discrete optimization problem. It uses the mode goodness ratio heuristic that uses a Gaussian mixture model to prioritize search subregions. The algorithm searches for effective paths through the parameter space at different levels of resolution. The task difficulty map is a spatial representation of sensor detection probability and defines areas of different levels of difficulty [15].

4.4.6 RESOURCE MANAGER FOR A TEAM OF AUTONOMOUS AIRCRAFT

Knowledge of meteorological properties is fundamental to many decision processes. It is useful if related measurement processes can be conducted in a fully automated fashion. The first type of cooperation that the autonomous aircraft may exhibit is to support each other if there is evidence that an interesting physical phenomenon has been discovered. The second type of cooperation that the aircraft can exhibit through their control algorithm is when an aircraft is malfunctioning or may be malfunctioning. If an aircraft internal diagnostic indicates a possible malfunction, then it will send out an omnidirectional request to the other aircraft for help to complete its task. Each autonomous aircraft will calculate its priority for providing help. The autonomous aircraft send their priority for providing a help message back to the requesting aircraft. The requester subsequently sends out a message informing the group of the ID of the highest priority aircraft. The high priority aircraft then proceeds to aid the requester. The support provided by the helping aircraft can take on different forms. If the requester suspects a malfunction in its sensors, the helper may measure some of the same points originally measured by the autonomous aircraft in doubt. This will help establish the condition of the requester's sensors. If additional sampling indicates the requester is malfunctioning and represents a liability to the group it will return to base. In this case, the supporter may take over the mission of the requester [180].

Whether or not the supporter samples all the remaining sample points of the requester, subsequently abandoning its original points, depends on the sample points' priorities. If it is established that the requester is not malfunctioning or the requester can still contribute to the mission's success, it may remain in the field to complete its current mission [3].

4.4.6.1 Routing with refueling depots for a single aircraft

A single autonomous aircraft routing problem with multiple depots is considered. The aircraft is allowed to refuel at any depot. The objective of the problem is to find a path for the UAV such that each target is visited at least once by the aircraft, the fuel constraint is never violated along the path for the UAV and the total fuel required by the UAV is a minimum. An approximation algorithm for the problem is developed and a fast construction and improvement heuristic is proposed for the solution [189].

As small autonomous aircraft typically have fuel constraints, it may not be possible for them to complete a surveillance mission before refueling at one of the depots. For example, in a typical surveillance mission, an aircraft starts at a depot and is required to visit a set of targets. It is assumed that the fuel required to travel a given path for the UAV is directly proportional to the length of the path. To complete this mission, the aircraft might have to start at one depot, visit a subset of targets and then reach one of the depots for refueling before starting a new path. If the goal is to visit each of the given targets once, the UAV may have to repeatedly visit some depots for refueling before visiting all the targets. In this scenario, the following problem arises:

Problem 4.17. ***Routing Problem:*** *Given a set of targets and depots and a UAV where the aircraft is initially stationed at one of the depots, find a path for the UAV such that each target is visited at least once, the fuel constraint is never violated along the path for the UAV and the travel cost for the aircraft is a minimum. The travel cost is defined as the total fuel consumed by the aircraft as it traverses its path.*

The main difficulty in this problem is mainly combinatorial [113]. As long as a path of minimum length can be efficiently computed from an origin to a destination of the autonomous aircraft, the motion constraints of the autonomous aircraft do not complicate the problem. The UAV can be modeled as a Dubins aircraft. If the optimal heading angle is specified at each target, the problem of finding the optimal sequence of targets to be visited reduces to a generalization of the traveling salesman problem [155].

The autonomous aircraft must visit each target at a specified heading angle. As a result, the travel costs for the autonomous aircraft may be asymmetric. Symmetry means that the cost of traveling from target A with the heading χ_A and arriving at target B with heading χ_B may not equal the cost of traveling from target B with heading χ_B and arriving at target A with the heading χ_A.

Definition 4.14. *An α approximation algorithm for an optimization problem is an algorithm that runs in polynomial time and finds a feasible solution whose cost is at most α times the optimal cost for every instance of the problem.*

This guarantee α is also referred to as the approximation factor of the algorithm. This approximation factor provides a theoretical upper bound on the quality of the solution produced by the algorithm for any instance of the problem. These upper bounds are known a priori. The bound provided by the approximation factor is generally conservative.

Let $\mathbb{T}$ denote the set of targets and $\mathbb{D}$ represent the set of depots. Let $s \in \mathbb{D}$ be the depot where the UAV is initially located. The problem is formulated on the complete directed graph $G = (\mathbb{V}, E)$ with $\mathbb{V} = \mathbb{T} \bigcup \mathbb{D}$. Let f_{ij} represent the amount of fuel required by the aircraft to travel from vertex $i \in \mathbb{V}$ to vertex $j \in \mathbb{V}$. It is assumed that the fuel costs satisfy the triangle inequality, i.e., for all distinct $i, j, k \in \mathbb{V}$, $f_{ij} + f_{jk} \geq f_{ik}$. Let L denote the maximum fuel capacity of the aircraft. For any given target $i \in \mathbb{I}$, it is assumed that there are depots d_1 and d_2 such that $f_{d_1 i} + f_{i d_2} \leq aL$, where $0 < a < 1$ is a fixed constant. It is also assumed that it is always possible to travel from one depot to another depot (either directly or indirectly, by passing through some intermediate depots, without violating the fuel constraints). Given two distinct depots d_1 and d_2, let $\ell'_{d_1 d_2}$ denote the minimum fuel required to travel from d_1 to d_2. Then, let β be a constant such that $\ell'_{d_1 d_2} \leq \beta \ell'_{d_1 d_2}$ for distinct d_1 and d_2. A tour for the aircraft is denoted by a sequence of vertices $T = (s, v_1, v_2, \dots, v_p, s)$ visited by the aircraft, where $v_i \in \mathbb{V}$ for $i = 1, \dots, p$. A tour visiting all the targets can be transformed to a tour visiting all the targets and the initial depot and vice versa.

Problem 4.18. *The objective of the problem is to find a tour* $T = (s, v_1, v_2, \dots, v_p, s)$ *such that* $\mathbb{T} \subseteq \{v_1, v_2, \dots, v_p\}$*, the fuel required to travel any subsequence of vertices of the tour* $(d_1, t_1, t_2, \dots, t_k, d_2)$ *starting at a depot* d_1 *and ending at the next visit to a depot* d_2*, while visiting a sequence of targets* $(t_1, t_2, \dots, t_k) \in \mathbb{T}$ *must be at most equal to* L*, i.e.,*

$$f_{d_1 t_1} + \sum_{i=1}^{k-1} f_{t_i t_{i+1}} + f_{t_k d_2} \leq L \tag{4.213}$$

The travel cost $f_{s v_1} + \sum_{i=1}^{p-1} f_{v_i v_{i+1}} + f_{v_p s}$ is a minimum. Let x_{ij} denote an integer decision variable which determines the number of directed edges from vertex i to vertex j in the network, $x_{ij} \in \{0, 1\}$, if either i or vertex j is a target.

The collection of edges chosen by the formulation must reflect the fact that there must be a path from the depot to every target. Flow constraints are used to formulate these connectivity constraints. In these flow constraints, the aircraft collects $|T|$ units of a commodity at the depot and delivers one unit of commodity at each target as it travels along its path. p_{ij} denotes the amount of commodity flowing from vertex i to vertex j and r_i represents the fuel left in the aircraft when the ith target is visited.

Problem 4.19. *The problem can be formulated as mixed integer linear programming as follows:*

$$min \sum_{(i,j)\in\mathbb{E}} f_{ij}x_{ij} \tag{4.214}$$

subject to degree constraints:

$$\sum_{i\in\mathbb{V}/\{k\}} x_{ik} = \sum_{i\in V/\{k\}} x_{ki} \qquad \forall k \in \mathbb{V} \tag{4.215}$$

$$\sum_{i\in\mathbb{V}/\{k\}} x_{ik} = 1, \forall k \in \mathbb{T} \tag{4.216}$$

Capacity and flow constraints:

$$\sum_{i\in\mathbb{V}/\{s\}} (p_{si} - p_{is}) = |T| \tag{4.217}$$

$$\sum_{j\in\mathbb{V}/\{i\}} (p_{ji} - p_{ij}) = 1 \quad \forall i \in \mathbb{T} \tag{4.218}$$

$$\sum_{j\in\mathbb{V}/\{i\}} (p_{ji} - p_{ij}) = 0 \quad \forall i \in \mathbb{D}/\{s\} \tag{4.219}$$

$$0 \leq p_{ij} \leq |T|x_{ij} \quad \forall i,j \in \mathbb{V} \tag{4.220}$$

Fuel constraints:

$$-M(1 - x_{ij}) \leq r_j - r_i + f_{ij} \leq M(1 - x_{ij}) \quad \forall i,j \in \mathbb{T} \tag{4.221}$$

$$-M(1 - x_{ij}) \leq r_j - L + f_{ij} \leq M(1 - x_{ij}) \quad \forall i \in \mathbb{D}, \forall j \in \mathbb{T} \tag{4.222}$$

$$-M(1 - x_{ij}) \leq r_i - f_{ij} \quad \forall i \in \mathbb{T}, \forall j \in \mathbb{D} \tag{4.223}$$

$$x_{ij} \in \{0,1\}, \forall i,j \in \mathbb{V}; \textit{either } i \textit{ or } j \textit{ is a target} \tag{4.224}$$

$$x_{ij} \in \{0,1,2,\ldots,|T|\}, \forall i,j \in \mathbb{D} \tag{4.225}$$

Equation (4.215) states that the in-degree and out-degree of each vertex must be the same and equation (4.216) ensures that each target is visited once by the aircraft. These equations allow for the aircraft to visit a depot any number of times for refueling. The constraints (4.217) to(4.220) ensure that there are $|T|$ units of commodity shipped from one depot and the aircraft delivers exactly one unit of commodity at each target. In equations (4.221) to (4.225), M denotes a large constant and can be chosen to be equal to $L + max_{i,j\in\mathbb{V}} f_{ij}$. If the UAV is traveling from target i to target j, equations (4.221) ensure that the fuel left in the aircraft after reaching target j is $r_j = r_i - f_{ij}$. If the UAV is traveling from depot i to target j, equation (4.222) ensures that the fuel left in the aircraft after reaching target j is $r_j = L - f_{ij}$. If the UAV is directly traveling from any target to a depot constraint, (4.224) must be at least equal to the amount required to reach the depot.

An approach to trajectory generation for autonomous aircraft is using **mixed integer linear programming (MILP)** and a modification of the A^* algorithm to optimize paths in dynamic environments particularly having pop-ups with a known future probability of appearance. Each pop-up leads to one or several possible evasion maneuvers, characterized with a set of values used as decision-making parameters in an integer linear programming model that optimizes the final route by choosing the most suitable alternative trajectory, according to the imposed constraints such as fuel consumption and spent time. The decision variables are the UAV state variables, i.e., position and speed. The constraints are derived from a simplified model of the UAV and the environment where it has to fly [166].

4.4.6.2 Routing with refueling depots for multiple aircraft

The multiple autonomous aircraft routing problem can also be considered. Given a set of heterogeneous aircraft, find an assignment of targets to be visited by each UAV along with the sequence in which it should be visited so that each target is visited at least once by an autonomous aircraft, all the UAV return to their respective depots after visiting the targets and the total distance traveled by the collection of autonomous aircraft is minimized;

Problem 4.20. *Let there be n targets and m aircraft located at distinct depots, let $\mathbb{V}(T)$ be the set of vertices that correspond to the initial locations of the aircraft (targets) with the m vertices, $\mathbb{V} = \{V_1, \dots, V_m\}$, representing the aircraft (i.e., the vertex i corresponds to the ith aircraft) and $T = \{T_1, \dots, T_n\}$ representing the targets. Let $\mathbb{V}^i = \mathbb{V}_i \cup T$ be the set of all the vertices corresponding to the ith aircraft and let $C^i : \mathbb{E}^i \rightarrow \mathbb{R}^+$ denote the cost function with $C^i(a,b)$ representing the cost of traveling from vertex a to vertex b for aircraft i. The cost functions are considered to be asymmetric i.e., $C^i(a,b) \neq C^i(b,a)$. An aircraft either does not visit any targets or visits a subset of targets in T. If the ith aircraft does not visit any target then its tour $TOUR_i = \emptyset$ and its corresponding tour $C(TOUR_i) = 0$. If the ith aircraft visits at least one target, then its tour may be represented by an ordered set $\{V_i, T_{i_1}, \dots, T_{i_{r_i}}, V_i\}$ where $T_{i_\ell}, \ell = 1, \dots, r_i$ corresponds to r_i distinct targets being visited in that sequence by the ith aircraft. There is a cost $C(TOUR_i)$ associated with a tour for the ith aircraft visiting at least one target defined as:*

$$C(TOUR_i) = C^i(V_i, T_{i_1}) + \sum_{k=1}^{r_i - 1} C^i(T_{i_k}, T_{i_{k+1}}) + C^i(T_{i_{r_1}}, V_i) \tag{4.226}$$

Find tours for the aircraft so that each target is visited exactly once by some aircraft and the overall cost defined by $\sum_{i \in V} C(TOUR_i)$ is minimized.

The approach is to transform the routing problem into a single asymmetric traveling salesman problem (ATSP) and use the algorithms available for the asymmetric traveling salesman problem to address the routing problem [140].

In the generalized traveling salesman problem, a major issue is that its mathematical formulation involves both continuous and integer decision variables [69]. To solve the problem, it is necessary to determine the following topics which minimize the mission completion time:

1. the order of the visit of the points.
2. the number of takeoffs and the number of points that have to be visited between each takeoff and landing.
3. the continuous path that the aircraft has to follow.

The **multi criteria decision analysis** (MCDA) technique is a process that allows to make decisions in the presence of multiple potentially conflicting criteria [187]. Common elements in the decision analysis process are a set of design alternatives, multiple decision criteria and preference information representing the attitude of a decision maker in favor of one criterion over another, usually in terms of weighting factors. Because of different preferences and incomplete information, uncertainty always exists in the decision analysis process [188].

To effectively select the most appropriate decision analysis method for a given decision problem, an approach consisting of the following steps can be proposed:

1. Define the problem,
2. define evaluation criteria,
3. calculate appropriateness index,
4. evaluate decision analysis method,
5. choose the most suitable method,
6. conduct sensitivity analysis.

An integrated approach based on graph theory for solving the deadlock problem in the cooperative decision-making and control is presented in [54]. The vehicle team can contain a group of fixed-wing UAV with different operational capabilities and kinematic constraints. Because of heterogeneity, one task cannot be performed by arbitrary vehicles in the heterogeneous group. The task assignment problem is described as a combinatorial optimization problem. Each assignment that allocates multiple vehicles to perform multiple tasks on multiple targets is a candidate solution. The execution time that the mission takes to be accomplished is chosen as the objective function to be minimized. A vehicle that performs multiple tasks on targets needs to change its path, waiting for other if another target needs to change its path, waiting for another vehicle that executes a former or simultaneous task has not finished or arrived. This creates risks of deadlocks. Two or more vehicles may fall into a situation of infinite waiting due to shared resources and precedence constraints among various tasks. A task-precedence graph of solutions is constructed and analyzed for detecting deadlocks. In addition, the topological sort of tasks is used for the path elongation of vehicles. Thus, deadlock-free solutions are obtained and the path coordinate is done [35].

The focus is to find a method to manage the non-deadlock condition and the time constraint. Consequently, the combinatorial optimization problem could be processed. All initial assignments are encoded according to a scheme that makes candidate solutions satisfy the constraints. Each feasible assignment of tasks is a feasible solution of the combinatorial optimization problem. After an initial assignment is generated, it must be checked whether it encodes deadlocks because the non-deadlock condition is a prerequisite for the subsequent process. An initial assignment is first processed into two groups according to two types of task relation. The first subgraph, the task executing subgraph, is derived by the vehicle-based task group. The second subgraph, the task constraint subgraph, is derived by the target tasks [71].

4.5 CONCLUSION

In the first part of this chapter, path and trajectory planning is presented. Trim is concerned with the ability to maintain flight equilibrium with controls fixed. Then an algorithm for 2D and 3D open-loop path planning is derived for the system presented in the previous section. Then, the Zermelo's navigation problem is considered in the sequel. Parametric paths are investigated, depending on the initial and final configurations. Smoother paths can be obtained by asking for the continuity of the derivatives of the path curvature and torsion. Maneuvers should be kept only to join two trim flight paths. Finally, some parametric curves such as polynomials, Pythagorean hodograph and η^3 splines are presented.

In the second part of the chapter, guidance and collision/obstacle avoidance topics are investigated into static and dynamic environments. In the author's opinion, there is no algorithm better than another. Depending on the mission, some have a better performance than the others. Only practitioners can choose the algorithm suitable for their case study.

Mission planning is presented in the last part of this chapter: route optimization, fuzzy planning, coverage problem and resource manager are topics of this important subject.

REFERENCES

1. Acevedo, J. J.; Arrue, B. C.; Diaz-Banez, J. M.; Ventura, I.; Maza, I.; Ollero, A. (2014): *One-to one coordination algorithm for decentralized area partition in surveillance missions with a team of aerial robots*, Journal of Intelligent and Robotic Systems, vol. **74**, pp. 269–285.
2. Ahuja, R.K.; Magnanti, T.L; Orlin, J.B. (1993): *Network Flows*, Prentice-Hall, Englewood Cliffs, NJ.
3. Alighanbari, M.; Bertuccelli, L. F.; How, J. P. (2006): *A robust approach to the UAV task assignment problem*, IEEE Conference on Decision and Control, San Diego, CA, pp. 5935–5940.

4. Anderson, R. P.; Milutinovic, D. (2014): *A stochastic approach to Dubins vehicle tracking problems*, IEEE Transactions on Automatic Control, vol. **59**, pp. 2801–2806.
5. Arsie, A.; Frazzoli, E. (2007): *Motion planning for a quantized control system on SO(3)*, 46th IEEE conference on Decision and Control, New Orleans, pp. 2265–2270.
6. Atkins, E.; Moylan, G.; Hoskins, A. (2006): *Space based assembly with symbolic and continuous planning experts*, IEEE Aerospace Conference, Big Sky, MT, DOI 10.1109/AERO.2006.1656009.
7. Aurenhammer, F. (1991): *Voronoi diagrams, a survey of fundamental geometric data structure*, ACM Computing Surveys, vol. **23**, pp. 345–405.
8. Avanzini, G. (2004): *Frenet based algorithm for trajectory prediction*, AIAA Journal of Guidance, Control and Dynamics, vol. **27**, pp. 127–135.
9. Avis, D.; Hertz, A.; Marcotte, O. (2005): *Graph Theory and Combinatorial Optimization*, Springer–Verlag.
10. Babel, L. (2013): *Three dimensional route planning for unmanned aerial vehicles in a risk environment*, Journal of Intelligent and Robotic Systems, vol. **71**, pp. 255–269, DOI 10.1007/s10846-012-9773-7.
11. Bakolas, E.; Tsiotras, P.(2010): *Zermelo–Voronoi Diagram: A dynamic partition problem*, Automatica, vol. **46**, pp. 2059–2067.
12. Bakolas, E.; Tsiotras, P. (2012): *Feedback navigation in an uncertain flowfield and connections with pursuit strategies*, AIAA Journal of Guidance, Control and Dynamics, vol. **35**, pp. 1268–1279.
13. Banaszuk, A.; Fonoberov, V. A.; Frewen, T. A.; Kobilarov, M.; Mathew, G.; Mezic, I; Surana, A. (2011): *Scalable approach to uncertainty quantification and robust design of interconnected dynamical systems*, IFAC Annual Reviews in Control, vol. **35**, pp. 77–98.
14. Bertsimas, D.; VanRyzin, G. (2011): *The dynamic traveling repairman problem*, MIT Sloan paper 3036-89-MS.
15. Bertuccelli, L. F.; Pellegrino, N.; Cummings, M. L. (2010): *Choice Modeling of Relook Tasks for UAV Search mission*, American Control Conference, pp. 2410–2415.
16. Bertsekas, D. P. (1995): *Dynamic Programming and Optimal Control*, One, Athena Scientific.
17. Bestaoui, Y.; Dahmani, H.; Belharet, K. (2009): *Geometry of translational trajectories for an autonomous aerospace vehicle with wind effect*, 47th AIAA Aerospace Sciences Meeting, Orlando, FL, paper AIAA 2009–1352.
18. Bestaoui, Y.: (2009) *Geometric properties of aircraft equilibrium and nonequilibrium trajectory arcs*, **Robot Motion and Control 2009**, Kozlowski, K. ed., Springer, Lectures Notes in Control and Information Sciences, Springer, pp. 1297–1307.
19. Bestaoui, Y.; Lakhlef, F. (2010): *Flight Plan for an Autonomous Aircraft in a Windy Environment*, **Unmanned Aerial Vehicles Embedded Control**, Lozano, R. (ed.), Wiley, ISBN-13-9781848211278.
20. Bestaoui, Y. (2011): *Collision avoidance space debris for a space vehicle*, IAASS conference, Versailles, France, **ESA Special Publication**, vol. **699**, pp. 74–79.
21. Bestaoui, Y. (2012): *3D flyable curve for an autonomous aircraft*, ICNPAA

World Congress on Mathematical Problems in Engineering, Sciences and Aerospace, AIP, Vienna, pp. 132–139.
22. Bestaoui,Y.; Kahale, E. (2013): *Time optimal trajectories of a lighter than air robot with second order constraints and a piecewise constant velocity wind*, AIAA Journal of Information Systems, vol. **10**, pp. 155–171, DOI 10.2514/1.55643.
23. Bestaoui Sebbane Y. (2014): *Planning and Decision-making for Aerial Robots*, Springer, Switzerland, ISCA 71.
24. Bethke, B.; Valenti, M.; How, J. P. (2008): *UAV task assignment, an experimental demonstration with integrated health monitoring*, IEEE Journal on Robotics and Automation, vol. **15**, pp. 39–44.
25. Bhatia, A.; Maly, M.; Kavraki, L.; Vardi, M. (2011): *Motion planing with complex goals*, IEEE Robotics and Automation Magazine, vol. **18**, pp. 55–64.
26. Bicho, E.; Moreira, A.; Carvalheira, M.; Erlhagen, W. (2005): *Autonomous flight trajectory generator via attractor dynamics*, Proc. of IEEE/RSJ Intelligents Robots and Systems, vol. **2** , pp. 1379–1385.
27. Bijlsma, S.J. (2009): *Optimal aircraft routing in general wind fields* , AIAA Journal of Guidance, Control, and Dynamics, Vol. **32**, pp. 1025–1029.
28. Bloch, A.M. (2003): *Non Holonomics Mechanics and Control*, Springer-Verlag, Berlin.
29. Boizot, N.; Gauthier, J. P. (2013): *Motion planning for kinematic systems*, IEEE Transactions on Automatic Control, vol. **58**, pp. 1430–1442.
30. Boukraa, D.; Bestaoui, Y.; Azouz, N. (2008): *Three dimensional trajectory generation for an autonomous plane*, Inter. Review of Aerospace Engineering, vol. **4**, pp. 355–365.
31. Braunl, T. (2008): *Embedded Robotics*, Springer.
32. Brooks, A.; Makarenko, A.; Williams, S.; Durrant-Whyte (2006): *Parametric POMDP for planning in continuous state spaces*, Robotics and Autonomous Systems, vol. **54**, pp. 887–897.
33. Bryson, A. E.; Ho, Y. C. (1975): *Applied Optimal Control: Optimization, Estimation and Control*, Taylor and Francis, New York.
34. Budiyono, A.; Riyanto, B.; Joelianto, E. (2009): *Intelligent Unmanned Systems: Theory and Applications*, Springer.
35. Busemeyer, J. R. , Pleskac, T. J (2009): *Theoretical tools for aiding dynamic decision-making*, Journal of Mathematical Psychology, vol. **53**, pp. 126–138.
36. Calvo, O.; Sousa, A.; Rozenfeld, A.; Acosta, G. (2009): *Smooth path planning for autonomous pipeline inspections*, IEEE Multi-conference on Systems, Signals and Devices, pp. 1–9, IEEE, DOI 978-1-4244-4346-8/09.
37. Campbell, S.; Naeem, W., Irwin, G. W. (2012): *A review on improving the autonomy of unmanned surface vehicles through intelligent collision avoidance maneuvers*, Annual Reviews in Control, vol. **36**, pp. 267–283.
38. Chakravarthy, A.; Ghose, D. (2012):*Generalization of the collision cone approach for motion safety in 3D environments*, Autonomous Robots, vol.**32**, pp. 243–266.
39. Chang, D. E. (2011): *A simple proof of the Pontryagin maximum principle on manifolds*, Automatica, vol. **47**, pp. 630–633.
40. Chavel, I. (ed.) (1984): *Eigenvalues in Riemannian Geometry*, Academic Press.
41. Choset, H.; Lynch, K.; Hutchinson, S.; Kantor, G.; Burgard, W.; Kavraki,

L.; Thrun, S. (2005): *Principles of Robot Motion, Theory, Algorithms and Implementation*, The MIT Press.
42. Chryssanthacopoulos, J.; Kochender, M. J. (2012): *Decomposition methods for optimized collision avoidance with multiple threats*, AIAA Journal of Guidance, Control and Dynamics, vol. **35**, pp. 368–405.
43. Clelland, J.N.; Moseley, C.; Wilkens, G. (2009): *Geometry of control affine systems*, Symmetry, Integrability and Geometry Methods and Applications (SIGMA), vol. **5**, pp. 28–45.
44. Cook, W. J. (2012): *In Pursuit of the Traveling Salesman: Mathematics at the Limits of Computation*, Princeton University Press.
45. Cormen, T. H. (2012): *Introduction to Algorithms*, The MIT Press, Cambridge.
46. Cotta, C.; Van Hemert, I. (2008): *Recent Advances in Evolutionary Computation for Combinatorial Optimization*, Springer.
47. Cruz, G. C.; Encarnacao, P. M. (2012): *Obstacle avoidance for unmanned aerial vehicles*, Journal of Intelligent and Robotics Systems, vol. **65**, pp. 203–217.
48. Dadkhah, N.; Mettler, B. (2012) (2012): *Survey of motion planning literature in the presence of uncertainty: considerations for UAV guidance*, Journal of Intelligent and Robotics Systems, vol. **65**, pp. 233–246.
49. Dai, R.; Cochran, J. E. (2010): *Path planning and state estimation for unmanned aerial vehicles in hostile environments*, AIAA Journal of Guidance, Control and Dynamics, vol. **33**, pp. 595–601.
50. Dantzig, G.; Fulkerson R.; Johnson, S. (1954): *Solution of a large-scale traveling salesman problem*, Journal of the Operations Research Society of America, vol. **2**, pp. 393–410.
51. Dantzig, G B.; Ramser, J.H. (1959): *The Truck dispatching problem*, Management Science, vol. **6**, pp. 80–91.
52. De Filippis, L.; Guglieri, G. (2012): *Path planning strategies for UAV in 3D environments*, Journal of Intelligent and Robotics Systems, vol. **65**, pp. 247–264.
53. Delahaye, D.; Puechmurel, S.; Tsiotra,s P.; Feron, E. (2013): *Mathematical models for aircraft design: a survey*, Air Traffic Management and Systems, Springer, Japan, pp. 205–247.
54. Deng, Q.; Yu, J.; Mei, Y. (2014): *Deadlock free consecutive task assignment of multiple heterogeneous unmanned aerial vehicles*, AIAA Journal of Aircraft, vol. **51**, pp. 596–605.
55. Devasia, S. (2011): *Nonlinear minimum-time control with pre- and post-actuation*, Automatica, vol. **47**, pp. 1379–1387.
56. Dicheva, S.; Bestaoui, Y. (2011): *Route finding for an autonomous aircraft*, AIAA Aerospace Sciences Meeting, Orlando, FL, USA, paper AIAA 2011–79.
57. Doshi-Velez, F.; Pineau, J.; Roy, N. (2012): *Reinforcement learning with limited reinforcement: Using Bayes risk for active learning in POMDP*, Artificial Intelligence, vol **187**, pp. 115–132.
58. Dubins, L. E. (1957): *On curves of minimal length with a constraint on average curvature and with prescribed initial and terminal positions and tangents.* American Journal of Mathematics, vol. **79**, pp. 497–517.
59. Eele, A.; Richards, A. (2009): *Path planning with avoidance using nonlinear branch and bound optimization*, AIAA Journal of Guidance, Control and

Dynamics, vol. **32**, pp. 384–394.
60. Enes, A.; Book, W. (2010): *Blended shared control of Zermelo's navigation problem*, American Control conference, Baltimore, MD, pp. 4307-4312.
61. Farault, J. (2008): *Analysis on Lie Groups: an Introduction*, Cambridge Studies in Advanced Mathematics.
62. Farouki, R. T. (2008): *Pythagorean Hodograph Curves*, Springer.
63. Foka, A.; Trahanias, P. (2007): *Real time hierarchical POMDP for autonomous robot navigation*, Robotics and Autonomous Systems, vol. **55**, pp. 561–571.
64. Foo, J.; Knutzon, J.; Kalivarapu, V.; Oliver, J.; Winer, E. (2009): *Path planning of UAV using B-splines and particles swarm optimization*, AIAA Journal of Aerospace Computing, Information and Communication, vol. **6**, pp. 271–290.
65. Fraccone, G. C.; Valenzuela-Vega, R.; Siddique, S.; Volovoi, V. (2013): *Nested modeling of hazards in the national air space system*, AIAA Journal of Aircraft, vol. **50**, pp. 370–377, DOI: 10.2514/1.C031690.
66. Fraichard, T.; Scheuer, A. (2008): *From Reeds and Shepp's to continuous curvature paths*, IEEE Transactions on Robotics, vol. **20**, pp. 1025–1035.
67. Frederickson, G.; Wittman, B. (2009): *Speedup in the traveling repairman problem with unit time window*, arXiv preprint arXiv:0907.5372.[cs.DS].
68. Funabiki, K.; Ijima, T.; Nojima, T. (2013): *Method of trajectory generation for perspective flight path display in estimated wind condition*, AIAA Journal of Aerospace Information Systems, Vol. **10**, pp. 240–249. DOI: 10.2514/1.37527.
69. Garone, E.; Determe, J. F.; Naldi, R. (2014): *Generalized traveling salesman problem for carrier-vehicle system*, AIAA Journal of Guidance, Control and Dynamics, vol. **37**, pp. 766–774.
70. Girard, A. R.; Larba, S. D.; Pachter, M.; Chandler, P. R. (2007): *Stochastic dynamic programming for uncertainty handling in UAV operations*, American Control Conference, pp. 1079–1084.
71. Goel, A.; Gruhn, V. (2008): *A general vehicle routing problem*, European Journal of Operational Research, vol. **191**, pp. 650–660.
72. Goerzen, C.; Kong, Z.; Mettler, B. (2010): *A survey of motion planning algorithms from the perspective of autonomous UAV guidance*, Journal of Intelligent and Robotics Systems, vol. **20**, pp. 65–100.
73. Greengard, C.; Ruszczynski, R. (2002): *Decision-making under Uncertainty: Energy and Power*, Springer.
74. Guerrero, J. A.; Bestaoui, Y. (2013): *UAV path planning for structure inspection in windy environments*, Journal of Intelligent and Robotics Systems, vol. **69**, pp. 297–311.
75. Habib, Z.; Sakai, M. (2003): *Family of G^2 cubic transition curves*, IEEE. Int. Conference on Geometric Modeling and Graphics. pp. 117–122.
76. Hameed, T.A. (2014): *Intelligent coverage path planning for agricultural robots and autonomous machines on three dimensional terrain*, Journal of Intelligent and Robotics Systems, vol. **74**, pp. 965–983.
77. Holdsworth, R. (2003): *Autonomous in flight path planning to replace pure collision avoidance for free flight aircraft using automatic dependent surveillance broadcast*, PhD Thesis, Swinburne Univ., Australia.
78. Holt, J.; Biaz, S.; Aj, C. A. (2013): *Comparison of unmanned aerial system collision avoidance algorithm in a simulated environment*, AIAA Journal of

Guidance, Control and Dynamics, vol. **36**, pp. 881–883.
79. Holzapfel, F.; Theil, S.(eds) (2011): *Advances in Aerospace Guidance, Navigation and Control*, Springer.
80. Hota, S.; Ghose, D. (2014): *Optimal trajectory planning for unmanned aerial vehicles in three-dimensional space*, AIAA Journal of Aircraft, vol. **51**, pp. 681–687.
81. Hota, S., Ghose, D. (2014): *Time optimal convergence to a rectilinear path in the presence of wind*, Journal of Intelligent and Robotic Systems, vol. **74**, pp. 791–815.
82. Howlett, J. K.; McLain, T., Goodrich, M. A. (2006): *Learning real time A* path planner for unmanned air vehicle target sensing*, AIAA Journal of Aerospace Computing, Information and Communication, vol. **23**, pp. 108–122.
83. Hsu, D.; Isler, V.; Latombe, J. C.; Lin, M. C. (2010): *Algorithmic Foundations of Robotic*, Springer.
84. Hutter, M. (2005): *Universal Artificial Intelligence, Sequential Decisions based on Algorithmic Probability*, Springer.
85. Huynh, U.; Fulton, N. (2012): *Aircraft proximity termination conditions for 3D turn centric modes*, Applied Mathematical Modeling, vol. **36**, pp. 521–544.
86. Ibe, O.; Bognor, R. (2011): *Fundamentals of Stochastic Networks*, Wiley.
87. Igarashi, H.; Loi. K (2001): *Path-planning and navigation of a mobile robot as a discrete optimization problems*, Art Life and Robotics, vol. **5**, pp. 72–76.
88. Innocenti, M.; Pollini, L.; Turra, D. (2002): *Guidance of unmanned air vehicles based on fuzzy sets and fixed waypoints*, AIAA Journal on Guidance, Control and Dynamics, vol. **27**, pp. 715–720.
89. Itoh, H.; Nakamura, K. (2007): *Partially observable Markov decision processes with imprecise parameters*, Artificial Intelligence, vol. **171**, pp. 453–490.
90. Jaklic, G.; Kozak, J.; Krajnc, M.; Vitrih, V.; Zagar, E. (2008): *Geometric Lagrange interpolation by planar cubic Pythagorean hodograph curves*, Computer Aided Design, vol **25**, pp. 720–728.
91. Jardin, M. R.; Bryson, A. E. (2001): *Neighboring optimal aircraft guidance in winds*, AIAA Journal of Guidance, Control and Dynamics, vol. **24**, pp. 710–715.
92. Jardin, M. R.; Bryson, A. E. (2012): *Methods for computing minimum time paths in strong winds*, AIAA Journal of Guidance, Control and Dynamics, vol. **35**, pp. 165–171.
93. Jarvis, P. A.; Harris, R.; Frost, C. R. (2007): *Evaluating UAS Autonomy Operations Software In Simulation*, AIAA Infotech@Aerospace Conference and Exhibit, DOI 10.2514/6.2007-2798.
94. Jennings, A. L.; Ordonez, R.; Ceccarelli, N. (2008): *Dynamic programming applied to UAV way point path planning in wind*, IEEE International Symposium on Computer-Aided Control System Design, San Antonio, TX, pp. 215–220.
95. Jiang, Z.; Ordonez, R. (2008): *Robust approach and landing trajectory generation for reusable launch vehicles in winds.* 17th IEEE International Conference on Control Applications, San Antonio, TX, pp. 930–935.
96. Johnson, B.; Lind, R. (2011): *3-dimensional tree-based trajectory planning with highly maneuverable vehicles*, 49th AIAA Aerospace Sciences Meeting, paper AIAA 2011–1286.
97. Jonsson, A. K. (2007): *Spacecraft Autonomy: Intelligence Software to increase*

crew, spacecraft and robotics autonomy, AIAA Infotech@Aerospace Conference and Exhibit, DOI 10.2514/6.2007-2791, paper AIAA 2007–2791.

98. Jurdjevic, V. (2008): *Geometric Control Theory*, Cambridge studies in advanced mathematics.
99. Kaelbling, L.; Littman, M.; Cassandra, A. (1998): *Planning and acting in partially observable stochastic domains*, Artificial Intelligence, vol. **101**, pp. 99–134.
100. Kalyanam, K.; Chandler, P.; Pachter, M.; Darbha, S. (2012): *Optimization of perimeter patrol operations using UAV*, AIAA Journal of Guidance, Control and Dynamics, vol. **35**, pp. 434–441.
101. Kalyanam, K.; Park, M.; Darbha, S.; Casbeer, D.; Chandler, P.; Pachter, M. (2014): *Lower bounding linear program for the perimeter patrol optimization*, AIAA Journal of Guidance, Control and Dynamics, vol. **37**, pp. 558–565.
102. Kampke, T.; Elfes, A. (2003): *Optimal aerobot trajectory planning for wind based opportunity flight control*, IEEE/RSJ Inter. Conference on Intelligent Robots and Systems , Las Vegas, NV, pp. 67–74.
103. Kang, Y.; Caveney, D. S.; Hedrick, J. S. (2008): *Real time obstacle map building with target tracking*, AIAA Journal of Aerospace Computing, Information and Communication, vol.**5**, pp. 120–134.
104. Khatib, O. (1985): *Real time obstacle avoidance for manipulators and mobile robots*, IEEE Int. Conference on Robotics and Automation, pp. 500–505.
105. Kothari, M.; Postlethwaite, I.; Gu, D. W. (2014): *UAV path following in windy urban environments*, Journal of Intelligent and Robotic Systems, vol. **74**, pp. 1013–1028.
106. Kluever, C.A. (2007): *Terminal guidance for an unpowered reusable launch vehicle with bank constraints*, AIAA Journal of Guidance, Control, and Dynamics, Vol. **30**, pp. 162–168.
107. Kuwata, Y.; Schouwenaars, T.; Richards, A.; How, J. (2005): *Robust constrained receding horizon control for trajectory planning*, AIAA Conference on Guidance, Navigation and Control, DOI 10.2514/6.2005-6079.
108. Lam, T. M. (ed) (2009): *Aerial Vehicles*, In-Tech, Vienna, Austria.
109. Laumond, J. P. (1998):*Robot Motion Planning and Control*, Springer.
110. Laugier, C.; Chatila, R. (eds) (2007): *Autonomous Navigation in Dynamic Environments*, Springer.
111. Lavalle, S. M. (2006): *Planning Algorithms*, Cambridge University Press.
112. Lavalle, S. M. (2011): *Motion planning*, IEEE Robotics and Automation Magazine, vol. **18**, pp. 108–118.
113. Lawler, E. L.; Lenstra, J. K; Rinnoy Kan, A. H.; Shmoys, D.B (1995): *A Guided Tour of Combinatorial Optimization*, Wiley.
114. Lee, J.; Kwon, O.; Zhang, L.; Yoon, S. (2014): *A selective retraction based RRT planner for various environments*, IEEE Transactions on Robotics, vol. **30**, pp. 1002–1011, DOI 10.1109/TRO.2014.2309836.
115. Lekkas, A. M.; Fossen, T. I. (2014): *Integral LOS path following for curved paths based on a monotone cubic Hermite spline parametrization*, IEEE Transactions on Control System Technology, vol. **22**, pp. 2287–2301, DOI 10.1109/TCST.2014.2306774.
116. LeNy, J.; Feron, E.; Frazzoli, E. (2012): *On the Dubins traveling salesman problem*, IEEE Transactions on Automatic Control, vol. **57**, pp. 265–270.

117. Li, Z.; Canny, J. F. (1992): *Non holonomic Motion Planning*, Kluwer Academic Press, Berlin.
118. Li, B.; Xu, C.; Teo, K. L.; Chu, J. (2013): *Time optimal Zermelo's navigation problem with moving and fixed obstacles*, Applied Mathematics and Computation, vol. **224**, pp. 866–875.
119. Lin, L.; Goodrich, M. A. (2014): *Hierarchical heuristic search using a Gaussian mixture model for UAV coverage planning*, IEEE Transactions on Cybernetics, vol. **44**, pp. 2532–2544, DOI 10.1109/TCYB.2014.2309898.
120. Liu, Y.; Saripelli, S. (2014): *Path planning using 3D Dubins curve for unmanned aerial vehicles*, Int. Conference on Unmanned Aircraft System, pp. 296–304, DOI 978-1-4799-2376-2.
121. Littman, M. (2009): *A tutorial on partially observable Markov decision process*, Journal of Mathematical Psychology, vol. **53**, pp. 119–125.
122. Ludington B.; Johnson, E., Vachtsevanos, A. (2006): *Augmenting UAV autonomy GTMAX*, IEEE Robotics and Automation Magazine, vol. **21** pp. 67-71.
123. Macharet, D.; Neto, A. A.; Campos, M. (2009): *On the generation of feasible paths for aerial robots in environments with obstacles*, IEEE/RSJ Int. conference on Intelligent Robots and Systems, pp. 3380–3385.
124. Maggiar, A.; Dolinskaya, I. S. (2014): *Construction of fastest curvature constrained path in direction dependent media*, AIAA Journal of Guidance, Control and Dynamics, vol. **37**, pp. 813–827.
125. Marigo, A., Bichi, A. (1998): *Steering driftless nonholonomic systems by control quanta*, IEEE Inter. conference on Decision and Control, vol. **4**, pp. 466–478.
126. Masoud, A. A. (2012): *A harmonic potential approach for simultaneous planning and control of a generic UAV platform*, Journal of Intelligent and Robotics Systems, vol. **65**, pp. 153–173.
127. Matsuoka, Y., Durrant-Whyte, H., Neira, J. (2010): *Robotics, Science and Systems*, The MIT Press.
128. Mattei, M.; Blasi, L. (2010): *Smooth flight trajectory planning in the presence of no-fly zones and obstacles*, AIAA Journal of Guidance, Control and Dynamics, vol. **33**, pp. 454–462, DOI 10.2514/1.45161.
129. Matveev, A. S.; Teimoori, H.; Savkin, A. (2011): *Navigation of a uni-cycle like mobile robot for environmental extremum seeking*, Automatica, vol. **47**, pp. 85–91.
130. McGee, T.; Hedrick, J. K. (2007): *Optimal path planning with a kinematic airplane model*, AIAA Journal of Guidance, Control and Dynamics, vol. **30**, pp. 629–633, DOI 10.2514/1.25042.
131. Miele, A.; Wang, T.; Melvin, W. (1986): *Optimal takeoff trajectories in the presence of windshear*, Journal of Optimization, Theory and Applications, vol. **49**, pp. 1–45.
132. Miele, A.; Wang, T; Melvin, W. (1989): *Penetration landing guidance trajectories in the presence of windshear* AIAA Journal of Guidance, Control and Dynamics, vol. **12**, pp. 806–814.
133. Missiuro, P.; Roy, N. (2006): *Adaptive probabilistic roadmaps to handle uncertain maps*, IEEE Int. Conference on Robotics and Automation, pp. 1261–1267, Orlando, FL.
134. Mufalli, F.; Batta, R.; Nagi R. (2012): *Simultaneous sensor selection and rout-*

ing of unmanned aerial vehicles for complex mission plans, Computers and Operations Research, vol. **39**, pp. 2787–2799.
135. Mujumda, A., Padhi, R. (2011): *Evolving philosophies on autonomous obstacles/collision avoidance of unmanned aerial vehicles*, AIAA Journal of Aerospace Computing, Information and Communication, vol. **8**, pp. 17–41.
136. Musial, M. (2008): *System Architecture of Small Autonomous UAV*, VDM.
137. Naldi, R.; Marconi, L. (2011): *Optimal transition maneuvers for a class of V/STOL aircraft*, Automatica, vol. **47**, pp. 870–879.
138. Nelson, R.; Barber, B.; McLain, T.; Beard, R. (2007): *Vector Field Path Following for Miniature Air Vehicle.* IEEE Transactions on Robotics, vol. **23**, pp. 519–529,.
139. Ng, H. K.; Sridhar, B.; Grabbe, S. (2014): *Optimizing trajectories with multiple cruise altitudes in the presence of winds*, AIAA Journal of Aerospace Information Systems, vol. **11**, pp. 35–46.
140. Oberlin, P.; Rathinam, S.; Darbha, S. (2010):*Todays traveling salesman problem*, IEEE Robotics and Automation Magazine, vol. **17**, pp. 70–77.
141. Obermeyer, K.; Oberlin, P.; Darbha, S. (2012): *Sampling based path planning for a visual reconnaissance unmanned air vehicle*, AIAA Journal of Guidance, Control and Dynamics, vol. **35**, pp. 619–631.
142. Oikonomopoulos, A. S.; Kyriakopoulos, K. J.; Loizou, S. G. (2010): *Modeling and control of heterogeneous nonholonomic input-constrained multi-agent systems*, 49th IEEE Conference on Decision and Control, pp. 4204–4209.
143. Okabe, A. R.; Boots, B.; Sugihara, K.; Chiu, S.N. (2000): *Spatial Tessalations: Concepts and Applications of Voronoi Diagrams*, Wiley.
144. Patsko, V. S.; Botkin, N. D.; Kein, V. M.; Turova, V. L.; Zarkh, M. A. (1994): *Control of an aircraft landing in windshear*, Journal of Optimization Theory and Applications, vol. **83**, pp. 237–267.
145. Patterson, T.; McClean, S.; Morrow, P.; Parr, G. (2012): *Modeling safe landing zone detection options to assist in safety critical UAV decision-making*, Procedia Computer Science, vol. **10**, pp. 1146–1151.
146. Pavone, M.; Frazzoli, E.; Bullo, F. (2011): *Adaptive and distributive algorithms for vehicle routing in a stochastic and dynamic environment*, IEEE Transactions on Automatic Control, vol. **56**, pp. 1259–1274.
147. Peng, R.; Wang, H.; Wang. Z.; Lin, Y. (2010): *Decision-making of aircraft optimum configuration utilizing multi dimensional game theory*, Chinese Journal of Aeronautics, vol. **23**, pp. 194–197.
148. Persiani, F.; De Crescenzio, F.; Miranda, G.; Bombardi, T.; Fabbri, M.; Boscolo, F. (2009):*Three dimensional obstacle avoidance strategies for uninhabited aerial systems mission planning and replanning*, AIAA Journal of Aircraft, vol. **46**, pp. 832–846.
149. Pettersson, P. O.; Doherty, P. (2004): *Probabilistic road map based path planning for an autonomous helicopter*, Journal of Intelligent and Fuzzy Systems: Applications in Engineering and Technology, vol. **17**, pp. 395–405.
150. Petres, C.; Pailhas, Y.; Pation, P.; Petillot, Y.; Evans, J.; Lame, D. (2007): *Path planning for autonomous underwater vehicles*, IEEE Transactions on Robotics, vol. **23**, pp. 331–341.
151. Phillips, J. M.; Bedrossian, N.; Kavraki, L. E. (2004): *Guided expansive space trees: a search strategy for motion and cost constrained state spaces*, IEEE Int.

Conference on Robotics and Automation, vol. **5**, pp. 3968–3973.

152. Piazzi, A.; Guarino Lo Bianco, C.; Romano, M. (2007): η^3 *Splines for the smooth path generation of wheeled mobile robot*, IEEE Transactions on Robotics, vol. **5**, pp. 1089–1095.
153. Plaku, E.; Hager, G. D. (2010): *Sampling based motion and symbolic action planning with geometric and differential constraints*, IEEE Int. Conference on Robotics and Automation, pp. 5002–5008.
154. Poggiolini, L.; Stefani, G. (2005): *Minimum time optimality for a bang-singular arc: second order sufficient conditions*, IEEE 44th Conference on Decision and Control, pp. 1433–1438.
155. Powell, W. B. (2011): *Approximate Dynamic Programming: Solving the Curse of Dimensionality*, Halsted Press, New York.
156. Prasanth, R. K.; Boskovic, J. D.; Li, S. M.; Mehra, R. (2001): *Initial study of autonomous trajectory generation for UAV*, IEEE Inter. conference on Decision and Control, Orlando, Fl, pp. 640–645.
157. Prats, X.; Puig, V.; Quevedo, J.; Nejjari, F. (2010): *Lexicographic optimization for optimal departure aircraft trajectories*, Aerospace Science and Technology, vol. **14**, pp. 26–37.
158. Puterman, M. L. (2005): *Markov Decision Processes Discrete Stochastic Dynamic Programming*, Wiley.
159. Qu, Y.; Zhang, Y.; Zhang, Y. (2014): *Optimal flight planning for UAV in 3D threat environment*, Int. Conference on Unmanned Aerial Systems, DOI 978-1-4799-2376-2.
160. Rabier, P. J.; Rheinboldt, W. C. (2000): *Nonholonomic Motion of Rigid Mechanical Systems from a DAE Viewpoint*, SIAM press.
161. Richards, A.; Schouwenaars, T.; How, J.; Feron, E (2002): *Spacecraft trajectory planning with avoidance constraints using mixed integer linear programming*, AIAA Journal of Guidance, Control and Dynamics, vol. **25**, pp. 755–764.
162. Rosen, K.H. (2013): *Discrete Mathematics*, McGraw Hill.
163. Rysdyk, R. (2007): *Course and heading changes in significant wind*, AIAA Journal of Guidance, Control and Dynamics, vol. **30**, pp. 1168–1171.
164. Ruchti J.; Senkbeil, R.; Carroll, J.; Dickinson, J.; Holt, J.; Biaz, S. (2014): *Unmanned aerial system collision avoidance using artificial potential fields*, AIAA Journal of Aerospace Information Systems, vol. **11**, pp. 140–144.
165. Rupniewski, M. W.; Respondek, W. (2010): *A classification of generic families of control affine systems and their bifurcations*, Mathematics, Control, Signals and Systems, vol. **21**, pp. 303–336.
166. Ruz, J.J.; Arevalo, O.; Pajares, G.; Cruz, J. M. (2007): *Decision-making along alternative routes for UAV in dynamic environments*, IEEE conference on Emerging Technologies and Factory Automation, pp. 997–1004, DOI-1-4244-0826-1.
167. Sabo C., Cohen K., Kumar M., Abdallah S.: (2009): *Path planning of a fire-fighting aircraft using fuzzy logic*, AIAA Aerospace Sciences Meeting, Orlando, FL, paper AIAA 2009-1353.
168. Sadovsky, A.V. (2014): *Application of the shortest path problem to routing terminal airspace air traffic*, AIAA Journal of Aerospace Information System, vol. **11**, pp. 118–130.
169. Samad, T.; Gorinevsky, D.; Stoffelen, F. (2001): *Dynamic multi-resolution*

route optimization for autonomous aircraft, IEEE Int. Symposium on Intelligent Control, pp. 13–18.

170. Santamaria, E.; Pastor, E.; Barrado, C.; Prats, X.; Royo, P.; Perez, M. (2012): *Flight plan specification and management for unmanned aircraft systems*, Journal of Intelligent and Robotic Systems, vol. **67**, pp. 155–181.
171. Sastry, S. (1999): *Nonlinear Systems, Analysis, Stability and Control*, Springer, Berlin.
172. Savla, K.; Frazzoli, E.; Bullo, F. (2008): *Traveling salesperson problems for the Dubbins vehicle*, IEEE Transactions on Automatic Control, vol. **53**, pp. 1378–1391.
173. Schmitt, L.; Fichter, W. (2014): *Collision avoidance framework for small fixed wing unmanned aerial vehicles*, AIAA Journal of Guidance, Control and Dynamics, vol. **37**, pp. 1323–1328.
174. Schouwenaars, T.; Valenti, M.; Feron, E.; How, J.; Roche, E. (2006): *Linear programming and language processing for human/unmanned-aerial-vehicle team missions*, AIAA Journal of Guidance, Control, and Dynamics, vol. **29**, no. 2, pp. 303–313.
175. Seibel, C. W.; Farines, J. M.; Cury, J. E. (1999): *Towards hybrid automata for the mission planning of Unmanned Aerial Vehicles*, **Hybrid Systems V**, Antsaklis, P. J. (ed), Springer-Verlag, pp. 324–340.
176. Selig, J. M. (1996): *Geometric Methods in Robotics*, Springer.
177. Sennott, L.I. (1999): *Stochastic Dynamic Programming and the Control of Queuing Systems*, Wiley.
178. Shah, M. Z.; Samar, R.; Bhatti, A. I. (2015): *Guidance of air vehicles: a sliding mode approach*, IEEE transactions on Control Systems Technology, vol. **23**, pp. 231–244, DOI 10.1109/TCST.2014.2322773.
179. Sinha, A.; Tsourdos, A.; White, B. (2009): *Multi-UAV coordination for tracking the dispersion of a contaminant cloud in an urban region*, European Journal of Control, vol. **34**, pp. 441–448.
180. Smith, J. F.; Nguyen, T. H. (2006): *Fuzzy logic based resource manager for a team of UAV*, Annual Meeting of the IEEE Fuzzy Information Processing Society, pp. 463–470.
181. Soler, M.; Olivares, A.; Staffetti, E. (2014): *Multiphase optimal control framework for commercial aircraft 4D flight planning problems*, AIAA Journal of Aircraft, vol. **52**, pp. 274–286, DOI 10.2514/1C032677.
182. Song, C.; Liu, L.; Feng, G.; Xu, S. (2014): *Optimal control for multi-agent persistent monitoring*, Automatica, vol. **50**, pp. 1663–1668.
183. Sridharan, M.; Wyatt, J.; Dearden, R. (2010): *Planning to see: a hierarchical approach to planning visual action on a robot using POMDP*, Artificial Intelligence, vol. **174**, pp. 704–725.
184. Stachura, M.; Frew, G. W. (2011): *Cooperative target localization with a communication aware- Unmanned Aircraft System*, AIAA Journal of guidance, control and dynamics, **34**, pp. 1352–1362.
185. Stranders, R.; Munoz, E.; Rogers, A.; Jenning, N. R. (2013): *Near-optimal continuous patrolling with teams of mobile information gathering agents*, Artificial Intelligence, vol. **195.**, pp. 63–105.
186. Sujit, P. B.; Saripalli, S.; Sousa, J. B. (2014): *Unmanned aerial vehicle path following*, IEEE Control System Magazine, vol. **34**, pp. 42–59.

187. Sun, L. G.; de Visser, C. C.; Chu, Q. P.; Falkena, W. (2014): *Hybrid sensor-based backstepping control approach with its application to fault-tolerant flight control*, AIAA Journal of Guidance, Control and Dynamics, vol. **37**, pp. 59–71.
188. Sun, X.; Gollnick, V.; Li, Y.; Stumpf, E. (2014): *Intelligent multi criteria decision support system for systems design*, AIAA Journal of Aircraft, vol. **51**, pp. 216–225.
189. Sundar, K.; Rathinam, S. (2014): *Algorithms for routing an unmanned aerial vehicle in the presence of refueling depots*, IEEE Transactions on Automation Science and Engineering, vol. **11**, pp. 287–294
190. Sussman, H. J. (1997): *The Markov Dubins problem with angular acceleration control*, IEEE 36th Conference on Decision and Control, pp. 2639–2643.
191. Sydney, N.; Paley, D. A. (2014): *Multiple coverage control for a non stationary spatio-temporal field*, Automatica, vol. **50**, pp. 1381–1390.
192. Tang, J.; Alam, S.; Lokan, C.; Abbass, H.A. (2012): *A multi-objective approach for dynamic airspace sectorization using agent based and geometric models*, Transportation Research part C, vol. **21**, pp. 89–121.
193. Techy L. (2011): *Optimal navigation in planar true varying flow: Zermelo's problem revisited*, Intelligent Service Robotics, vol. **4**, pp. 271–283.
194. Temizer, S. (2011): *Planning under uncertainty for dynamic collision avoidance*, PhD Thesis, MIT, MA, USA.
195. Tewari, A. (2011): *Advanced Control of Aircraft, Spacecrafts and Rockets*, Wiley Aerospace Series.
196. Toth, P.; Vigo, D. (2002): *The Vehicle Routing Problem*, SIAM, Philadelphia.
197. Trumbauer, E.; Villac, B. (2014): *Heuristic search based framework for onboard trajectory redesign*, AIAA Journal of Guidance, Control and Dynamics, vol. **37**, pp. 164–175.
198. Turnbull, O.; Richards, A.; Lawry, J.; Lowenberg, M. (2006): *Fuzzy decision tree cloning of flight trajectory optimization for rapid path planning*, IEEE Conference on Decision and Control, San Diego, CA, pp. 6361–6366.
199. VanDaalen, C. E.; Jones, T. (2009): *Fast conflict detection using probability flow*, Automatica, vol. **45**, pp. 1903–1909.
200. Vanderberg, J.P. (2007): *Path planning in dynamic environments*, PhD thesis, Univ. of Utrecht, The Netherlands.
201. Vazirani, V. (2003): *Approximation Algorithms*, Springer-Verlag.
202. Wang, H. F.; Wen, Y. P. (2002): *Time-constrained Chinese postman problems*, International Journal of Computers and Mathematics with Applications, vol. **44**, pp. 375–387.
203. Wang, X.; Wei, G.; Sun, J. (2011): *Free knot recursive B Spline for compensation of nonlinear smart sensors*, Measurement, vol. **44**, pp. 888–894.
204. Wang, Y.; Wang, S.; Tan, M.; Zhou, C.; Wei, Q. (2014): *Real time dynamic Dubins-helix method for 3D trajectory smoothing*, IEEE Transactions on Control System Technology, vol. **23**, pp. 730–736, DOI 10.1109/TCST.2014.2325904.
205. Watkins, A. S. (2007): *Vision based map building and trajectory planning to enable autonomous flight through urban environments*, PhD Thesis, Univ. of Florida, Gainesville .
206. Weiss, A.; Petersen, C.; Baldwin, M.; Scott, R.; Kolmanovsky, I. (2014): *Safe positively invariant sets for spacecraft obstacle avoidance*, AIAA Jour-

nal of Guidance, Control and Dynamics, vol. **38**, pp. 720–732, DOI : 10.2514/1.G000115.
207. Wilkins, D. E.; Smith, S. F.; Kramer, L. A.; Lee, T.; Rauenbusch, T. (2008): *Airlift mission monitoring and dynamic rescheduling*, Engineering Application of Artificial Intelligence, vol. **21**, pp. 141–155.
208. Williams, P. (2005): *Aircraft trajectory planning for terrain following incorporating actuator constraints*, AIAA Journal of Aircraft, vol. **42**, pp. 1358–1362.
209. Wu, P. (2009): *Multi-objective mission flight planning in civil unmanned aerial systems*, PhD thesis, Queensland Univ. of Technology (Australia).
210. Yakimenko, O.A. (2000): *Direct method for rapid prototyping of near optimal aircraft trajectory*, AIAA. Journal of Guidance, Control and Dynamics, vol. **23**, pp. 865–875.
211. Yang, I.; Zhao, Y. (2004): *Trajectory planning for autonomous aerospace vehicles amid known obstacles and conflicts*, AIAA Journal of Guidance, Control and Dynamics, vol. **27**, pp. 997-1008.
212. Yanmaz, E.; Costanzo, C.; Bettstetter, C.; Elmenreich, W. (2010): *A discrete stochastic process for coverage analysis of autonomous UAV networks*, IEEE GLOBECOM Workshop, pp. 1777–1782.
213. Yanushevsky, R.(2011): *Guidance of Unmanned Aerial Vehicles*, CRC Press.
214. Yokoyama, N. (2012): *Path generation algorithm for turbulence avoidance using real time optimization*, AIAA Journal of Guidance, Control and Dynamics, vol. **36**, pp. 250–262.
215. Zhao, Y.; Tsiotras, P. (2013): *Time optimal path following for fixed wing aircraft*, AIAA Journal of Guidance, Control and Dynamics, vol. **36**, pp. 83–95.
216. Zhang, J.; Zhan, Z.; Liu Y., Gong Y. (2011): *Evolutionary computation meets machine learning: a survey*, IEEE Computational Intelligence Magazine, vol. **6**, pp.68–75.
217. Zou, Y.; Pagilla, P. R.; Ratliff R. T. (2009): *Distributed formation flight control using constraint forces*, AIAA Journal of Guidance, Control and Dynamics, vol. **32**, pp. 112–120.

5 Flight Safety

ABSTRACT

Situational awareness is used for low level control and for flight and mission planning which constitute the high level control elements. Data are coming from different kinds of sensors, each one being sensitive to a different property of the environment. Data can be integrated to make the perception of an autonomous aircraft more robust. This allows to obtain new information otherwise unavailable. Due to the uncertainty and imprecise nature of data acquisition and processing, individual informational data sources must be appropriately integrated and validated. An integrated navigation system is the combination of an on board navigation solution providing position, velocity and attitude as derived from accelerometers and gyro-inertial sensors. This combination is accomplished with the use of different filters, presented in the first part of the chapter. In autonomous aircraft, the on board control system must be able to interpret the meaning of the system health information and decide on the appropriate course of action. This requires additional processing on board the unmanned aircraft to process and interpret the health information and requires that the health monitoring and aircraft control systems be capable of integration. In the second part of the chapter, integrated system health monitoring is investigated, presenting some diagnostic tools and approaches. As there are uncertainties in the information, usually several scenarios are considered and a trade-off solution is offered. In the third part of the chapter, fault tolerant flight control is considered for LTI and LPV formulations followed by model reference adaptive control. The last part of the chapter concerns fault-tolerant planner detailing trim state discovery, reliability analysis, safety analysis of obstacle avoidance and finally risk measurement.

5.1 INTRODUCTION

Selecting the best set of decisions or actions in order to maximize the obtained cumulative reward for the autonomous aircraft is an important problem in many application domains. The challenge in using these controllers and planners in the real world arises from uncertainties that are a result of modeling errors or inadequacies of available mathematical models as well as a dynamic environment with meteorological phenomena.

An autonomous flight controller must possess the most adaptive capability that can be imagined for any control system [13]: robustness and failure tolerance.

1. **Robustness** of the control law:

   a. Taking into account the uncertainties/dispersions, coverage of domains,
   b. Guaranteeing compliance with specifications (frequency and time) at a given probability,
   c. Classic robustness/performance trade-off to be managed.
2. **Failure tolerance**
   a. Can be ensured through sophisticated sensor redundancy and rapid detection/correction of actuator abnormal behavior,
   b. Minimum capacity of algorithms to handle double failures in order to give the autonomous aircraft a chance.

The limitations of this approach can be:

1. Couplings between axes,
2. Simplified equations for controller logic with respect to complexity of physics/reality,
3. Interaction between control function and other functions and subsystems.

Flight control system is validated through:

1. **Ground system studies**
   a. Definition of algorithms
   b. A priori validation of flight envelope with mathematical models
   c. A posteriori validation using the unmanned aircraft
2. **Electrical system and embedded software studies**
   a. Encoding on on board computer (real time)
   b. Inputs/outputs coupled with electrical hardware
   c. Equipment could be backed up to improve reliability

An integrated **vehicle health management** system is executed in an environment that includes different sensor technologies and multiple information systems that include different data models. **Integrated vehicle health management** (IVHM) involves data processing that captures data related to aircraft components, monitoring parameters, assessing current or future health conditions and providing recommended maintenance actions [112].

Early applications of system heath monitoring focused on developing health and usage monitoring systems to improve safety by providing real measures of equipment use and insure maintenance was performed after the prescribed number of operating hours. Maintenance nowadays can be performed based on the actual condition of the equipment.

Remark 5.1. *While health monitoring systems developed for manned platforms can be applied for unmanned aircraft, the unmanned systems often have lower limits on size, weight and power requirements than manned systems making direct transition of the technology impractical. Another challenge in transitioning health monitoring technology from manned to unmanned aircraft* *is the absence of an operator on board to interpret the meaning of the health information and take appropriate action [126].*

Advanced fault detection and diagnosis (FDD) techniques limit the impacts of flight control system failures [21]. Some uncertainties with sufficient data to determine the precise distributions and distribution parameters can be described within a probabilistic framework whereas some lacking sufficient data can be described within a non probabilistic framework [141].

decision-making and estimation are central in unmanned aircraft, as the need often arises to make inferences about the state of the environment, based on information which is at best limited, if not downright misleading.

Remark 5.2. *Some studies assessing the safety of UAS operations use* ***uniform traffic densities****. Some others are related to quantitatively establishing the boundary of* well-clean *for sense and avoid systems. In [66], a distributed traffic model is constructed using actual traffic data collected over a one-year period to enable a probabilistic approach to risk assessment.*

5.2 SITUATION AWARENESS

The situation awareness system is capable of detecting problems, finding solutions and reacting to them. Eight subfunctions have been categorized by the FAA into detect, track, evaluate, prioritize, declare, determine action, command and execute modules. The category is defined by awareness sensors, awareness data fusion, self-separation declaration/collision avoidance declaration and self-separation reaction/collision reaction [8].

3D perception and representation of the environment is an important basis for autonomous intelligent functionality in unstructured environments [18, 86]. To be able to interact properly with its environment, a smart autonomous aircraft has to know the environment and its state in that environment. Using several sensors and knowing the map of the environment, localization algorithms allow to compute the position and orientation of the aircraft [114]. Any localization problem involves four main concepts: the environment, the map of the environment, the configuration of the aircraft and the measurements of the environment. Each measurement of the environment is associated to an equation or a set of equations linking the map, the configuration of the aircraft and the measurement. The problem of localization can be formulated as a **constraint satisfaction problem**(CSP). A constraint satisfaction problem can be seen as a set of equations or constraints involving variables to be determined such as the configuration of the aircraft. Each of these variables is known to belong to a known set called domain or search space. In this case, each constraint can be considered as a representation of the information [106].

A technique to assist in system analysis is **Markov analysis**. It is useful when investigating systems where a number of states may be valid and are also interrelated. The question of whether a system is airworthy is not a simple mathematical calculation but depends upon the relative states of the parts of the system [78, 79]. The problem of making decisions in the presence of uncertainties has also been studied in the planning literature. The typical approach there is to formulate these problems in the **Markov decision processes** (MDP) framework and search for an optimal policy.

Remark 5.3. *If the situation awareness system is combining measurements from different technologies, their respective data should be aligned in time, properly compensated for uncertainties that differ by technology or dimension and given appropriate weights [7].*

5.2.1 FILTERS

Estimation of states and parameters for dynamical systems, in general, is performed in the Bayesian framework, where uncertainty is represented as the **probability density function** (PDF) [33]. Developing an **estimation scheme** first requires a suitable description of the measurements and the dynamics equations describing the estimator state-time evolution [53].

Dead reckoning is the estimation of an aircraft location based on its estimated speed, direction and time of travel with respect to a previous estimate. The Kalman filter is used to estimate the new configuration. Integrated navigation systems are used in this context. The primary sensing methods available are air data, magnetic, inertial and radar sensors. Each of the on board sensors' families have their own attributes and associated strengths and shortcomings. on board sensors are also used in conjunction with external navigation aids and systems to achieve the optimum performance for the navigation system. The quest algorithm has been the most widely and most frequently implemented algorithm for three axis attitude estimation. It is also an important component of many attitude Kalman filters [26].

The state dynamics can be modeled as a discrete time system, allowing for the implementation of several standard filtering approaches such as the approaches presented in this section. In this section, the general Bayesian framework for linear and nonlinear filtering is reviewed.

5.2.1.1 Classical Kalman filter

For the linear Gaussian system, it is possible to get exact analytical expressions for the evolving sequence of moments, which characterizes the probability density function completely. This is the Kalman filter approach. It is an effective procedure for combining noisy sensor outputs to estimate the state of a system with uncertain dynamics. Uncertain dynamics include unpredictable disturbances of the aircraft, caused by the winds or unpredictable changes in the sensor or actuator parameters [44]. The classical **Kalman filter** (KF) maintains two types of variables:

1. An **estimated state vector**,

2. A **covariance matrix** which is a measure of the estimation uncertainty.

The equations used to propagate the covariance matrix model and manage uncertainty taking into account how sensor noise and dynamic uncertainty contribute to uncertainty about the estimated system state are presented next.

By maintaining an estimate of its own estimation uncertainty and the relative uncertainty in the various sensor outputs, the Kalman Filter is able to combine all sensor information optimally in the sense that the resulting estimate minimizes any quadratic loss function of estimation error, including the mean squared value of any linear combination of state estimation errors. The Kalman filter gain is the optimal weighting matrix for combining new sensor data with a priori estimate to obtain a new estimate [128].

The Kalman filter is a two step process: **prediction** and **correction**. The correction step makes correction to an estimate based on new information obtained from sensor measurement. In the prediction step, the estimate $\hat{X}$ and its associated covariance matrix of estimation uncertainty $\mathbf{P}$ are propagated from time epoch to another. This is the part where the aircraft linearized dynamics is included. The state of the aircraft is a vector of variables that completely specify enough of the initial boundary value conditions for propagating the trajectory of the dynamic process forward in time and the procedure for propagating that solution forward in time is called *state prediction.*

Algorithm 12 Essential Kalman Filter Equations

1. Predictor (Time Updates)
   a. Predicted state vector

$$\hat{X}_k^- = \Phi_k \hat{X}_{k-1}^+$$

   b. covariance matrix

$$\mathbf{P}_k^- = \Phi_k \mathbf{P}_{k-1}^+ \Phi_k^T + \mathbf{Q}_{k-1}$$

2. Corrector (Measurement Updates)
   a. Kalman gain

$$\bar{\mathbf{K}}_k = \mathbf{P}_k^- \mathbf{H}_k^T \left(\mathbf{H}_k \mathbf{P}_k^- \mathbf{H}_k^T + \mathbf{R}_k \right)^{-1}$$

   b. Corrected state estimator

$$\hat{X}_k^+ = \hat{X}_k^- + \bar{\mathbf{K}}_k \left(Z_k - \mathbf{H}_k \hat{X}_k^- \right)$$

   c. Corrected covariance matrix

$$\mathbf{P}_k^+ = \mathbf{P}_k^- - \bar{\mathbf{K}}_k \mathbf{H}_k \mathbf{P}_k^-$$

The model for propagating the covariance matrix of estimation uncertainty is derived from the model used for propagating the state vector. The following formulation is used:

$$\begin{aligned} X_k &= \Phi_k X_{k-1} + \vartheta_{k-1} \\ Z_k &= \mathbf{H}_k X_k + \nu_{k-1} \end{aligned} \tag{5.1}$$

The noise process ϑ_k, ν_k are white, zero mean uncorrelated and have known covariance matrices, respectively, $\mathbf{Q}_k, \mathbf{R}_k$. In the following derivation, $\hat{X}_k^-$ represents the a priori estimate, $\hat{X}_k^+$ represents the posteriori estimate, $\mathbf{P}_k^-$ represents the a priori covariance and $\mathbf{P}_k^+$ represents the posteriori covariance.

Algorithm 12 presents the main steps of the Kalman filter.

The **Joseph formula** is a general covariance update equation valid not only for the Kalman filter but for any linear unbiased estimator under standard Kalman filtering assumptions [136]. The Joseph formula is given by:

$$\mathbf{P}^+ = (\mathbf{I}_{n\times n} - \mathbf{KH})\mathbf{P}^-(\mathbf{I}_{n\times n} - \mathbf{KH})^T + \mathbf{KRK}^T \tag{5.2}$$

where $\mathbf{I}_{n\times n}$ is the identity matrix, $\mathbf{K}$ is the gain, $\mathbf{H}$ is the measurement mapping matrix and $\mathbf{P}^-$ and $\mathbf{P}^+$ are the pre and post measurement update estimation error covariance matrices respectively.

The optimal linear unbiased estimator or equivalently the optimal linear minimum mean square error estimate or Kalman filter often uses a simplified covariance update equation such as

$$\mathbf{P}^+ = (\mathbf{I}_{n\times n} - \mathbf{KH})\mathbf{P}^- \tag{5.3}$$

or

$$\mathbf{P}^+ = \mathbf{P}^- - \mathbf{K}(\mathbf{HP}^-\mathbf{H}^T + \mathbf{R})\mathbf{K}^T \tag{5.4}$$

Remark 5.4. *While these alternative formulations require fewer computations than the Joseph formula, they are only valid when* $\mathbf{K}$ *is chosen as the optimal Kalman gain.*

5.2.1.2 Extended Kalman filter

For a nonlinear system exhibiting Gaussian behavior, the system is linearized locally about the current mean value and the covariance is propagated using the approximated linear dynamics. This approach is used in extended Kalman filter (EKF). The aircraft's position and attitude is provided in six degrees of freedom by the flight control computer using a differential GPS sensor, a magnetometer and an inertial measurement unit. The raw data of these sensors can be integrated by an extended Kalman filter to provide an accurate solution in all six degrees of freedom. The extended Kalman filter is also widely used in attitude estimation.

The problem of nonlinear filtering requires the definition of dynamical and measurement models. It is assumed that the dynamic state $X(t) \in \mathbb{R}^n$ at time

t evolves according to the continuous-time stochastic model:

$$\dot{X} = f(X(t), t) + \mathbf{G}(X(t), t)\vartheta(t) \tag{5.5}$$

where $f: \mathbb{R}^n \times \mathbb{R} \to \mathbb{R}^n$, $G: \mathbb{R}^n \times \mathbb{R} \to \mathbb{R}^{n \times m}$ and $\vartheta(t)$ is an m-dimensional Gaussian with white noise process with covariance matrix $\mathbf{Q}(t)$. In particular in equation (5.5), the function f encodes the deterministic force components of the dynamics, such as gravity, lift, drag ... while the process noise term models the stochastic uncertainties. In many applications, a discrete-time formulation of the dynamical model is used:

$$X_{k+1} = f_k(X_k) + \mathbf{G}_k(X_k)\vartheta_k \tag{5.6}$$

where $X_k = X(t_k)$, $f: \mathbb{R}^n \to \mathbb{R}^n$, $\mathbf{G}_k: \mathbb{R}^n \to \mathbb{R}^{n \times m}$ and ϑ_k is an m-dimensional zero-mean Gaussian white noise sequence with covariance matrix $\mathbf{Q}_k$.

Algorithm 13 Extended Kalman Filter Equations

1. The extended Kalman filter update equations are as follows [128]:

$$\hat{X}_{k+1}^- = f\left(\hat{X}_k, U_k, 0\right)$$

$$\mathbf{P}_{k+1}^- = \mathbf{A}_k \mathbf{P}_k \mathbf{A}_k^T + \mathbf{W}_k \mathbf{Q}_k \mathbf{W}_k^T$$

2. and the measurement update equations are:

$$\mathbf{K}_k = \mathbf{P}_k^- \mathbf{H}_k^T \left(\mathbf{H}_k \mathbf{P}_k^- \mathbf{H}_k^T + \mathbf{V}_k \mathbf{R}_k \mathbf{V}_k^T\right)^{-1}$$

$$\hat{X}_k = \hat{X}_k^- + \mathbf{K}_k \left(Z_k - h(\hat{X}_k^-, 0)\right)$$

$$\mathbf{P}_k = (\mathbf{I}_{n \times n} - \mathbf{K}_k \mathbf{H}_k)\mathbf{P}_k^-$$

3. where

$$X_{k+1} = f_k(X_k, U_k, W_k)$$

$$Z_k = h_k(X_k, V_k)$$

4. and

$$\mathbf{A}_k = \frac{\partial f}{\partial X}(\hat{X}_k, U_k, 0)$$

$$\mathbf{W}_k = \frac{\partial f}{\partial W}(\hat{X}_k, U_k, 0)$$

$$\mathbf{H}_k = \frac{\partial h}{\partial X}(\hat{X}_k, 0),$$

$$\mathbf{V}_k = \frac{\partial h}{\partial W}(\hat{X}_k, 0)$$

A sequence of measurements $Z_k = z_1, \dots, z_k$ is related to the corresponding kinematic states X_k via measurement functions $h_k : \mathbb{R}^n \rightarrow \mathbb{R}^p$ according to the discrete-time measurement model

$$Z_k = h_k(X_k) + \nu_k \tag{5.7}$$

In this equation, ν_k is a p-dimensional zero-mean Gaussian white noise sequence with covariance matrix $\mathbf{R}_k$. More general filter models can be formulated from measurement models with non-Gaussian or correlated noise terms and sensor biases.

In the Bayesian approach to dynamic state estimation, the posterior probability density function of the states is constructed based on information of a prior state, and received measurements are constructed. Encapsulating all available statistical information, the posterior probability density function $p(X_k|Z_k)$ may be regarded as the complete solution to the estimation problem and various optimal state estimates can be computed from it. Analytical solutions to the filter prediction and correction steps are generally intractable and are only known in a few restrictive cases. In practice, models are nonlinear and states can be non-Gaussian; only an approximate or suboptimal algorithm for the Bayesian state estimator can be derived.

The extended Kalman filter equations are presented in algorithm 13.

Remark 5.5. *This approach performs poorly when the nonlinearities are high, resulting in an unstable estimator. However, the error in mean and covariance can be reduced if the uncertainty is propagated using nonlinear dynamics for a minimal set of sample points called* ***sigma points****. The PDF of the states, characterized by sigma points, captures the posterior mean and covariance accurately to the third order in the Taylor series expression for any nonlinearity with Gaussian behavior. This technique has resulted in the unscented Kalman filter (UKF). The aforementioned filters are based on the premise of Gaussian PDF evolution. If the sensor updates are frequent, then the extended Kalman filter yield satisfactory results.*

5.2.1.3 Unscented Kalman filter

The unscented Kalman filter employs a set of deterministically selected points in order to compute the mean and covariance of Y as well as the cross variance between U and Y. The EKF and UKF are used extensively in air tracking; they represent state uncertainties by a covariance matrix and this may not be adequate in all situations. Higher-order versions of the UKF (also called Gauss-Hermite filters) make use of efficient Gauss-Hermite quadrature rules. The unscented Kalman filter selects its points based on moment matching. The filter formulation is based on standard attitude vector measurement using a gyro based for attitude propagation [60].

Remark 5.6. *For nonlinear systems, the unscented filter uses a carefully selected set of sample points to map the probability distribution more accurately* *than the linearization of the standard extended Kalman filter (EKF) leading to faster convergence from inaccurate initial condition particularly in attitude estimation problems.*

The unscented Kalman filter is derived for discrete time nonlinear equation, where the model is given by:

$$\begin{aligned} X_{k+1} &= f(X_k, k) + \mathbf{G}_k \varrho_k \\ \tilde{Z}_k &= h(X_k, k) + \nu_k \end{aligned} \tag{5.8}$$

where X_k is the $n \times 1$ state vector and $\tilde{Z}_k$ is the $m \times 1$ measurement vector. A continuous time mode can always be expressed in the form of (5.8) through an appropriate numerical integration scheme. The process noise ϱ_k and measurement error noise ν_k are assumed to be zero mean Gaussian noise processes with covariances given by $\mathbf{Q}_k$ and $\mathbf{R}_k$, respectively. The update equations are first rewritten as:

$$\begin{aligned} \hat{X}_k^+ &= \hat{X}_k^- + \mathbf{K}_k \vartheta_k \\ \hat{\mathbf{P}}_k^+ &= \hat{\mathbf{P}}_k^- - \mathbf{K}_k \mathbf{P}_k^{VV} \mathbf{K}_k^T \end{aligned} \tag{5.9}$$

where $\hat{X}_k^-$ and $\hat{\mathbf{P}}_k^-$ are the pre-update state estimate and covariance, respectively. $\hat{X}_k^+$ and $\hat{\mathbf{P}}_k^+$ are the post-update state estimate and covariance. The covariance of ϑ_k is denoted by $\mathbf{P}_k^{VV}$. The innovation ϑ_k is given by

$$\vartheta_k = \tilde{Z}_k - \hat{Z}_k^- = \tilde{Z}_k - h(\hat{X}_k^-, k) \tag{5.10}$$

The gain $\mathbf{K}_k$ is computed by

$$\mathbf{K}_k = \mathbf{P}_k^{XY} (\mathbf{P}_k^{VV})^{-1} \tag{5.11}$$

where $\mathbf{P}_k^{XY}$ is the cross correlation matrix between $\hat{X}_k^-$ and $\hat{Z}_k^-$.

The unscented filter uses a different propagation from the standard EKF. Given an $n \times n$ covariance matrix $\mathbf{P}$, a set of $2n$ sigma points can be generated from the columns of the matrix $\pm\sqrt{(n+\lambda)\mathbf{P}}$ where $\sqrt{\mathbf{M}}$ is the shorthand notation for a matrix $\mathbf{Z}$ such that $\mathbf{Z}\mathbf{Z}^T = \mathbf{M}$.

The set of points is zero mean, but if the distribution has mean μ, then simply adding μ to each of the points yields a symmetric set of $2n$ points having the desired mean and covariance. Because of the symmetric nature of this set, its odd central moments are zero and so its first three moments are the same as the original Gaussian distribution. The scalar λ is a convenient parameter for exploiting knowledge (if available) about the higher moments of the given distribution [44].

The transformation process is represented by the following algorithm 14 [60].

Remark 5.7. *In scalar systems, for example, for* $n = 1$*, a value of* $\lambda = 2$ *leads to errors in the mean and variance that are sixth order. For higher dimensions*

Algorithm 14 Unscented Kalman Filter

1. The set of sigma points are created.
2. The transformed set is given by:

$$\chi_i(k+1|k) = f(\chi_i(k|k), U_k, k)$$

3. The predicted mean is computed as:

$$\hat{X}(k+1|k) = \sum_{i=0}^{2^n} \mathbf{W}_i \chi_i(k+1|k)$$

4. The predicted covariance is computed as:

$$\mathbf{P}(k+1|k) =$$
$$= \sum_{i=0}^{2^n} \mathbf{W}_i \left(\chi_i(k+1|k) - \hat{X}(k+1|k) \left(\chi_i(k+1|k) - \hat{X}(k+1|k) \right)^T \right)$$

5. Instantiate each of the prediction points through the observation model

$$Z_i(k+1|k) = h\left(\chi_i(k+1|k), U(k|k)\right)$$

6. The predicted observation is calculated by:

$$\hat{Z}(k+1|k) = \sum_{i=0}^{2^n} \mathbf{W}_i Z_i(k+1|k)$$

7. Since the observation noise is additive and independent, the innovation covariance is:

$$\mathbf{P}_{vv}(k+1|k) = \mathbf{R}(k+1) +$$
$$+ \sum_{i=0}^{2^n} \mathbf{W}_i \left(Z_i(k|k-1) - \hat{Z}(k+1|k) \left(Z_i(k|k-1) - \hat{Z}(k+1|k) \right)^T \right)$$

8. Finally, the cross correlation matrix is determined by:

$$\mathbf{P}_{XZ}(k+1|k) =$$
$$= \sum_{i=0}^{2^n} \mathbf{W}_i \left(\chi_i(k|k-1) - \hat{X}(k+1|k) \left(Z_i(k|k-1) - \hat{Z}(k+1|k) \right)^T \right)$$

---

systems, choosing $\lambda = 3 - n$ minimizes the mean squared error up to the fourth order. However, caution should be exercised when λ is negative because a possibility exists that the predicted covariance can become non positive semi-definite. When $n + \lambda$ tends to zero, the mean tends to that calculated by the truncated second order filter.

Another approach can be used that allows for scaling of the sigma points, which guarantees a positive semi-definite covariance matrix.

This unscented Kalman filtering has several advantages over the extended Kalman filter including

1. The expected error is lower than the extended Kalman filter.
2. This filter can be applied to a non differentiable function.
3. It avoids the derivation of Jacobian matrices.
4. It is valid to higher order expansions than the standard extended Kalman filter.

5.2.1.4 Monte Carlo filter

Recently, simulation based sequential filtering methods, using **Monte Carlo** simulations, have been developed to tackle nonlinear systems with non-Gaussian uncertainty, while Kalman filters assume in general that the error in sensors have a Gaussian probability density function. The Monte Carlo estimator makes no assumption about the distribution of errors. Many different versions of the aircraft state vector are maintained. When a new measurement is available, it is tested versus the old versions. The best fitting states are kept and are randomly perturbed to form a new generation of states. Collectively, these many possible states and their scores approximate a probability density function for the state to be estimated. Monte Carlo methods are often used when simulating systems with a large number of coupled degrees of freedom within significant uncertainty in inputs.

Monte Carlo methods involve representing the probability density function of the states using a finite number of samples. The filtering task is obtained by recursively generating properly weighted samples of the state variable using importance sampling. Monte Carlo filters are based on sequential Monte Carlo methods. In the particle filter, ensemble numbers or particles are propagated using nonlinear system dynamics. These particles with proper weights, determined from the measurements, are used to obtain the state estimate.

Monte Carlo methods estimate a probability distribution of a system's output response from uncertain input parameters [18]. A typical calculation step of Monte Carlo methods to obtain the model output statistics is as follows:

1. Uncertain input parameters for an analytical or numerical system model are randomly sampled from their respective probability distribution function.
2. Multiple simulation runs are conducted using each corresponding output for each case. The probability distribution of a user-defined output metric can then be generated while estimating various statistics such as mean and variance.

In the standard Monte Carlo method, since random sampling of the input parameter distributions is required, the number of simulation runs must be large enough to ensure representation of the entire input parameter range and also to converge to the output distribution.

Statistical techniques can provide predicted path coordinates under uncertainty. Relevant output statistics such as mean, variances and covariance can also be calculated. Based on these statistics, the motion path can be augmented with ellipsoids defined by the variances and covariance. The ellipsoids indicate confidence levels for the predicted position on the path. Given a sufficient sample size n from the Monte Carlo method of motion path coordinates $X_i = (x_i, y_i, z_i)^T$, a sample mean vector $\bar{X} = (\bar{x}, \bar{y}, \bar{z})^T$ can be defined as:

$$\bar{x} = \frac{1}{n}\sum_{i=1}^{n} x_i \qquad \bar{y} = \frac{1}{n}\sum_{i=1}^{n} y_i \qquad \bar{z} = \frac{1}{n}\sum_{i=1}^{n} z_i \tag{5.12}$$

The sample covariance matrix S is then determined as:

$$S = \frac{1}{n-1}\sum_{i=1}^{n}\left(X_i - \bar{X}\right)\left(X_i - \bar{X}\right)^T = \\ = \begin{pmatrix} s_x^2 & rs_{xy} & rs_{xz} \\ rs_{xy} & s_y^2 & rs_{yz} \\ rs_{xz} & rs_{yz} & s_z^2 \end{pmatrix} \tag{5.13}$$

where s_x, s_y, s_z are the sample standard deviations, s_{xy}, s_{xz}, s_{yz} the sample covariance and r the sample correlation index.

The equation for a **confidence ellipsoid** can be formulated by the following equation:

$$C^2 = \left(X - \bar{X}\right)^T S^{-1}\left(X - \bar{X}\right) \tag{5.14}$$

where $C = \sqrt{-2\ln(1-P)}$ and P is the probability that determines the confidence level of the predicted position.

The principal semi-axes a_x, a_y of the confidence ellipse in the $x - y$ plane for a given probability P are obtained from:

$$a_x = Cs'_x \qquad a_y = Cs'_y \qquad a_z = Cs'_z \tag{5.15}$$

where s'_x, s'_y are expressed by:

$$s'_x = \sqrt{s_x^2 + s_y^2 + \frac{\sqrt{(s_x^2 - s_y^2)^2 + 4r^2 s_x^2 s_y^2}}{2}} \tag{5.16}$$

$$s'_y = \sqrt{s_x^2 + s_y^2 - \frac{\sqrt{(s_x^2 - s_y^2)^2 + 4r^2 s_x^2 s_y^2}}{2}} \tag{5.17}$$

The orientation of the confidence ellipse with respect to the $x - y$ coordinate is defined by the inclination angle

$$\alpha = \frac{1}{2}\arctan\left(\frac{2rs_x s_y}{s_x^2 - s_y^2}\right) \tag{5.18}$$

5.2.1.5 Particle filter

The aircraft nonlinear dynamical system with states $X \in \mathbb{R}^n$ and outputs $\tilde{Z} \in \mathbb{R}^m$ is given by:

$$\begin{aligned} \dot{X} &= g(X, \Delta) \\ \tilde{Z} &= h(X) + \nu \end{aligned} \tag{5.19}$$

where ν is the measurement noise and Δ is a vector of random parameters. Let $p(\Delta)$ be the probability density functions of Δ and $\mathbb{D}_\delta$ be the domain of Δ. Discrete measurement updates are available at times $t_0, t_1, \ldots, t_k$. If the system has only initial condition uncertainty, then $\Delta = X$. Parameter estimation can be included in this framework by suitably augmenting the system.

Particle filters are based on the importance sampling theorem, where random samples are drawn from a given distribution at time t_{k-1}. Each sample point is assigned a weight that is determined from the distribution function. These sample points are taken as initial conditions for the dynamical system and evolved using (5.19) to time t_k. Depending on their locations, the prior density function is obtained using the corresponding likelihood function in a Bayesian setting. Based on the new density function, a new set of sample points are generated and the process repeats.

However, due to particle degeneracy, particle filters require a large number of ensembles for convergence, leading to higher computational effort. This problem is tackled through re-sampling. Particle filters with the re-sampling technique are commonly known as **bootstrap filters**. However, bootstrap filters introduce other problems like loss of diversity among particles if the re-sampling is not performed correctly.

Recently developed techniques have combined importance sampling and **Markov chain Monte Carlo** (MCMC) methods to generate samples to get better estimates of states and parameters. Several other methods, like regularized particle filters and filters involving the Markov chain Monte Carlo move step, have been developed to improve sample diversity. The main problem with particle filters is the determination of the weights for each sample. This greatly affects the accuracy. At the same time, for large scale problems, the exponential growth in the number of samples makes this method computationally prohibitive. The problem of determining weights can be resolved by using the Frobenius–Perron (FP) operator.

The **Frobenius–Perron operator** is used to predict evolution of uncertainty in the nonlinear system and obtain the prior probability density function in the estimation process. The Frobenius–Perron operator on the **Liouville equation** predicts evolution of uncertainty in a more computationally efficient manner than Monte Carlo.

The problem of determining uncertainty in state due to parametric system uncertainty, for a nonlinear system, can be solved using the Frobenius–Perron operator [62]. The definition of the operator in continuous time for the dynamical system

$$\dot{X} = f(X) \quad X \in \mathbb{R}^n \quad f : \mathbb{X} \to \mathbb{X} \tag{5.20}$$

is given by the **Liouville equation**:

$$\frac{\partial P}{\partial t} + \sum_{i=1}^{n} \frac{\partial P f_i(X)}{\partial X_i} \tag{5.21}$$

where $f(X) = (f_1(X), \dots, f_n(X))$ and $p(t, X) = P_t p(t_0, x)$. The symbol P_t represents the continuous time Frobenius–Perron operator and $p(t_0, X)$ is the density function at initial time $t = t_0 \geq 0$.

The Frobenius–Perron operator is a **Markov operator** and has the following properties:

1. **Continuity**:

$$lim_{t \to t_0} \| P_t p(t_0, X) - p(t_0, X) \| = 0 \tag{5.22}$$

2. **Linearity**:

$$P_t \left(\lambda_1 p_1(t_0, X) + \lambda_2 p_2(t_0, X) \right) = \lambda_1 P_t p_1(t_0, X) + \lambda_2 P_t p_2(t_0, X) \tag{5.23}$$

such that $\lambda_1, \lambda_2 \in \mathbb{R}$ and $p_1(t_0, X), p_2(t_0, X) \in \mathbb{L}^1$.

3. **Positivity**:

$$P_t p(t_0, X) \geq 0 \text{ if } p(t_0, X) \geq 0 \tag{5.24}$$

4. **Norm preserving**:

$$\int_X P_t p(t_0, X) \mu(dX) = \int_X p(t_0, X) \mu(dX) \tag{5.25}$$

such that $p(t_0, X) \in \mathbb{L}^1$ where $\mathbb{L}^1$ is a set of functions, $f : \mathbb{X} \to \mathbb{R}$, satisfying $\int_{\mathbb{X}} |f(X)| \mu(dX) < \infty$ and μ is a measure defined on $\mathbb{X}$.

These properties ensure that the initial probability density function $p(t_0, X)$ evolves continuously over time while satisfying properties of probability density functions.

A method of characteristics for solving a first order linear partial differential equation (PDE) that can be easily solved uses the method of characteristics. Thus relation (5.21) can be written as:

$$\frac{\partial P}{\partial t} + \sum_{i=1}^{n} \frac{\partial p}{\partial X_i} f_i(X) + p \sum_{i=1}^{n} \frac{\partial f_i(X)}{\partial X_i} \tag{5.26}$$

Defining

$$g(X, p) = -p \sum_{i=1}^{n} \frac{\partial f_i(X)}{\partial X_i} \tag{5.27}$$

the following form is obtained

$$\frac{\partial P}{\partial t} + \sum_{i=1}^{n} \frac{\partial p}{\partial X_i} f_i(X) = g(X, p) \tag{5.28}$$

which is in the standard form. Assuming $g(X, p) \neq 0$, (5.28) can be solved by solving $(n+1)$ coupled ordinary differential equations by:

$$\frac{dX_1}{dt} = f_1(X) \quad \dots \quad \frac{dX_n}{dt} = f_n(X) \quad \frac{dp}{dt} = g(X, p) \tag{5.29}$$

These equations trace out a trajectory in the $(n+1)$ dimensional space spanned by $(X_1, \dots, X_n, p)$. To make the solution unique, the value of $p(t, X)$ has to be specified at a given point $X(t_0)$ at time t_0. The evolution of $p(t, X)$ over $\mathbb{R} \times \mathbb{X}$ can be determined by specifying $p(t, X)$ over several points in $\mathbb{X}$. These points are obtained by sampling the density function at t_0. The equation (5.29) determines the evolution of p along $X(t)$.

An important point is accuracy in prediction of uncertainty. Since this method requires a selection of points in the state space, this approach is similar to Monte Carlo. The main difference is that in the Frobenius–Perron operator, the value of the density function is determined along the trajectory and a final time; the value of the density function is known at certain discrete locations. These locations are values of the state vector at that final time. The value of the density function over the domain is then determined using interpolation [11].

Remark 5.8. *Process noise cannot be addressed in this framework. This is a limitation of the Frobenius–Perron operator, or the Liouville equation. To include process noise, the Fokker–Planck–Kolmogorov equation must be solved.*

A nonlinear state estimation algorithm that combines the Frobenius–Perron operator theory with the Bayesian estimation theory is presented in [33]. Let the aircraft and measurement model be given by (5.19). It is assumed that measurements are available at discrete times $t_0, \dots, t_k$. At a given time t_k, let X_k, Y_k and p_k be the state, measurement and probability density functions. Let $p_k^-(.)$ and $p_k^+(.)$ denote the prior and posterior density functions at time t_k. The estimation algorithm is described next.

1. **STEP 1: Initialization:** To begin, the domain $\mathbb{D}_\Delta$ of the initial random variable X_0 is discretized. From the discretized domain, N samples are chosen at random, based on probability density functions $P_0(X_0)$ of the random variable X_0. Let the samples be represented by $X_{0,i}$ for $i = 1, 2, \dots, N$ and $P_0(X_0, i)$ be the value of $P_0(X)$ at the sample points. The following steps are then performed recursively starting from $k = 1$.

2. **STEP 2: Propagation:** With the initial states at the $(k-1)^{th}$ step as $(X_{k-1,i}P_{k-1}(X_{k-1,i}))$, equation (5.29) is integrated for each grid over the interval $[t_{k-1}, t_k]$ to get $\left(X_{k,i}\bar{P}_k(X_{k,i})\right)^T$. $P_k(X_{k,i})$ obtained by integration are the prior probability density functions values for $X_{k,i}$.
3. **STEP 3: Update:** First the likelihood function $P(\tilde{Y}_k|X_k = X_{k,i})$ is determined for each grid point i, using Gaussian measurement noise and the sensor model in (5.19). It is defined as

$$\ell(\tilde{Y}_k|X_k = X_{k,i}) = \frac{1}{\sqrt{(2\pi)^m|\mathbf{R}|}} \exp^{-0.5\left(\tilde{Y}_k - h(X_{k,i})\right)^T \mathbf{R}^{-1}\left(\tilde{Y}_k - h(X_{k,i})\right)} \tag{5.30}$$

where $|\mathbf{R}|$ is the determinant of the covariance matrix of measurement noise. The probability density functions of the states are constructed next for each grid point i using classical Bayes rule. It is defined as the density function of the states given current measurement, i.e.,

$$P_k^+(X_{k,i}) = P(X_k = X_{k,i}|\tilde{Y}_k) = \\ = \frac{\ell(\tilde{Y}_k|X_k=X_{k,i})\mathbf{P}_k^-(X_k=X_{k,i})}{\sum_{j=1}^N \ell(\tilde{Y}_k|X_k=X_{k,i})\mathbf{P}_k^-(X_k=X_{k,j})} \tag{5.31}$$

4. **STEP 4: Getting the state estimate:** The state estimate for the kth step is then computed, depending on the desired computation. The following are commonly used criteria:
   a. For the maximum likelihood estimate, maximize the probability that $X_{k,i} = \hat{X}_k$. This results in $\hat{X}_k$ =mode of $\mathbf{P}^+(X_{k,i})$.
   b. For the minimum variance estimate,

$$\hat{X}_k = \operatorname{argmin} \sum_{i=1}^N \|X_k - X_{k,i}\|^2 \mathbf{P}_k^+(X_{k,i}) = \sum_{i=1}^N X_{k,i}\mathbf{P}_k^+(X_{k,i}) \tag{5.32}$$

   the estimate is the mean of $X_{k,i}$.
   c. For the minimum error estimate, minimize the maximum $|X - X_{k,i}|$. This results in $\hat{X}$ =median of $\mathbf{P}_k^+(X_{k,i})$.
5. **STEP 5: Re-sampling:** The degeneracy of sample points can be detected by looking at values of $X_{k,i}$ for which $\mathbf{P}^+(X_{k,i}) < \epsilon$ for which $\epsilon << 1$ and is prespecified. Existing methods for re-sampling can be used to generate new points and the corresponding posterior density $\mathbf{P}_k^+(X_{k,i})$ serve as initial states.

5.2.1.6 Wind estimation

Wind is one of the major contributors to uncertainties while flying, particularly in continuous descent operation [57]. The **true airspeed** (TAS) V_a is determined via the air data computer (ADC) and pitot-static system for input of impact pressure, static pressure and total air temperature. The ground speed is calculated by the inertial reference unit and GPS. The **flight management system** (FMS) predicts trajectory parameters, such as time and fuel estimates, by performing calculations that require estimates of the wind field along the trajectory toward the base to improve the accuracy of these calculations. An error in estimating the wind field will result in an error of the predicted ground speed, deceleration and flight path angle. In turn, the flight path angle error affects the predicted ground speed and predicted vertical trajectory. Hence, the accuracy of the trajectory, both temporal and spatial, is greatly influenced by the wind estimation error. The **aircraft meteorological data relay**(AMDAR) combines the observed atmospheric data that contains position (latitude and longitude), altitude, time, temperature, horizontal wind direction and speed, turbulence, humidity and icing, phase of flight, roll and pitch angles and the aircraft identifier.

The use of **automatic dependent surveillance-broadcast** (ADS-B) for meteorological monitoring is explored in [67]. Although originally developed for surveillance, the data that the system provides can be used to estimate wind, pressure and temperature profiles. The ground speed vector is the sum of the airspeed and wind vectors:

$$V_g = V_a + W \tag{5.33}$$

V_g is the aircraft speed relative to the ground, V_a the aircraft speed relative to the air and W the speed of the wind relative to the ground. Wind can be estimated from a series of observations of the aircraft ground-speed vector at different track angles [93, 105].

To estimate the wind from the ADS-B data of one aircraft, the North and East components of the aircraft true airspeed can be written as:

$$V_{a_x} = \|V_g\| \sin \chi_g - \|W\| \sin \chi_w \tag{5.34}$$

$$V_{a_y} = \|V_g\| \cos \chi_g - \|W\| \cos \chi_w \tag{5.35}$$

where χ_g and χ_w are the angles of the ground speed and wind vector with respect to the North. It follows that during a turn at constant airspeed and wind vector:

$$\|V_a\|^2 - \left(\|V_g\| \cos \chi_g - \|W\| \cos \chi_w\right)^2 - \left(\|V_g\| \sin \chi_g - \|W\| \sin \chi_w\right)^2 = 0 \tag{5.36}$$

Equation (5.36) is treated as a nonlinear least squares problem with measurement and solution vector:

$$X = \left(\|V_g\|, \chi_g\right)^T \qquad w_w = \left(\|V_a\|, \|W\|, \chi_w\right)^T \tag{5.37}$$

that minimizes

$$\sum_{i=0}^{t} \|Z_i - \Gamma(w_w, X_i)\|^2 \tag{5.38}$$

where Z_i and $\Gamma(w_w, x_i)$ are equal to the right and left parts of relation (5.36), respectively. To solve this nonlinear least squares problem recursively, the following algorithm 15 can be proposed.

Algorithm 15 Wind Estimation Algorithm

For $k = 0, 1, \ldots, n$, the solution is:

$$W_{k+1} = W_k + \mathbf{K}_k \left(Z_k - \Gamma(W_k, X_k) \right)$$

where

$$\mathbf{K}_k = \mathbf{P}_k \mathbf{H}_k^T \left(\mathbf{H}_k \mathbf{P}_k \mathbf{H}_k^T + \mathbf{R}_k \right)^{-1}$$

$$\mathbf{P}_{k+1} = (\alpha + 1) \left(\mathbf{P}_k - \mathbf{K}_k \mathbf{H}_k \mathbf{P}_k + \epsilon \mathbf{I}_{n \times n} \right)$$

$$\mathbf{H}_k = \frac{\partial \Gamma(W, X)}{\partial W} |_{W = \hat{W}, X = X_k}$$

where $\alpha > 0, \epsilon > 0$ and $\mathbf{P}_0$ and $\mathbf{R}_t$ are symmetric positive definite covariance matrices.

5.2.2 AERIAL SIMULTANEOUS LOCALIZATION AND MAPPING

To be autonomous, aircraft must know the environment in which they are to move. Autonomous systems have to sense, understand and remember at least those parts of the environment that are in the vicinity [4]. A main requirement for autonomous aircraft is to detect obstacles and to generate environmental maps for sensor data. One approach to represent the environment is the use of grid-based maps. They allow a fusion of data from different sensors, including noise reduction and simultaneous pose estimation, but they have large memory requirements. Further, they do not separate single objects. A second approach, called feature-based maps, focuses on individual objects. An early work uses lines to represent the world in 2D. Later approaches uses planar or rectangular surfaces for 3D modeling but mostly to rebuild the world with details and possible texture mapping. A suitable model for autonomous behavior is the velocity obstacle paradigm that can be added with the introduced specifications on how to measure the obstacles.

5.2.2.1 Problem formulation

There is a need to have different map types for different aircraft tasks. Because map-based aircraft localization and mapping are interdependent, both problems are solved simultaneously. The field of mapping is divided into metric and topological approaches. Metric maps capture the geometric properties of the environment while topological maps describe the connectivity of different places by means of nodes and graphs. In practice, metric maps are finer grained than topological maps, but higher resolution comes at a computational burden. Metric maps can be discretized based on the probability of space occupation. The resulting mapping approaches are known as occupancy-grid mapping. In contrast, the metric maps of geometric elements retain positions and properties of objects with specific geometric features.

In many cases, it is advantageous to use grid-based maps for sensor fusion and feature based polygonal metric maps for local planning, e.g., in order to avoid obstacles. Additionally, non-metric topological maps are most suitable for global search tasks like route planning. In a complex scenario, an aircraft must deal with all of these different maps and keep them updated. These tasks need the usual information exchange. The approach presented in [4] combines grid maps and polygonal obstacle representations and tackles the problem of large environments by using small grid maps that cover only essential parts of the environment for sensor fusion. Characteristic features are recognized; their shapes are calculated and inserted to a global map that takes less memory and is easily expandable. This map is not restricted to the sensor environment and is used for path planning and other applications [96]. Aerial mapping is an active research area. It addresses the problem of acquiring spatial models of the environment through aircraft on board sensors. Thrun in [121] has presented a survey of a mobile robot mapping algorithm with a focus on indoor environments. However, aircraft navigation occurs outdoor in a 3D dynamic environment and the situation is quite different.

To achieve real autonomy, an aircraft must be able to localize itself in the environment it is exploring [12]. When a map is not available, the aircraft should build it, while at the same time localizing itself within it. This problem, known as simultaneous localization and mapping (SLAM) can be cast as an estimation problem. A crucial issue for a SLAM algorithm is to close loops, i.e., to recognize places that have already been visited. The difficulty of the SLAM problem increases with the size of the environment.

Since both the aircraft pose and the map are uncertain, the mapping problem and the induced problem of localizing the robot with respect to the map are considered [80]. The fundamental principle used in aircraft mapping and localization is Bayesian filtering. The filter is used to calculate the aircraft pose and map posterior probability distribution, given all the control and measurements.

Remark 5.9. *If unmanned aircraft could relate the information obtained by their sensors to their motor/propulsion actions, they would detect incoherence and would be able to react to unexpected situations.*

Some SLAM methods are incremental and allow real time implementation whereas others require several passes through the whole of the perceived data. Many incremental methods employ Kalman filters to estimate the map and the aircraft localization and generate maps that describe the position of landmarks, beacons and certain objects in the environment. An alternative family of methods is based on Dempster's expectation maximization algorithm, which tries to find the most probable map by means of a recursive algorithm. These approaches solve the correspondence problem between sensorial measurement and objects in the real world [104, 107].

Mapping dynamic environments is a considerable problem since many realistic applications for aircraft are in non-static environments. Although Kalman filter methods can be adapted for mapping dynamic environments by assuming landmarks that move slowly over time, and, similarly, occupancy grid maps may consider some motion by reducing the occupancy over time, map generation in dynamic environments has been poorly explored. Smart autonomous aircraft address key problems of intelligent navigation, such as navigation in dynamic environments, navigation in modified environments [7].

An aircraft can be tasked to explore an unknown environment and to map the features it finds, without the use of infrastructure based localization systems such as GPS or any a priori terrain data. Simultaneous localization and mapping allows for the simultaneous estimation of the location of the aircraft as well as the location of the features it sees [20]. One key requirement for SLAM to work is that it must re-observe features and this has two effects:

1. The improvement of the location estimate of the feature.
2. The improvement of the location estimate of the platform because of the statistical correlations that link the platform to the features.

SLAM is generally implemented as a statistical filter, where the prediction and estimation of the location of point features is concurrent with the prediction and estimation of the pose and velocity of the aircraft. A feature can be any object in the environment which can be represented by a point in 3D space. The SLAM prediction stage involves the propagation of the aircraft and feature models and their corresponding uncertainties and mainly relies on inertial navigation. The SLAM estimation step occurs when there is an observation of a feature on the ground; the observation is used to improve both the location estimate of the feature and the location estimate of the aircraft because of the correlations in the SLAM structure that links the platform to the feature. The effect that observations have towards correcting the aircraft location estimates is dependent on the order on which features are observed and the trajectory of the aircraft. As the aircraft explores unknown terrain, initializing new features into the map, it becomes necessary to return to the well-known regions of the map in order to reduce the growth in localization errors. This process is referred to as closing the loop in which the uncertainty in both the aircraft localization and map position estimates is reduced via the correlations made between the aircraft and map states in the SLAM filter. Additionally, the maneuvers the aircraft takes during feature observations affect the accuracy in localization estimates. The aircraft's control strategy has an effect on the accuracy of the filter estimates.

The estimation process minimizes a cost function J_{est} to obtain the optimal estimate of all landmarks l_j, the past trajectories $X_{0:c}$ during $[t_0, t_c]$, the internal parameters ρ_{int} and the past process noises $\nu_{0:c}$. The cost function J_{est} is the negative log of the posterior $p\left(X_{0:c}, \nu_{0:c}, l_{1,K}, \rho_{int} | \{Z_{ij}, U_{0:c-1}\}\right)$ which provides the maximum a posteriori (MAP) estimate [120]. The assumption is made that the states and parameters have a Gaussian prior, i.e., $X_0 \propto \aleph\left(\hat{X}_0, \mathbf{P}_X^{-1}\right)$ and $\rho_{int} \propto \aleph\left(\hat{\rho}_{int}, \mathbf{P}_{int}^{-1}\right)$, and that uncertainties are also Gaussian $\nu_i \propto \aleph\left(0, \mathbf{P}_{\nu_i}^{-1}\right)$ and $\vartheta_i \propto \aleph\left(0, \mathbf{P}_{\vartheta_i}^{-1}\right)$. The cost function J_{est} has the following form:

$$J_{est} = \left\| X_0 - \hat{X}_0 \right\|_{\mathbf{P}_X}^2 + \left\| \rho_{int} - \hat{\rho}_{int} \right\|_{\mathbf{P}_{int}}^2 + \left\| \nu \right\|_{\mathbf{P}_{\nu_i}}^2 + \sum_{ij} \left\| h_{ij}(X_i, l_j) - Z_{ij} \right\|_{\mathbf{P}_{\vartheta_{ij}}}^2 \tag{5.39}$$

subject to the dynamic constraints:

$$\dot{X} = f\left(X(t), U(t), \nu(t), \rho_{int}, t\right) \qquad \text{Dynamics} \tag{5.40}$$

$$Z_{ij} = h_{ij}\left(X(t_i), l_j\right) + \vartheta_{ij} \qquad \text{Measurements} \tag{5.41}$$

and as state and input constraints $X(t) \in \mathbb{X}, U(t) \in \mathbb{U}$, where f and h_{ij} are nonlinear dynamics and measurement functions, respectively.

The control process minimizes another cost function J_{mpc} to compute the optimal control policy $U_{c:N-1}$ and the corresponding future trajectory $X_{c+1:N}$:

$$J_{mpc} = \tilde{J}_{X_N} + \sum_{i=c}^{N-1} L_i\left(X_i, U_i, l_{1:K}, \rho_{int}\right) \tag{5.42}$$

where $\tilde{J}_{X_N}$ is the terminal cost and $L_i\left(X_i, U_i, l_{1:K}, \rho_{int}\right)$ is the stage-wise cost function.

Remark 5.10. *Factor graphs are used as a common framework to represent both estimation and control problems. Traditional techniques to solve nonlinear factor graphs without nonlinear constrained factors typically apply nonlinear optimization methods. The general dynamics and kinematics can be represented as a differential equation over the state Lie-group manifold. In factor graphs, they are equivalent to constrained factors at every time step. Then sequential quadratic programming method can be used to solve this graph optimization [120].*

For SLAM to be justified as a localization technique, it must be demonstrated that the aircraft state errors can be constrained using SLAM alone, without the need for external data such as from GPS. The amount of information contained in the probability distribution of the SLAM estimates when differing control actions are taken by the aircraft is considered. Information refers to the degree to which the probability mass in a distribution is concentrated to a small volume of the distribution state space: a property measured by the **entropy** of the distribution (the compactness of the probability distribution).

Information measures are used as a utility function for determining aircraft control actions that improve localization system performance. There are several practical limitations to using information measures for planning in SLAM:

1. In the case of an aircraft, the available actions to optimize over are large as the aircraft is capable of maneuvering in six degrees of freedom.
2. As the number of features N in the map grows, the computational complexity of evaluating the information measures grows in the order of $O(N^2)$. This growth in computational complexity can be mitigated to some degree by computing the information utility of proposed paths using approximations.

When also considering the requirement for high-rate control of the aircraft, the complexity involved in computing the information utility must be reduced before a real-time planner will be practically feasible. The information gain for every feasible trajectory is not evaluated in [20], instead the path planner evaluates the information gain involved with simple straight and level flight trajectories that involve traveling to and making observations of each feature in the map. Further control of the aircraft then reduces to what trajectories should be flown when the observation of the feature is taking place. This component of the problem is tackled by undertaking an observability analysis of inertial SLAM and evaluating several behavior based decision rules based on this analysis. The decision rule trajectories are designed to perform aircraft motions that excite the direction of locally unobservable modes in the system, thus maximizing the observability of the states over multiple time segments.

5.2.2.2 Inertial SLAM algorithm

The inertial SLAM algorithm is formulated using an extended Kalman filter in which map feature locations and the aircraft's position, velocity and attitude are estimated using relative observations between the aircraft and each feature [12, 20, 37].

The estimated state vector $\hat{X}(k)$ contains the 3D aircraft position $p^n = (x^n, y^n, z^n)$, velocity v^n, Euler angles $\eta_2^n = (\phi^n, \theta^n, \psi^n)^T$ and the 3D feature locations m_i^n in the environment

$$\hat{X}(k) = \begin{pmatrix} p^n(k) \\ v^n(k) \\ \eta_2^n(k) \\ m_1^n(k) \\ m_2^n(k) \\ \vdots \\ m_N^n(k) \end{pmatrix}$$

$i = 1, \ldots, n$; the superscript n indicates the vector is referenced in a local-level navigation frame. The state estimate $\hat{X}(k)$ is predicted forward in time from $\hat{X}(k-1)$ via the process model:

$$\hat{X}(k) = F\left(\hat{X}(k-1), U(k), k\right) + W(k) \tag{5.43}$$

where F is the nonlinear state transition function at time k, $U(k)$ is the system input at time k and $W(k)$ is uncorrelated, zero-mean aircraft process noise errors of covariance $\mathbf{Q}$. The process model is the standard six degrees of freedom inertial navigation equations which predict the position, velocity and attitude of the aircraft. An inertial frame mechanization is implemented:

$$\begin{pmatrix} p^n(k) \\ v^n(k) \\ \eta_2^n(k) \end{pmatrix} = \begin{pmatrix} p^n(k-1) + v^n(k)\Delta t \\ v^n(k-1) + \left(\mathbf{R}_b^n(k-1)f^b(k) + g^n\right)\Delta t \\ \eta_2^n(k-1) + \mathbf{J}_b^n(k-1)\omega^b(k)\Delta t \end{pmatrix} \tag{5.44}$$

where f^b and ω^b are, respectively, the body frame referenced aircraft accelerations and rotation rates which are provided by the inertial sensors on the aircraft and g^n is the acceleration due to gravity as shown in chapter 2. The direction cosine matrix $\mathbf{R}_b^n$ is given by equation (2.4) and the rotation rate transformation J_b^n between the body and navigation frames has been defined in equation (2.25). Feature locations are assumed to be stationary and thus the process models for the position of the ith feature is given as:

$$m_i^n(k) = m_i^n(k-1) \tag{5.45}$$

An on board sensor makes range and bearing observations $Z_i(k)$ to the ith feature. Such observations can be made using either radar or by using a combination of a vision camera and laser range finder.

The SLAM algorithm requires that point feature can be extracted and associated from the observation sensor data. Features in this sense are points in the sensor data that are distinct and recognizable or else points in the sensor data that appear to correlate well with a given feature model or template that is specified off-line. The sensor processing algorithm on board the aircraft may be provided with a visual model and/or a model of the shape of interest that is likely to be in the environment and the feature extraction will attempt to find areas in the sensor data that correlate with the properties of the model. Data association of extracted features from subsequent frames can be performed using a simple matching of the properties of the sensor data corresponding to feature or for more generic features by using innovation gating.

The observation $Z_i(k)$ is related to the estimated states using:

$$Z_i(k) = H_i\left(p^n(k), \eta_2^n(k), m_i^n(k), k\right) + \nu(k) \tag{5.46}$$

Here, $H_i(\ldots, k)$ is a function of the feature location, aircraft position and Euler angles and $\nu(k)$ is uncorrelated, zero-mean observation noise errors of covariance R. The observation model is given by:

$$Z_i(k) = \begin{pmatrix} \rho_i \\ \varphi_i \\ v_i \end{pmatrix} = \begin{pmatrix} \sqrt{(x^s)^2 + (y^s)^2 + (z^s)^2} \\ \arctan\left(\frac{y^s}{x^s}\right) \\ \arctan\left(\frac{-z^s}{\sqrt{(x^s)^2+(y^s)^2}}\right) \end{pmatrix} \tag{5.47}$$

where ρ_i, φ_i and v_i are the observed range, azimuth and elevation angles to the feature and x_s, y_s, z_s the Cartesian coordinates of p^s_{ms}, the relative position of the feature with respect to the sensor, measured in the sensor frame. p^s_{ms} is given by:

$$p^s_{ms} = \mathbf{R}^s_b \mathbf{R}^b_n \left(m^n_i - p^n - \mathbf{R}^n_b p^b_{sb}\right) \tag{5.48}$$

where $\mathbf{R}^s_b$ is the transformation matrix from the body frame to the sensor frame and p_{sb} is the sensor offset from the aircraft center of mass, measured in the body frame, known as the lever arm.

The estimation process is recursive as presented in the following algorithm.

1. **Prediction**: the aircraft position, velocity and attitude are predicted forward in time in between feature observations with data provided by the inertial sensors. The state covariance $\mathbf{P}$ is propagated forward

$$\mathbf{P}(k|k-1) = \nabla F_X(k)\mathbf{P}(k-1|k-1)\nabla F_X^T(k) + \nabla F_w(k)\mathbf{Q}\nabla F_w^T(k) \tag{5.49}$$

   where ∇F_X and ∇F_w are the Jacobians of the state transition function with respect to the state vector $\hat{X}(k)$ and the noise input $W(k)$, respectively.
2. **Feature initialization**: When the first range/bearing observation of a particular feature is obtained, its position is calculated using the initialization function $G_1\left(\hat{X}(k)\right), G_2\left(\hat{X}(k)\right)$ which is given as:

$$G_1 \longrightarrow m^n_i = p^n + \mathbf{R}^n_b p^b_{sb} + \mathbf{R}^n_b \mathbf{R}^b_s p^s_{ns} \tag{5.50}$$

$$G_2 \longrightarrow p^s_{ns} = \begin{pmatrix} \rho_i \cos\varphi_i \cos v_i \\ \rho_i \sin\varphi_i \cos v_i \\ -\rho_i \sin v_i \end{pmatrix} \tag{5.51}$$

   The state vector covariances are then augmented to include the new feature position:

$$\hat{X}_{aug}(k) = \begin{pmatrix} \hat{X}(k) \\ m^n_i(k) \end{pmatrix} \tag{5.52}$$

   and

$$\mathbf{P}_{aug}(k) = \begin{pmatrix} I & \mathbf{0} \\ \nabla G_x & \nabla G_z \end{pmatrix} \begin{pmatrix} \mathbf{P}(k) & 0 \\ 0 & \mathbf{R}(k) \end{pmatrix} \begin{pmatrix} \mathbf{I}_{n\times n} & 0 \\ \nabla G_x & \nabla G_z \end{pmatrix}^T \tag{5.53}$$

where ∇G_x and ∇G_z are the Jacobians of the initialization with respect to the state estimate $\hat{X}(k)$ and the observation $Z_i(k)$, respectively. The position of this feature becomes correlated to both the pose and velocity of the aircraft and the position of other features of the map.

3. **Update**: Once a feature has been initialized into the state vector consisting of the aircraft pose and velocity and the position of this feature and other features in the environment. The state estimate is updated as:

$$\hat{X}(k|k) = \hat{X}(k|k-1) + \mathbf{W}(k)\nu(k) \tag{5.54}$$

where the gain matrix $\mathbf{W}(k)$ and innovation $\nu(k)$ are calculated as:

$$\nu(k) = Z_i(k) - \mathbf{H}_i\left(\hat{X}(k|k-1)\right) \tag{5.55}$$

$$\mathbf{W}(k) = \mathbf{P}(k|k-1)\nabla H_X^T(k)S^{-1}(k) \tag{5.56}$$

$$\mathbf{S}(k) = \nabla H_X(k)\mathbf{P}(k|k-1)\nabla H_X^T(k) + R \tag{5.57}$$

where $\nabla H_x^T(k)$ is the Jacobian of the observation function with respect to the predicted state vector $\hat{X}(k|k-1)$. The state covariance $\mathbf{P}(k|k)$ is updated after the observation using the covariance update

$$\mathbf{P}(k|k) = \mathbf{P}(k|k-1) - \mathbf{W}(k)\mathbf{S}(k)\mathbf{W}^T(k) \tag{5.58}$$

Once a feature leaves the field of view of the sensor, its position remains in the state vector and continues to be updated via its correlations to other visible features in the state vector.

The evolution of the probability distributions in the extended Kalman filter is a function of the state X, due to the linearization of the process and observation models. When the value of the state can be controlled to some degree, the evolution of the extended Kalman filter probability distributions can be controlled in order to minimize entropy. An action can be defined as a set of controlled states and observations to be made n steps into the future:

$$a \in \{X(k), Z(k), X(k+1), Z(k+1), \ldots, X(k+n), Z(k+n)\} \tag{5.59}$$

In the case of the aircraft in a localization and mapping task, an action consists of a set of observations of different features to be made as well as the position, velocity and attitude trajectories of the aircraft over a finite time horizon [54]. The utility for each possible action that can be made is specified by the entropic information gain $I[X,a]$ which is defined as the difference between the entropies of the distributions about the estimated states before and after taking the action

$$I[X,a] = h(X) - h(X|a) = -\frac{1}{2}log\left(\frac{|\mathbf{P}(X|a)|}{|\mathbf{P}(X)|}\right) \tag{5.60}$$

where $h(X)$ and $\mathbf{P}(X)$ are the prior entropy and covariance and $h(X|a)$ and $\mathbf{P}(X|a)$ are the entropy and covariance of the state X subsequent to taking action a (i.e., taking a particular aircraft trajectory and making observations of features along the way). The entropic information gain is a number which is negative for a loss and positive for a gain in information. The advantage of entropy and entropic information gain as utility measures in a control problem is that they represent the whole informativeness of a multivariate distribution in a scalar value, hence simplifying the control problem to:

$$a^* = \text{argmax} \ (I[X,a]) \tag{5.61}$$

where a^* is the best control action. This scalar measure, however, can pose a disadvantage in the sense that the distribution of the information across states may be uneven. For SLAM purposes, however, the scalar measure is sufficient for determining overall information gain.

5.2.2.3 Sense-and-avoid setup

Visual control is an active field of research [28]. The localization techniques for UAV have obtained many promising performances which use GPS, motion capture system (MCS), laser, camera, kinect RGBD sensor. Collision avoidance, also referred to as sense-and-avoid problem, has been identified as one of the most significant challenges facing the integration of aircraft into the airspace. Hence, the term sense relates to the use of sensor information to automatically detect possible aircraft conflicts and the term avoid relates to the automated control actions used to avoid any detected/predicted collisions. The on board single or multiple sensors can provide the sense-and-avoid capability for aircraft. However, the camera sensor is the best on board candidate for UAV which can be used in collision avoidance applications. In particular, UAV requires this collision avoidance ability in the event of failures, for example, GPS has dropped out, INS generated the drift, the software/hardware of UAV has faults suddenly. For applications that need environmental sensing capabilities, a vision system separated from the flight controller can be installed on the aircraft. This configuration can use only cameras because they are lightweight, passive and have low power consumption. A dedicated computer is used to process image information. This results in improved speed and no influence on the real time behavior of the flight control computer. For interaction between image-based results and flight control, data exchange is provided via a local network. A mission planning and automatic control system can be installed on the flight control computer. It calculates trajectories around obstacles, considers changes due to actual image-based map updates and instructs the aircraft to fly these paths.

Since obstacle mapping and other image-based algorithms require flight information, a navigation solution provides the global position and attitude of the aircraft. A limited look-ahead discrete event supervisor controller can also be used for the sense-and-avoid problem. The controlled UAV and the approaching uncontrolled intruding aircraft that must be avoided are treated as a hybrid system. The UAV control decision-making is discrete while the embedded flight model dynamics of the aircraft are continuous. The technique is to compute the controller in an on line fashion on a limited time horizon [35]. It is assumed that there exists a sensor package on the UAV that is capable of identifying the position of the intruding aircraft in 3D. Its heading and speed are initially unknown. In addition to accounting for parametric and state unknowns, it is also necessary to account for uncontrollable actions that influence the aircraft's flight path. For maximum safety, it is assumed that at every moment the intruder is attempting to collide with the UAV then there can be no worse scenario.

The basic method to interpret data from depth image sequences follows the classical approach with occupancy grids. For obstacle mapping, these grids allow sensor fusion and reduce sensor noise. In this case, a world centric 3D grid represents the map. Each cell consists of a value describing the presence of obstacles. Higher values refer to a higher probability of the cell being occupied. The map is created incrementally by starting with an empty grid and writing the actual sensor information with each new depth image. Occupancy grid maps are used to interpret data from depth image sequences. In addition to that, feature maps are built out of these occupancy grids to store global obstacle information in a compressed way and to be an input for applications that use the map.

For this reason, it is a primary requirement to determine the required level of detailing to represent objects in a feature map [10]. For obstacle avoidance applications, the important criteria are [14]

1. Small details are not needed.
2. An identification of planes for texture projection is not needed.
3. Real-time capabilities are more important than centimeter accuracy.

These criteria imply that it is sufficient to mark a box around an object and simply avoid this area. In a real scenario like a city, objects can be modeled as prisms. The world is split into layers in different heights and each layer has its own polygonal $2D$ obstacle map. This is sufficient for a lot of applications like flight trajectory planning in urban scenarios. The mapping process works as presented in the following algorithm 16.

This approach uses different map types. The occupancy grid map for sensor inputs (layer 0) is partitioned into zones and each zone is searched for obstacles separately (layer 1). Next, a map with separate obstacles but without zone separation is generated (layer 2) and finally the prism shapes are extracted out of them (layer 3). Grid resolution and zone sizes are user-defined but may not change over time when processing image series.

Extracting objects from the grid maps, object features are detected by segmenting the global map into occupied and free areas applying a threshold. A single object is a cluster of connected occupied cells of the 3D array. These

Algorithm 16 : Mapping Algorithm

1. Create an occupancy grid around the aircraft's position if not existing in the map.
2. If a new grid is allocated or the grid is extended, check whether previously stored features can be inserted.
3. Insert the actual sensor data information to the grid.
4. Find clusters of occupied grid cells and mark them as single obstacle features.
5. Find out which obstacle features are new and which are updates of objects that have already been identified in the previous loop cycle. Pre-existing objects may also be removed.
6. Calculate the shape of each new or updated feature.
7. To insert the next sensor data, go back to step 1.

objects are recognized with a flood fill algorithm. By saving the minimal and maximal values of the coordinates of cells belonging to the object, the bounding box is calculated and put into the global feature map. For each object, the binary cell shape is stored in addition to the bounding box so that this shape can be re-inserted. Compression with an octree structure is applicable in this approach.

5.2.2.4 Monocular visual–inertial SLAM using fuzzy logic controllers

The uncertainty, inaccuracy, approximation and incompleteness problems widely exist in real controlling technique. Hence, the model-free control approach often has the good robustness and adaptability in the highly nonlinear, dynamic, complex and time-varying UAV dynamics. The fuzzy logic controller mainly consists of three different types of parameters:

1. **Scaling Factors**: (SF) are defined as the gains for inputs and outputs. Their adjustment causes macroscopic effects to the behavior of the fuzzy controller, i.e., affecting the whole rule table.
2. **Membership Function**: (MF) its modification leads to medium size changes, i.e changing one row/column of the rule tables.
3. **Rule weight**: (RW) also known as the certainty grade of each rule, its regulation brings macroscopic modifications for the fuzzy logic controller, i.e., modifying one unit of the rule tables.

The **cross-entropy** (CE) method involves an iterative procedure where each iteration can be broken down in two phases:

1. In the first stage, a random data sample (e.g., scaling factors or membership function of fuzzy controller) is generated according to a specified mechanism,

2. then the parameters of the random mechanism are updated based on the data in order to produce a better sample in the next iteration.

Fusion of vision and inertial measurement unit can be classified into three different categories:

1. **Correction**: where it uses the results from one kind of sensor to correct or verify the data from another sensor.
2. **Colligation**: where one uses some variables from the inertial data together with variables from the visual data.
3. **Fusion**: to efficiently combine inertial and visual data to improve pose estimation.

Fuzzy logic controller has a good robustness and adaptability to control the UAV orientation. This fuzzy logic controller is PID type with

1. Three inputs: the angle error estimation between the angle reference and the heading of the UAV, the derivative and the integral values of this estimated error.
2. One output: the command in degrees/seconds to change the heading.

The product t-norm is used for rules conjunction and the defuzzification method is:

$$Y = \frac{\sum_{i=1}^{M} \overline{Y}^l \sum_{i=1}^{N} \left(\mu_{X_i^l}(X_i)\right)}{\sum_{i=1}^{M} \sum_{i=1}^{N} \left(\mu_{X_i^l}(X_i)\right)} \tag{5.62}$$

where N and M represent the number of input variables and total number of rules, respectively, $\mu_{X_i^l}(X_i)$ denotes the membership function of the lth rule for the ith input variable and $\overline{Y}^l$ represents the output of the lth rule [24].

5.2.3 GEOLOCATION

5.2.3.1 Cooperative geolocation with articulating cameras

Geolocation is the tracking of a stationary or moving point of interest using visual cameras for payloads. It is the process of using sensor data to develop statistical estimates of a point of interest on the ground.The geolocation system for an autonomous aircraft requires the complex integration of several hardware components (camera, UAV, GPS, attitude sensors) and software components (camera image processing, inner-loop and path planning control, and estimation software) to develop accurate estimation of the object being tracked. The sensor biases and the unknown point of interest state are estimated in a decentralized manner, while using the solution from the on board navigation system to save significant computation. The joint estimation problem is solved for multiple UAV cooperating in a decentralized fashion such that the UAV share information on the point of interest state and model only their local biases. This decentralized formulation saves computation as well as communication, while giving geolocation accuracy that is comparable with the centralized case. Further, this decentralized approach allows for effective cooperation not only among autonomous aircraft with potentially different biases, but among different sensors altogether.

For the application of a vision sensor on a UAV based on its position and orientation, the camera points through a gimbal's payload mounted inside the UAV at the point of interest on the ground, while the aircraft is moving and the point of interest is potentially moving; the camera gimbals must adjust their angles to point at the point of interest. The camera is directed at the point of interest, such that it always remains within the field of view of the camera [127]. The objective of geolocation is then to estimate the position (2D or 3D) of the point of interest from the aircraft, gimbals and camera measurements. Uncertainties in the aircraft position and orientation, gimbal angles, camera specifications and measurements, and disturbances complicate this problem. The most accurate estimator couples the UAV navigation (NAV), attitude (ATT), camera gimbals (GIM) and point of interest states in a single estimator, which requires full UAV and gimbals models and a model for the point of interest. However, this estimator requires very high computations, memory and communications in the case of multiple UAV. Fortunately, most autonomous aircraft use a navigation system with estimators that provide estimates and covariances for both the ATT and NAV states. In addition, the GIM states can be directly measured. Therefore a geolocation estimator can be developed that estimates the point of interest state only, saving computation and memory [129].

An **extended information filter** (EIF) is developed which uses the navigation system to solve the cooperative geolocation problem and makes explicit the assumptions about the estimates of the UAV state. The state to be estimated, $X_{k,POI}$, is the state of the point of interest, with discrete-time dynamics governed by:

$$X_{k+1} = f(X_k, W_k) = f_{POI}\left(X_{k,POI}, \vartheta_{k,POI}\right) \tag{5.63}$$

where the disturbance $\vartheta_{k,POI}$ is zero mean white Gaussian noise with covariance $\mathbf{Q}_{k,POI}$ and the subscript k denotes time step t_k. Assume there are N UAV with states ψ^j_{k+1} for $j = 1, \dots, N$ composed of UAV position $\psi^j_{k+1,NAV}$, UAV attitude $\psi^j_{k+1,ATT}$ for $j = 1, \dots, N$ and camera attitude $\psi^j_{k+1,GIM}$ for $j = 1, \dots, N$ written in vector form as:

$$\psi^j_{k+1} = \begin{pmatrix} \psi^j_{k+1,NAV} \\ \psi^j_{k+1,ATT} \\ \psi^j_{k+1,GIM} \end{pmatrix} \tag{5.64}$$

The unmanned aircraft are further assumed to have on board navigation system and measurements of the camera's gimbal angles, which give an estimate of the UAV state $\hat{\psi}^j_{k+1}$. A simple model can be used

$$\psi^j_{k+1} = \hat{\psi}^j_{k+1} + \eta^j_{k+1} \tag{5.65}$$

where the UAV state estimate error η^j_{k+1} is zero-mean, Gaussian and white with covariance ${}^{\eta}\mathbf{R}^j_{k+1}$. This model is known to be incorrect because the statistics are not white, but correlated through the navigation filter. Many times, the errors due to autocorrelation are small. Biases may also exist which can have a significant effect on accuracy. Measurements of the point of interest are made on each autonomous aircraft using:

$$Z^j_{k+1} = h^j\left(X_{k+1}, \eta^j_{k+1}, \nu^j_{k+1}\right) = h_{SCR}\left(X_{k+1,POI}, \hat{\psi}^j_{k+1} + \eta^j_{k+1}, \nu^j_{k+1,SCR}\right) \tag{5.66}$$

where the sensor noise ν^j_{k+1} is zero-mean, white Gaussian noise with covariance ${}^{\nu}\mathbf{R}^j_{k+1}$. The process noise, sensor noises and navigation system noises $(\vartheta_k, \nu^j_{k+1}, \eta^j_{k+1})$ are assumed to be uncorrelated with each other. The measurement function in (5.66) is a complicated nonlinear function of the point of interest state and the UAV state. The cooperative geolocation problem can now be solved with an extended information filter as follows: The information matrix $\mathbf{Y}_k$ and information state Y_k are defined based on the state estimate error covariance $\mathbf{P}_k$ and state estimate $\hat{X}_k$ as

$$\mathbf{Y}_k = \mathbf{P}_k^{-1} \tag{5.67}$$

$$Y_k = \mathbf{Y}_k \hat{X}_k \tag{5.68}$$

The extended information filter algorithm can be written for N autonomous aircraft as in [129].

5.2.3.2 Geolocation with bias estimation

The assumption that the errors in the estimate of the UAV state are zero-mean, white and Gaussian is not accurate in the practical case for two reasons: correlated outputs of the navigation filter and biases in the outputs. Sensor biases are a significant source of error for geolocation.

Series of sensed points of interest (SPOI) are the line of sight intersection of the camera with the ground as computed based on the estimates of the autonomous aircraft state (NAV, ATT, GIM). Series of sensed points of interest moves in a roughly circular pattern around the true point of interest location. The period of this oscillation corresponds directly to the UAV orbit about the point of interest and is due to non-zero mean errors (biases) in the autonomous aircraft state estimate. The sensor biases are explicitly modeled and both the sensor biases and the unknown point of interest location are jointly estimated. The biases b_k^j are now modeled explicitly as a part of the UAV navigation system output and camera gimbals measurement as:

$$\psi_k^j = \hat{\psi}_k^j + b_k^j + \eta_k^j \tag{5.69}$$

where the model of the bias state b_k^j used here is

$$b_{k+1}^j = b_k^j + \vartheta_{k,b}^j \tag{5.70}$$

Autocorrelation of the UAV state estimate error could be taken into account in the same way as the biases by adding autocorrelation states $\mu_k^{j,m}$ in (5.69) as

$$\psi_k^j = \hat{\psi}_k^j + b_k^j + \eta_k^j + \mu_k^{j,1} + \cdots + \mu_k^{j,n_\mu} \tag{5.71}$$

where each of the autocorrelation terms correspond to a different frequency of autocorrelation. The autocorrelation terms can be modeled as

$$\mu_{k+1}^{j,n_\mu} = a^{j,m}\mu_k^{j,n_\mu} + \vartheta_{k,\mu_{j,m}}^j \tag{5.72}$$

where the parameter $a^{j,m}$ is chosen to capture the appropriate autocorrelation frequency. The navigation system and camera gimbals measurements are used directly, while only the biases b^j are estimated recursively, with the point of interest state $X_{k,POI}$. This saves significant computation while effectively improving the estimate of the UAV state and thus improving geolocation [23].

5.3 INTEGRATED SYSTEM HEALTH MONITORING

There are several flight critical components and systems for the UAV: actuators, control surfaces, engines, sensors, flight computers and communication devices [98]. Faults and failures in such components may compromise UAV flights. The consequence of a control surface fault is reduced performance and possibly instability depending on the effectiveness of the health management system. Sensors in the feedback loop are subject to both hard-over failures which are catastrophic but relatively easy to detect. Hard-over failures are typically detected and identified by a sensor with built-in testing. Soft failures include a small bias in measurements, slow drifting of measurements, a combination of the two, loss of accuracy and freezing of the sensor to certain values.

An environmental hazard can cause damage to a UAV. The platform itself may be damaged as well as the flight critical components and systems. In the case of a fixed-wing UAV, control surface damage can change the dynamics, translating in a modified control input-to-state matrix and as an additional nonlinear term representing asymmetrical post-damage dynamics. A degradation in the performance of UAV sensors and actuators may be the result of poor weather or other adverse environmental effects. Forces of nature have a greater impact on small vehicles than on large aircraft. Constraints on available on board power, as well as allowed payload mass and volume, indirectly limit the complexity of control laws that can be embedded in the small UAV. The aerodynamic scale and available control authority make it difficult for small UAV to counteract the effects of wind. For UAV equipped with GPS receivers, examples of faults include jamming of GPS data and the multi-path effect of reflections causing delays. These in turn result in inaccurate positions. All these factors necessarily restrict the choice and numbers of actuators and sensors.

Definition 5.1. *A **fault** means an unpermitted deviation of characteristic properties or parameters of the system from the standard condition, and a **failure** means a permanent interruption of partial systems.*

Remark 5.11. *Manned and unmanned aircraft share many common degraders and can use similar analytical techniques to detect and track faults in components and subsystems. The physical implementation of the health monitoring system may differ in unmanned and manned systems, however, due to differences in platform size, weight and electrical power. A key requirement is the integration of the health management system with an autonomous aircraft controller [17]. The health monitoring system and the aircraft control computer must be able to exchange information to realize higher levels of autonomy. The implementation of integrated system health management typically involves the collection, processing and monitoring of data from sensors and signals to determine the state of health of the platform. The weight, size and power requirements for health monitoring are important constraints in unmanned systems because they represent a high percentage of the operating margins for those parameters [39].*

Generic, object-oriented fault models, built according to causal-directed graph theory, can be integrated into an overall software architecture dedicated to monitoring and predicting the health of mission critical systems [134]. Processing over the generic faults is triggered by event detection logic that is defined according to the specific functional requirements of the system and its components. Once triggered, the fault models provide an automated way for performing both upstream **root cause analysis** (RCA) and for predicting downstream effects or impact analysis. The methodology has been applied to integrated system health management (ISHM) implementations [43].

Remark 5.12. *The real time **flight envelope** monitoring system uses a combination of three separate detection algorithms to provide a warning at present number of degrees prior to stall. real time aircraft control surface hinge moment information could be used to provide a robust and reliable prediction of aircraft performance and control authority degradation. For a given airfoil section with a control surface, aileron, rudder or elevator, the control surface hinge moment is sensitive to the aerodynamic characteristics of the section. As a result, changes in the aerodynamics in the section due to angle of attack or environmental effects such as icing, heavy rain, surface contaminants or* *bird strike will affect the control surface hinge moment. These changes include both the magnitude of the hinge moment and its sign in a time-averaged sense and the variation of the hinge moment with time. The system attempts to take the real time hinge moment information from the aircraft control surfaces and develop a system to predict aircraft envelope boundaries across a range of conditions, alerting the flight management system to reduction in aircraft controllability and flight boundaries.*

By applying genetic optimization and goal seeking algorithms on the aircraft equipment side, a war game can be conducted between a system and its model as in [15]. The end result is a collection of scenarios that reveals any undesirable behaviors of the system under test. It leverages advances in state and model-based engineering, which are essential in defining the behavior of an autonomous system. It can also use goal networks to describe test scenarios [89].

A widely used approach is the **fault detection and isolation** (FDI) technique which is based on **redundancy management system**.

Definition 5.2. ***Fault detection*** *refers to the decision of whether or not a fault occurs, and* ***fault isolation*** *is defined as the process of finding and excluding the faulty component.*

Generally, the fault detection and isolation technique is categorized into **hardware redundancy** and **analytic redundancy** management.

Remark 5.13. *The triple or quadruple hardware redundancies generally used for large conventional aircraft are not suitable for small unmanned aircraft because small inexpensive systems are limited by cost and payload space.*

The problem of **model-based fault detection and isolation** (FDI) has been widely studied in recent years. Fundamentally, the objective of all model-based fault detection and isolation approaches is to monitor residuals constructed by comparing the measured system output to a predicted output synthesized using a model of the expected healthy aircraft behavior. If the model is accurate, then the measurement predictions should all be close to the corresponding actual sensor signals. Faults in the actuators, sensors or the aircraft itself become manifested through discrepancies that become abnormally large in a statistically significant sense. Known nominal aircraft models can form the basis for **analytic redundancy-based designs** that simultaneously diagnose both sensor and actuator faults.

In [130], a recursive strategy for online detection of actuator faults on a unmanned aircraft subject to accidental actuator faults is presented. It offers necessary flight information for the design of fault tolerant mechanisms to compensate for the resultant side effect when faults occur. The proposed fault detection strategy consists of a bank of unscented Kalman filters with each one detecting a specific type of actuator faults and estimating corresponding velocity and attitude information.

An undetected fault in a system can have catastrophic effects. A combination of a fault detection scheme and a control system is also known as a **fault tolerant flight control system** (FTFCS). They are used to detect, identify and accommodate for any type of failure that may occur on board the aircraft [106].

One approach to fault detection and identification in actuators and sensors is based on **multiple model methods** [59]. These methods have been extended to detect faults, identify the fault pattern and estimate the fault values. Such methods typically use Kalman filter or EKF or UKF filters in conjunction with multiple hypothesis testing and have been reported to be effective for bias type faults such as aircraft control surfaces getting stuck at unknown values or sensors (e.g., rate gyros) that develop unknown constant varying biases. A basic requirement for these methods is that the inputs and sensors should be identifiable.

Multiple sensor fault detection, isolation and accommodation (SFDIA) schemes are particularly important if the measurements of a failed error sensor are used in a control system. Since the aircraft control laws require sensor feedback to set the current dynamic state of the aircraft, even slight sensor inaccuracies, if left undetected and accommodated for, can lead to closed-loop instability. Moreover, nowadays, condition monitoring is heading towards a more informative process, where practitioners need a health management system where the fault can also be accommodated [131].

The majority of the model based multiple sensor fault detection, isolation and accommodation schemes rely on **linear time invariant** (LTI) models. Unfortunately, in nonlinear time-varying systems, such as aircraft, LTI models can sometimes fail to give satisfactory results [106]. In general most FDI methods can be divided into two groups, both initially designed to be robust when applied to the real aircraft:

1. One that makes use of an aircraft model,
2. One that does not make use of an aircraft model.

There is a set of performance criteria

1. Fault detection time,
2. False alarm rate,
3. Number of undetected faults,
4. Ability to isolate the fault.

Definition 5.3. ***Reliability*** *is the probability that an item will perform its intended function for a specified time interval under stated conditions.* ***Availability*** *is the probability that a given system is performing successfully at time t, independently of its previous states.* ***Average availability*** *measures the up-time as a proportion of a given working cycle.*

From the various techniques to quantify reliability, **fault tree analysis** (FTA) is one of the most commonly used. Other techniques as **failure modes and effects analysis** (FMEA), **Markov analysis, Monte Carlo** approaches can be combined with fault tree analysis to empower the outcome. Reliability is a complex theory. Different concepts exist applicable to different systems, depending on its features. Failure rate λ is widely used as a measure of reliability. It gives the number of failures per unit time from a number of components exposed to failure. It is frequently measured in failures per 10^9 hours and these units are denoted as FIT.

Availability reaches a steady state condition when time tends to infinity which is affected only by failure and repair rates. Moreover, one must suppose that the probability distribution function (PDF) of failure is exponential.

$$A(\infty) = \frac{\mu}{\mu + \lambda_f} \tag{5.73}$$

μ is the repair rate and λ_f is the failure rate.

5.3.1 DIAGNOSTIC TOOLS AND APPROACHES

Model-based diagnostic and prognostic techniques depend upon models of the aircraft. In the presence of uncertainties, modeling errors can decrease system sensitivity to faults, increase the rate of occurrence of false alarms and reduce the accuracy of failure prognoses. Robust designs have been presented that explicitly address the presence of aircraft model uncertainty. In addition, **adaptive** designs have been presented that assume known nominal aircraft dynamics and also consider aircraft with unknown system parameters [88]. Diagnostic applications are built around three main steps:

1. Observation,
2. Comparison,
3. Diagnosis.

A **sensor failure model** may be not specifically considered in the design; rather it is stated that any sensor anomaly will cause the associated output estimation error not to converge, and thus a failure to be detected and isolated. Moreover, the design can support diagnosis of arbitrary combinations of sensor faults rather than requiring an output estimator for every sensor to be monitored.

Model-based approaches have proven fruitful in the design and implementation of intelligent systems that provide automated diagnostic functions [95]. A wide variety of models are used in these approaches to represent the particular domain knowledge, including **analytic state-based models, input-output transfer function models, fault propagation models** and **qualitative and quantitative physics-based models** [61].

If the modeling begins in the early stages of UAV development, engineering models such as **fault propagation models** can be used for testability analysis to aid definition and evaluation of instrumentation suites for observation of system behavior. **Analytical models** can be used in the design of monitoring algorithms that process observations to provide information for the second step in the process. The expected behavior of the smart autonomous aircraft is compared to the actual measured behavior. In the final diagnostic, step reasoning about the results of the comparison can be performed in a variety of ways, such as dependency matrices, graph propagation, constraint propagation and state estimation [100, 116, 137].

The three primary components of the diagnosis engine for run time analysis are described next.

1. **Hybrid observer** employs a filter combined with a hybrid automaton scheme for tracking nominal system behavior. The diagnosis engine architecture includes a hybrid observer that tracks continuous system behavior and mode changes while taking into account measurement noise and modeling errors. When observer output shows statistically significant deviations from behavior predicted by the model, the fault detector triggers the fault isolation scheme.
2. **Fault detection and symbol generation**: the fault detector continually monitors the measurement residual $R(k) = Y(k) - \hat{Y}(k)$ where Y is the measured value and $\hat{Y}$ is the expected system output, determined by the hybrid observer. Since the system measurements are typically noisy and the system model imperfectly known, so that the prediction system is not perfect, the fault detector employs a statistical testing scheme based on the *Z-test* for **robust fault detection**. A fault signature is defined in terms of magnitude and higher order derivative changes in a signal.
3. The **fault isolation** engine uses a **temporal causal graph (TCG)** as the diagnosis model. The temporal causal graph captures the dynamics of the cause-effect relationships between system parameters and observed measurements. All parameter changes that can explain the initial measurement deviations are implicated as probable fault candidates. Qualitative fault signatures generated using the temporal causal graph are used to track further measurement deviations. An inconsistency between the fault signature and the observed deviation results in the fault candidate being dropped. As more measurements deviate the set of fault candidates become smaller.

For hybrid diagnosis, to accommodate the mode changes that may occur during the diagnostic analysis, the fault isolation procedure is extended by two steps:

1. **qualitative roll-back**,
2. **qualitative roll-forward**.

Fault identification uses a parameter scheme: a mixed simulation-and-search optimization scheme is applied to estimate parameter deviations in the system model. When more than one hypothesis remains in the candidate set, multiple optimization are run simultaneously and each one estimates one scalar degradation parameter value. The parameter value that produces the least square error is established as the true fault candidate.

Control surface fault diagnosis is essential for time detection of maneuvering and stability risks for an unmanned aircraft. An approach where a basic generic model is applied and necessary parameters in residual generators are identified on the flight is presented in [19]. Initial estimates of parameters are known from off-line analysis of previous flights. Self-tuning residual generators are combined with change detection to obtain timely fault diagnosis. The parameter convergence is investigated as well as detection properties for the suggested combination of identification and change detection techniques. Self-tuning can be employed to obtain adaptation to flight parameters while retaining the capability to detect faults. Change detection is applied to residuals alone and to residuals combined with parameter estimate.

Definition 5.4. *A* ***runaway*** *is an unwanted control surface deflection that can persist until the moving surface stops. A runaway can have various speeds and is mainly due to electronic component failure, mechanical breakage or a flight control computer malfunction.*

Depending on the runaway dynamic, this fault could either result in an undesired pitch maneuver or could require a local structural load augmentation in the aircraft.

Definition 5.5. *A* ***jamming*** *is a generic system failure case that generates a mechanical control surface stuck at its current position. A negative effect of surface jamming is the resulting increased drag, which leads to increased fuel consumption.*

Remark 5.14. *Employing dissimilar hardware and redundancy-based techniques is the standard industrial practice and provides a high level of robustness and good performance compliant with the certification requirements for manned aircraft. Fault detection is mainly performed by cross-checks or consistency checks [39]. The current industrial practice for control surface jamming and runaway detection in manned aircraft consists mainly of consistency checks between two redundant signals. If the difference between the signals computed in the two flight control computer channels is greater than a given threshold during a given time, the detection is confirmed. The whole procedure can be divided into two steps: residual generation and residual evaluation.*

Alarms are triggered when the signal resulting from the comparison exceeds a given threshold during a given time window or confirmation time. The basic idea of another fault detection and diagnosis is to integrate a dedicated Kalman filter between residual generation and decision-making blocs.

5.3.1.1 Sensor selection and optimization

Traditionally, in flight systems, sensors are primarily selected through an ad hoc heuristic process, where various domain groups are polled as to what sensors they require in the system. Although safety and reliability are just as important as system performance, the sensors are primarily selected based on **control requirements** and performance assessment, rather than on health monitoring and management. To be incorporated into the design process, accepted methodologies and procedures must be established that provide justification of the selected sensors within the constraints of the aircraft requirements. To precisely quantify the benefits of sensors in flight systems, heuristic sensor selection approaches must give way to systematic techniques that are based on accepted selection criteria and performance metrics that measure their value to the systems [75].

The goal of the **sensor selection** process is to provide a suite of sensors that fulfill specified performance requirements within a set of system constraints. These performance requirements are defined as the **figures of merit** (FOM) of the system:

1. **Observability** addresses how well the sensor suite will provide information about the given system process, which parameters that are directly observed and which parameters can be inferred.
2. **Sensor reliability/sensor fault robustness** addresses sensor reliability and how sensor availability impacts the overall sensor suite performance.
3. **Fault detectability/fault discriminability** specifically addresses whether the sensor suite can detect and discriminate system failures.
4. **Cost** can include development, purchase and maintenance costs for the sensors as well as resource and communication costs.

Sensor selection is based on a set of criteria called the objective function which is an algorithmic representation of established **figures of merit** and **system constraints**. Accurate in-flight detection of sensor uncertainties and failures is crucial in safety- and mission-critical systems of an autonomous aircraft. For example, if a sensor is indicating that a parameter is out of the normal operating range, there are two real possibilities and the proper reaction of the control system is different in both cases:

1. The sensor has failed in such a way that it is providing erroneous measurements and in fact the monitored process is functioning normally. Given a model-based diagnostic system integrated with the control system, equipment operation can continue as planned.
2. The measured parameter is truly out of the normal operating range. The control system needs to take action, or perhaps uses an alternative or virtual sensor.

Accelerometers are commonly used for health monitoring of aircraft.

They are used to sense vibration signatures and shock within turbine engines in order to identify problems during operation [68]. They are used to sense vibration signatures and shock within turbine engines in order to identify problems during operation. By employing physics understanding of these systems, it is possible to identify potential failures at a much earlier time thereby providing an opportunity to take corrective action. For the specific case of an aircraft health monitoring system designed to conduct corrective action based on feedback from accelerometers (e.g., an automated shut down of an engine), the reliability of the sensor must be ensured.

The **self-diagnosis accelerometer** (SDA) system actively interrogates piezoelectric accelerometers to verify the proper operation of the accelerometers. By interrogating the sensor, it is possible to identify the following failure modes: a physically damaged sensor, electrical disconnection, as well as sensor detachment/loosening from the structure. The interrogation of the sensor is accomplished by deriving the piezoelectric crystal with a frequency swept sinusoidal voltage and monitoring the response from the crystal so that the frequency of a particular resonance is determined. The sensor's resonant frequency which depends on the electromechanical properties of the accelerometer is far above the specified operational frequency range. Furthermore, the resonant frequency of the sensor is sensitive to the mounting torque of the sensor. Other failures such as physical damage to the sensor crystal and electrical disconnection are also both identifiable when interrogating the sensor with this type of input signal [75].

Information-driven sensor planning utilizes information theoretic functions to assess the value of a set of sensor measurements and to decide the best sensor mode or measurement sequence [72]. Because target classification can be reduced to the problem of estimating one or more random variables from partial or imperfect measurements, the value of future measurements may be represented by their expected information value. Information theoretic functions, such as information entropy, have been proposed to assess the information value of sensor measurements.

5.3.1.2 Hazard alerting

Smart autonomous aircraft operations are conducted to perform some mission that could be disrupted by making unplanned maneuvers. Therefore, it is important that sense-and-avoid systems distinguish threatening from non threatening traffic or other hazards, and call for a maneuver only when required for safety [8]. The threat declaration function then must balance two primary requirements:

1. To determine that a hazard poses a threat such that some maneuver is required,
2. To minimize these declarations for targets that are actually non-threatening.

Hazard alerting systems alert the autonomous aircraft to potential future undesirable events so that action may be taken to mitigate risks. One way to develop a hazard alerting system based on probabilistic models is by using a threshold based approach, where the probability of the undesirable event without mitigation is compared against a threshold. Another way to develop such a system is to model the system as a **Markov decision process** (MDP) and solve for the hazard alerting strategy that maximizes expected utility [27].

Threshold based approaches to the hazard-alerting system developed involve using explicit threshold criteria to decide when to alert and what alert to issue. They typically require computing the probability of the undesirable event assuming no alert is ever issued and the event probability when following the guidance provided by each alert indefinitely [77]. The probabilities can be computed using analytic, Monte Carlo, numerical approximation or dynamic programming methods. Threshold based systems can be divided into two categories:

1. **Constant threshold** system uses thresholds that remain fixed from scenario to scenario.
2. **Dynamic threshold** system uses thresholds that vary as the hazard situation evolves.

5.3.1.3 Fault tree analysis

Fault tree analysis provides a hierarchical and visual representation of combinations of possible events affecting a system in a set of odd behaviors previously defined. The basis of fault tree analysis lies in **Boolean algebra** for simple analysis and it is extended to probability and statistics analysis allowing more complex models. Fault tree analysis is an up-bottom technique oriented to represent failure modes of the system. The basic rules to build a fault tree are:

1. Describe the top event accurately and the boundaries of the system behavior.
2. Divide the problem into scenarios as technical failures, human errors or software and develop them as deep as necessary.
3. Use consistent nomenclature for all devices and events; otherwise the analysis can be incorrect.

Analysis of common cause failure in redundant control systems can be done using fault trees. **Redundancy** in control systems (sensors, control units, input/output cards, communication systems and actuators) is critical. The actual implementation of redundancy can follow different strategies. Control system components have different **failure probability** and different sensitivity to external common disturbances. Redundancy implies an extra cost and therefore an accurate reliability analysis to discover critical components is needed in order to assist the design of such facilities. Fault tree modeling can be used to analyze control reliability. A modular decomposition of the system at different levels and the identification of interactions among them is used to build the tree. The failure rate associated to each module or component is used in the reliability analysis of the whole system. Reliability is hardly affected by **common cause failures** (CCF), i.e., events that affect more than one component at the same time. Common cause failures are in general unpredictable. They might be the result of an event that affects a common area where several components are located. The analysis of common cause failures is an important task in order to ensure the continuous operation of critical processes [43, 45].

Fault tree analysis construction ends with a system failure probability model where the combination of elements that lead the system to failure is explicitly represented. A basic model consists in connecting logical gates (AND and OR are the basic ones) that describe elements and dependencies among subsystems. There are different ways to assess the failure probability of the basic events

$$\Pr_{and} = \Pr\left(a_1 \cap .. \cap a_n\right) = \prod_{i=1}^{n} \Pr(a_i) \tag{5.74}$$

$$\Pr_{or} = \Pr\left(a_1 \cup .. \cup a_n\right) = \sum_{i=1}^{n} \Pr(a_i) - \prod_{i=1}^{n} \Pr(a_i) \tag{5.75}$$

$\Pr(a_i)$ is the probability of element a_i.

In the construction of the fault tree, the top node represents the failure mode and collects the influence of all the components and subsystems through branches composed by logical gates. When branches cannot be further developed those elements placed at the bottom are called basic events. They must be sufficiently detailed to relate them to those devices from which failure and repair rates are available or can be calculated. Previous formulas are used to derive the final availability of the whole system and analyze the influence of each component and event. Those elements affected by the same adverse influences (stressors), typically those that are placed together or submitted to the same ambient conditions, are candidates to be affected by common cause failures.

Remark 5.15. *In redundant studies, the consideration of common cause failure (CCF) is very important because the total reliability without considering them can be too optimistic.*

A phased mission describes a situation when the requirements for success change throughout the time of operation; therefore the causes of failure during the mission also change. The consecutive and distinct periods in the mission performing different tasks are known as phases, performed in sequence. In order for the mission to be successful, each of the phases must be completed successfully; therefore, the mission fails if at least one phase fails [100]. A significant factor in the decision-making process of autonomous systems is the mission failure probability during the future phases. There are two types of prediction required: before the start of the mission and while the mission is in progress.

A prognostic tool is expected to provide accurate information in a short time so that the decision-making process would be well-informed and appropriate decisions would be made before a catastrophic event. Fault tree analysis is suitable when describing a non-repairable system where component failures are treated independently. Binary decision diagrams can be developed as an alternative logic function representation. Phased missions are used to define the behavior of a system during different parts of the mission and to perform the analysis. The following characteristics determine a phased mission:

1. Every mission consists of many consecutive phases performed in sequence.
2. Since a different task is to be performed in each phase there are different failure criteria in each phase.
3. For a mission to be successful all phases must be completed successfully.

The phased mission is represented by a number of fault trees, each of them expressing the conditions leading to a failure of a phase. Let F_i express the logical expression for the failure conditions being met in phase i and Ph_i express the logical expression for mission failure in phase i; then

$$\begin{aligned} Ph_1 &= \bar{F}_1 \\ &\vdots \\ Ph_i &= \overline{F}_1\overline{F}_2\ldots\overline{F}_{i-1}\overline{F}_i \end{aligned} \tag{5.76}$$

Since the conditional phase failures are mutually exclusive events the total mission failure probability is simply the sum of mission phase failure probabilities. Fault tree analysis can be used to quantify the probability of mission failure during mission phase i, Q_i:

$$Q_i = \Pr(Ph_i) \tag{5.77}$$

Since the mission fails if at least one of its phases fails, the logical expression for mission failure is given by:

$$Ph_{mission} = Ph_1 + Ph_2 + \cdots + Ph_i \tag{5.78}$$

The total mission failure probability is obtained by adding these mission phase failure probabilities:

$$Q_{mission} = Q_1 + Q_2 + \cdots + Q_n \tag{5.79}$$

where n is the total number of phases in the mission. Once the mission is underway, Q_i is updated taking into account the success of the previous phases.

The updated phase failure probability, $Q_{j|\bar{k}}$, is the probability of failure in phase j given that phase k was successfully completed:

$$Q_{j|\bar{k}} = \frac{Q_j}{1 - \sum_{i=1}^{k} Q_i} \tag{5.80}$$

Then the overall mission failure probabilities calculated by adding the phase failure probabilities of the future phases:

$$Q_{mission|\bar{k}} = \sum_{j=k+1}^{n} Q_{j|\bar{k}} \tag{5.81}$$

If the probability of mission failure becomes too high then the future mission is considered to be too risky and an alternative mission configuration is used.

The probability of collision between two objects provides a quantitative measure of the likelihood that the objects will collide with each other [31].

5.3.1.4 Failure modes and effects analysis

Fault/failure detection, isolation and recovery (FDIR) methods are usually designed for flight-critical components such as actuators and sensors, using detailed models. At some point during normal operation, a fault may begin to develop in a component. As the fault progresses, it may begin to impact the performance and operation of the platform [99]. If the fault is not detected and corrected, it will eventually lead to a functional failure and possible collapse of the platform. In autonomous aircraft, the diagnose-repair-replace step may involve execution of diagnostic procedures, putting the platform in some sort of sleep or safe mode, then switching to a redundant backup subsystem if the platform was designed and built with the necessary redundancy. Finally, aspects of the mission may be re-planned or re-scheduled to accommodate the break in execution or to accommodate reduced functionality resulting from the fault or failure. One objective of integrated system health monitoring is to detect faults after initiation, but before they impact operation. Once a fault is detected, the health of the platform can be assessed. System health monitoring enables **automated diagnosis** and response to a developing fault.

In the case of **aircraft maintenance**, there exist three criteria or inputs that affect the failure time and cause the model to lose the other inputs. The related forecast equation involves four main estimation functions, exponential, logarithmic, linear and power functions

$$\begin{aligned}\hat{Y} = &\left(\alpha_1 \exp\left(-\beta_1 X_1\right) + \gamma_1 \ln\left(-\beta_1 X_1\right) + \delta_1 X_1 + \epsilon_1 X_1^{\zeta_1}\right) \\ &+ \left(\alpha_2 \exp\left(-\beta_2 X_2\right) + \gamma_2 \ln\left(-\beta_2 X_2\right) + \delta_2 X_2 + \epsilon_2 X_2^{\zeta_2}\right) \\ &+ \left(\alpha_3 \exp\left(-\beta_3 X_3\right) + \gamma_3 \ln\left(-\beta_3 X_3\right) + \delta_3 X_3 + \epsilon_3 X_3^{\zeta_3}\right)\end{aligned} \tag{5.82}$$

where X_1 is the total lifetime of takeoffs, X_2 is the previous year failure rate, X_3 is the failure time and $\hat{Y}$ is the estimated failure time. The forecast equation has been built as the linear combination of four different estimation functions for each input. Genetic algorithms can be used for the parameter estimate of a forecast equation [1, 42]. A genetic algorithm has been built that aims to forecast equation parameters in a way to minimize **sum square errors** (SSE) which is represented as

$$SSE = \sum_{i=1}^{n} (Y_i - \hat{Y}_i)^2 \tag{5.83}$$

where Y is the real failure time, n is the number of data and $\hat{Y}$ is the estimated failure time.

5.3.2 RISK-BASED SENSOR MANAGEMENT

A risk-based optimal sensor management scheme has to be developed for integrated detection and estimation under limited sensor resources in the presence of uncertainties. The objective is to effectively detect and satisfactorily estimate every unknown state of interest within a mission domain while minimizing the risk associated with the sensing decisions.

Definition 5.6. ***Integrity risk*** *is defined as the probability of the state estimate error exceeding predefined bounds of acceptability.*

An expression can be derived that relates the measurement noise and disturbance input autocorrelation functions to the state estimate error vector [65].

Detection and estimation are integrated into a single-risk based framework, which facilitates the optimal resource allocation across multiple tasks that are competing for the same limited sensory resources. A single or multiple sensors are used to perform detection of discrete random variables concurrently with the estimation of some other continuous random variables. Bayesian sequential detection and its extension to Bayesian sequential estimation are used to address the problem. Two costs can be taken into account for risk evaluation: the cost of making an erroneous detection or estimation decision without taking further observations, and the cost of taking more observations to decrease the risk associated with the detection and estimation decisions. The **Renyi information divergence** is introduced as a measure of the relative information loss, which is used to define the observation cost, in making a suboptimal sensor allocation decision.

The Bayesian sequential detection method can be used to address the detection problem. It is a sequential hypothesis testing for a stationary discrete random variable, which allows the number of observations to vary to achieve optimal decisions [126]. The risk analysis for integrated decision and estimation requires the comparison of expected information gains for a hybrid mix of discrete (for detection) and continuous (for estimation) random variables.

Renyi information measure can be used to model the information gained by making a certain sensor allocation decision. The relative information loss in making a suboptimal allocation decision is used to define the observation cost.

5.3.2.1 Bayesian sequential detection

The existence state X is modeled as a discrete-time, time-independent Markov chain, where the transitional probability matrix is given by the identity matrix. It is equal to one if a process exists within a given region and zero if no process exists.

Since the existence state is binary, a sensor model with a **Bernoulli distribution** is employed. At time k, the sensor is assumed to give the output $Y_k = 1$: a positive observation, i.e., process exists, or $Y_k = 0$, a negative observation or no process exists. The sensor model is given by the following general conditional probability matrix:

$$\mathbf{B} = \begin{pmatrix} \Pr(Y_k = 0|X = 0) & \Pr(Y_k = 0|X = 1) \\ \Pr(Y_k = 1|X = 0) & \Pr(Y_k = 1|X = 1) \end{pmatrix} \tag{5.84}$$

where $\Pr(Y_k = i|X = l), i, l = 0, 1$ describes the probability of measuring $Y_k = i$ given that the existence state is $X = l$. Let β be the probability of the sensor measurement being equal to the true existence state, i.e the detection probability, hence $\Pr(Y_k = 0|X = 0) = \Pr(Y_k = 1|X = 1) = \beta$ and $\Pr(Y_k = 0|X = 1) = \Pr(Y_k = 1|X = 0) = -\beta$.

As measurements are made with time, the probability of process existence is updated according to Bayes' rule. This probability is given by $\Pr(X = 1; k) = p_k$ and $\Pr(X = 0; k) = 1 - p_k$. Hence, it suffices to only update p_k as measurements are collected.

The update equation is as follows:

$$\hat{p}_k = \left\{ \begin{array}{ll} \frac{\beta \bar{p}_k}{(1-\beta)(1-\bar{p}_k)+\beta\bar{p}_k} & \text{if } Y_k = 1 \\ \frac{(1-\beta)\bar{p}_k}{\beta(1-\bar{p}_k)+(1-\beta)\bar{p}_k} & \text{if } Y_k = 0 \end{array} \right\} \tag{5.85}$$

These equations constitute the **belief prediction** and **update equations**.

The goal of Bayesian sequential detection is to determine the actual state of process existence X with minimum risk given a sequence of observations up to time t. Each decision made can be associated to a Bayes risk, which considers two types of costs

1. The expected cost of making an erroneous decision,
2. The expected cost of taking future new observations for a possibly better decision. The observation cost c_{obs} may include sensing energy, potential information loss, financial cost.

The sensor will stop and make a decision regarding process existence when the risk of making a decision with the current data is lower than the risk of taking another measurement to increase the probability of making a correct decision [125, 126, 141].

Decision Cost Assignment: The following hypotheses are considered: H_0,the null hypothesis that $X = 0$ and H_1, the alternative hypothesis that $X = 1$. Define the cost of accepting hypothesis H_i when the actual existence state is $X = j$ as $C(i,j)$. Using a **uniform cost assignment** (UCA), the decision cost matrix is given by

$$C(i,j) = \left\{ \begin{array}{ll} 0 & \textbf{if } i = j \\ c_d(\tau) & \textbf{if } i \neq j \end{array} \right\} \tau \geq 0 \tag{5.86}$$

where $i = 0, 1$ correspond to accepting, respectively, H_0 and H_1 and $j = 0, 1$ correspond, respectively, to the true state of existence $X = 0$ and $X = 1$ $c_d(\tau) \geq 0$ is the cost of making the wrong decision at time $\tau \geq 0$ indicating the number of observations.

Detection decision-making: At time t, after taking an observation $Y_t, \widehat{p}_t$ and $\bar{p}_{t+1}$ are updated using recursive Bayes rule. If a decision regarding process existence is made without taking any further future observations, i.e., the observation number $\tau = 0$, a Bayes risk r is defined as the expected cost of accepting the wrong hypothesis over all possible realization of X conditioned on all previous observations.

If the detection probability is close to 0.5, i.e., the sensor returns a true or false observation with equal probability, more observations need to be taken before an optimal decision with minimum risk can be reached.

5.3.2.2 Bayesian sequential estimation

Bayesian risk analysis tools are developed for sequential Bayesian estimation. Consider a linear system for a continuous random variable, which satisfies the discrete time Markov chain model:

$$\begin{array}{c} X_{k+1} = \mathbf{F}_k X_k + \nu_k \\ Y_k = \mathbf{H}_k X_k + \vartheta_k \end{array} \tag{5.87}$$

where the first equation defines the evolution of the process state sequence $X_k \in \mathbb{R}^n, k \in \mathbb{N}, \mathbf{F}_k \in \mathbb{R}^{n \times n}$ is the process state matrix, $\nu_k \in \mathbb{R}^n, k \in \mathbb{N}$ is the **independent identically distributed Gaussian process noise** (IID) sequence with zero mean and positive semi-definite covariance $\mathbf{Q}_k \in \mathbb{R}^{n \times n}, Y_k \in \mathbb{R}^m, k \in \mathbb{N}$ is the measurement sequence, $\mathbf{H}_k \in \mathbb{R}^{n \times n}$ is the output matrix and $\vartheta_k \in \mathbb{R}^m, k \in \mathbb{N}$ is the independent identically distributed Gaussian measurement noise sequence with zero mean and positive definite covariance $\mathbf{R}_k \in \mathbb{R}^{m \times m}$. The initial condition for the process state is assumed Gaussian with mean $\bar{X}_0$ and positive definite covariance $\mathbf{P}_0 \in \mathbb{R}^{n \times n}$. The initial process state, process noise and measurement noise are assumed to be all uncorrelated.

As an optimal filter for linear Gaussian systems, the Kalman filter allows constructing a suitable estimator. At time k, the process state and error covariance matrix prediction equations are given:

$$\begin{aligned} \bar{X}_k &= \mathbf{F}_{k-1}\hat{X}_{k-1} \\ \bar{\mathbf{P}} &= \mathbf{Q}_{k-1} + \mathbf{F}_{k-1}\hat{\mathbf{P}}_{k-1}\mathbf{F}_{k-1}^T \end{aligned} \tag{5.88}$$

where $\hat{X}_{k-1}$ is the process state estimate update at time k given measurements up to time $k-1$. The posterior state estimate is given by:

$$\hat{X}_k = \bar{X}_k + \mathbf{K}_k\left(Y_k - \mathbf{H}_k\bar{X}_k\right) \tag{5.89}$$

The only decision to be made is whether to accept the estimate as the true state and hence stop making additional measurements or take at least one more measurement. Hence, the list of decisions are

1. Accept the estimate and stop taking measurements,
2. Take one more measurement.

Let $X_k^e(Y_k)$ be an estimator, i.e., computed estimate, of the actual process state X_k based on observation Y_k. The cost of accepting the estimate $X_k^e(Y_k)$ given the actual process state X_k is defined as $C\left(X_k^e, X_k\right)$.

The function $C\left(X_k^e, X_k\right) = c_e(\tau)\left\|X_k^e - X_k\right\|^2$ is a quadratic cost with $c_e(\tau) > 0$ being some τ-dependent cost value and $\tau > 0$ indicating the number of future observations:

$$C\left(X_k^e, X_k\right) = \left\{ \begin{array}{ll} 0 & \textbf{If } \|X_k^e - X_k\| \leq \epsilon \\ c_e(\tau) & \textbf{If } \|X_k^e - X_k\| > \epsilon \end{array} \right\} \tag{5.90}$$

where ϵ is some preset interval. The updated Kalman filter estimate $\widehat{X}$ can be used.

For estimation decision-making, at time k, after making a measurement Y_k, if the sensor decides not to take any more measurements, as in the sequential detection approach, the Bayes risk is defined as the expected value, over all possible realizations of the process state, conditioned on all previous measurements, of the cost of choosing the estimate $\widehat{X}_t$.

5.4 FAULT TOLERANT FLIGHT CONTROL

In general, a fault tolerant flight control system is required to perform failure detection, identification and accommodation for sensor and actuator failures [97]. Active fault tolerant flight control schemes can be broadly classified into projection based and online controller redesign based approaches. Currently, there is a shift from robust passive fault tolerant flight control towards active methods relying on switching gain scheduled or linear parameter-varying (LPV) methods [94]. Much research has been done to find a generic flight control system, which can be applied to any aircraft and trained in flight and a fault-tolerant flight control system able to handle failures and to account for modeling errors [91]. Significant research effort has focused on control methods to recover stability with a damaged aircraft. The existing reconfigurable control design methods fall into one of the following approaches: linear quadratic, pseudo-inverse, gain scheduling/linear parameter varying, adaptive control/model following, eigenstructure assignment, multiple model feedback linearization or dynamic inversion, robust control, sliding mode control, intelligent control [109, 111, 137]. Adaptive flight control methods adapt to system uncertainty caused by aircraft damage or failure, which includes aerodynamic changes, structural degradation, engine damage and reduced flight control effectiveness [132]. The robust control problem appeared with the objective of reducing the difference between the dynamic behavior of real systems and their models [142].

Conventional control systems and deterministic optimization techniques provide a means to deal with uncertainty within certain boundaries. The introduction of artificial intelligence methodologies in order to implement, enhance and improve the UAV autonomous functionalities seems to be a natural evolution of more conventional feedback control theory: neural networks, fuzzy logic, genetic algorithm, reinforcement learning, temporal logic, knowledge based systems, constraint satisfaction problem. **Adaptive fuzzy control** (AFC) has been proposed for dynamical systems subject to model uncertainties and external disturbances. Neuro-fuzzy approximations are used to perform identification of the unknown system dynamics and become components of nonlinear controllers that finally make the system's state vector converge to the reference set-points. The adaptation gains in the learning performed by the neuro-fuzzy approximators are finally determined through Lyapunov stability analysis. A **neural network** control architecture incorporating a direct adaptive control with dynamic inversion can accommodate damage or failures over a range of failure conditions. This section presents mostly conventional control systems.

5.4.1 LINEAR TIME INVARIANT FORMULATION

For robust control for linear time invariant systems, there are two fundamental approaches: the robust control considering dynamic uncertainty and the robust control considering parametric uncertainty.

The main idea of **admissible model matching** (AMM) fault tolerant control approach is that instead of looking for a controller that provides an exact (or best) matching to a given single behavior after the fault appearance, a family of closed-loop behaviors that are acceptable is specified [81]. Consider the following linear time invariant aircraft model:

$$\dot{X}(t) = \mathbf{A}X(t) + \mathbf{B}U(t) \tag{5.91}$$

A classical state feedback control law is considered:

$$U(t) = -\mathbf{K}X(t) \tag{5.92}$$

In the model matching problem, relations (5.91) to (5.92) result in the closed-loop behavior that follows the reference model:

$$\dot{X}(t) = (\mathbf{A} - \mathbf{BK})\,X(t) = \mathbf{M}X(t) \tag{5.93}$$

where $\mathbf{M}$ is chosen to be stable. A set of system matrices that are acceptable are considered and the flight tolerant controller tries to provide a closed-loop behavior inside the set. Let $\mathbb{M}_a$ be a set of matrices such that any solution of

$$\dot{X}(t) = \mathbf{M}X(t) \qquad \mathbf{M} \in \mathbb{M}_a \tag{5.94}$$

has an acceptable dynamic behavior. The set of reference models $\mathbb{M}_a$ is defined off-line by the designer.

Moreover, if for the nominal system operation a state feedback $\mathbf{K}_m$ that satisfies some nominal control specifications has been obtained, then

$$\dot{X}(t) = (\mathbf{A}_m - \mathbf{B}_m\mathbf{K}_m)\,X(t) = \mathbf{M}^* X(t) \tag{5.95}$$

and $\mathbf{M}^*$ is known as the reference model.

For a given fault $\mathbf{A}_f, \mathbf{B}_f$, the goal of the fault accommodation is to find a feedback gain $\mathbf{K}_f$ that provides an admissible closed-loop behavior $(\mathbf{A}_f - \mathbf{B}_f\mathbf{K}_f) \in \mathbb{M}_a$.

A characterization of $\mathbb{M}_a$ can be specified by a set of d inequality constraints:

$$\mathbb{M}_a = \{\mathbf{M} : \Phi\,(m_{ij}, i = 1, \dots n, j = 1 \dots n) \leq 0\} \tag{5.96}$$

where $m_{ij}, i = 1, \dots n, j = 1 \dots n$ are the entries of matrix $\mathbf{M}$ and $\Phi : \mathbb{R}^{n \times n} \rightarrow \mathbb{R}^d$ is a given vector function.

5.4.2 LINEAR PARAMETER VARYING FORMULATION

The aircraft model being a strongly nonlinear system, it can be modeled as an LPV system, as shown in chapter 2. This is the basis of the development of model-based methods using LPV models. There are two commonly used approaches [124]:

1. The **fault estimation** method where the estimated fault is used as the fault indicating signal.
2. **Residual generation** method where the residuals are generated in order to be robust against modeling errors and unknown inputs.

5.4.2.1 Short-period dynamics

An uncertain quasi-LPV model for the short-period dynamics is provided in [90]:

$$\begin{aligned}\begin{pmatrix} \dot{q} \\ \Delta\dot{\alpha} \end{pmatrix} = &\begin{pmatrix} 0 & \frac{1}{I_{yy}} M_\alpha(\alpha)_{unc} \\ 1 + \frac{\cos\alpha}{mV_0} Z_q(\alpha)_{unc} & \frac{\cos\alpha}{mV_0} Z_q(\alpha)_{unc} \end{pmatrix} \begin{pmatrix} q \\ \Delta\alpha \end{pmatrix} + \\ &+ \begin{pmatrix} \frac{1}{I_{yy}} M_{\delta_e}(\alpha)_{unc} \\ \frac{\cos\alpha}{mV_0} Z_{\delta_e}(\alpha)_{unc} \end{pmatrix} \Delta\delta_e \end{aligned} \tag{5.97}$$

The parameters q and α represent the pitch rate and the angle of attack, respectively, and δ_e is the elevator deflection. I_{yy} and m are the moment of inertia and the mass of the aircraft, respectively.

The model is valid for an equilibrium point characterized by the speed V_0 and altitude z_0. The signals $\Delta\alpha = \alpha - \alpha_0$ and $\Delta\delta_e = \delta_e - \delta_{e0}$ represent large deviations from the equilibrium values α_0, δ_{e0}. The uncertainty is represented in the form of weighted additive uncertainty on the aerodynamics coefficient. The following nonlinear dynamic inversion control law is designed as:

$$\Delta\delta_e = \left(\frac{M_{\delta_e}(\alpha)}{I_{yy}}\right)^{-1}\left(U_\theta - \frac{M_\alpha(\alpha)_{unc}}{I_{yy}}\Delta\alpha\right) \tag{5.98}$$

The external signal U_θ is the commanded pitch acceleration.

5.4.2.2 General formulation

5.4.2.2.1 Controller design

The aircraft model can be described by the following LPV representation as follows:

$$\dot{X}(t) = \mathbf{A}(\varpi)X(t) + \mathbf{B}(\varpi)U(t) \tag{5.99}$$

$$Y(t) = \mathbf{C}(\varpi)X(t) + \mathbf{D}(\varpi)U(t) \tag{5.100}$$

In [81], additionally to the schedule of the parameters with the operating conditions $p(t)$, they are scheduled with the fault estimation $\hat{f}(t)$ provided by a fault, detection and isolation module. Then the scheduling of the linear parameter varying parameters can be expressed as follows:

$$\varpi_t = \varpi(p(t), \hat{f}(t)) \tag{5.101}$$

where the fault estimation $\hat{f}(t)$ is in the set of tolerated fault $\hat{f}(t) \in \mathbb{F}_t$ and operating point $p(t) \in \mathbb{P}$. The time-varying parameter vector ϖ of polytopic LPV system varies on a polytope. The state space matrices range in a polytope of matrices defined by the convex hull of a finite number of matrices N:

$$\begin{pmatrix} \mathbf{A}(\varpi_t) & \mathbf{B}(\varpi_t) \\ \mathbf{C}(\varpi_t) & \mathbf{D}(\varpi_t) \end{pmatrix} \in Co\left\{\begin{pmatrix} \mathbf{A}_j(\varpi_j) & \mathbf{B}_j(\varpi_j) \\ \mathbf{C}_j(\varpi_j) & \mathbf{D}_j(\varpi_j) \end{pmatrix}, j = 1, \ldots, N\right\}$$

$$\begin{pmatrix} \mathbf{A}(\varpi_t) & \mathbf{B}(\varpi_t) \\ \mathbf{C}(\varpi_t) & \mathbf{D}(\varpi_t) \end{pmatrix} = \sum_{j=1}^{N} \alpha_j(\varpi_t)\begin{pmatrix} \mathbf{A}_j(\varpi_j) & \mathbf{B}_j(\varpi_j) \\ \mathbf{C}_j(\varpi_j) & \mathbf{D}_j(\varpi_j) \end{pmatrix} \tag{5.102}$$

with $\alpha_j(\varpi_t) \geq 0$, $\sum_{j=1}^{N} \alpha_j(\varpi_t) = 1$ and $\varpi_j = \varpi(p_j, f_j)$ is the vector of parameters corresponding to the jth model. Each jth model is called a vertex system. Consequently, the LPV fault representation can be expressed as follows:

$$\dot{X}(t) = \sum_{j=1}^{N} \alpha_j(\varpi_t)\left(\mathbf{A}_j(\varpi_j)X(t) + \mathbf{B}_j(\varpi_j)U(t)\right) \tag{5.103}$$

$$Y(t) = \sum_{j=1}^{N} \alpha_j(\varpi_t)\left(\mathbf{C}_j(\varpi_j)X(t) + \mathbf{D}_j(\varpi_j)U(t)\right) \tag{5.104}$$

where $\mathbf{A}_j, \mathbf{B}_j, \mathbf{C}_j, \mathbf{D}_j$ are the state space matrices defined for the jth model. The state space matrices of system (5.103) to (5.104) are equivalent to the interpolation between LTI models, that is:

$$\mathbf{A}(\varpi_t) = \sum_{j=1}^{N} \alpha_j(\varpi_t)\left(\mathbf{A}_j(\varpi_j)\right) \tag{5.105}$$

and analogously for the other matrices. There are several ways of implementing (5.102) depending on how $\alpha_j(\varpi_t)$ functions are defined. They can be defined with a barycentric combination of vertexes.

Considering the system to be controlled is the LPV fault representation (5.103); a state feedback control law is written as follows:

$$U(t) = -\mathbf{K}_{\varpi}X(t) = -\sum_{j=1}^{N} \alpha_j(\varpi_t)\left(\mathbf{K}_j(\varpi_j)X(t)\right) \tag{5.106}$$

such that the closed-loop behavior satisfies:

$$\varpi_{\mathbf{t}} = \sum_{j=1}^{N} \alpha_j(\varpi_t)\left(\mathbf{M}_j(\varpi_j)\right) \tag{5.107}$$

where

$$\mathbf{M}_j(\varpi_j) = \mathbf{A}_j(\varpi_j) - \mathbf{B}_j(\varpi_j)\mathbf{K}_j(\varpi_j) \tag{5.108}$$

The admissibility condition can be expressed using LMI; the following set of LMI regions should be solved if and only if a symmetric matrix $\mathbf{X}_j$ exists for all $j \in [1 \dots N]$ such that:

$$\left(\mathbf{A}_j - \mathbf{B}_j\mathbf{K}_j\right)\mathbf{X}_j + \mathbf{X}_j\left(\mathbf{A}_j^T - \mathbf{K}_j^T\mathbf{B}_j^T\right) + 2\alpha\mathbf{X}_j < 0 \tag{5.109}$$

$$\begin{pmatrix} -r\mathbf{X}_j & \left(\mathbf{A}_j - \mathbf{B}_j\mathbf{K}_j\right)\mathbf{X}_j \\ \mathbf{X}_j\left(\mathbf{A}_j^T - \mathbf{K}_j^T\mathbf{B}_j^T\right) & -r\mathbf{X}_j \end{pmatrix} < 0 \tag{5.110}$$

In [103], the problem of designing an LPV state feedback controller for uncertain LPV systems that can guarantee some desired bounds on the H_∞ and H_2 performances and that satisfies some desired constraints on the closed-loop location is considered. The H_∞ performance is convenient to enforce robustness to model uncertainty and to express frequency domain specifications such as bandwidth and low frequency gain. The H_2 performance is useful to handle stochastic aspects such as measurement noise and random disturbances. By constraining the poles to lie in a prescribed region a satisfactory transient response can be ensured [103].

A common issue in LPV controller design method based on robust control theory is the conservatism of the designed controllers. As a means to reduce the conservatism and to improve the closed-loop performance achieved by a single LPV controller, a switching LPV controller design method can use the multiple parameter dependent Lyapunov functions. The moving region of the gain scheduling variables is divided into subregions and each subregion is associated with one local LPV controller [47].

5.4.2.2.2 *Observer based approach*

The aircraft equations of motion can be described as:

$$\dot{X}(t) = \mathbf{A}(\varpi)X(t) + \mathbf{B}(\varpi)U(t) + \sum_{j=1}^{m} L_j(\varpi)f_j(t) \tag{5.111}$$

$$Y(t) = \mathbf{C}(\varpi)X(t) + \mathbf{D}(\varpi)U(t) + \sum_{j=1}^{m} M_j(\varpi)f_j(t) \tag{5.112}$$

There are different failure signals f_j affecting the system; the matrices $\mathbf{A}, \mathbf{B}, \mathbf{C}, \mathbf{D}$ are parameter dependent; L_j are the directions of the faults acting on the input, most often on the actuators, while M_j are the output fault directions most often acting on the sensors. In a particular FDI filter synthesis problem, the goal is to detect a subset of these faults and be insensitive to the rest of them [123].

The following approach has been used for the LPV aircraft actuator model and nonlinear elevator in [123]. The observer based approach is to estimate the outputs of the system from the measurements by using a **Luenberger** observer, assuming a deterministic setting or a **Kalman** filter in the stochastic case. Then the weighted output estimation error is used as a residual. It is desired to estimate the output, a linear parameter varying function of the state, i.e., $\mathbf{C}(\varpi)X(t)$, using a functional or generalized LPV Luenberger-like observer with the following observer:

$$\dot{Z}(t) = \mathbf{F}(\varpi)Z(t) + \mathbf{K}(\varpi)Y(t) + \mathbf{J}(\varpi)U(t) \tag{5.113}$$

$$\tilde{W}(t) = \mathbf{G}(\varpi)Z(t) + \mathbf{R}(\varpi)Y(t) + \mathbf{S}(\varpi)U(t) \tag{5.114}$$

$$\hat{Y}(t) = \tilde{W}(t) + \mathbf{D}(\varpi)U(t) \tag{5.115}$$

$$R(t) = \mathbf{Q}\left(Y(t) - \hat{Y}(t)\right) = \mathbf{Q}_1(\varpi)Z(t) + \mathbf{Q}_2(t)Y(\varpi) + \mathbf{Q}_3(\varpi)U(t) \tag{5.116}$$

where $Z(t) \in \mathbb{R}^q$ is the state vector of the functional observer and $\mathbf{F}, \mathbf{K}, \mathbf{J}, \mathbf{R}, \mathbf{G}, \mathbf{S}\ \mathbf{D}, \mathbf{Q}, \mathbf{Q}_1, \mathbf{Q}_2, \mathbf{Q}_3$ are matrices with appropriate dimensions. The output $\tilde{W}(t)$ of this observer is said to be an estimate of $\mathbf{C}(\varpi)X(t)$, in an asymptotic sense in the absence of faults. The residual $R(t)$ is generated based on the states of the observer, where the $\mathbf{Q}_i$ entries are free parameters, but have to satisfy the following set of equations:

$$eig\left(\mathbf{F}(\varpi)\right) < 0 \tag{5.117}$$

$$\mathbf{TA}(\varpi) - \mathbf{F}(\varpi)\mathbf{T} = \mathbf{K}(\varpi)\mathbf{C} \tag{5.118}$$

$$\mathbf{J}(\varpi) = \mathbf{TB}(\varpi) - \mathbf{K}(\varpi)\mathbf{D} \tag{5.119}$$

$$\mathbf{Q}_1(\varpi)\mathbf{T} + \mathbf{Q}_2(\varpi)\mathbf{C} = 0 \tag{5.120}$$

$$\mathbf{Q}_3(\varpi) + \mathbf{Q}_2(\varpi)\mathbf{D} = 0 \tag{5.121}$$

where $\mathbf{T}$ is a coordinated transformation matrix constant if $\mathbf{C}, \mathbf{D}$ are also constant. It can be seen that the residual depends solely on faults in the asymptotic sense, given a stable estimator dynamics.

In [101], a fault tolerant control strategy is presented that compensates the effects of time-varying or constant actuator faults by designing an **adaptive polytopic observer (APO)** which is able to estimate both the states of the system and the magnitude of the actuator faults. Based on the information provided by this adaptive polytopic observer, a state feedback control law is derived in order to stabilize the system.

5.4.3 SLIDING MODE APPROACH

Sliding mode theory has found applications in the field of fault tolerant control which exploits robustness properties with respect to matched uncertainties. Sliding mode techniques have robustness properties with respect to the matched uncertainty [3]. Most applications of sliding mode to fault tolerant control deal with actuator faults as these faults occurring in the input channels are naturally classified as matched uncertainties. Another property is associated with the ability of sliding mode observer to reconstruct unknown signals in the system (e.g., faults) while simultaneously providing accurate state estimation. Sliding mode observers are able to reconstruct unknown signals in the system (e.g., faults) while simultaneously providing accurate state estimation. This reconstruction capability is advantageous in comparison to traditional residual based fault tolerant control since the fault reconstruction signals provide further information with regard to the shape and size of the unknown fault signals. The scheme presented in [3] is from a sensor fault tolerant control point of view and uses the existing controller in the closed-loop system. A sliding mode observer is used to reconstruct sensor fault and then use this signal to correct the corrupted measurement before it is used by an existing controller. Only a small modification to the feedback loop is required to implement this scheme and the observer designed independently. Sliding mode control cannot directly deal with total actuator failure because the complete loss of effectiveness of a channel destroys the regularity of the sliding mode and a unique equivalent can no longer be determined [46]. The integral switching function aims to eliminate the reaching phase present in traditional sliding mode control so that the sliding mode will exist from the time instant the controller is switched online.

Most of the technologies for LTI based FDI design can be directly extended to LPV models to deal with normal changes in the system, e.g., a change of operating conditions. Although LPV representations are better than LTI models in terms of modeling the behavior of a real system and can deal with changes in operating conditions, plant-model mismatches or uncertainty are still present [2]. Consider an uncertain affine LPV aircraft model subject to actuator faults represented by:

$$\begin{aligned} \dot{X}(t) &= \mathbf{A}(\varpi)X(t) + \mathbf{B}(\varpi)U(t) + \mathbf{D}(\varpi)f_i(t) + \mathbf{M}\zeta(t,y,u) \\ Y_m(t) &= \mathbf{C}(\varpi)X(t) + d(t) \end{aligned} \tag{5.122}$$

The inputs $U(t)$ and output measurements Y_m are available for the FDI scheme. The proposed observer has the structure:

$$\begin{aligned} \dot{\hat{X}}(t) &= \mathbf{A}(\varpi)\hat{X}(t) + \mathbf{B}(\varpi)U(t) - \mathbf{G}_l(\varpi)e_{ym}(t) + \mathbf{G}_m\nu(t,y,u) \\ \hat{Y}(t) &= \mathbf{C}\hat{X}(t) \end{aligned} \tag{5.123}$$

where $\mathbf{G}_l(\varpi), \mathbf{G}_m \in \mathbb{R}^{n\times p}$ are the observer gain matrices and $\nu(t)$ represents a discontinuous switched component to introduce a sliding motion. The output estimation error is:

$$e_{ym}(t) = \hat{Y}(t) - Y_m(t) = \mathbf{C}e(t) - d(t) \tag{5.124}$$

where $e(t) = \hat{X}(t) - X(t)$. The state estimation error output is:

$$\dot{e}(t) = \mathbf{A}(\varpi)e(t) - \mathbf{G}_l(\varpi)e_{ym}(t) + \mathbf{G}_m\nu(t,y,u) - \mathbf{D}(\varpi)f_i(t) - \mathbf{M}\zeta(t,y,u) \tag{5.125}$$

The objective of the observer is to force the output estimation error $e_{ym}(t)$ to zero in finite time and then a sliding mode is said to have been attained on the sliding surface

$$\mathbb{S} = \{e \in \mathbb{R}^n : e_{ym}(t) = 0\} \tag{5.126}$$

Ideally the measurements corruption term $d(t) = 0$ in which case relation (5.126) corresponds to a surface in which the output of the observer exactly follows the plant output. The idea is that during sliding, the signal $\nu(t)$ must take on average to maintain sliding.

Parameters in LPV systems can be viewed as parametric uncertainty or parameters which can be measured in real time during flight [41]. An H_∞ filter is designed in [50] for an LPV system. Conditions of existence of parameter-dependent Lyapunov function are formulated via LMI constraints and an algorithm for LPV filter gain based on the solutions to the LMI conditions is presented.

5.4.4 DIRECT MODEL REFERENCE ADAPTIVE CONTROL

Flight control systems are designed to tolerate various faults in sensors and actuators during flight; these faults must be detected and isolated as soon as possible to allow the overall system to continue its mission [43]. **Fault tolerant flight control** (FTFC) aims to increase the survivability of a failed aircraft by reconfiguring the control laws rather than by means of hardware redundancy, only. There are many control approaches possible in order to achieve fault tolerant flight control [79]. An important aspect of these algorithms is that they should not only be robust but even adaptive in order to adapt to the failure situation. Currently, much research is performed in the field of indirect adaptive control, where the adaptation is more extensive than only tuning the PID control gains. One of these new indirect control possibilities is **adaptive model predictive control** (AMPC) which deals with inequality constraints. These constraints are a good representation for actuator faults.

5.4.4.1 Linear approach

Adaptive control algorithms are able to deal with changes in the system's dynamics due to possible systems components' faults and failures. Actuator and sensor faults have been implicated in several aircraft loss of control accidents and incidents. **Direct model reference adaptive control** (MRAC) methods have been suggested as a promising approach for maintaining stability and controllability in the presence of uncertainties and actuator failures without requiring explicit fault detection, identification and controller reconfiguration. In addition to actuator faults, a sensor fault is unknown sensor bias which can develop during operation in one or more sensors such as rate gyros, accelerometers, altimeters [58].

Consider a linear time-invariant aircraft model, subject to actuator failures and sensor biases, described by:

$$\dot{X}(t) = \mathbf{A}X(t) + \mathbf{B}U(t) \qquad Y(t) = X(t) + \beta \tag{5.127}$$

where $\mathbf{A} \in \mathbb{R}^{n\times n}$, $\mathbf{B} \in \mathbb{R}^{n\times m}$ are the system and input matrices (assumed to be unknown and uncertain), $X(t) \in \mathbb{R}^n$ is the system state and $U(t) \in \mathbb{R}^m$ is the control input. The state measurement $Y(t) \in \mathbb{R}^n$ includes an unknown constant bias $\beta \in \mathbb{R}^n$, which may be present at the outset or may develop or change during operation. In addition to sensor bias, some of the actuators $U(t) \in \mathbb{R}^m$ (for example control surface or engines in aircraft flight control) may fail during operation [58]. Actuator failures are modeled as:

$$U_j(t) = \overline{U}_j \qquad t \geq t_j, j \in \mathbb{J}_p \tag{5.128}$$

where the failure pattern $\mathbb{J}_p = \{j_1, j_2, \dots, j_p\} \subseteq \{1, 2, \dots, m\}$ and the failure time of occurrence t_j are all unknown. Let $V(t) = (V_1, V_2, \dots, V_m)^T \in \mathbb{R}^m$ denote the applied (commanded) control input signal.

In the presence of actuator failures, the actual input vector $U(t)$ to the system can be described as:

$$U(t) = V(t) + \sigma(\overline{U} - V(t)) = (\mathbf{I}_{m\times m} - \sigma)V(t) + \sigma\overline{U} \tag{5.129}$$

where

$$\overline{U} = \left[\overline{U}_1, \overline{U}_2, \dots, \overline{U}_m\right]^T \qquad \sigma = diag\left(\sigma_1, \sigma_2, \dots, \sigma_m\right) \tag{5.130}$$

where σ is a diagonal matrix (failure pattern matrix) whose entries are piecewise constant signals: $\sigma_i = 1$ if the ith actuator fails and $\sigma_i = 0$ otherwise.

The actuator failures are uncertain in value $\overline{U}_j$ and time of occurrence t_j. The objective is to design an adaptive feedback control law using the available measurement $Y(t)$ with unknown bias β such that the closed-loop signal boundedness is ensured and the system state $X(t)$ tracks the state of a reference model described by:

$$\dot{X}_m(t) = \mathbf{A}_m X_m(t) + \mathbf{B}_m R(t) \tag{5.131}$$

where $X \in \mathbb{R}^n$ is the reference model state, $\mathbf{A}_m \in \mathbb{R}^{n\times n}$ is Hurwitz, $\mathbf{B}_m \in \mathbb{R}^{n\times m_r}$ and $R(t) \in \mathbb{R}^{n_r}, (1 \leq m_r \leq m)$ is a bounded reference input used in aircraft operation. The reference model $(\mathbf{A}_m, \mathbf{B}_m)$ is usually based on the nominal plant parameters and is designed to capture the desired closed-loop response to the reference input. For example, the reference model may be designed using optimal and robust control methods such as linear quadratic regulator (LQR), H_2, H_∞.

This paragraph considers the single reference input case $m_r = 1, \mathbf{B}_m \in \mathbb{R}^n$; the actuators are assumed to be similar (for example segments of the same control surface), that is, the columns b_i of the B matrix can differ only by unknown scalar multipliers. It is also assumed that:

$$b_i = \mathbf{B}_m/\alpha_i \qquad i = 1, \dots, m \tag{5.132}$$

for some unknown (finite and nonzero) α_i values whose signs are assumed known. The multiplier α_i represents the uncertainty in the effectiveness of the ith actuator due to modeling uncertainties and/or reduced effectiveness (example caused by partial loss of an aircraft control surface). The objective is to design an adaptive control law that will ensure closed-loop signal boundedness and asymptotic state tracking, i.e., $lim_{t\to\infty}(X(t) - X_m(t)) = 0$ despite system state uncertainties, actuator failures and sensor bias faults.

The adaptive controller should synthesize the control signal $U(t)$ capable of compensating for actuator failures and sensor bias faults automatically. It is assumed that the following matching conditions hold: there exist gains $\mathbf{K}_1 \in \mathbb{R}^{n\times m}, k_2, k_3 \in \mathbb{R}^m$ such that

$$\begin{aligned} \mathbf{A}_m &= \mathbf{A} + \mathbf{B}(\mathbf{I}_{m\times m} - \sigma)\mathbf{K}_1^T \\ \mathbf{B}_m &= \mathbf{B}(\mathbf{I}_{m\times m} - \sigma)k_2 \\ \mathbf{B}\sigma\overline{U} &= -\mathbf{B}(\mathbf{I}_{m\times m} - \sigma)(\mathbf{K}_1^T\beta + k_3) \end{aligned} \tag{5.133}$$

The first two matching conditions are typical MRAC matching conditions which address actuator failures without sensor bias, whereas the third condition contains a modification because of sensor bias. At least, one actuator must be functional for the matching condition to be satisfied. The reference model $(\mathbf{A}_m, \mathbf{B}_m)$ represents the desired closed-loop characteristics and it is usually designed using an appropriate state feedback for the nominal plant $(\mathbf{A}, \mathbf{B})$. Thus the matching conditions are always satisfied under nominal conditions. For the adaptive control scheme, only $\mathbf{A}_m$ and $\mathbf{B}_m$ need to be known. Because $\mathbf{A}_m$ is a Hurwitz matrix, there exist positive definite matrices: $\mathbf{P} = \mathbf{P}^T, \mathbf{Q} = \mathbf{Q}^T \in \mathbb{R}^{n \times n}$, such that the following Lyapunov inequality holds:

$$\mathbf{A}_m^T \mathbf{P} + \mathbf{P}\mathbf{A}_m \leq -\mathbf{Q} \tag{5.134}$$

The sensor measurements available for feedback have unknown biases as in equation (5.127). Let $\hat{\beta}(t)$ denote an estimate of the unknown sensor bias. Using $\hat{\beta}$, the corrected state $\overline{X}(t) \in \mathbb{R}^n$ is defined as:

$$\overline{X} = Y - \hat{\beta} = X + \tilde{\beta} \qquad \tilde{\beta} = \beta - \hat{\beta} \tag{5.135}$$

An adaptive control law can be designed as:

$$U = \hat{\mathbf{K}}_1^T Y + \hat{k}_2 R + \hat{k}_3 \tag{5.136}$$

where $\hat{\mathbf{K}}_1(t) \in \mathbb{R}^{n \times m}$, $\hat{k}_2, \hat{k}_3 \in \mathbb{R}^m$ are the adaptive gains. Therefore, the closed-loop corrected state equation is:

$$\begin{aligned} \dot{\overline{X}} = \left(\mathbf{A} + \mathbf{B}(\mathbf{I}_{m \times m} - \sigma)\mathbf{K}_1^T\right) X + \\ \mathbf{B}(\mathbf{I}_{m \times m} - \sigma)\left(\tilde{K}_1^T Y + k_2 R + \tilde{k}_3\right) + \mathbf{B}(\mathbf{I}_{m \times m} - \sigma) k_2 R + \\ + \mathbf{B}(\mathbf{I}_{m \times m} - \sigma)\mathbf{K}_1^T \beta + \mathbf{B}(\mathbf{I}_{m \times m} - \sigma) k_3 + \dot{\tilde{\beta}} + \mathbf{B}\sigma\overline{U} \end{aligned} \tag{5.137}$$

where $\tilde{\mathbf{K}}_1 = \hat{\mathbf{K}}_1 - \mathbf{K}_1$, $\tilde{k}_2 = \hat{k}_2 - k_2$, $\tilde{k}_3 = \hat{k}_3 - k_3$. Using equation (5.133) in equation (5.137):

$$\dot{\overline{X}} = \mathbf{A}_m \overline{X} + \mathbf{B}_m R + \mathbf{B}(\mathbf{I}_{m \times m} - \sigma)\left(\tilde{\mathbf{K}}_1^T Y + \tilde{k}_2 R + \tilde{k}_3\right) - \mathbf{A}_m \tilde{\beta} + \dot{\tilde{\beta}} \tag{5.138}$$

A measurable auxiliary error signal $\hat{e}(t) \in \mathbb{R}^n$ is defined as $\hat{e} = \overline{X} - X_m$ denoting the state tracking error. Using equations (5.131) to (5.138), the closed-loop auxiliary error system can be expressed.

Theorem 5.1

For the system given by equations (5.127), (5.129), (5.131) and the adaptive controller given by relation (5.136), the gain adaptation laws can be proposed as:

$$\begin{aligned}\dot{\hat{\mathbf{K}}}_{1j} &= -sign(\alpha_j)\Gamma_{1j}\mathbf{B}_m^T\mathbf{P}\hat{e}\\ \dot{\hat{k}}_{2j} &= -sign(\alpha_j)\gamma_{2j}\mathbf{B}_m^T\mathbf{P}\hat{e}\\ \dot{\hat{k}}_{3j} &= -sign(\alpha_j)\gamma_{3j}\mathbf{B}_m^T\mathbf{P}\hat{e}\end{aligned} \tag{5.139}$$

$\Gamma_{1j} \in \mathbb{R}^{n\times n}$ being a constant symmetric positive definite matrix, γ_{2j}, γ_{3j} constant positive scalars, $\mathbf{P}$ defined in (5.134) and the bias estimation law selected as:

$$\dot{\hat{\beta}} = -\eta \mathbf{P}^{-1}\mathbf{A}_m^T\mathbf{P}\hat{e} \tag{5.140}$$

where $\eta \in \mathbb{R}$ is a tunable positive constant gain, guarantees that $\hat{e} \to 0$ and all the closed-loop signals including $e(t)$, the adaptive gains and bias estimate are bounded. ■

To demonstrate this theorem, studies were performed on a fourth order longitudinal dynamics model of an aircraft in a wings-level cruise condition with known nominal trim conditions in [58].

Recently, **virtual sensors** and **virtual actuators** for linear systems have been proposed as a fault reconfiguration approach. A sensor fault tolerant control combines a set-based FDI module with **controller reconfiguration** (CR) based on the use of virtual sensors. The detection mechanism is based on the separation between pre-computed invariant sets and transition sets where appropriate residual variables jump when faults occur. The closed-loop system is then reconfigured by means of a virtual sensor which is adapted to the identified fault [83].

Virtual sensors for linear systems have been proposed as a **fault accommodation** approach. In particular, this approach of virtual sensors can be extended for fault tolerant control to linear parameter varying systems. The main idea of this fault tolerant control method is to reconfigure the control loop such that the nominal controller could still be used without need of retuning it. The aircraft with the faulty sensor is modified adding the virtual sensor block that masks the sensor fault and allows the controller to see the same plant as before the fault. The linear parameter varying virtual sensor is designed using LMI regions and taking into account the effect of the fault and the operating point. This approach requires to approximate the linear parameter varying model in a polytopic way guaranteeing the desired specifications at the expense of introducing some conservatism. As a benefit, controller design can be reduced to solve a convex optimization problem [82].

In [109], an application of robust gain scheduling control concept is presented. It uses a liner parameter varying control synthesis method to design fault tolerant controllers. To apply the robust linear parameter varying control synthesis method, the nonlinear dynamics must be represented by a linear parameter varying model which is developed using the function substitution method over the entire flight envelope. The developed linear parameter varying model associated with the aerodynamic coefficients' uncertainties represents nonlinear dynamics including those outside the equilibrium manifold. Passive and active fault tolerant control are designed for the longitudinal equations in the presence of elevator failure.

Instead of identifying the most likely model of the faulty plant, models that are not compatible with the observations are discarded. This method guarantees that there will not be false alarms as long as the model of the non faulty aircraft remains valid; i.e., if the assumption regarding the bounds on the exogenous disturbances are not violated and the model of the dynamics of the aircraft is valid then no fault is declared. Moreover, one need not compute the decision threshold used to declare whether or not a fault has occurred [102].

Remark 5.16. *The regulation of aircraft trajectory in the vicinity of obstacles must focus on critical outputs which vary as a function of the aircraft's position relative to obstacles. Aircraft mismanagement relative to its environment can lead to grave consequences.*

In [34], the possibility of addressing the issue of critical regulation is addressed via gain scheduling techniques, aimed at modifying the control parameters and possibly controller architecture as the conditions surrounding the plan evolve. The output depends on parameter vector which is a function of the aircraft environment. Parameter dependent control laws are explored to meet the condition of closed-loop system stability by looking simultaneously for parameter dependent Lyapunov functions. The use of parameter dependent Lyapunov function in the context of generalized LMI based H_∞ control results in analysis and synthesis conditions in the form of parametrized LMI [25].

5.4.4.2 Nonlinear approach

An alternative indirect adaptive nonlinear control approach allows a **reconfigurable control** routine placing emphasis on the use of physical models to be developed and thus producing internal parameters which are physically interpretable at any time. This technique can deal with control surface failures as well as structural damage resulting in aerodynamic changes. Reconfiguring control is implemented by making use of **adaptive nonlinear inversion** for autopilot control. The adaptativity of the control setup is achieved with a real time identified physical model of the damaged aircraft. In failure situations, the damaged aircraft model is identified by the two-step method in real time and this model is then provided to the model based adaptive **nonlinear dynamic inversion** (NDI) routine into a modular structure which allows flight control reconfiguration on line.

Three important modules of this control setup can be discussed:

1. aerodynamic model identification,
2. adaptive nonlinear control and

3. control allocation.

Aircraft often use redundant control inputs for tracking flight paths. A system with more inputs than degrees of freedom is an overactuated system. The reference signals can be tracked with a certain combination of existing control effectors [30]. **Control allocation** is able to distribute the virtual control law requirements to the control effectors in the best possible manner while accounting for their constraints [56]. Even under the condition that control effectors are damaged, the control reallocation can be implemented without redesigning the control law for fault-tolerant control. The general approaches of **control allocation** include interior-point algorithms, weighed pseudoinverse, linear programming, quadratic programming, sequential least-squares and quantized control allocation approach. Most previous works study linear control allocation by programming algorithms, which can be iteratively conducted to minimize the error between the commands produced by virtual control law and the moments produced by practical actuator combination.

The performance of the fault tolerant flight control system can be augmented with different control allocation (CA) methods. These methods can be used to distribute the desired control forces and moments over the different control effectors available, i.e., control surfaces and engines which makes the control system more versatile when dealing with in-flight failures. **Control allocation** is especially important when some dynamic distribution of the control commands is needed towards the different input channels. The main assumption in nonlinear dynamic inversion is that the plant dynamics are assumed to be perfectly known and therefore can be canceled exactly. However, in practice, this assumption is not realistic, not only with respect to system uncertainties but especially to unanticipated failures for the purpose of fault tolerant flight control.

In order to deal with this issue, robust control can be used as an outer loop control or neural networks to augment the control signal [52]. However, another solution is the use of a real time identification algorithm, which provides updated model information to the dynamics inversion controller. These augmented structures are called **adaptive nonlinear dynamics inversion** (ANDI). Three consecutive inversion loops can be implemented: a body angular rate loop, an aerodynamic angle loop and a navigation loop which can be placed in a cascaded order based upon the timescale separation principle. The navigation loop tracks the course, the flight path, the angle of attack and the throttle setting.

A high performance flight control system based on the nonlinear dynamic inversion principle requires highly accurate models for aircraft aerodynamics. In general, the accuracy of the internal model determines to what degree the aircraft nonlinearities can be canceled. In [122], a control system is presented that combines nonlinear dynamic inversion with multivariate spline-based control allocation. Three control allocation strategies that use expressions for the analytic Jacobian and Hessian of the multivariate spline models are presented.

A major advantage of nonlinear dynamic inversion is that gain scheduling is avoided through the entire flight envelope. Furthermore, the simple structure allows easy and flexible design for all flying modes. It can be augmented with a control allocation module in the case that an aircraft has redundant or cross-coupled control effectors.

In a first stage, a number of different control allocation methods can be compared using a simplified aircraft model, yielding two methods: **weighted least square** (WLS) and **direct control allocation** (DCA). Weighted least square is in theory able to find a feasible solution in all attainable virtual control. This is due to the fact that it is able to free up previously saturated actuators. Direct control allocation is an alternative approach to optimization based control allocation. Instead of optimizing some criterion, Direct control allocation produces a solution based on geometric reasoning. The direct control allocation method looks for the largest virtual control that can be produced in the direction of the desired virtual control. High-performance aircraft are designed with significantly overlapping control moment effectiveness at a particular flight condition and so the resultant control allocation is underdetermined [87]. In the context of solving the three moment constrained control allocation, much of the existing literature has focused on efficient numerical optimization methods that are practically implementable in the flight software, relaxing the requirement that the moments are globally linear in the controls and the explicit consideration of the actuator dynamics and fault tolerance.

5.4.5 BACKSTEPPING APPROACH

An incremental type sensor-based backstepping control approach is presented in [116], based on singular perturbation theory and Tikhonov theorem. This Lyapunov function based method used measurements of control variables and less model knowledge and it is not susceptible to the model uncertainty caused by fault scenario. A flight controller combines nonlinear dynamic inversion with the sensor-based backstepping control approach.

An aircraft under many post-failure circumstances can still achieve a certain level of flight performance with the remaining valid control effectors. However, as a consequence of the structural actuator failures, the control authority or the safe flight envelope of the aircraft is inevitably limited. In the sensor-based backstepping control approach, a singular perturbation theory based control approximation is adopted. To apply the **singular perturbation theory**, the system dynamics of the control plant need to have the timescale separation property. In the aircraft system, the actuator system can be viewed as a subsystem cascaded to the body angular dynamic system. Because the actuator dynamics are much faster than the body rate dynamics, the time scale separation property of the aircraft is guaranteed. The following expression holds for the body angular rate aerodynamics:

$$\begin{pmatrix} \dot{p} \\ \dot{q} \\ \dot{r} \end{pmatrix} = -\mathbf{I}^{-1} \left(\begin{pmatrix} p \\ q \\ r \end{pmatrix} \times \left(\mathbf{I} \begin{pmatrix} p \\ q \\ r \end{pmatrix} \right) \right) + \\ -\frac{1}{2}\rho V^2 S \mathbf{I}^{-1} \begin{pmatrix} bC_{ls} \\ \bar{c}C_{ms} \\ bC_{ns} \end{pmatrix} + \frac{1}{2}\rho V^2 S \mathbf{I}^{-1} \mathbf{M}_{CA} U \tag{5.141}$$

where $U = (\delta_a, \delta_e, \delta_r)^T$, $\mathbf{M}_{CA} = \begin{pmatrix} b & 0 & 0 \\ 0 & \bar{c} & 0 \\ 0 & 0 & b \end{pmatrix} \mathbf{M}_E$, where $\mathbf{M}_{CA}$ is the control allocation matrix, $\mathbf{M}_E$ is the control effectiveness matrix, U is the vector consisting of all the control inputs and C_{ls}, C_{ms}, C_{ns} are the non dimensional moments contributed by all of the current states.

Rewriting relation (5.141), a simplified formulation of the aircraft motion equation is derived:

$$\dot{X} = f(X) + gU \qquad X = (p, q, r)^T \qquad g = \frac{1}{2}\rho V^2 S \mathbf{I}^{-1} \mathbf{M}_{CA} \tag{5.142}$$

and

$$f(X) = -\left(\mathbf{I}^{-1} \left(\begin{pmatrix} p \\ q \\ r \end{pmatrix} \times \left(\mathbf{I} \begin{pmatrix} p \\ q \\ r \end{pmatrix} \right) \right) - \frac{1}{2}\rho V^2 S \mathbf{I}^{-1} \begin{pmatrix} bC_{ls} \\ \bar{c}C_{ms} \\ bC_{ns} \end{pmatrix} \right) \tag{5.143}$$

The error is defined as $e = X - X_r$ where $X_r = (p_r, q_r, r_r)^T$. To design a single loop body rate backstepping controller, the control Lyapunov function $\tilde{V}$ is chosen as follows:

$$\tilde{V}(e) = \frac{1}{2}e^2 + \frac{1}{2}\mathbf{K}\lambda^2 \qquad \text{where } \dot{\lambda} = e \tag{5.144}$$

where $\mathbf{K}$ is a diagonal matrix of controller gains and λ is an integral term introduced to remove the tracking errors caused by the internal dynamics.

Using relations (5.144) to (5.143), the following expression can be derived for the desired state of the control system $e = X_{des} - X_r$ and:

$$\dot{\tilde{V}} = e\dot{e} + \mathbf{K}\lambda e = e\left(\dot{X}_{des} - \dot{X}_r + \mathbf{K}\lambda \right) \tag{5.145}$$

To stabilize this system, $\dot{X}_{des}$ can be selected as:

$$\dot{X}_{des} = -\mathbf{P}e - \mathbf{P}(X - X_r) + \dot{X}_r - \mathbf{K}\lambda \tag{5.146}$$

where $\mathbf{P}$ is a positive diagonal matrix to stabilize the system such that $\dot{e} = -c(X - X_r)$ and the equivalent inputs are computed as $R_{red} = \mathbf{M}_{CA}u$.

$$\epsilon \dot{U}_{red} = -sign\left(\frac{\partial \dot{X}}{\partial U_{red}} \right) \dot{X} - \dot{X}_{des} \tag{5.147}$$

where ϵ is a turning parameter with a small positive value. From relation (5.141), the following formulation can be obtained:

$$\epsilon \dot{U}_{eq} = -sign\left(\frac{1}{2}\rho V^2 \mathbf{S}\mathbf{I}^{-1}\right)\dot{X} + \mathbf{P}(X - X_r) - \dot{X}_r + \mathbf{K}\lambda \tag{5.148}$$

and

$$U_{eq_k} = U_{eq_{k-1}} + \int_{(k-1)T}^{kT} \dot{U}_{eq} dt \tag{5.149}$$

According to relation (5.146), the control input U can be solved using a control allocation algorithm if $\mathbf{M}_{CA}$ is available.

5.4.6 CONTROL PROBABILISTIC METHODS

The development of algorithms for aircraft robust dynamic optimization considering uncertainties is relatively limited compared to aircraft robust static optimization. Modeling uncertainty and propagating it through computational models is a key step in robust optimization considering uncertainties. The **Monte Carlo** (MC) method is a natural choice for this because of its ease of implementation. An alternative method known as the **polynomial chaos** (PC) expansion scheme can provide accuracy comparable to the Monte Carlo method at a significantly lower computational cost. In [70], an approach for dynamic optimization considering uncertainties is developed and applied to robust aircraft trajectory optimization. The non intrusive polynomial chaos expansion scheme can be employed to convert a robust trajectory optimization problem with stochastic ordinary differential equation into an equivalent deterministic trajectory optimization problem with deterministic ordinary differential equation.

Robust optimization has attracted increasing attention due to its ability to offer performance guarantees for optimization problems in the presence of uncertainty. Robust control design requires the construction of a decision such that the constraints are satisfied for all admissible values of some uncertain parameter. An alternative approach is to interpret robustness in a probabilistic sense, allowing for constraint violation for low probability, formulating chance constrained optimization problems. Randomization of uncertainty offers an alternative way to provide performance guarantees without assumption on the probability distribution. Typically, it involves sampling the uncertainty and substituting the chance constraint with a finite number of hard constraints, corresponding to the different uncertainty realization [74].

A method based on probability and randomization has emerged to synergize with the standard deterministic methods for control of systems with uncertainty [22]. The study of robustness of complex systems began based on a deterministic description of the uncertainty acting on the system to be controlled [44, 142]. However, these deterministic methodologies were affected by serious computational problems, especially for uncertainty entering in a nonlinear fashion into the control system. For this reason, approximation and relaxation techniques have been proposed. Only upper and lower bounds of the robustness margins can be determined.

The starting point of probabilistic and randomized methods is to assume that the uncertainty affecting the system has a probabilistic nature. The objective is then to provide probabilistic assessments/characteristics on the system. A given performance level is robustly satisfied in a probabilistic sense if it is guaranteed against most possible uncertainty outcomes. The risk of a system property being violated can be accepted for a set of uncertainties having small probability measure. Such systems may be viewed as being practically robust from an engineering point of view.

Remark 5.17. *The probabilistic approach is not limited to control problems, but is useful in a wide range of related areas such as robust optimization and general engineering design, where decisions that work satisfactorily well in an uncertain or adversarial environment should be devised.*

One of the advantages of the probabilistic approach for control is to use **classical worst case** bounds of robust control together with probabilistic information, which is often neglected in a deterministic context. The probabilistic approach has also connections with adaptive control methods. The interplay of probability and robustness leads to innovative concepts such as the **probabilistic robustness margin** and the **probability degradation function**. However, assessing probabilistic robustness may be computationally hard, since it requires the computation of multi-dimensional probability integrals. These integrals can be evaluated exactly only in very special cases of limited practical interest [22]. This computational problem can be resolved by means of randomized techniques, which have been used to tackle difficult problems that are too hard to be treated via exact deterministic methods. Specific examples include the Monte Carlo method in computational physics and the Las Vegas techniques in computer science.

In the context of unmanned aircraft, **uncertainty randomization** requires the development of specific techniques for generating random samples of the structured uncertainty acting on the system. The probability is estimated using a finite number of random samples. Since the estimated probability is itself a random quantity, this method always entails a risk of failure, i.e., there exists a non zero probability of making an erroneous estimation. The resulting algorithms are called **randomized algorithms**; (RA) i.e., algorithms that make random choices during execution to produce a result. Randomized algorithms have low complexity and are associated with robustness bounds which are less conservative than the classical ones, obviously at the expense of a probabilistic risk.

In the probabilistic approach, the system is not fixed a priori but depends on some parameters (for instance, parameters defining the controller) that need to be determined in order to make the system behave as desired. A randomized algorithm for design should be able to determine these parameters to guarantee the desired system specifications up to a given level of probability. For example, in order to account for unknown kinematic target, the target motion can be assumed to be robustly defined by planar Brownian motion. Stochastic problems in the control of Dubins vehicles typically concentrate on variants of the traveling salesman problem and other routing problems, in which the target location is unknown or randomly generated. The goal of the work developed in [6] is to develop a feedback control policy that allows a UAV to optimally maintain a nominal standoff distance from the target without full knowledge of the current target position or its future trajectory. The feedback control policy is computed off-line using a Bellman equation discretized through an approximating Markov chain. The tracking aircraft should achieve and maintain a nominal distance to the target. If the target is also modeled as a Dubins vehicle with a Brownian heading angle, the knowledge of the target's heading angle would provide the UAV with an indicator of the target's immediate motion and an appropriate response.

Let a performance function $f(q)$ be defined for a generic uncertain dynamic system, where $q \in \mathbb{Q}$ is a vector of random uncertain parameters and $\mathbf{Q} \subset \mathbb{R}^\ell$ is a given uncertainty domain. Two probabilistic analysis problems can be defined:

Problem 5.1. ***Reliability estimation:*** *Given $\Gamma > 0$, estimate the reliability of the estimation of the specification $f(q) \leq \Gamma$, that is, evaluate:*

$$R_l = \Pr\{f(q) \leq \Gamma\} \tag{5.150}$$

Problem 5.2. ***Performance level estimation:*** *Given $\epsilon \in [0, 1]$, estimate a performance level Γ such that $f(q) \leq \Gamma$ holds with reliability at $1 - \epsilon$, that is, to find Γ such that:*

$$\Pr\{f(q) \leq \Gamma\} \geq 1 - \epsilon \tag{5.151}$$

Remark 5.18. *The availability of samples drawn from this probability is needed and the stated results hold irrespective of the probability measure. This is important, since in some practical situations, the samples can be directly acquired through measurements or experiments on the real UAV. In other cases, when samples are not directly available, they must be generated and hence a probability measure on the uncertain set has to be assumed. In this situation, the UAV reliability R_l depends on the specific choice of this measure. In extreme cases, the probability may vary between zero and one, when considering different measures.*

Without any guidelines on the choice of the measure, the obtained probability estimate may be meaningless. In many cases of interest, with bounded uncertainty set $\mathbb{Q}$, the uniform probability distribution possesses such worst case properties, and the distribution is often used in practice, when little is known about the actual distribution. In other cases, when the uncertainty is concentrated around the mean, the traditional Gaussian distribution or truncated Gaussian can be used [53]. For general sets, asymptotic sampling techniques that progressively reach a steady-state distribution that coincides with the desired target distribution are usually employed: **Markov chain Monte Carlo methods** (MCMC). These methods do not produce independent identically distributed samples at steady state and also the time needed for such algorithms in order to reach steady state is not known in advance. Hence, explicit analytic results for finite sample complexity are not generally available [5].

5.5 FAULT TOLERANT PLANNER

5.5.1 ARTIFICIAL INTELLIGENCE PLANNERS

Artificial intelligence planners are useful for converting domain knowledge into situational-relevant plans of action but execution of these plans cannot generally guarantee hard real time response [89]. Planners and plan-execution systems have been extensively tested on problems such as mobile robot control and various dynamic system simulations, but the majority of architectures used today employ a best-effort strategy. As such, these systems are only appropriate for soft real time domain in which missing a task deadline does not cause total system failure [36].

To control mobile robots, for example, **soft real time plan** execution succeeds because the robot can slow down to allow increased reaction times or even stop moving should the route become too hazardous [76]. For autonomous aircraft flight, **hard real time response** is required, and fault tolerance is mandated. Moreover, a planner is desired, particularly for selecting reactions to anomalous or emergency situations. Violating response timing constraints in safety-critical systems may be catastrophic. Developing explicit mechanisms for degrading system performance if resource shortage does not allow all constraints to be met is a major real time research issue [108].

The interface of a real time resource allocator with an artificial intelligence planner to automatically create fault-tolerant plans that are guaranteed to execute in hard real time, is described in [48]. The planner produces an initial plan and task constraint set required for failure avoidance. This plan is scheduled by the resource allocator for the nominal *no-fault case* as well as for each specified *internal* fault condition, such as a processor or communication channel failure. If any resource is overutilized, the most costly task is determined using a heuristic. The costly task is fed back to the planner, which invokes dynamic backtracking to replace this task, or, if required, generate a more schedulable task set. This combination of planning and resource allocation takes the best elements from both technologies: plans are created automatically and are adaptable, while plan execution is guaranteed to be tolerant to potential faults from a user-specified list and capable of meeting hard real time constraints. This work is improved by adding fault-tolerance and the capability to reason about multiple resource classes and instances of each class, so that all aspects relevant to plan execution are explicitly considered during the planning phase.

A heuristic is presented in [40] to combine pertinent information from planning and resource allocation modules. The resulting *bottleneck* task information is used to guide planner backtracking when scheduling conflicts arise. An algorithm is described which incorporates this heuristic and a fault condition list to develop a set of fault-tolerant plans which will execute with hard real time safety guarantees. The utility of this algorithm can be illustrated with the example of an aircraft agent that must be tolerant to a single-processor failure during plan execution then describe related work in the development of planning and plan-execution agent architectures.

During planning, the world model is created incrementally based on initial states and all available transitions. The planner builds a state-transition network from initial to goal states and selects an action (if any) for each state based on the relative gain from performing the action. The planner backtracks if the action selected for any state does not ultimately help achieve the goal or if the system cannot be guaranteed to avoid failure. The planner minimizes memory and time usage by expanding only states produced by transitions from initial states or their descendants and includes a probability model which promotes a best first state space search as well as limiting search size via removal of *highly improbable* states. Planning terminates when the goal has been reached while avoiding failure states. During plan construction, action transition timing constraints are determined such that the system will be guaranteed to avoid all **temporal transitions to failure** (TTF), any one of which would be sufficient to cause catastrophic system failure. The temporal model allows computation of a minimum delay before each temporal transitions to failure can occur; then the deadline for each preemptive action is set to this minimum delay. After building each plan, all preemptive actions (tasks) are explicitly scheduled such that the plan is guaranteed to meet all such deadlines, thus guaranteeing failure avoidance. The establishment of Bayesian networks relies mainly on knowledge from experts who give causal relationship between each node according to their experience [38].

Smart autonomous aircraft path planning systems should meet feasibility, optimality and real time requirements.

5.5.2 TRIM STATE DISCOVERY

Automatic trajectory planners require knowledge of aircraft flight envelopes to ensure their solution is feasible. Existing flight management systems presume a nominal performance envelope, requiring on-line identification or characterization of any performance degradation. Given predictable failures such as control surface jam, stability and controllability can be evaluated off-line to establish a stabilizable trim state database that can be used in real time to plan a feasible landing trajectory. In less predictable cases such as structural damage, performance can only be determined on line. Damage or failures that significantly impact performance introduce pressure to respond quickly and accurately. Although adaptive control methods may maintain stability and controllability of a damaged aircraft, in many situations the flight envelope unavoidably contracts. The result may be the need for the increased thrust or larger actuator deflections to compensate for the reduced performance. As saturation limits are reached, reference commands are no longer possible, necessitating alteration of the flight plan in addition to the control law. Aircraft dynamics are nonlinear; thus it is typically difficult to extend stability and controllability results in a manner that identifies the full operating envelope. To make valid sequence decisions when damage/failure occurs, an emergency flight management system is proposed in [133].

Knowledge of the safe maneuvering envelope is of vital importance to prevent loss of control aircraft accidents [51]. In this section, determination of the safe maneuvering envelope is addressed in a reachability framework. The **forwards and backwards reachable sets** for a set of initial trim conditions are obtained by solving a Hamilton–Jacobi partial differential equation through a semi-Lagrangian method.

Definition 5.7. *The **safe maneuvering set** is defined as the intersection between the forwards and backwards reachable sets for a given set of a priori known safe states. The **flight envelope** describes the area of altitude and airspeed where an airplane is constrained to operate.*

The flight envelope boundaries are defined by various limitations on the performance of the airplane, for example, available engine power, stalling characteristics, structural considerations. A common way to present the flight envelope is the diagram which relates the altitude, velocity and possibly other variables at which the aircraft can safely fly. The boundaries defined on the flight envelope are adequate during normal operation of aircraft.

Remark 5.19. *The main problem with the conventional definition of flight envelope is that only constraints on quasi-stationary states are taken into account. Additionally, constraints posed on the aircraft state by the environment are not part of the conventional definition. The aircraft's dynamic behavior can pose additional constraints on the flight envelope.*

Ther safe maneuver envelope is the part of the state space for which safe operation of the aircraft can be guaranteed and external constraints will not be violated. It is defined by the intersection of three envelopes:

1. **Dynamic envelope**: constraints posed on the envelope by the dynamic behavior of the aircraft, due to its aerodynamics and kinematics. It is the region of the aircraft's state space in which the aircraft can be safely controlled and no loss of control events can occur.

2. **Structural envelope**: constraints posed by the airframe and payload, defined through maximum accelerations and loads.
3. **Environmental envelope**: constraints due to the environment in which the aircraft operates.

Reachable set analysis is a useful tool in safety verification of systems.

Definition 5.8. ***Reachable Set:*** *The reachable set describes the set that can be reached from a given set of initial conditions within a certain amount of time, or the set of states that can reach a given target set within a certain time.*

The dynamics of the system can be evolved backwards and forwards in time resulting in the backwards and forwards reachable set. For a forwards reachable set, the initial conditions are specified and the set of all states that can be reached along trajectories that start in the initial set is determined. For the backwards reachable sets, a set of target states are defined and a set of states from which trajectory states that can reach that target set are determined. The dynamics of the system are given by:

$$\dot{X} = f(X, U, d) \tag{5.152}$$

where $X \in \mathbb{R}^n$ is the state of the system, $U \in \mathbb{U} \subset \mathbb{R}^m$ is the control input and $d \in \mathbb{D} \subset \mathbb{R}^q$ a disturbance input.

Definition 5.9. *The* ***backwards reachable set*** $\mathbb{S}(\tau)$ *at time* $\tau, 0 \leq \tau \leq t_f$ *of* (5.152) *starting from the target set* $\mathbb{S}_0$ *is the set of all states* $X(\tau)$*, such that there exists a control input* $U(t) \in \mathbb{U}, \quad \tau \leq t \leq t_f$*, for all disturbance inputs* $d(t) \in \mathbb{D}, \tau \leq t \leq t_f$ *for which some* $X(t_f)$ *in* $\mathbb{S}_0$ *are reachable from* $X(\tau)$ *along a trajectory satisfying* (5.152).

Definition 5.10. *The* ***forwards reachable set*** $\mathbb{V}(\tau)$ *at time* $\tau, 0 \leq \tau \leq t_f$ *of* (5.152) *starting from the initial set* $\mathbb{V}_0$ *is the set of all states* $X(\tau)$*, such that there exists a control input* $U(t) \in \mathbb{U}, \tau \leq t \leq t_f$*, for all disturbance inputs* $d(t) \in \mathbb{D}, \tau \leq t \leq t_f$ *for which some* $X(0)$ *in* $\mathbb{V}_0$ *are reachable from* $X(\tau)$ *along a trajectory satisfying* (5.152).

If the initial target set is known to be safe, then all states that are part of both the forwards and backwards reachable sets can be considered safe as well.

To cope with in-flight damage/failure, the planner must guide the aircraft through a sequence of feasible trim states leading to a safe landing. Before transitioning from the current proven equilibrium state to a new state, the feasibility of the new state should be predicted; otherwise the aircraft may transition outside its operating envelope. The process of trim state discovery computes a path trim state space with 3D coordinates $(V_T, \dot{\gamma}, \dot{\chi})$ translated to physical 3D space to verify required altitude and airspace constraints are met.

Trim state discovery can be used to characterize unknown failures, for which envelope constraint information cannot be obtained prior to the discovery process. This requires use of a dynamic path planning strategy in which envelope constraints, as they are approached, are modeled as obstacles for the state path planner.

Problem 5.3. ***Trim State Discovery:*** *Given an initial stable damaged aircraft position in trim state space* $s_0 = (V_0, \dot{\chi}_0, \gamma_0)$ *and an ideal final approach trim state* s_{app}, *generate a continuous path* P_T *in trim state space from* s_0 *to* s_{app} *where* P_T *is a sequence of continuous trim states and transitions. All trim states in* P_T *must be stabilizable in the presence of disturbances.*

With the above formulation, trim state discovery is mapped to motion planning with obstacle avoidance. Every step (local transition) through trim-state space involves two phases: exploration and exploitation.

Failure or damage events that degrade performance pose significant risk to aircraft in flight. Adaptive control and system identification may stabilize a damaged aircraft, but identified models may be valid only near each local operating point. A guidance strategy designed to discover a set of feasible flight states sufficient to enable a safe landing given an unknown degradation event is presented in [133]. The aircraft is progressively guided through a sequence of trim states that are stabilizable given local envelope estimates. The proposed guidance strategy progressively explores trim state space rather than 3D physical space, identifying a set of trim states. A potential field method is adapted to steer exploration through trim states' space, modeling envelope constraints as obstacles and desirable trim states as attractors.

Design of maneuvers for carefree access of an aircraft to its complete flight envelope including **post-stall regimes** is useful for devising recovery strategies from an accident scenario. Maneuvers for an aircraft can be efficiently designed if a priori knowledge of its maneuverability characteristics is available to the control designers. Different types of agility metrics that characterize aircraft maneuvering capabilities based on different criteria have been proposed. Another approach to define maneuverability characteristics is based on computing attainable equilibrium sets. This approach involves a $2D$ section of **attainable equilibrium sets** of a particular maneuver using an inverse trimming formulation [63].

Construction of maneuvers based on attainable equilibrium sets involves accessing desired aircraft states in the attainable equilibrium set from a normal flying condition, such as a level flight trim condition. Computing an attainable equilibrium set for a given aircraft model and developing control algorithm to switch aircraft states between different operating points lying within the accessible region defined by attainable equilibrium sets are thus essential for aircraft maneuver design. For aircraft models, use of nonlinear control design techniques based on **dynamic inversion** (DI) or **sliding mode control** (SMC) have been proposed for control prototyping to design maneuvers.

A procedure for computing a 2D **attainable equilibrium region** (AER) section is based on the assumption that the boundary of the attainable equilibrium region is defined by saturation of one or more control surfaces. Therefore, a point lying on the boundary of an attainable equilibrium region is initially located using a continuation method. Using the trim point on the boundary as a starting condition, a separate continuation procedure is carried out with the saturated control fixed at its limit value to obtain the envelope containing equilibrium points.

To compute the attainable equilibrium region, the **extended bifurcation analysis** (EBA) method uses one control input as a continuation parameter. It consists of two stages:

1. In **stage 1**, a trim point lying on the attainable equilibrium region boundary is computed simultaneously solving the following set of state and constraint equations:

$$\dot{X} = f(X, U, p) = 0 \qquad g(X) = 0 \tag{5.153}$$

   where $X = (V, \alpha, \beta, p, q, r, \phi, \theta)^T \in \mathbb{X} \subset \mathbb{R}^8$ is the vector of state variables, $U \in \mathbb{R}$ is a continuation parameter and p denotes a bounded set of free control parameters, $g(X)$ represent the desired constraint. To solve the augmented system of (5.153) using the **extended bifurcation analysis**, it is necessary that an equal number of constraints, for example 3, as the number of free parameters are imposed so that the continuation problem becomes well posed. The continuation run is terminated when any of the free control parameters saturate or the continuation parameter itself reaches its deflection limit. Continuation results obtained for system (5.153) provide the augmented state vector for the trim point lying on the attainable equilibrium region boundary along with information about the saturated control. Therefore this trim point can be used as a starting point for computing the attainable equilibrium region boundary.
2. In **stage 2**, a continuation procedure consisting of multiple runs is carried out to trace the boundary of attainable region. Thus the system (5.154) is solved:

$$\dot{X} = f(X, U, p_1, p_2, p_s) = 0 \qquad g(X) = 0 \tag{5.154}$$

   where p_1, p_2 are the two free control parameters, p_s is the saturated control and $g(X)$ is a 2D vector representing the specified constraints. Continuation for system (5.154) is carried out until the continuation parameter reaches its deflection limit or any of the two free control parameters saturate. In case a free control parameter saturates before the limit of continuation parameter is reached, the fixed and free control parameters are interchanged and the continuation procedure continues. However, if the continuation parameter reaches its limit before any of the free control parameters attain its maximum/minimum value, then the fixed and continuation parameters are exchanged and the continuation process repeats. The solution of each continuation run of stage 2 is appended with the results of the subsequent runs. The starting point for any continuation run in stage 2 is automatically available from the solution of the preceding run. The continuation procedure is carried out until a bounded attainable zone is obtained.

In [9], an approach is presented for safe landing trajectory generation of an airplane with structural damage to its wing in proximity to local terrain. A damaged airplane maneuvering flight envelope is estimated by analyzing the local stability and flying qualities at each trim condition. The safety value index to prioritize choice of trim conditions for post-damage flight is chosen to be the trim state distance of a flight envelope boundary. A potential field strategy is used to rapidly define emergency landing trajectories. The damage to an aircraft can cause a shift in the center of gravity which complicates the equations of motion. The damaged aircraft center of gravity variation is calculated with respect to the center of gravity position of the undamaged aircraft. The new equations of motion according to the new center of gravity location can be derived. The damaged airplane trim states are derived using the airplane nonlinear equations of motion:

$$\dot{X} = f(X, U) \tag{5.155}$$

$X = (V, \alpha, \beta, p, q, r, \phi, \theta) \in \mathbb{R}^8, U = (\delta_a, \delta_e, \delta_r, T) \in \mathbb{R}^4$ and f is the vector of nonlinear six degrees of freedom equations. The desired trim conditions represent a constant speed V, desired turn rate $\dot{\chi}$ and desired flight path angle γ for a reference altitude z. One method for deriving these trim conditions is to solve the nonlinear constrained optimization problem that minimizes the following cost function:

$$J_{trim}(X, U) = \frac{1}{2}\dot{X}^T \mathbf{Q} \dot{X} \tag{5.156}$$

where $\mathbf{Q}$ represents the weighting matrix describing each state's priority with respect to maintaining the trim state.

A safety value index relating trim state distance (in respective flight envelope dimensions) to the flight envelope boundary can be defined. This metric is calculated based on the normalized distance of each trim state to the flight envelope boundary. The task of the flight planner is to identify an appropriate sequence of trim states from the initial airplane state where damage occurs to the desired landing site position and heading. A trajectory consists of a sequence of states or motion primitives that can be divided into the trim trajectories and the maneuver. A neighborhood set is defined for each trim condition and maneuvers are guaranteed based on the neighboring states of each trim. Motion planning takes place in discrete space by considering a fixed time step for trim and maneuver trajectory segments.

5.5.3 OBSTACLE/COLLISION AVOIDANCE SYSTEMS

5.5.3.1 Safety analysis of collision/obstacle avoidance system

Collision avoidance systems are critical to the safety of airspace. Quantification of the overall (system level) safety impact of collision avoidance system requires an understanding of the relevant interactions among the various layers of protection against collisions, as well as the frequencies and patterns of encounters that can lead to collisions [125]. One can divide the overall problem of estimating the risk of collision into three steps:

1. Determine the conflict frequency,
2. Given the conflict, determine the chances of resolving it by a deployed collision avoidance system,
3. Determine collision chances, given mid-air collision which is the failure of the collision avoidance system to resolve a conflict.

The concept of conflict probability is proposed for collision avoidance between two aircraft and collision risk reduction [92]. The collision probability calculation involves the relative trajectory between the objects and the combined relative position-error probability density. The volume for conflict probability is defined to be a cylindrically shaped volume much larger than the actual size of the aircraft. The cylindrical height of a conflict volume is aligned vertically. The circular cross section of the cylinder lies in the horizontal plane containing the North-South and the East-West directions. The position error covariances of the two aircraft are assumed Gaussian and are combined to form the relative position error covariance matrix, which is centered on the primary aircraft. The conflict volume is centered on the secondary aircraft. The probability that the primary aircraft will penetrate the conflict volume is the conflict probability. Conflict probability can be used as a metric.

From the system **reliability** and safety modeling standpoint, a collision avoidance system relies on **time redundancy** because there are several consecutive attempts to detect and resolve a conflict. This time redundancy is supplemented by functional redundancy because the time before the conflict is separated into distinct phases or layers with the conflict resolution task assigned to distinct subsystems. This functional separation is motivated by the increased urgency of the task combined with less uncertainty about the conflict. Therefore, as a general rule, as time progresses, conflict resolution should be less complex in order to facilitate reliability and can be simpler as it deals with less uncertainty. In addition, increasing the diversity of the protective layers provides some protection against **common cause failures** that can defeat the intended redundancy.

Combining structural and time-redundancy is well recognized as providing a more efficient means of protection than each type of redundancy alone in other applications, such as in designing a fault tolerant computer system to negotiate the effects of transient faults. Although detection becomes more efficient as time progresses, there is potential for **accumulation of failures**. If dynamic interactions are confined to a single layer of protection, then a hierarchical analysis is possible as advocated in the context of a sense-and-avoid system: an inner loop that includes a collision encounter model and relies on Monte Carlo simulation combined with an outer-loop analysis based on fault trees. However, if different layers share common failure modes, neglecting this coupling in the fault tree analysis can lead to non conservative risk estimates [125].

Sampling-based approaches have several advantages for complex motion planning problems including efficient exploration of high dimensional configuration spaces, paths that are dynamically feasible by construction and trajectory-wise (e.g., non-enumerative) checking of possible complex constraints. The RRT algorithm has been demonstrated as a successful planning algorithm for UAV [64]. However, it does not explicitly incorporate uncertainty. The RRT algorithm can be extended to an uncertain environment incorporating uncertainty into the planned path. The tree can be extended with different conditions sampled from a probabilistic model as with particle filters. Each vertex of the tree is a cluster of the simulated results. The likelihood of successfully executing an action is quantified and the probability of following a full path is then determined. Another approach identifies a finite-series approximation of the uncertainty propagation in order to reduce model complexity and the resulting number of simulations per node. In [64], a chance constrained RRT algorithm is used to handle uncertainty in the system model and environmental situational awareness. The chance constrained formulation is extended to handle uncertain dynamics obstacles.

Path planning of smart autonomous aircraft with known and unknown obstacles is considered as one of the key technologies in unmanned vehicle systems. The fault tree analysis method was applied to the **traffic alert and collision avoidance (TCAS)** for safety analysis while **failure modes and effects analysis** (FMEA) was also used. Functional failure analysis (FFA) was performed for safety analysis of UAV operation including collision avoidance.

Two critical hazards were defined in this analysis: midair collision and ground impact. The **Markov decision process (MDP)** solver can be proposed to generate avoidance strategies optimizing a cost function that balances flight-plan deviation with anti-collision. The minimum distance from the aircraft to an obstacle during collision avoidance maneuver is chosen as the criterion for the performance assessment. To successfully perform collision avoidance maneuvers, the minimum distance to the obstacle d_{min} must be greater than the radius of the obstacle r_n including a safe margin. The worst case analysis in the presence of all the possible uncertainties is cast as a problem to find the combinations of the variations where the minimum distance to the obstacle d_{min} appears [113].

Problem 5.4. *Initial robustness analysis can be carried out by solving the following optimization problem:*

$$d_{min} = min(d(t)) \tag{5.157}$$

subject to $P_L \leq P \leq P_U$ *where* P *is the uncertain parameters set,* P_L *and* P_U *are the lower and upper bounds of* P*,* $d(t)$ *is the distance to the obstacle,* T *is the collision avoidance maneuver during the period and* d_{min} *is the minimum distance to the obstacle.*

Several methods can be used to solve this problem such as **sequential quadratic programming (SQP)**, **genetic algorithms** or **global optimization**. Genetic algorithms can be applied to the UAV collision avoidance system to find the global minimum. The uncertain parameter set is considered here as the genetic representation, i.e., the chromosome. Each of the uncertainties corresponds to one gene. A binary coded string is generated to represent the chromosome where each of the uncertain parameters lies between the lower and upper bounds. The selection function of a roulette wheel can be used for this study.

In [72], an approach for building a potential navigation function and roadmap based on the information value and geometry of the targets is presented. The method is based on a potential function defined from conditional mutual information that is used to design a switched feedback control law, as well as to generate a PRM for escaping local minima, while obtaining valuable sensor measurements. Shaffer in [108] presents a method of re-planning in case of failure of one of the fins while Atkins in [10] presents a planning technique for loss of thrust.

5.5.3.2 Informative motion planning

Informative motion planning consists in generating trajectories for dynamically constrained sensing agents. The aircraft platforms used as mobile agents to traversing the operating environment are typically subject to nonholonomic and differential dynamic constraints. Obstacles in the operating environment can both constrain the aircraft motion and occlude observations. Finally, the limitations inherent in the available sensing mechanism (e.g., narrow field of view) can further limit the informativeness of agent plans. The **information-rich rapidly exploring random tree** (IRRT) algorithm is proposed in [69] as an on-line solution method that by construction accommodates very general constraint characterizations for the informative motion planning problem. The method IRRT extends the RRT by embedding information collection, as predicted using the **Fisher information matrix** (FIM) at the tree expansion and path selection levels, thereby modifying the structure of the growing feasible plan collection and biasing selection toward information-rich paths. As the IRRT is a sampling-based motion planner, feasible solutions can be generated on-line.

Several solution strategies to effect information-rich path planning can be considered. Analytical solutions often use the Fisher information matrix to quantify trajectory information collection in an optimal control framework. Solutions seek to maximize, for example, a lower bound on the determinant of the Fisher information matrix at the conclusion of the trajectory. While analytical solutions often have a simple form and perform optimally for low dimension unconstrained problems, they typically are difficult to scale to complicated scenarios. Another approach is the heuristic path shape that performs well in steady state. For aircraft with side-mounted cameras, circular trajectories with optima radii at a fixed altitude and varying altitude can be proposed. While these heuristically constrained trajectories may capture the physical and geometrical intuition of bearings-only target tracking, the gap between anticipated and realized informativeness of the motion plan can become arbitrarily large when operating under realistic dynamic and sensing constraints. The partially observable Markov decision process framework is a way of solving problems of planning under uncertainty with observations but there are also tractability issues. Belief space planning can be considered for both the target tracking problem and its inverse; that of localizing a vehicle through sensor measurements of perfectly known targets in a previously mapped environment. Partially observable Markov decision process solutions are currently intractable for vehicle models with complex dynamics [16].

5.5.4 GEOMETRIC REINFORCEMENT LEARNING FOR PATH PLANNING

Smart autonomous aircraft often fly in a complicated environment. Many threats such as hills, trees, other aircraft can be fatal in causing the autonomous aircraft to crash [138]. These threats can only in general be detected within a limited range from a single aircraft. However, by sharing information with other autonomous aircraft, these threats can be detected over a longer distance. Furthermore, an effective path for navigation should be smooth, provide an escape route and must be computationally efficient [55]. In previous work on path planning for a single autonomous aircraft, Voronoi graph search and visibility graph search have been proven to be effective only in a simple environment. They are not real time and also lead to fatal failure when the map information is not entirely available, such as when obstacles are not detected [32].

Evolutionary algorithms can be used as a candidate to solve path planning problems and provide feasible solutions within a short time [29, 84]. A radial basis functions artificial neural network (RBF-ANN) assisted differential evolution (DE) algorithm is used to design an off-line path planner for autonomous aircraft coordinated navigation in known static environments [115, 116, 118]. The **behavior coordination and virtual** (BCV) goal method proposes a real time path planning approach based on the coordination of the global and local behaviors. A fuzzy logic controller (FLC) for controlling the local behavior is designed to achieve the threat avoidance [49].

Path planning of multiple autonomous aircraft concentrates on **collaborative framework**, **collaborative strategies** and **consistency**. The Voronoi graph search and the A^* or Dijkstra algorithms plan a global path for multiple aircraft to simultaneously reach the target in an exhaustive procedure [73].

Path planning of multiple autonomous aircraft can also be addressed from the perspective of reinforcement learning. Q-learning is a way to solve the path planning problem. The basic idea of Q-learning is to obtain the optimal control strategy from the delayed rewards according to the observed state of the environment in a learning map and to make a control strategy to select the action to achieve the purpose. But the method is actually designed for the entire map of the environment known to planners. Q-learning fails to use the geometric distance information which is a very valuable element for path planning when only partial information of the map is available. Moreover for multiple autonomous aircraft, the shared information from other aircraft cannot be well exploited as there are a lot of unnecessary calculations to propagate from one point to another in Q-learning. Also some special points such as the start and target points are not well considered.

An algorithm based on reinforcement learning using the geometric distance and risk information from detection sensors and other autonomous aircraft can be proposed. It builds a general path planning model. By dividing the map into a series of lattices, path planning is formulated as the problem of optimal path planning. A continuous threat function can be used to simulate the real situation. To reduce the complexity of calculation, the parameters of control are finely modulated to control the size of the map. Moreover, the algorithm is generalized to multiple autonomous aircraft by using the information shared from other aircraft, which provides an effective solution for path planning and avoids local optimum. In this approach, the autonomous aircraft detect threatening objects in real time and share the information with each other. Collisions are avoided by the virtual aircraft created from an aircraft which is considered as a new obstacle for the other aircraft. Further, the reward matrix can be changed in terms of an exposure risk function. The target planner gets the final path according to all known threat information and aircraft real time positions [138].

Modeling of probabilistic risk exposure to obstacles can be presented as follows. It is necessary for an autonomous aircraft to keep a certain distance from regions of high risk to ensure safe flying. So the measure of probabilistic risk exposure to obstacles can be seen as a continuous distribution function. For example, consider the case where an obstacle is at position (x_i, y_i, z_i); the measure of the risk is denoted by F_i, in which the parameters are related to the dimension of the planning space. In the 3D space,

$$F_i(x,y,z) = \frac{1}{\sqrt{2\pi}\sigma_i}\exp\left(-\frac{d_i^2}{2\sigma_i}\right) \tag{5.158}$$

where

$$d_i = \sqrt{(x-x_i)^2 + (y-y_i)^2 + (z-z_i)^2} \tag{5.159}$$

σ_i is an adujstable parameter.

The probabilistic risk of the area where aircraft could not fly over can be represented as a very big value. Furthermore, when more than one obstacle exists on the map, the probabilistic risk at position (x, y, z) can be calculated as:

$$F(x,y,z) = 1 - \prod_{i=1}^{M}[1 - f_i(x,y,z)] \tag{5.160}$$

Cooperative and geometric learning algorithm is executed when a new threatening object is detected and the weight matrix $\mathbf{Q}$ is updated. This weight matrix describes the relationship between different points in the map. It is designed to measure any two points on the map in terms of risk and geometric distance [140]. The risk matrix $\mathbf{Q}$ and time or length matrix $\mathbf{T}$ on a path can be computed by:

$$\mathbf{Q} = \int_{\mathbb{C}} F_i(x,y,z)ds \qquad \mathbf{T} = \int_{\mathbb{C}} V ds \tag{5.161}$$

where $\mathbb{C}$ is the point set of a given path and V is the speed of the autonomous aircraft. In order to find the next point for the current aircraft configuration, the relationship weight matrix based on the geometric distance and integral risk measure is considered.

The key idea of the cooperative and geometric learning algorithm is to calculate the cost matrix $\mathbf{G}$ which can be used to find the optimal path from a starting point to a target point in terms of distance and integral risk. Each element in the matrix $\mathbf{G}$ is defined to be the sum of cost from its position to a target point.

A hybrid approach to the autonomous motion control in cluttered environments with unknown obstacles combining the optimization power of evolutionary algorithm and the efficiency of reinforcement learning in real time is presented in [73]. This hybrid navigation approach tends to combine the high level efficiency of deliberative model-based schemes and the low level navigation ability of reactive schemes.

5.5.5 PROBABILISTIC WEATHER FORECASTING

Weather conditions such as thunderstorms, icing, turbulence and wind have great influences on UAV safety and mission success. It is therefore important to incorporate weather forecasting into path planning. The recent development in numerical weather prediction makes possible the high resolution ensemble forecasting. In ensemble forecasting, different weather models with model inputs, initial conditions and boundary conditions are being slightly changed for each run. Each single run contains a different number of ensemble members and generates a prediction spectrum. This allows to build an objective and stochastic weather forecast that supports statistical post-processing [119]. Based on this information, a probabilistic weather map can be constructed. With continuous ensemble forecasting updating at a rate of once per time unit, this analysis provides an online 4D weather map. In the probabilistic 3D weather map, the path can be optimally planned. The problem is defined as follows.

Problem 5.5. *Given a UAV in an area of operation described by nonuniform grids, each waypoint assigned with a probability of adverse weather that is updated periodically, find a path from start to destination with minimum cost on defined terms and meeting the constraints on mission failure risk. The cost function is defined as:*

$$Min \ \left(J = w_{time} T_{time} + w_{wea} W_{wea}\right) \tag{5.162}$$

subject to:

$$R_{mission} < R_{critical} \qquad T_{time} < T_{max} \tag{5.163}$$

where w_{time}, w_{wea} *are the weighting factors on mission duration and weather condition, respectively, with:* $w_{time} + w_{wea} = 1$*,* $R_{mission}$ *is the risk of the mission and* $R_{critical}$ *is the critical risk level defined by users.* T_{time} *is the mission duration and* T_{max} *is the maximum mission duration allowed and* W_{wea} *is the weather condition along the flight route.*

Incremental search algorithm makes an assumption about the unknown space and finds a path with the least cost from its current location to the goal. When a new area is explored, the map information is updated and a new route is replanned if necessary. This process is repeated until the goal is reached or it turns out that the goal cannot be reached (due to obstacles, for instance). When the weather map is updated, the weather in cells close to the UAV is more certain than those in the grids far away from the UAV. In this sense, the weather uncertainty in grids is proportional to its distance to the UAV. When the uncertainties become larger, the weather condition can be regarded as unknown. Therefore the weather map is not completely known [71].

The mission risk evaluation and management can be improved by integrating an uncertainty factor in path planning. The uncertainty factor for a grid denoted as $U(x)$ can be defined as a Gaussian function:

$$U_{un}(X) = 1 - \exp\left(-\frac{(X - X_0)^2}{2\sigma^2}\right) \tag{5.164}$$

where σ is an adjustable parameter, X_0 is the UAV current location and X are the centers of the grids. Everytime the weather forecasting is updated, this uncertainty factor is recalculated to obtain a new set of uncertainty so that the impact of adverse weather to mission success is also updated. The probability of adverse weather in each grid is then weighted by the uncertainty factor:

$$P_{ad-un} = P_{ad}(i)\left(1 - U_{un}(i)\right) \tag{5.165}$$

where P_{ad-un} is the probability of adverse weather adjusted by the uncertainty factor, $P_{ad}(i)$ is the probability of adverse weather in i^{th} grid and $U_{un}(i)$ is the probability of adverse weather adjusted by the uncertainty factor.

To evaluate the mission risk of a planned path, the probability of adverse weather in each grid cell needs to be converted to the probability of UAV failure as it traverses the cell. The **Weibull distribution** can be used to calculate the probability of failure. The inputs consist of the probability of adverse weather occurring in each cell along the path and the time that the UAV spends flying through each of the cells. In the study presented by [139], the Weibull scale factor is calculated as:

$$\alpha = \frac{\mu_{fail}}{\Gamma\left(1 + \frac{1}{\beta}\right)} \tag{5.166}$$

where $\Gamma(.)$ is the Gamma function and μ_{fail} is the average time to failure for the aircraft in each cell. Then a Weibull distribution can be established to calculate the probability of UAV failure.

5.5.6 RISK MEASUREMENT

The determination of the minimum number of aircraft needed for the mission subject to fuel constraints, risk, cost and importance of various points for sampling is important. The measurement space includes sample points and the desirable neighborhoods that surround them. The sample points or the desirable neighborhoods are where the autonomous aircraft will make measurements. Risk refers to turbulent regions. The measurement space also includes forbidden points and the undesirable neighborhoods that surround them [11, 141]. The forbidden points are points of turbulence and other phenomena that could threaten the aircraft. The undesirable neighborhoods surrounding them also represent various degrees of risk. The planning algorithm automatically establishes the order in which to send the aircraft taking into account its value, on board sensor payload, on board resources such as fuel, computer CPU and memory. The priority of sample points and their desirable neighborhoods are taken into account. The planning algorithm also calculates the optimal path around undesirable regions routing the aircraft to or at least near the points to be sampled [65, 126].

Remark 5.20. *In the planning phase to establish likely positions, human experts can be consulted. The experts provide subjective probabilities of the* *points of interest. The points on the sampling grid are the sample points. Sample points arising from the highest probability hypothesis positions have priority 1, sample points associated with lower probability hypothesis positions, priority 2.*

Each sample point is surrounded by what are referred to as desirable neighborhoods. Depending on local weather, topography. the desirable neighborhoods are generally concentric closed balls with a degree of desirability assigned to each ball. The desirable region need not have spherical geometry. The notion of a desirable neighborhood is inspired by the fact that a sample point may also be a forbidden point and at least part of the sample point's desirable neighborhood falls within the forbidden point's undesirable neighborhood; the aircraft may only sample within a desirable neighborhood that is consistent with its risk tolerance [66, 100].

A forbidden point and the undesirable neighborhoods containing the point generally represent a threat to the aircraft. The threat may take the form of high winds, turbulence, icing conditions, mountains. The undesirable neighborhoods around the forbidden point relate to how spatially extensive the threat is. A method of quantifying the risk and incorporating it into the path assignment algorithm can use fuzzy logic to quantify how much risk a given neighborhood poses for an autonomous aircraft. This quantitative risk is then incorporated into the aircraft cost for traveling through the neighborhood. Once the cost is established, an optimization algorithm is used to determine the best path for the aircraft to reach its goal.

When determining the optimal path for the autonomous aircraft to follow, both the planning algorithm and the control algorithm running on each aircraft take into account forbidden points and the undesirable neighborhood around each forbidden point. The path planning and control algorithms will not allow the aircraft to pass through a forbidden point. Depending on its risk tolerance, an aircraft may pass through various neighborhoods of the forbidden point, subsequently experiencing various degrees of risk.

Remark 5.21. *Both the concepts of risk and risk tolerance are based on human expertise and employ rules carrying a degree of uncertainty. This uncertainty is born of linguistic imprecision, the inability of human experts to specify a crisp assignment for risk. Owing to this uncertainty, risk and risk tolerance are specified in terms of fuzzy logic.*

Risk is represented as a fuzzy decision tree. The risk tree is used to define forbidden points and the undesirable neighborhoods surrounding the forbidden points. The best path algorithm is actually an optimization algorithm that attempts to minimize a cost function to determine the optimal trajectory for each autonomous aircraft to follow, given a priori knowledge. The cost function for the optimization algorithm takes into account various factors associated with the aircraft properties and mission. Two significant quantities that contribute to the cost are the effective distance between the initial and final proposed positions of the aircraft and the risk associated with travel.

Aircraft are assigned as a function of their abilities to sample high priority points first. The planning algorithm determines flight paths by assigning as many high priority points to a path as possible, taking into account relative distances including sampling and non-sampling velocity, risk from forbidden points and fuel limitations. Once flight paths are determined, the planning algorithm assigns the best aircraft to each path using the fuzzy logic decision tree for path assignment. The planning algorithm must assign aircraft to the flight paths determined by an optimization procedure: the aircraft path assignment problem. The planning algorithm makes this assignment using a fuzzy logic-based procedure [135].

Remark 5.22. *The **fuzzy degree of reliability** experts assign to the sensors of aircraft (i)* A_i *is denoted as* $\mu_{sr}(A_i)$*. This is a real number between 0 and 1 with 1 implying the sensors are very reliable and 0 that they are totally unreliable. Likewise* $\mu_{nsr}(A_i)$ *is the fuzzy degree of reliability of other non-sensor system on board the aircraft (i). This fuzzy concept relates to any non-sensor system such as propulsion, computers, hard disk, deicing system. The value of aircraft (i) is denoted* $V(A_i)$*. The amount of fuel that aircraft (i) has at time t is denoted* $fuel(A_i, t)$*. All the aircraft participating in a mission are assumed to leave base at time* $t = t_0$*.*

Given the fuzzy grade of membership, it is necessary to defuzzify, i.e., make definite aircraft path assignments. The set of all possible permutations of the aircraft's possible paths is considered. An assignment benefit is calculated. In order to formalize the priorities, a multi-objective control problem must be addressed. To formalize objectives and priorities, a scalar objective function and a set of bounds of increasing difficulties are assigned to each objective. The bounds can be listed in a priority table where the first priority objectives are found in the first column and so on. The columns thus correspond to different intersections of level sets of the objective function [85].

5.6 CONCLUSION

The smart autonomous aircraft must be able to overcome environmental uncertainties such as modeling errors, external disturbances and an incomplete situational awareness. This chapter tackles flight safety, presenting first situation awareness then integrated system health monitoring. The benefit of integrating systems health monitoring with the command and control system in unmanned aircraft is that it enables management of asset health by matching mission requirements to platform capability. This can reduce the chance of mission failures and loss of the platform due to a faulty component. Fault tolerant control and fault tolerant planners are the last part of this chapter.

REFERENCES

1. Altay, A.; Ozkam, O.; Kayakutlu, G. (2014): *Prediction of aircraft failure times using artificial neural networks and genetic algorithms*, AIAA Journal of Aircraft, vol. **51**, pp. 45–53.
2. Alwi, H.; Edwards, C. (2010): *Robust actuator fault reconstruction for LPV system using sliding mode observers*, IEEE Conference on Decision and Control, pp. 84–89.
3. Alwi, H.; Edwards, C.; Menon, P. P. (2012): *Sensor fault tolerant control using a robust LPV based sliding mode observer*, IEEE conference on Decision and Control, pp. 1828–1833.
4. Andert, F.; Goormann, L. (2009): *Combining occupancy grids with a polygonal obstacle world model for autonomous flights*, **Aerial Vehicles**, Lam, T. M. (ed.), In Tech, pp. 13–28.
5. Anderson, R.; Bakolas, E.; Milutinovic, D.; Tsiotras, P. (2012): *The Markov-Dubins problem in the presence of a stochastic drift field*, IEEE Conference on Decision and Control, pp. 130–135.
6. Anderson, R.; Milutinovic, D. (2014): *A stochastic approach to Dubins vehicle tracking problems*, IEEE Transactions on Automatic Control, vol. **59**, pp. 2801–2806, DOI : 10.1109/TAC.2014.2314224.
7. Angelov, P.; Filev, D. P.; Kasabov, N. (2010): *Evolving Intelligent Systems*, IEEE Press.
8. Angelov, P. (2012): *Sense and Avoid in UAS–Research and Applications*, Wiley.
9. Asadi, D.; Sabzehparvor, M.; Atkins, E. M.; Talebi, H. A. (2014): *Damaged airplane trajectory planning based on flight envelope and motion primitives*, AIAA Journal of Aircraft, vol. **51**, pp. 1740–1757.
10. Atkins, E. M.; Abdelzaher, T. F.; Shin, K. G.; Durfee, E. H. (2001): *Planning and resource allocation for hard real time fault tolerant plan execution*, Autonomous Agents and Multi-Agents Systems, Vol. **4**, pp.57–78.
11. Banaszuk, A.; Fonoberov, V. A.; Frewen, T. A.; Kobilarov, M.; Mathew, G.; Mezic, I.; Surana, A. (2011): *Scalable approach to uncertainty quantification and robust design of interconnected dynamical systems*, IFAC Annual Reviews in Control, vol. **35**, pp. 77–98.
12. Benedettelli, D.; Garulli, A.; Giannitraponi, A. (2010): *Multi-robot SLAM using M-space feature representation*, IEEE Conference on Decision and Control (CDC), pp. 3826–3831.
13. Benosman, M.; Lum, K. (2010): *Passive actuators fault tolerant control for affine nonlinear systems*, IEEE Transactions on Control System Technology, vol. **18**, pp. 152–163.
14. Berger, J.; Barkaoui, M.; Boukhtouta, A. (2006): *A hybrid genetic approach for airborne sensor vehicle routing in real time reconnaissance missions*, Aerospace Science and Technology, vol. **11**, pp. 317–326.
15. Bertsekas, D. P.; Tsitsiklis, J. N. (1996): *Neuro-Dynamic Programming*, One, Athena Scientific.
16. Bestaoui Sebbane Y. (2014): *Planning and Decision-making for Aerial Robots*, Springer, Switzerland.
17. Bethke, B.; Valenti, M.; How, J. P. (2008): *UAV task assignment, an experimental demonstration with integrated health monitoring*, IEEE Journal on Robotics and Automation, vol. **15**, pp. 39–44.

18. Birk, A.; Vaskevicius, N.; Pathak, K.; Schwerfeger, S.; Poppinga, J.; Bulow, H. (2009): *3D perception and modeling*, IEEE Robotics and Automation Magazine, vol. **16**, pp. 53–60.
19. Blanke, M.; Hansen, S. (2013): *Towards self-tuning residual generators for UAV control surface fault diagnosis*, IEEE Conference on Control and Fault-tolerant Systems (SYSTOL), Nice, France, pp. 37–42.
20. Bryson, M.; Sukkarieh, S. (2008): *Observability analysis and active control for airborne SLAM*, IEEE Transactions on Aerospace and Electronic Systems, vol. **44**, pp. 261–278.
21. Burkholder, J. O.; Tao, G. (2011): *Adaptive detection of sensor uncertainties and failures*, AIAA Journal of Guidance, Control and Dynamics, vol. **34**, pp. 1065–1612.
22. Calafiore, G. C.; Dabbene, F., Tempo, R. (2011): *Research on probabilistic methods for control system design*, Automatica, vol. **47**, pp. 1279–1295.
23. Campbell, M. E.; Wheeler, M. (2010): *Vision based geolocation tracking system for uninhabited vehicle*, AIAA Journal of Guidance, Control and Dynamics, vol. **33**, pp. 521–532.
24. Catena, A.; Melita, C.; Muscato, G. (2014): *Automatic tuning architecture for the navigation control loop of UAV*, Journal of Intelligent and Robotic Systems, vol. **73**, pp. 413–427.
25. Chen, B.; Nagarajaiah, S. (2007): *Linear matrix inequality based robust fault detection and isolation using the eigenstructure assignment method*, AIAA Journal of Guidance, Control and Dynamics, vol. **30**, pp. 1831–1835.
26. Cheng, Y.; Shusterr, M. D. (2014): *Improvement to the implementation of the Quest algorithm*, AIAA Journal of Guidance, Control and Dynamics, vol. **37**, pp. 301–305.
27. Chryssanthacopoulos, J. P.; Kochenderfer, M. J. (2012): *Hazard alerting based on probabilistic models*, AIAA Journal of Guidance, Control and Dynamics, vol. **35**, pp. 442–450.
28. Corke, P. (2011): *Robotics, Vision and Control*, Springer.
29. Cotta, C.; Van Hemert, I. (2008): *Recent Advances in Evolutionary Computation for Combinatorial Optimization*, Springer.
30. Cui, L.; Yang, Y. (2011): *Disturbance rejection and robust least-squares control allocation in flight control system*, AIAA Journal of Guidance, Control and Dynamics, vol. **34**, pp. 1632–1643.
31. DeMars, K. J.; Cheng, Y.; Jah, M. R. (2014): *Collision probability with Gaussian mixture orbit uncertainty*, AIAA Journal of Guidance, Control and Dynamics, vol. **37**, pp. 979–984.
32. Doshi-Velez, F.; Pineau, J.; Roy, N. (2012): *Reinforcement learning with limited reinforcement: Using Bayes risk for active learning in POMDP*, Artificial Intelligence, vol. **187**, pp. 115–132.
33. Dutta, P.; Bhattacharya, R. (2011): *Hypersonic state estimation using the Frobenius Perron operator*, AIAA Journal of Guidance, Control and Dynamics, vol. **34**, pp. 325–344.
34. Farhood, M.; Feron, E. (2012): *Obstacle-sensitive trajectory regulation via gain scheduling and semi definite programming*, IEEE Transactions on Control System Technology, vol. **20**, pp. 1107–1113.
35. Fatemi, M.; Millan, J.; Stevenson, J.; Yu, T.; Young, S. (2008): *Discrete event*

control of an unmanned aircraft, Int. Workshop on Discrete Event Systems, pp. 352–357.
36. Forrest, L.J.; Kessler, M.L.; Homer D.B (2007): *Design of a human-interactive autonomous flight manager for crewed lunar landing*, AIAA conference Infotech@aerospace, Rohnert Park, CA, AIAA 2007–2708.
37. Fu, C.; Olivares-Mendez, M.; Suarez-Fernandez, R.; Compoy, P. (2014): *Monocular visual-inertial SLAM based collision avoidance strategy for fail-safe UAV using fuzzy logic controllers*, Journal of Intelligent and Robotic Systems, vol. **73**, pp. 513–533.
38. Ge, Y.; Shui, W. (2008): *Study on algorithm of UAV threat strength assessment based on Bayesian networks*, IEEE Int. conference In Wireless Communications, Networking and Mobile Computing, pp. 1–4, DOI 978-1-4244-2108-41.
39. Ghearghe, A.; Zolghadri, A.; Cieslak, J.; Goupil, P.; Dayre, R.; Le Berre, H. (2013): *Model-based approaches for fast and robust fault detection in an aircraft control surface servoloop*, IEEE Control System Magazine, vol. **33**, pp. 20–30.
40. Ghosh, P.; Conway, B.A (2012): *Numerical trajectory optimization with swarm intelligence and dynamic assignment of solution structure*, AIAA Journal of Guidance, Control and Dynamics, vol. **35**, pp. 1178–1191.
41. Girard, A. R.; Larba, S. D.; Pachter, M.; Chandler, P. R. (2007): *Stochastic dynamic programming for uncertainty handling in UAV operations*, American control conference, pp. 1079–1084.
42. Goldberg, D. E. (2008): *Genetic Algorithms*, Addison Wesley publishing company.
43. Goupil P. (2011): *Airbus state of the art and practices on FDI and FTC in flight control systems*, Control Engineering Practice, vol. **19**, pp. 524–539.
44. Greval, M.; Andrews, A. (2008): *Kalman Filtering*, Wiley.
45. Halder, B.; Sarkar, N. (2009): *Robust Nonlinear Fault Detection and Isolation of Robotic Systems*, VDM.
46. Hamayoun, M. T.; Alwi, H.; Edwards, C. (2012): *An LPV fault tolerant control scheme using integral sliding modes*, IEEE Conference on Decision and Control, pp. 1840–1845.
47. Hamifzadegan, M.; Nagamine, R. (2014): *Smooth switching LPV controller design for LPV system*, Automatica, vol. **50**, pp. 1481–1488.
48. Hantos, P. (2011): *Systems engineering perspectives on technology, readiness assessment in software intensive system development*, AIAA Journal of Aircraft, vol. **48**, pp. 738–748.
49. Haykin, S. (2009): *Neural Networks and Learning Machines*, Pearson Education.
50. He, X.; Zhao, J. (2012): *Parameter dependent H_∞ filter design for LPV system and an autopilot application*, Applied Mathematics and Computation, vol. **218**, pp. 5508–5517.
51. Holzapfel, F.; Theil, S. (eds) (2011): *Advances in Aerospace Guidance, Navigation and Control*, Springer.
52. Horn, J.; Schmidt, E.; Geiger, B. R.; DeAngelo, M. (2012): *Neural network based trajectory optimization for UAV*, AIAA Journal of Guidance, Control and Dynamics, vol. **35**, pp. 548–562.
53. Horwood, J.; Aragon, N.; Poore, A. (2011): *Gaussian sum filters for space surveillance: theory and simulation*, AIAA Journal of Guidance, Control and

Dynamics, vol. **34**, pp. 1839–1851.

54. Jefferies, M. E.; Yeap, W. K. (2010): *Robotics and Cognitive Approach to Spatial Mapping*, Springer.
55. Jensen, R.; Shen, Q. (2008): *Computational Intelligence and Feature Selection*, IEEE Press.
56. Johansen, T. A.; Fossen, T. I. (2013): *Control allocation: a survey*, Automatica, vol. **49**, pp. 1087–1103.
57. de Jong, P. M.; van der Laan, J. J., Veld, A.C.; van Paassen, M.; Mulder, M. (2014): *Wind profile estimation using airborne sensors*, AIAA Journal of Aircraft, vol. **51**, pp. 1852–1863, DOI 10.2514/1.C032550.
58. Joshi, S. M.; Patre, P. (2014) :*Direct model reference adaptive control with actuator failures and sensor bias*, AIAA Journal of Guidance, Control and Dynamics, vol. **37**, pp. 312–317.
59. Joshi, S. M.; Gonzalez, O. R.; Upchurch, J. M. (2014): *Identifiability of additive actuator and sensor faults by state augmentation*, AIAA Journal of Guidance, Control and Dynamics, vol. **37**, pp. 941–946.
60. Julier, S. J.; Uhlmann, J. K. (1997): *A new extension of the Kalman filter to nonlinear systems*, Int. Symp. on Aerospace/Defense Sensing, Simulation and Controls, vol. **3**, pp. 182–193.
61. Kelley, T.; Avery, E.; Long, L.; Dimperio, E. (2009): *A hybrid symbolic and subsymbolic intelligent system for mobile robots*, AIAA Infotech @ aerospace conference, paper AIAA 2009–1976.
62. Khammash, M.; Zou, L.; Almquist, J.A.; Van Der Linden, C (1999): *Robust aircraft pitch axis control under weight and center of gravity uncertainty*, 38th IEEE Conference on Decision and Control, vol. **2**, pp. 190–197.
63. Khatic, A. K.; Singh, J.; Sinha, N. K. (2013): *Accessible regions for controlled aircraft maneuvering*, AIAA Journal of Guidance, Control and Dynamics, vol. **36**, pp. 1829–1834.
64. Khotari, M.; Postlethwaite, I. (2012): *A probabilistically robust path planning algorithm for UAV using rapidly exploring random trees*, Journal of Intelligent and Robotic Systems, vol. **71**, pp. 231–253, DOI 10.1007/s10846-012-9776-4.
65. Langel, S. E.; Khanafseh, S. M.; Pervan, B. S. (2014): *Bounding integrity risk for sequential state estimators with stochastic modeling uncertainty*, AIAA Journal of Guidance, Control and Dynamics, vol. **37**, pp. 36–46.
66. Lee, H. T.; Meyn, L. A.; Kim, S. Y. (2013): *Probabilistic safety assessment of unmanned aerial system operations*, AIAA Journal of Guidance, Control and Dynamics, vol. **36**, pp. 610–616.
67. Leege, A. M.; van Paassen, M.; Mulder, M. (2013): *Using automatic dependent surveillance broadcast for meteorological monitoring*, AIAA Journal of Aircraft, vol. **50**, pp. 249–261.
68. Lekki, J.; Tokars, R.; Jaros, D.; Riggs, M. T.; Evans, K. P.; Gyekenyesi, A. (2009): *Self diagnostic accelerometer for mission critical health monitoring of aircraft and spacecraft engines*, 47th AIAA Aerospace Sciences Meeting, AIAA 2009–1457.
69. Levine, D.; Luders, B., How, J. P. (2013): *Information theoretic motion planning for constrained sensor networks*, AIAA Journal of Aerospace Information Systems, vol. **10**, pp. 476–496.
70. Li, X.; Nair, P. B.; Zhang, Z., Gao, L.; Gao, C. (2014): *Aircraft robust trajectory*

optimization using nonintrusive polynomial chaos, AIAA Journal of Aircraft, vol. **51**, pp. 1592–1603, DOI 10.2514/1.C032474.
71. Lin, L.; Goodrich, M. A. (2014): *Hierarchical heuristic search using a Gaussian mixture model for UAV coverage planning*, IEEE Transactions on Cybernetics, vol. **44**, pp. 2532–2544, DOI 10.1109/TCYB.2014.2309898.
72. Lu, W.; Zhang, G.; Ferrari, S. (2014): *An information potential approach to integrated sensor path planning and control*, IEEE Transactions on Robotics, vol. **30**, pp. 919–934, DOI 11.1109/TRO.2014.2312812.
73. Maravall, D.; De Lope, J.; Martin, J. A. (2009): *Hybridizing evolutionary computation and reinforcement learning for the design of almost universal controllers for autonomous robots*, Neurocomputing, vol. **72**, pp. 887–894.
74. Margellos, K.; Goulart, P.; Lygeros, J. (2014): *On the road between robust optimization and the scenario approach for chance constrained optimization problems*, IEEE Transactions on Automatic Control, vol. **59**, pp. 2258–2263, DOI 10.1109/TAC.2014.2303232.
75. Maul, W. A.; Kopasakis, G.; Santi, L. M.; Sowers, T. S.; Chicatelli, A. (2007): *Sensor selection and optimization for health assessment of aerospace systems*, AIAA Infotech@ Aerospace, AIAA 2007–2849.
76. Melingui, A.; Chettibi, T.; Merzouki, R.; Mbede, J. B. (2013): *Adaptive navigation of an omni-drive autonomous mobile robot in unstructured dynamic environment*, IEEE Int. conference Robotics and Biomimetics, pp. 1924–1929.
77. Miller, S. A.; Harris, Z.; Chong, E. (2009): *Coordinated guidance of autonomous UAV via nominal belief state optimization*, American Control Conference, St. Louis, MO, pp. 2811–2818.
78. Moir, A.; Seabridge, A. (2006): *Civil Avionics Systems*, Wiley.
79. Moir, A.; Seabridge, A. (2007): *Military Avionics Systems*, AIAA Press.
80. Montemerlo, M.; Thrun, S. (2006): *Fast SLAM: a Scalable Method for the Simultaneous Localization and Mapping Problems in Robotics*, Springer.
81. Montes de Oca, S.; Puig, V.; Theilliol, D.; Tornil-Sin, S. (2010): *Fault tolerant control design using LPV admissible model matching with H_2/H_∞ performance, application to a 2 dof helicopter*, IEEE Conference on Control and Fault-tolerant Systems, Nice, France, pp. 251–256.
82. Montes de Oca, S.; Rotondo, U.; Nejjari, F.; Puig, V. (2011): *Fault estimation and virtual sensor FTC approach for LPV systems*, IEEE Conference on Decision and Control, Orlando, FL, pp. 2251–2256.
83. Nazari, R.; Seron, M. M.; De Dona, J. A. (2013): *Fault-tolerant control of systems with convex polytopic linear parameter varying model uncertainty using virtual sensor-based controller reconfiguration*, Annual Reviews in Control, vol. **37**, pp. 146–153.
84. Nicolos, I. K.; Valavanis, K. P.; Tsourveloudis, N. T.; Kostaras, A. N. (2003): *Evolutionary algorithm based offline/online path planner for UAV navigation*, IEEE Transactions on Systems, Man and Cybernetics, vol. **33**, pp. 898–912.
85. Ogren, P.; Robinson, J. (2011): *A model based approach to modular multi-objective robot control*, Journal of Intelligent and Robotic Systems, vol. **63**, pp. 257–282, DOI 10.1007/s10846-010-9523-7.
86. Ollero, A. (2003): *Control and perception techniques for aerial robotics*, 7th IFAC Symposium on Robot Control, Wroclaw, Poland.
87. Orr, J. S.; Slegers, N. J. (2014): *High-efficient thrust vector control allocation*,

AIAA Journal of Guidance, Control and Dynamics, vol. **37**, pp. 374–382.

88. Palade, V.; Danut Bocaniala, C.; Jain, L.C. (eds) (2006): *Computational Intelligence in Fault Diagnosis*, Springer.
89. Panella, I. (2008): *Artificial intelligence methodologies applicable to support the decision-making capability onboard unmanned aerial vehicles*, IEEE Bio-inspired, Learning and Intelligent Systems for Security Workshop, pp. 111–118.
90. Papagoergiou, C.; Glover, K. (2014):*Robustness analysis of nonlinear dynamic inversion control laws with application to flight control*, 43th IEEE Conference on Decision and Control, pp. 3485–3490, DOI. 0-7803-8682-5.
91. Pashilkar, A. A.; Sundararajan, N.; Saratchandran, P. (2006): *A fault-tolerant neural aided controller for aircraft auto-landing*, Aerospace Science and Technology, vol. **10**, pp. 49–61.
92. Patera, R. P. (2007): *Space vehicle conflict avoidance analysis*, AIAA Journal of Guidance, Control and Dynamics, vol. **30**, pp. 492–498.
93. Patsko, V. S.; Botkin, N. D.; Kein, V. M.; Turova, V.L.; Zarkh, M.A. (1994): *Control of an aircraft landing in windshear*, Journal of Optimization Theory and Applications, vol. **83**, pp. 237–267.
94. Peni, T.; Vanek, B.; Szabo, Z.; Bokor J. (2015): *Supervisory fault tolerant control of the GTM UAV using LPV methods*, International Journal of Applied Mathematics and Computer Science, vol. **25**, pp. 117–131.
95. Poll, S.; Patterson-Hine, A.; Camisa, J.; Nishikawa, D.; Spirkovska, L.; Garcia, D.; Lutz, R. (2007): *Evaluation, selection and application of model-based diagnosis tools and approaches*, AIAA Infotech@Aerospace, AIAA 2007–2941.
96. Pongpunwattana, A.; Rysdyk, R. (2007): *Evolution-based dynamic path planning for autonomous vehicles*, Innovations in Intelligent Machines, Springer, Berlin Heidelberg, pp. 113–145.
97. Postlethwaite, I.; Bates, D. (1999): *Robust integrated flight and propulsion controller for the Harriet aircraft*, AIAA Journal of Guidance, Control and Dynamics, vol. **22**, pp. 286–290.
98. Rabbath, C. A.; Lechevin, N. (2010): *Safety and Reliability in Cooperating Unmanned Aerial Systems*, World Scientific.
99. Reichard, K.; Crow, E.; Rogan, C. (2007): *Integrated system health management in unmanned and autonomous systems*, AIAA Infotech @aerospace conference, AIAA 2007–2962.
100. Remenyte-Prescott, R.; Andrews, J. D.; Chung, P. W. (2010): *An efficient phased mission reliability analysis for autonomous vehicles*, Reliability Engineering and Systems Safety, vol. **95**, pp. 226–235.
101. Rodrigues, M.; Hamdi, H.; Braiek, N. B.; Theilliol D. (2014): *Observer-based fault tolerant control design for a class of LPV descriptor system*, Journal of the Franklin Institute, vol. **351**, pp. 3104–3125.
102. Rosa, P.; Silvestre, C. (2013): *Fault detection and isolation of LPV system using set-valued observers: an application to fixed wing aircraft*, Control Engineering Practice, vol. **21**, pp. 242–252.
103. Rotondo, D.; Nejjari, F.; Ping, V. (2014): *Robust state feedback control of uncertain LPV system: an LMI based approach*, Journal of the Franklin Institute, vol. **35**, pp. 2781–2803.
104. Rutkowski, L. (2008): *Computational Intelligence: Methods and Techniques*, Springer Verlag.

105. Rysdyk, R. (2007): *Course and heading changes in significant wind*, AIAA Journal of Guidance, Control and Dynamics, vol. **30**, pp. 1168–1171.
106. Samy, I.; Postlethwaite, I.; Gu. D.W. (2011): *Survey and application of sensor fault detection and isolation schemes*, Control Engineering Practice, vol. **19**, pp. 658–674.
107. Santamaria, E.; Barrado, C.; Pastor, E.; Royo, P.; Salami, E. (2012): *Reconfigurable automated behavior for UAS applications*, Aerospace Science and Technology, vol. **23**, pp. 372-386.
108. Shaffer, P. J.; Ross, I. M.; Oppenheimer, M. W.; Doman, D. B. (2007): *Fault tolerant optimal trajectory generator for reusable launch vehicles*, AIAA Journal of Guidance, Control and Dynamics, vol. **30**, pp. 1794–1802.
109. Shin, J.; Gregory, I. (2007): *Robust gain scheduled fault tolerant control for a transport aircraft*, IEEE Int. Conference on Control Applications, pp. 1209–1214.
110. Song, C.; Liu, L.; Feng, G.; Xu, S. (2014): *Optimal control for multi-agent persistent monitoring*, Automatica, vol. **50**, pp. 1663–1668.
111. Spooner, J. T.; Maggiore, M.; Ordonez ,R.; Passino, K. M. (2002): *Stable Adaptive Control and Estimation for Nonlinear Systems: Neural and Fuzzy Approximator Techniques*, Wiley.
112. Sreenuch, T.; Tsourdos, A.; Jennions, I. K. (2014): *Software framework for prototyping embedded integrated vehicle health management applications*, AIAA Journal of Aerospace Information Systems, vol. **11**, pp. 82–96.
113. Srikanthakumar, S.; Liu, C. (2012): *Optimization based safety analysis of obstacle avoidance system for unmanned aerial vehicle*, Journal of Intelligent and Robotic Systems, vol. **65**, pp. 219–231.
114. Stachura, M.; Frew, G. W. (2011): *Cooperative target localization with a communication aware-unmanned aircraft system*, AIAA Journal of Guidance, Control and Dynamics, vol. **34**, pp. 1352–1362.
115. Strang, G. (2007): *Computational Science and Engineering*, Wellesley-Cambridge Press.
116. Sun, L. G.; de Visser, C. C.; Chu, Q. P.; Falkena, W. (2014): *Hybrid sensor-based backstepping control approach with its application to fault-tolerant flight control*, AIAA Journal of Guidance, Control and Dynamics, vol. **37**, pp. 59–71.
117. Sun, B.; Ma, W. (2014): *Soft fuzzy rough sets and its application in decision-making*, Artificial Intelligence Review, vol. **41**, pp. 67–80.
118. Sutton, R. S.; Barto, A. G. (1998): *Reinforcement Learning*, The MIT Press.
119. Sydney, N.; Paley, D. A. (2014): *Multiple coverage control for a non stationary spatio-temporal field*, Automatica, vol. **50**, pp. 1381–1390.
120. Ta D.N.; Kobilarov M.; Dellaert F. (2014): *A factor graph approach to estimation and model predictive control on unmanned aerial vehicles*, Int. Conference on Unmanned Aircraft Systems (ICUAS), Orlando, FL, pp. 181–188.
121. Thrun, S.; Burgard, W.; Fox, D. (2006): *Probabilistic Robotics*, MIT Press.
122. Tol, H. J.; de Visser, C. C.; van Kamper, E.; Chu, Q. P. (2014): *Nonlinear multivariate spline based control allocation for high performance aircraft*, AIAA Journal of Guidance, Control and Dynamics, vol. **37**, pp. 1840–1862, DOI 10.2514/1.G000065.
123. Vank B., Edelmayer A., Szabo Z., Bohor J. (2014): *Bridging the gap between theory and practice in LPV fault detection for flight con-*

trol actuators, Control Engineering Practice, vol. **31**, pp. 171–182, DOI 10.1016/j.conengprac.2014.05.002.

124. Varrier, S.; Koening, D.; Martinez, J. J. (2014): *Robust fault detection for uncertain unknown inputs LPV system*, Control Engineering Practice, vol. **22**, pp. 125–134.
125. Volovoi, V.; Balueva, A.; Vega, R. (2014): *Analytical risk model for automated collision avoidance systems*, AIAA Journal of Guidance, Control and Dynamics, vol. **37**, pp. 359–363.
126. Wang, Y.; Hussein, I.; Erwin R. (2011): *Risk based sensor management for integrated detection and estimation*, AIAA Journal of Guidance, Control and Dynamics, vol. **34**, pp. 1767–1778.
127. Watkins, A. S. (2007): *Vision based map building and trajectory planning to enable autonomous flight through urban environments*, PhD Thesis, Univ. of Florida, Gainesville.
128. Welch, G.; Bishop, G. (2001): *An introduction to the Kalman filter*, Proceedings of the SISGRAPH conference, vol. **8**, pp. 1–41.
129. Whitacre, W.; Campbell, M.E. (2011): *Decentralized geolocation and bias estimation for UAV with articulating cameras*, AIAA Journal of Guidance, Control and Dynamics, vol. **34**, pp. 564–573.
130. Yang, X.; Mejias, L.; Gonzalez, F.; Warren, M.; Upcroft, B.; Arain, B. (2014): *Nonlinear actuator fault detection for small-scale UASs*, Journal of Intelligent and Robotic Systems, vol. **73**, pp. 557–572.
131. Yanmaz, E.; Costanzo, C.; Bettstetter, C.; Elmenreich, W. (2010): *A discrete stochastic process for coverage analysis of autonomous UAV networks*, IEEE GLOBECOM Workshop, pp. 1777–1782.
132. Yi, G.; Atkins E. M. (2010): *Trim state discovery for an adaptive flight planner*, AIAA aerospace sciences meeting, AIAA 2010–416.
133. Yi, G.; Zhong, J.; Atkins, E. M.; Wang, C. (2014): *Trim state discovery with physical constraints*, AIAA Journal of Aircraft, vol. **52**, pp. 90-106, DOI 10.2514/1.C032619.
134. Yoon, S.; Kim, S.; Bae, J., Kim, Y.; Kim, E. (2011): *Experimental evaluation of fault diagnosis in a skew configured UAV sensor system*, Control Engineering Practice, vol. **10**, pp. 158–173.
135. Zaheer, S; Kim, J. (2011): *Type 2 fuzzy airplane altitude control: a comparative study*, IEEE Int. Conference on Fuzzy Systems, Taipei, pp. 2170–2176.
136. Zanetti, R.; de Mars, K. J. (2013): *Joseph formulation of unscented and quadrature filters with applications*, AIAA Journal of Guidance, Control and Dynamics, vol. **36**, pp. 1860–1863.
137. Zhang, Y.; Jiang, J. (2008): *Bibliographical review on reconfigurable fault tolerant control systems*, Annual Reviews in Control, vol. **32**, pp. 229–252.
138. Zhang, B.; Mao, Z.; Liu, W.; Liu, J. (2015): *Geometric reinforcement learning for path planning of UAV*, Journal of Intelligent and Robotics Systems, vol. **77**, pp. 391–409, DOI 10.1007/s10846-013-9901-z
139. Zhang, B.; Tang, L.; Roemer, M. (2014): *Probabilistic weather forecasting analysis of unmanned aerial vehicle path planning*, AIAA Journal of Guidance, Control and Dynamics, vol. **37**, pp. 309–312.
140. Zhang, B.; Liu, W.; Mao, Z.; Liu, J.; Shen, L. (2014): *Cooperative and geometric learning algorithm (CGLA) for path planning of UAV with limited information*,

Automatica, vol. **50**, pp. 809–820.
141. Zhangchun, T.; Zhenzhou, L. (2014): *Reliability based design optimization for the structure with fuzzy variables and uncertain but bounded variables*, AIAA Journal of Aerospace Information, vol. **11**, pp. 412–422.
142. Zhou, K.; Doyle, J. C. (1998): *Essentials of Robust Control*, Prentice Hall.

6 General Conclusions

The study of smart autonomous aircraft is an innovative and ongoing part of research. A common perspective of autonomy is to segregate functions according to the nature of the tasks involved, such as those that are unique to the aircraft as compared to those that are applicable to the mission-level activities. Aircraft level autonomy includes vehicle stabilization and flight control, maneuvering flight and basic auto-land. Mission level autonomy encompasses functions such as auto-navigation, route planning, mission objective determination, flight plan contingencies, dynamic trajectory management and collision/obstacle avoidance.

In order to conduct this research, certain requirements for the design and simulation of concepts must be set. Accurate models of aircraft are necessary for the design of controllers and evaluation of the performance. The software architecture exhibits a multi-loop control structure in which an inner loop controller stabilizes the aircraft dynamics, while a guidance outer-loop controller is designed to control the vehicle kinematics, providing path-following capabilities. The problem of path following can be described as that of making an aircraft converge to and follow a desired spatial path, while tracking a desired velocity profile that may be path dependent. Collision/obstacle avoidance is another important component of the sense and avoid system, while the on board intelligence is to monitor their health and prevent flight envelope violations for safety. A UAV health condition depends on its aerodynamic loading, actuator operating status, structural fatigue. Any technique to maintain specific flight parameters within the operational envelope of an autonomous aircraft falls under envelope protection. Since UAV are expected to be operated more aggressively than manned counterparts, envelope protection is very important in smart autonomous aircraft and must be done automatically due to the absence of a pilot on board. Finally, it is important to ensure that the mission planning algorithm is deterministic in terms of computation time and solution cost. This book presented diverse methods useful for all these topics.

Autonomous aircraft sector growth is predicted to continue to rise and is described as the most dynamic growth part of the world aerospace industry this decade. Unmanned technologies will continue to improve in many different capability areas: such as increasingly data-intensive multi-sensor, multi-mission capabilities, learning, adapting and acting independently of human control. Autonomous mission performance demands the ability to integrate sensing, perceiving, analyzing, communicating, planning, decision-making and executing to achieve mission goals versus system functions. If mission learning is employed, smart autonomous aircraft can develop modified strategies for themselves by which they select their behavior. The future of smart autonomous aircraft is characterized as a movement beyond autonomous execution to autonomous mission performance. The difference between execution and performance is associated with mission outcomes that can vary even during a mission and requires deviation from the pre-programmed task. Software for autonomous aircraft is typically embedded, concurrent and must guarantee system properties such as safety, reliability and fault tolerance. There is a pressing need to regard the construction of new software applications as the composition of reusable building blocks. A key limitation that still remains for many tasks and applications is the ability of a machine to complete its decision cycle in real time. In this case, real time is defined by the task time constant that drives the decision process for the application at hand. This involves study and research in system theory, control theory, artificial intelligence, mission and flight planning and control of autonomous aircraft. Technology is evolving rapidly.

Acronyms

2D	Two-dimensional
3D	Three-dimensional
b-frame	Body-fixed frame
e-frame	Earth-fixed frame
gi-frame	Geocentric inertial fixed-frame
i-frame	Inertial-fixed frame
n-frame	Navigation frame
w-frame	Wind frame
ACO	Ant colony optimization
ADS-B	Automatic dependent surveillance broadcast
AER	Attainable equilibrium region
AFC	Adaptive fuzzy control
AFCS	Automatic flight control system
AGL	Above ground level
AIAA	American Institute of Aeronautics and Astronautics
AMF	Antecedent membership function
AMPC	Adaptive model predictive control
ANDI	Adaptive nonlinear dynamic inversion
ANFIS	Adaptive neural fuzzy inference system
ANN	Artificial neural network
APO	Adaptive polytopic observer
ATC	Air traffic control
ATSP	Asymmetric traveling salesman problem
BCV	Behavior coordination virtual
BIBS	Bounded input bounded state
BN	Bayesian network
BVP	Boundary value problem
C3	Command, control, communication
CA	Control allocation
CCF	Common cause failures
CE	Cross entropy
CEKM	Continuous enhanced Kernel–Mendel
CICS	Convergent input convergent state
CKM	Continuous Kernel–Mendel
CLF	Control Lyapunov function
CPT	Conditional probability table
CR	Controller reconfiguration
CSC	Circle straight circle
CSP	Constraint satisfaction problem

CT	Coordinated turn
CV	Constant velocity
DAG	Directed acyclic graph
DCA	Direct control allocation
DCM	Direction cosine matrix
DE	Differential evolution
DI	Dynamic inversion
DLM	Double-lattice method
DME	Distance measurement equipment
DMOC	Discrete mechanics and optimal control
DOF	Degrees of freedom
DSPP	Dubins shortest path problem
DSS	Decision support system
DTRP	Dynamic traveling repairman problem
DTSP	Dynamic traveling salesman problem
EA	Evolutionary algorithm
EASA	European aviation safety agency
EBA	Extended bifurcation analysis
EFCS	Electronic flight control system
EIF	Extended information filter
EKF	Extended Kalman filter
EKM	Enhanced Kernel–Mendel
ELOS	Equivalent level of safety
ENU	East North Up frame
EP	Evolutionary programming
ESD	Event sequence diagram
EST	Expansive search trees
ETSP	Euclidean traveling salesman problem
EVS	Enhanced vision system
FAA	Federal aviation authority
FAR	Federal aviation regulations
FCC	Flight control computer
FDD	Fault detection and diagnosis
FDI	Fault detection isolation
FDIR	Fault detection isolation and response
FFA	Functional fault analysis
FGS	Fuzzy guidance system
FIFO	First-in, first-out
FIM	Fisher information matrix
FIS	Fuzzy inference system
FL	Flight level
FLC	Fuzzy logic controller
FLF	Fuzzy Lyapunov function
FMEA	Failure modes and effects analysis

FOM	Figures of merit
FOTSP	Finite-one in set traveling salesman problem
FP	Frobenius–Perron
FRBS	Flight rule based system
FT	Fault tree
FTA	Fault tree analysis
FTFCS	Flight tolerant flight control system
FU	Footprint of uncertainty
GA	Genetic algorithm
GAS	Global asymptotic stability
GFS	Genetic fuzzy system
GIS	Geographic information system
GNC	Guidance, navigation and control
GNSS	Global navigation satellite system
GPS	Global positioning system
GVD	Generalized Voronoi diagram
HJB	Hamilton–Jacobi–Bellman
HJI	Hamilton–Jacobi–Isaacs
HOSMC	Higher order sliding mode control
IACR	Instantaneous acceleration center of rotation
IEEE	Institute of electrical and electronic engineers
IHU	Intelligent hardware units
IID	Independent identically distributed
ILS	Instrument landing system
IMU	Inertial measurement unit
INS	Inertial navigation system
IO	Input output
IOP	Input-to-output path
IRRT	Information-rich rapidly exploring random tree
IRU	Inertial reference unit
ISA	International standard atmosphere
ISHM	Integrated system health management
ISMC	Integral sliding mode control
ISR	Intelligence, surveillance, reconnaissance
IT2	Interval type 2
IVMH	Integrated vehicle health management
JAR	Joint aviation regulations
KF	Kalman filter
KM	Kernel–Mendel
LARC	Lie algebra rank condition
LASSO	Least absolute shrinkage and selection operation
LIFO	Last in first out
LLM	Local linear model
LMI	Linear matrix inequality

LOS	Line of sight
LPV	Linear parameter variant
LQE	Linear quadratic estimator
LQG	Linear quadratic Gaussian
LRU	Line replaceable unit
LS	Least squares
LTI	Linear time invariant
LTL	Linear temporal logic
LTV	Linear time variant
OLS	Orthogonal least squares
OODA	Observe, orient, decide, act
MA	Maneuver automaton
MAS	Multi-agent system
MBFS	Model based fuzzy system
MC	Monte Carlo
MCDA	Multi-criteria decision analysis
MCS	Motion capture system
MCMC	Markov chain Monte Carlo
MDL	Motion description language
MDP	Markov decision process
MF	Membership function
MFP	Mission flight planning
MILP	Mixed integer linear programming
MIMO	Multi-input multi-output system
MISO	Multi-input single-output
ML	Machine learning
MLD	Master logic diagram
MMAE	Multiple model adaptive estimation
MP	Mission priority
MPC	Model predictive control
MR	Mission risk
MRAC	Model reference adaptive control
MST	Minimum spanning trees
MTBF	Mean time between failure
MTTR	Mean time to repair
NAS	National air space
NDI	Nonlinear dynamic inversion
NDM	Naturalistic decision making
NED	North East Down frame
NOWR	Neighboring optimal wind routing
NP	Nonpolynomial
NT	Nie-Tau
OCP	Optimal control problem
OLS	Orthogonal least squares

OODA	Observe, orient, decide, act
OWP	Optimal wind routing
PC	Polynomial chaos
PD	Proportional derivative
PDC	Parallel distributed compensator
PDE	Partial differential equation
PDF	Probability distribution function
PID	Proportional integral derivative
PMP	Pontryagin maximum principle
PN	Proportional navigation
POI	Point of interest
POMDP	Partially observable Markov decision process
PP	Pole placement
PRA	Probabilistic risk assessment
PRM	Probabilistic road map
PSO	Particle swarm optimization
PVDTSP	Polygon visiting Dubins traveling salesman problem
PWM	Primary waypoints mission
QV	Quantization vector
RA	Randomized algorithm
RBF	Radial basis function
RCA	Root cause analysis
RCLF	Robust control Lyapunov function
RDM	Relative degree matrix
RF	Radio frequency
RGUAS	Robust global uniform asymptotic stability
RHC	Receding horizon control
RHTA	Receding horizon task assignment
RL	Reinforcement learning
RPDM	Recognition primed decision making
RRT	Rapidly exploring random tree
RS	Robust stabilizability
RT	Risk tolerance
RVM	Relevance vector learning mechanism
RW	Rule weight
SAA	Sense and avoid
SCP	Set covering problem
SDA	Self-diagnosis accelerometer
SDRE	State-dependent Riccati equation
SEC	Software enabled control
SF	Scaling factors
SFC	Specific fuel consumption
SFDIA	Sensor fault detection, isolation, accommodation
SGP	Set covering problem

SHM	System health management
SI	Swarm intelligence
SISO	Single-input single-output
SLAM	Simultaneous localization and mapping
SLHS	Stochastic linear hybrid system
SMC	Sliding mode control
SPOI	Sensed points of interest
SPP	Sensory graph plan
SQP	Sequential quadratic programming
SSE	Sum Square errors
STC	Self-tuning controller
SVD	Singular value decomposition
SVD-QR	Singular value decomposition and QR
TA	Terrain avoidance
TAA	Technical airworthiness authority
TCAS	Traffic alert and collision avoidance
TCG	Temporal causal graph
TF	Terrain following
TPBVP	Two-point boundary value problem
TR	Type reduction
TSFS	Takagi–Sugeno fuzzy system
TSK	Takagi–Sugeno–Kang
TSP	Traveling salesman problem
TTF	Temporal transition to failure
UAS	Unmanned aerial systems
UAV	Unmanned aerial vehicle
UB	Uncertainty bounds
UCA	Uniform cost assignment
UKF	Unscented Kalman filter
VFF	Virtual force field
VOR	VHF omni-directional range
VSC	Variable structure control
VTP	Virtual target point
WG	Waypoint generation
WLS	Weighted least squares
WSN	Wireless sensor network
ZVD	Zermelo–Voronoi diagram

Nomenclature

α	Angle of attack
β	Side-slip angle
χ	Heading angle
$\eta = \begin{pmatrix} \eta_1 \\ \eta_2 \end{pmatrix}$	6D vector position and orientation of the body-fixed frame expressed in the Earth-fixed frame
$\eta_1 = \begin{pmatrix} x \\ y \\ z \end{pmatrix}$	3D position of the body-fixed frame expressed in the Earth-fixed frame
$\eta_2 = \begin{pmatrix} \phi \\ \theta \\ \psi \end{pmatrix}$	Orientation of the body-fixed frame expressed in the Earth-fixed frame, expressed with Euler angles
γ	Flight path angle
Γ	Markov transition matrix
λ	Latitude
κ	Curvature of the trajectory
$\delta_e, \delta_a, \delta_r$	Deflection angles, respectively, of the elevator, aileron and rudder
ℓ	Longitude
μ_i	Validity parameter of the local model i
ϖ	Membership function
ν	Innovation vector
ω	Angle rate
ω_a	Solid body rotation speed of the air
ω_s	Spatial frequency
$\Omega = \begin{pmatrix} p \\ q \\ r \end{pmatrix}$	Angular velocity expressed in the fixed frame
ϕ	Roll angle
$\tilde{\lambda}$	Scale factor
σ	Bank angle
ψ	Yaw angle
ρ	Volume mass of the surrounding fluid
θ	Pitch angle
Θ	Darboux vector
τ	Torsion of the trajectory
$\mathbf{A}$	State space matrix
A_R	Aspect ratio

B(s)	Binormal vector
$\mathbf{B}$	Control space matrix
C	Side-force
$\mathbf{C}$	Measurement space matrix
C_c	Side-force aerodynamic force coefficient
C_D	Drag aerodynamic force coefficient
C_L	Lift aerodynamic force coefficient
C_l	Tangential aerodynamic torque coefficient
C_m	Normal aerodynamic torque coefficient
C_n	Lateral aerodynamic torque coefficient
d	Distance
D	Drag Force
E	Energy
F	Force acting on the vehicle
g	Acceleration of the gravity
G	Guard condition
h	Altitude
H	Hamiltonian function
$\mathbf{I}_{n*n}$	$n * n$ identity matrix
J	Cost function
$\mathbf{I}$	Inertia matrix
$\mathbf{K}$	Control gain matrix
L	Lift force
m	Mass
M	Moment acting on the vehicle
n, e, d	Unit vectors pointing, respectively, to North, East and Down
$N(s)$	Normal vector
$\mathbb{N}(s)$	Set of natural numbers
$\mathbf{O}_{n*m}$	n*m zero matrix
n_z	Load factor
O_x, O_y, O_z	Axes according, respectively, to x, y and z directions
P	Pressure or linear momentum
P_d	Detection probability
Prob	Probability
$\mathbf{Q}$	State weighting matrix
Q	Configuration
$\mathbf{R}$	3 *3 Rotation matrix
$\mathbb{R}$	Set of real numbers
$\mathcal{R}_a$	Aerodynamic frame
$\mathcal{R}_f$	Inertial fixed frame
$\mathcal{R}_m$	Body-fixed frame
$\mathbf{R}$	6 *6 Generalized rotation matrix
s	Curvilinear abscissa
S_{ref}	Characteristic or reference of wetted area

sfc	Specific fuel consumption
S_j	Surface of the panel j
$Sk(V)$	Skew matrix related to the vector V
$SO(3)$	Special orthogonal matrix group
T	Thrust
$T(s)$	Tangent vector
U	Control vector
$V = \begin{pmatrix} u \\ v \\ w \end{pmatrix}$	Linear velocity of the vehicle expressed in the body-fixed frame
$\mathbf{V}$	Generalized velocity of the vehicle $\mathbf{V} = (V, \Omega)^T$
$V = (u, v, w)^T$	Vectorial velocity field
V_a	Relative vehicle airspeed
$\tilde{V}$	Lyapunov function
W	3D wind speed in the inertial frame
W_g	Weight force
WP_n	n^{th} waypoint
$\mathbf{W_N}$	Spectral density matrix
$\tilde{W}$	White Gaussian process noise
X	State space vector
$X_u, X_w, Z_u, Z_w, \ldots$	Aerodynamic derivatives
Y	Output vector
Z	Measured output vector
$\mathbf{Z}$	Matrix Z
$\mathbb{Z}$	Set Z

Index